AF180688

Precambrian Geology of the Tobacco Root Mountains, Montana

Edited by

John B. Brady
Department of Geology
Smith College
Northampton, Massachusetts 01063
USA

H. Robert Burger
Department of Geology
Smith College
Northampton, Massachusetts 01063
USA

John T. Cheney
Department of Geology
Amherst College
Amherst, Massachusetts 01002
USA

Tekla A. Harms
Department of Geology
Amherst College
Amherst, Massachusetts 01002
USA

THE GEOLOGICAL SOCIETY OF AMERICA

Special Paper 377

3300 Penrose Place, P.O. Box 9140 ▪ Boulder, Colorado 80301-9140 USA

2004

Published by The Geological Society of America, Inc.
3300 Penrose Place, P.O. Box 9140, Boulder, Colorado 80301
www.geosociety.org

Printed in U.S.A.

GSA Books Science Editor: Abhijit Basu

Library of Congress Cataloging-in-Publication Data

Precambrian Geology of the Tobacco Root Mountains, Montana / edited by John B. Brady … [et al.].
p.cm. – (Special paper ; 377)
Includes bibliographical references.
ISBN 0-8137-2377-9 (pbk.)
1. Geology, Stratigraphic—Precambrain. 2. Geology—Montana—Tobacco Root Mountains.
I. Brady, John, 1948- II. Special papers (Geological Society of America) ; 377

QE653.P73338 2004
551.7'1'09786663—dc22

2003071045

Cover: Metamorphosed mafic dikes and sills (MMDS) that have intruded quartzofeldspathic gneisses of the Pony–Middle Mountain Metamorphic Suite are visible as nearly horizontal bands in a cliff face in the central Tobacco Root Mountains. The middle of the three prominent MMDS in the photo thins dramatically, as if caught in the act of intrusion. The photograph is taken from the northeast ridge of Spuhler Peak looking northwest toward previously unnamed, flat-topped, 10,080 foot Keck Mountain. In the foreground on the left are rocks of the Spuhler Peak Metamorphic Suite, which are typically more brown in color than the gray-weathering gneisses of the Pony–Middle Mountain Metamorphic Suite. The absence of the MMDS in the Spuhler Peak Metamorphic Suite suggest that the rocks of this suite were not in their present position at the time of intrusion of the MMDS.

10 9 8 7 6 5 4 3 2 1

Contents

Preface

OVERVIEW

The Tobacco Root Mountains occupy a unique niche in the Wyoming province of North America. Located near the northwestern margin of the province, the Precambrian rocks of the Tobacco Roots were subjected to an intense tectonothermal event during the collision of the Wyoming province with the Hearne province in the Early Proterozoic. This event overprinted earlier periods of deformation and metamorphism, but the unique lithologic packages present in the Tobacco Root Mountains aid in unraveling early from later events and in detailing many results stemming from this Early Proterozoic collision.

This Geological Society of America Special Paper is organized into three separate parts. In Chapters 2–6, an effort is made to review the geochemistry and petrology of the four major Precambrian rock sequences present in the Tobacco Root Mountains: the Indian Creek Metamorphic Suite, the Pony–Middle Mountain Metamorphic Suite, the Spuhler Peak Metamorphic Suite, and metamorphosed mafic dikes and sills. Each of these suites, or sequences, plays an important role in understanding the Precambrian geology of this region. This section focuses on what each rock suite represents in terms of original protolith and provides the foundation for understanding the extensive petrological and geochronological information that follows. In addition, Chapter 4 considers the meta-ultramafic rocks that are dispersed throughout all Precambrian exposures in the Tobacco Root Mountains. Chapter 6 presents petrologic observations and interprets them in terms of a metamorphic history and a pressure-temperature path for the four rock suites.

The second section (Chapters 7–9) presents the results of numerous and extensive radiometric age determinations from all four major rock suites. Each chapter details the results of a different approach; taken together, they provide a solid base for understanding the timing of the tectonothermal events that affected these rock suites and for separating the effects of the last major event in the Early Proterozoic. The final chapters (Chapters 10 and 11), comprising the third section, examine the structures and fabrics present in these Precambrian rocks and propose a sequence of events that correlates observed field relations and structures with age determinations and metamorphic history.

Included with this volume is a reprinted version of the Geologic Map of Southern Tobacco Root Mountains, Madison County, Montana, that was compiled by Charles J. Vitaliano and William S. Cordua and published by the Geological Society of America in 1979 as MC-31. The map is included as a pocket insert, and the text that accompanied the map is appended at the end of this volume. As this map was the foundation upon which our research was based and as it currently is out of print, we are pleased that it can be included here. This reprinting of MC-31 is made possible by a generous donation from a former student of Dr. Charles J. Vitaliano.

DEDICATION

The editors dedicate this volume to Sandra Glass, former program vice-president of the W.M. Keck Foundation. It was largely through the wisdom and foresight of Sandra Glass that the Keck Geology Consortium was born. Her insights into science education and research, especially (in our case) at undergraduate liberal arts colleges, provided the support that ultimately made possible our field seasons in Montana and, by extension, this volume.

ACKNOWLEDGMENTS

This volume is an outgrowth of three Keck Geology Consortium undergraduate summer research projects in the Tobacco Root Mountains and the resulting 32 undergraduate theses. Special thanks are due to the advisors of those theses for their careful supervision and guidance. We are grateful to the W.M. Keck

Foundation of Los Angeles and to the 12 colleges of the Keck Geology Consortium (Amherst College, Beloit College, Carleton College, Colorado College, College of Wooster, Franklin & Marshall College, Pomona College, Smith College, Trinity University, Washington and Lee University, Whitman College, and Williams College) for financial support of our field and laboratory work. We also wish to thank Henry H. Woodard and Cathryn A. Manduca, who served as Keck Geology Consortium coordinators during the time of our work in Montana. Their support and flexibility contributed substantially to the success of our project. We are very grateful for the generous assistance of the many reviewers, who are individually acknowledged in each chapter. Finally, we wish to thank GSA Science Editor Abhijit Basu and the GSA publications staff for their patience and assistance in making this publication a reality.

John B. Brady
H. Robert Burger
Northampton, Massachusetts

John T. Cheney
Tekla A. Harms
Amherst, Massachusetts

Geological Society of America
Special Paper 377
2004

General geology and tectonic setting of the Tobacco Root Mountains

H. Robert Burger
Department of Geology, Smith College, Northampton, Massachusetts 01063, USA

ABSTRACT

The Tobacco Root Mountains are located near the northwestern margin of the Archean Wyoming province. This position places the Archean rocks of the Tobacco Root Mountains approximately midway between the Great Falls tectonic zone, which many believe marks the collisional boundary between the Hearne and Wyoming provinces, and the Madison mylonite zone, 3-km-wide shear zone that is believed to represent major displacement during collision of these two Archean provinces.

The Archean of the Tobacco Root Mountains is dominated by quartzofeldspathic orthogneisses but contains substantial and unequivocal metasedimentary rocks that facilitate detailed structural mapping, an allochthonous sequence of metamorphosed mafic and ultramafic volcanic rocks, and mafic dikes and sills that crosscut early fabrics but also are metamorphosed and deformed. The unique location and compositional diversity of these Archean rocks supports new insights into the timing and tectonothermal effect of the collision of the Wyoming province with the Hearne province.

Geochronologic and geothermobarometric results support a collision between 1.78 and 1.72 Ga that produced a metamorphic episode at 5–6 kbar and 650–750 °C and involved partial melting. A sequence dominated by tholeiitic and komatiitic basalts and possible komatiites was brought into tectonic contact with the relatively autochthonous orthogneisses and metasedimentary rocks. The intensity of this tectonothermal event strongly reworked all Archean rocks, so that the dominant folds, fabrics, and mineralogy now observed in Tobacco Root rocks date from this time.

Keywords: Tobacco Root Mountains, Wyoming Province, Archean, Proterozoic, Big Sky orogeny.

THE WYOMING PROVINCE

Introduction

An Archean province of the North American craton as defined by Hoffman (1989, p. 448) is a "contiguous area of Archean continental lithosphere framed by Proterozoic orogens inferred to be suture zones." There are seven Archean provinces that constitute the North American craton: the Wyoming, Superior, Slave, Nain, Hearne, Rae, and Burwell. Amalgamation of all seven provinces to form the protocraton of Laurentia was complete by ca. 1.7 Ga. The provinces differ in the details of their internal segments with respect to crustal ages, lithologies, metamorphic grade, and assembly ages of the segments.

The Archean rocks in the Wyoming province (Fig. 1) are restricted to the cores of Laramide uplifts. Although these commonly provide excellent exposures, they reveal only ~5%–10% of the Archean basement of the province. Exposures nearest the margins of the province include those in the Black Hills (South Dakota) at the eastern margin, the northeastern Uintas (Utah) to the south, the Albion and Raft River Ranges (Utah and Idaho) and the Ruby Mountains (Nevada) to the southwest, the Ruby Range and the Highland Mountains (Montana) to the northwest, and the Little Rocky Mountains (Montana) at the northern boundary (Mueller et al., 1998a). For the most part, the actual margins of the Wyoming province are not exposed. The only exposed margin is the Cheyenne belt that marks the southeastern limit of the province. The Cheyenne belt margin cuts across the southern Laramie

Burger, H.R., 2004, General geology and tectonic setting of the Tobacco Root Mountains, *in* Brady, J.B., Burger, H.R., Cheney, J.T., and Harms, T.A., eds., Precambrian geology of the Tobacco Root Mountains, Montana: Boulder, Colorado, Geological Society of America Special Paper 377, p. 1–14. For permission to copy, contact editing@geosociety.org.

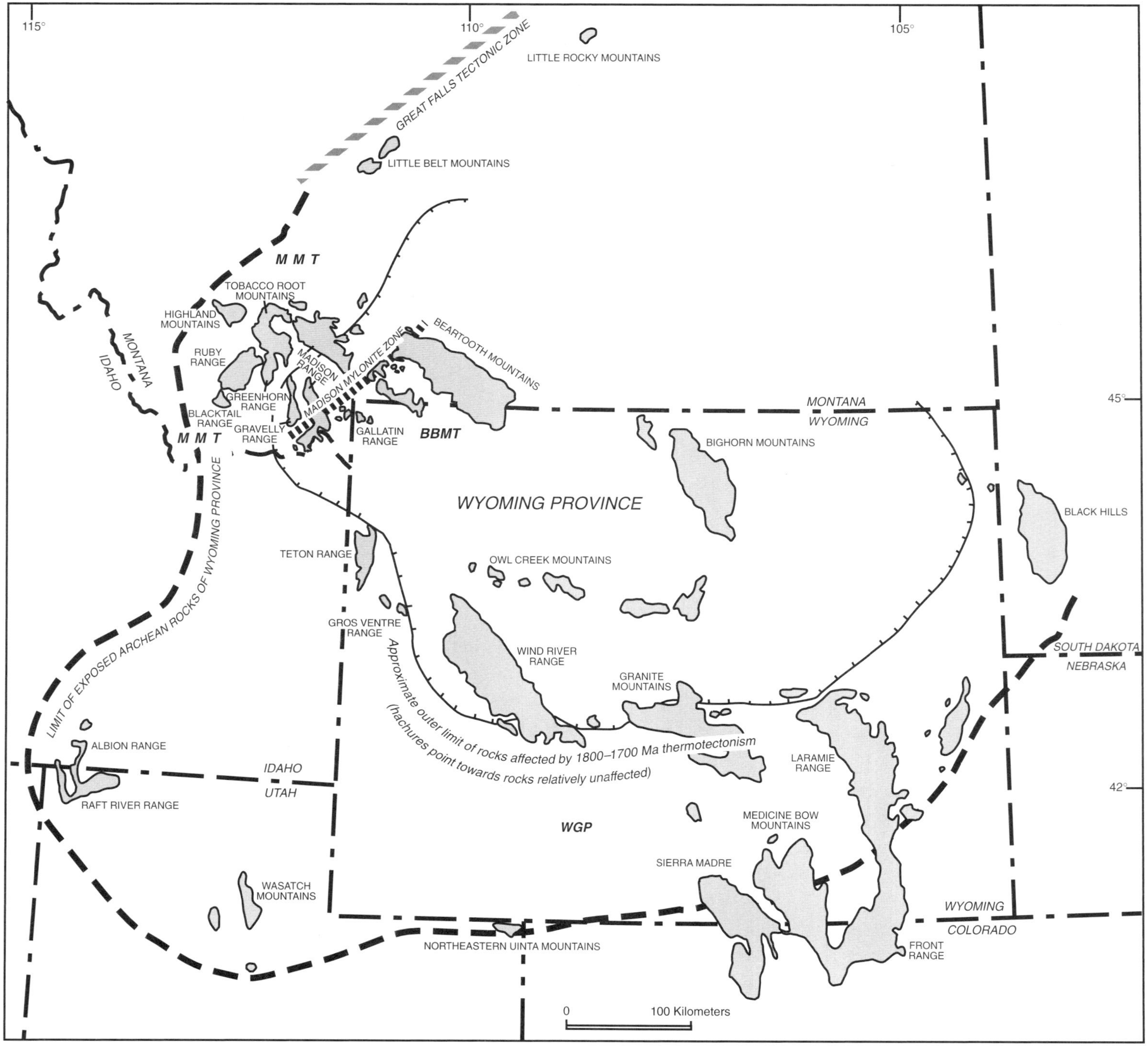

Figure 1. Outline map of the Wyoming province showing the distribution of Archean exposures and major boundaries referred to in this volume. MMT—Montana metasedimentary terrane; BBMT—Beartooth-Bighorn magmatic terrane; WGP—Wyoming greenstone province (Mogk et al., 1992a, 1992b).

Range and represents a major shear zone between Archean and Early Proterozoic rocks of the Wyoming province and late Early Proterozoic rocks to the south (Houston et al., 1993).

The Wyoming province can be divided into three subprovinces (Mogk et al., 1992a, 1992b) based on differences in metasedimentary sequences and the ages and compositions of metamorphic and magmatic rocks associated with the metasediments. These three subprovinces (Fig. 1) are the Montana metasedimentary terrane, the Beartooth-Bighorn magmatic terrane, and the Wyoming greenstone province. The Montana metasedimentary terrane, in which the Tobacco Root Mountains are located, is unique because of its carbonate-quartzite-pelite association even though quartzofeldspathic gneisses are the dominant lithology. At least some of the quartzofeldspathic gneisses have an igneous protolith, and magmatic ages for these range from 3.1 to 3.3 Ga (Mogk et al., 1992b; Mueller et al., 1993). The Beartooth-Bighorn magmatic terrane is dominated by tonalite-trondhjemite-granodiorite intrusives that date to 2.7–2.9 Ga, although large xenoliths of metasupracrustal rocks also are present (Mogk et al., 1992b). Greenstone belts containing metasedimentary and mafic to ultramafic metavolcanic rocks and intruded by abundant granitoids characterize the Wyoming greenstone province. The majority of these date from 2.5 to 2.7 Ga (Mueller and D'Arcy, 1990).

Tectonic Boundaries

The evidence is compelling that the Wyoming province existed as a small craton at least by 3.0 Ga and probably before that time. Analyses of detrital zircons document a major period of crustal growth between 3.2 and 3.4 Ga (Mueller et al., 1998b). Ages in this same range have been obtained from many igneous protoliths throughout the province, although these are most abundant in the Montana metasedimentary terrane. A question naturally arises concerning the relation among the Montana metasedimentary terrane, Beartooth-Bighorn magmatic terrane, and Wyoming greenstone province. Present information suggests that these subprovinces were separate geologic entities that were assembled in the Late Archean, most likely between 2.7 and 2.55 Ga (Mogk and Henry, 1988; Mogk et al., 1992b). However, as all subprovinces have distinctively high $^{207}Pb/^{204}Pb$ for a given $^{206}Pb/^{204}Pb$ ratio, they are believed to have originated within the Wyoming province and not to represent exotic terranes (Mueller et al., 1996b). A boundary between the Montana metasedimentary terrane and Beartooth-Bighorn magmatic terrane is present as a highly faulted and structurally complex zone in the North Snowy Block of the Beartooth Mountains (Mogk et al., 1992a, 1992b; Mueller et al., 1996a). Another boundary, a 0.5-km-wide ductile shear zone within the Montana metasedimentary terrane as exposed in the northern Madison Range, separates K-feldspar–rich gneisses that are common in the Tobacco Root Mountains and the western Madison Range from the dominantly plagioclase-rich gneisses with interlayered mafic amphibolites/granulites of the Gallatin Range and eastern Madison Range (Mogk et al., 1992a). It seems likely that many as yet unrecognized, and perhaps more subtle, tectonic boundaries exist throughout the Wyoming province. The likelihood of numerous tectonic boundaries suggests caution when attempting correlation from range to range based on general similarities among lithologies.

As noted, with the exception of the Cheyenne belt, the actual margins of the Wyoming province are not exposed, so the details of its assembly with the Hearne and Superior provinces are poorly understood. Although the collision zone between the Wyoming and Superior provinces is buried, recent work in the Black Hills demonstrates that thermotectonism related to this collision began at ca. 1770 Ma and ended at ca. 1715 Ma (Dahl et al., 1999a). The collision of the Hearne and Superior provinces during the Trans-Hudsonian orogeny was earlier, at ca. 1860–1790 Ma (Dahl et al., 1999a). There also are many dates from the northwestern part of the Wyoming province that suggest a possible collision between the Wyoming and Hearne provinces in the Early Proterozoic (Dahl et al., 1999a, 1999b; Peterman, 1981; Mueller et al., 1996b; Davis et al., 1995). Ranges with extensive exposures of Archean rocks that are located close to the northwestern margin of the Wyoming province and, therefore, are near a possible collision zone include the Ruby Range, the Highland Mountains, and the Tobacco Root Mountains. Other more limited exposures in possibly critical positions include those in the Little Rocky Mountains and the Little Belt Mountains.

It has been clear for more than 30 years (Giletti, 1966) that rocks in the northwestern part of the Montana metasedimentary terrane (Fig. 1) were subjected to a thermal event in the Early Proterozoic (1900–1700 Ma). Evidence that this event was a thermotectonic one and might be related to a collision between the Hearne and Wyoming provinces is much more recent (O'Neill et al., 1988; Erslev and Sutter, 1990), although the extent of reworking of Archean crust is, for the most part, still poorly documented. U/Pb dating of euhedral zircons from a gneiss dome in the Highland Mountains gives an age of ca. 1800 Ma (O'Neill et al., 1988). The structural configuration and metamorphic character of these rocks is demonstrated to date from that time. Erslev and Sutter (1990) present a convincing case, based on $^{40}Ar/^{39}Ar$ thermochronology of samples from the 3-km-thick Madison mylonite zone (Fig. 1), that this feature likely represents a major thrust zone on the margin of a major Early Proterozoic (1900–1800 Ma) orogen. Although the Madison mylonite zone is beyond the limit of rocks invariably containing evidence of the 1900–1700 Ma event and is substantially removed from the likely site of the Hearne-Wyoming collision zone, it does have the same general trend as a postulated collision zone site (the Great Falls tectonic zone). This strengthens the case for a major Early Proterozoic event that affected rocks well inboard of the northwestern margin of the Wyoming province.

The Great Falls tectonic zone is a northeast-trending geologic feature that can be traced from the Idaho batholith to just across the Canadian border. O'Neill and Lopez (1985) present evidence that this feature has exerted structural control over igneous intrusions, has experienced recurrent fault movements from at least the Middle Proterozoic to the Cenozoic, and probably represents a buried zone of crustal weakness. The Great Falls tectonic zone possesses a number of similarities to the northwestern, collisional margin of the Superior province, suggesting it may have a similar origin (O'Neill, 1998). If so, then the Great Falls tectonic zone marks the position of the boundary between the Hearne and Wyoming provinces. The evidence that this zone represents a fundamental structural discontinuity is further enhanced by its spatial correlation with gravity and magnetic anomalies. However, recent electromagnetic data (Boerner et al., 1998) are interpreted as demonstrating that the nature of the Great Falls tectonic zone is unlike that of other Paleoproterozoic collisional or accretionary collision zones in Laurentia. Boerner et al. (1998, p. 181) suggest the Great Falls tectonic zone is a reactivated intracontinental shear zone (similar to the Snowbird tectonic zone between the Rae and Hearne provinces) rather than a Proterozoic suture between Archean provinces.

An accretionary model was proposed by O'Neill (1998) to account for the observed distribution and character of Archean and Proterozoic basement in northwestern Montana and northern Idaho. His model interprets the Great Falls tectonic zone as the suture between the Hearne and Wyoming provinces and places the gneisses of the Highland Mountains within a metamorphic-plutonic terrane belt. Metamorphic basement in the Tobacco Root Mountains constitutes a huge crustal duplex structure at the margin of continental foreland and experiences only thermal metamorphic effects from the Early Proterozoic collision. Although O'Neill's model elegantly synthesizes many major regional features and relationships, it is clear from the results presented in this paper that, at the very least, the boundaries of the metamorphic-plutonic terrane belt must be extended to include Tobacco Root Archean rocks. With some refinement, this model certainly is worthy of careful consideration. Imaging by the Lithoprobe project in central Alberta (Ross et al., 1995) reveals crustal-scale thrust imbrication and crust-mantle boundary deflections that are dated to ca. 1.8 Ga and are roughly of the same scale and age as O'Neill's tectonic elements.

The southern part of the Hearne province, referred to as the Medicine Hat Block, is completely buried beneath the Mesozoic Cordilleran foreland basin. However, xenoliths in Eocene minette intrusions provide intriguing age information. Upper crustal xenoliths (hornblende-biotite tonalites, foliated and nonfoliated) provide exclusively Archean ages (2.6–3.3 Ga) from U/Pb analyses of zircons, but U/Pb results from zircons in lower crustal xenoliths (mafic and felsic granulites) are exclusively Early Proterozoic (1.7–1.8 Ga) (Davis et al., 1995). Nd isotopic data plus geothermobarometric information indicate that these Early Proterozoic ages are due to extensive reworking of Archean rocks under high-temperature conditions (11–13 kbar, 950 °C) (Davis et al., 1995, p. 332). These ages are essentially the same as those reported in this paper for a high-grade thermotectonic event that affected the Archean rocks of the Tobacco Root Mountains.

The Tobacco Root Mountains are located reasonably close to the Great Falls tectonic zone, are between this zone and the Madison mylonite zone, and are very close to the Highland Mountains. Therefore, the Precambrian geology of the Tobacco Root Mountains should provide important insights into the nature and extent of the Early Proterozoic event that affected the northwestern portion of the Wyoming province and the adjacent Medicine Hat Block of the Hearne province.

TOBACCO ROOT MOUNTAINS

Geographic and Geologic Setting

The Tobacco Root Mountains (Figs. 1 and 2) are a north-trending mountain range occupying an area of ~1300 km^2. Located in southwestern Montana, the range is bounded by 45° and 46° north latitude and bisected by 112° west longitude. The Tobacco Root Mountains have a topographic relief of almost 1700 m. Intermontane basins that flank the mountains to the east and west have an average elevation of 1500 m. Many peaks in the range have elevations exceeding 3100 m; the highest is Mount Jefferson with an elevation of 3181 m. Drainage from the steep western slopes enters the Jefferson and Ruby Rivers, whereas drainage from the more gentle eastern slopes flows into the Madison River. Towns next to the range include Whitehall to the north, Harrison, Norris, and Ennis to the east, Virginia City to the south, and Sheridan and Twin Bridges to the west.

A number of nearby ranges (Fig. 1) with substantial exposures of Precambrian rocks include (1) the Madison Range immediately to the east of the Tobacco Root Mountains, (2) the Gravelly and Greenhorn Ranges to the south, (3) the Ruby Range to the southwest, and (4) the Highland Mountains directly to the west.

The Tobacco Root Mountains possessed considerable relief by the Late Cretaceous–early Tertiary (70–60 Ma) (Schmidt, 1975). Today the mountains are essentially an east-tilted block uplift that is raised along a high-angle normal fault located on the western margin of the range. This present uplift began in the late Miocene (10 Ma) and is still taking place (Pardee, 1950).

Rocks exposed in the Tobacco Root Mountains span the Precambrian to the Quaternary, but Archean metamorphic rocks dominate. Paleozoic and Mesozoic rocks are most abundant in the northern portion of the range where they overlie the Archean rocks. Granitic to dioritic intrusives of the Late Cretaceous (77–72 Ma) Tobacco Root batholith (Vitaliano et al., 1980) occupy an area of ~310 km^2 in the central and eastern parts of the range. In the south, near Virginia City, Tertiary volcanics overlie the Archean rocks. Exposures of Archean rocks throughout much of the rest of the range eventually pass beneath Tertiary and Quaternary basin-fill deposits of the intermontane basins.

Early Studies

Early reconnaissance studies in the region that involved portions of the Tobacco Root Mountains include those by Hayden (1872), Peale (1896), and Winchell (1914). Mapping by Tansley and Schafer (Tansley et al., 1933) produced the first geologic

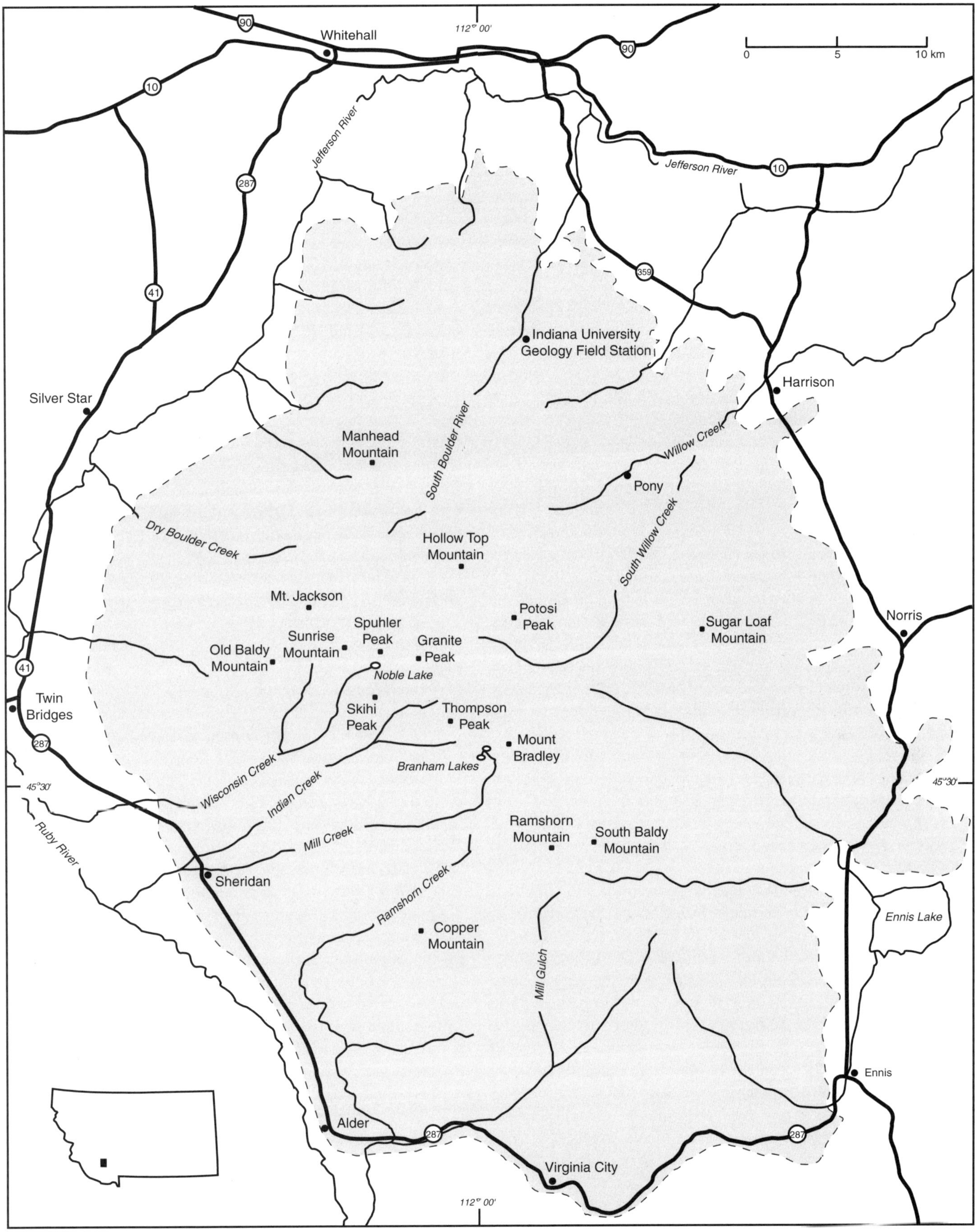

Figure 2. Outline map of the Tobacco Root Mountains area showing major towns, streams, peaks, and lakes, many of which are referred to in this volume. The outline of the Tobacco Root Mountains is based primarily on the 6000-foot contour, except in the vicinity of Virginia City.

map of the entire range, and many of their main lithologic elements and groupings are present on the most recent detailed map available (Vitaliano et al., 1979a; see reprinted map accompanying this volume). In their descriptions of the Archean in the Tobacco Root Mountains, Tansley and Schafer (Tansley et al., 1933, p. 8–9) used the term "Pony Series" to designate gneisses particularly well exposed near the town of Pony in the northeastern part of the range. Their description notes that quartzofeldspathic gneisses are the most abundant rock and that substantial amounts of hornblende gneiss also are present. Other rock types mentioned include amphibolite and mica schist.

In addition to "Pony Series," Tansley and Schafer (Tansley et al., 1933, p. 9–10) used "Cherry Creek Series" to refer to associations containing marble. The name "Cherry Creek" was first used by Peale (1896) to refer to a sequence containing marble exposed in the area of Cherry Creek in the Gravelly Range, located ~27 km southeast of Virginia City. Winchell (1914) correlated the Cherry Creek rocks with a somewhat similar sequence of marble-bearing rocks in the Tobacco Root Mountains. Tansley and Schafer do not mention his correlation, but they may have been influenced by it. It is interesting that they defined three groupings within the Cherry Creek rocks: (1) a sequence composed mainly of "garnetiferous schists and gneisses," (2) a sequence composed mainly of limestone with some quartzite and mica schist, and (3) a sequence dominated by quartzofeldspathic gneisses.

Tansley and Schafer believed the Cherry Creek rocks are younger than the Pony, primarily because they interpreted much of the Pony as igneous and the Cherry Creek as sedimentary because Pony gneisses were never seen to cut Cherry Creek rocks. Their reasons for attributing an igneous origin to many Pony rocks were based mainly on mineral composition, lack of persistence of some units, and abundant migmatite (Tansley et al., 1933, p. 9). Finally, they called attention to the dark dikes and sills containing oriented minerals that cut both Pony and Cherry Creek rocks. Two of their specific observations (the garnet-rich rocks in a portion of the Cherry Creek sequence and the dark dikes and sills) are returned to time and again by later investigators. These two observations are presented in this volume as perhaps the two most important keys that unlock the Precambrian history of the rocks in the Tobacco Root Mountains.

The next study directly concerned with the Precambrian rocks of the Tobacco Root Mountains was conducted by Levandowski (1956), who examined the geology and mineral deposits of an area between Sheridan and Alder (Cherry Creek rocks). This was followed by Reid (1957, 1963) who produced a reasonably detailed map that included a majority of the Precambrian rocks exposed in the northern one-third of the Tobacco Root Mountains. Although the areas mapped by Reid and Levandowski were once again mapped in the 1960s and 1970s (see below), it's useful to briefly examine several of Reid's observations in order to provide a context for later work.

Although Reid's map does not offer much more detail than that produced by Tansley and Schafer, with the exception of a series of northwest-southeast–trending faults added by Reid, he provides significantly more structural and petrologic detail. Reid substituted "metamorphics" for "Series," but continued to subdivide Precambrian rocks into Pony and Cherry Creek groups. The contact between these two groupings on Reid's map is in an almost identical position on Tansley and Schafer's map. Reid noted, as did Tansley and Schafer, that the bulk of the Pony metamorphic rocks are composed of quartzofeldspathic gneiss with subordinate biotite and hornblende gneiss, but he also added that thin amphibolites and quartzites are present. In addition, Reid observed that, although the Cherry Creek metamorphics contain marble, sillimanite schist, anthophyllite schist and coarse garnet amphibolite, they also contain all the major rock types that occur in the Pony (Reid, 1957, p. 5). He made special note of the fact that, where the distinctive metasedimentary units are not present, it is impossible to distinguish Cherry Creek from Pony (Reid, 1957, p. 6–7).

For the purpose of later discussion, it is important to note Reid's subdivision of Cherry Creek rocks. As did Tansley and Schafer, Reid called attention to what he called the "upper Cherry Creek" (although Tansley and Schafer placed these rocks at the bottom of their section). He describes the sequence located in the upper South Boulder–upper Wisconsin Creek divide near Noble and Spuhler Peaks to consist of a thick zone, dominantly amphibolite, overlain by coarse garnet amphibolite, anthophyllite schist, amphibolite, and sillimanite-bearing quartzite (Reid, 1957, p. 7).

In contrast to Tansley and Schafer, Reid believed the Cherry Creek to be older than the Pony, because wherever he found both sequences, the Cherry Creek was always structurally below the Pony. He apparently did not believe that rocks of such areal extent could be overturned.

Reid (1963, p. 297–301) proposed five periods of deformation and metamorphism to explain the relationships he observed. Some of these events, and especially some items of evidence, seem questionable today, but he did make several fundamentally important observations. Reid also recognized the importance of the metabasalt dikes and sills (the dark dikes and sills of Tansley and Schafer). He noted that they cut gneissic banding in both Pony and Cherry Creek rocks, are themselves deformed, and contain granulite facies assemblages. He clearly states that a pre-metabasalt high-grade metamorphism had occurred but was overprinted by the granulite-grade event that affected the metabasalts. As upper amphibolite facies assemblages dominate in these rocks and as he observed textural relationships such as sillimanite replacing kyanite, Reid proposed a subsequent amphibolite facies event.

Thus, at the end of the 1950s, basic aspects of the geology of Precambrian metamorphic rocks of the Tobacco Root Mountains included the following.

1. The rocks were affected by more than one tectonothermal event, but the exact number and sequence remained unresolved.

2. Upper amphibolite facies assemblages are most common, but granulite facies assemblages are also present in places.

3. The metamorphic rocks consist of two major groups. One contains a sequence of marble, quartzite, and iron formation (Cherry Creek Series), and the other does not (Pony Series).

However, minor amounts of quartzite and magnetite schist occur in the Pony, and the two groups cannot be differentiated where the obvious metasedimentary rocks are not present. The age relationship of the two groups remained unresolved.

4. The gneissic banding in both major groups is cut by metamorphosed mafic dikes and sills that themselves are deformed and contain granulite-grade assemblages.

5. A portion of Cherry Creek rocks, dominated by amphibolites, are unusual in composition and appearance.

Investigations during the 1960s and 1970s

During the 1960s and 1970s, five graduate students from Indiana University (Burger, 1966; Root, 1965; Hess, 1967; Cordua, 1973; and Hanley, 1975) and one from Harvard University (Gillmeister, 1971) mapped almost all Precambrian exposures in the Tobacco Root Mountains. Charles J. Vitaliano mapped the remaining unmapped Precambrian exposures and, working with William S. Cordua, compiled a map for the Tobacco Root Mountains (Fig. 3) that was published in 1979 (Vitaliano et al., 1979a, 1979b; see reprinted map and text accompanying this volume). This student mapping produced geologic maps at a scale of 1:24,000 as well as a wealth of detailed lithologic, petrologic, and geochemical information. The studies greatly extended our knowledge of the Precambrian history of these rocks but, in a sense and with a few exceptions, did not substantially revise the major conclusions of earlier studies nor resolve some of the more difficult relationships (e.g., the Pony–Cherry Creek relative age relationship).

Root (1965) mapped Precambrian exposures in the northeastern portion of the Tobacco Root Mountains. This region has less continuous outcrop than most of the rest of the range and is underlain by Tansley and Schafer's (Tansley et al., 1933) Pony metamorphics. Thus, distinct, continuous map units are uncommon and, where present, difficult to trace. Root concluded that the Precambrian rocks had experienced only one major period of intense deformation and metamorphism. He recorded mineral assemblages indicative of both upper amphibolite and lower granulite facies and ascribed these to one event.

Burger (1966, 1967, 1969) mapped the geology of the west-central Tobacco Root Mountains. This area principally contains Cherry Creek rocks, and the presence of marble greatly aided the mapping of a series of upright, subisoclinal folds. His structural studies indicated that the axes of these macroscopic folds and all other mesoscopic lineations are coincident with a mean orientation of 24, N10W. Burger mapped no earlier, clearly delineated folds with the exception of one area that suggests the presence of an earlier recumbent fold refolded by the dominant folds. As all folds deform gneissic banding, Burger proposed two major events: (1) an early high-grade tectonothermal event to produce the gneissic banding and (2) a later event that produced the dominant macroscopic folds and the upper amphibolite–lower granulite facies mineral assemblages. In addition, Burger mapped what he termed the "anthophyllite gneiss assemblage" and proposed a fault to separate the anthophyllite-gneiss assemblage from Cherry Creek rocks. Burger's (1966, p. 41) anthophyllite-gneiss assemblage appears to be equivalent to Reid's (1957, p. 7) upper Cherry Creek. Burger mapped the fault on the basis of (1) apparent regional truncation of Cherry Creek rocks, (2) the suggestion of lower-pressure metamorphic evolution for the anthophyllite-gneiss assemblage based on the abundance of cordierite, (3) a distinctive lithologic assemblage not present elsewhere in his map area, and (4) an apparent high-strain zone coincident with the mapped trace of the fault.

Hess (1967) mapped a zone of Precambrian rocks surrounding their contact with the Tobacco Root batholith in an effort to study the contact metamorphic effects of batholith intrusion as well as to gain a wide, regional perspective on the Precambrian rocks. He distinguished a series of recumbent folds with axes plunging to the northeast as well as north-south–trending folds with steep axial planes. Hess provided a detailed account of metamorphic mineral assemblages, recognizing and expanding upon the accounts of previous workers, and also described in great detail what he termed "orthoamphibolites" that are the metabasalt dikes and sills of Reid (1957). Hess, in contrast with others but in concert with Root (1965), proposed only one period of protracted deformation and metamorphism to explain the major structures and mineral assemblages in Tobacco Root Precambrian rocks. At the time Hess was finishing his thesis, the only date (1.6 Ga) that existed for these rocks in southwestern Montana had just been published by Giletti (1966). Although Hess recognized the significance of the metabasalt intrusions, he likely opted for a long, protracted event based on this one age.

Gillmeister (1971, 1972) extended Burger's (1966) mapping to the north. As did Burger, he mapped the anthophyllite-gneiss assemblage as a separate sequence and named it the Spuhler Peak Formation. Due to the nearly ubiquitous presence of a quartzite unit at the base of the Spuhler Peak Formation, Gillmeister interpreted the contact as an unconformity rather than a fault. He also argued that the fold regime of the Pony metamorphics and the Spuhler Peak Formation differs from that of the Cherry Creek metamorphics and placed a tectonic discontinuity between Pony and Cherry Creek rocks. Finally, and perhaps most importantly, Gillmeister pointed out that the metabasalt dikes and sills observed and noted by all previous workers are not present in the Spuhler Peak Formation even though they are common in Pony and Cherry Creek rocks near their contacts with the Spuhler.

Cordua (1973) mapped Precambrian rocks exposed in the southern Tobacco Root Mountains. A number of Cordua's observations merit discussion, as they are relevant to observations reported later in this volume. In at least three small areas Cordua mapped a number of early, recumbent, isoclinal folds. These are refolded by a series of north-plunging, overturned antiforms and synforms that plunge to the northwest in the western portion of Cordua's map area and to the northeast in the eastern portion. These variations in plunge are due to a later event of broad, open folds. Cordua's structural work suggests all mesoscopic lineations (fold axes and mineral lineations) in his area are of the same generation or were rotated into parallelism by a high-strain event.

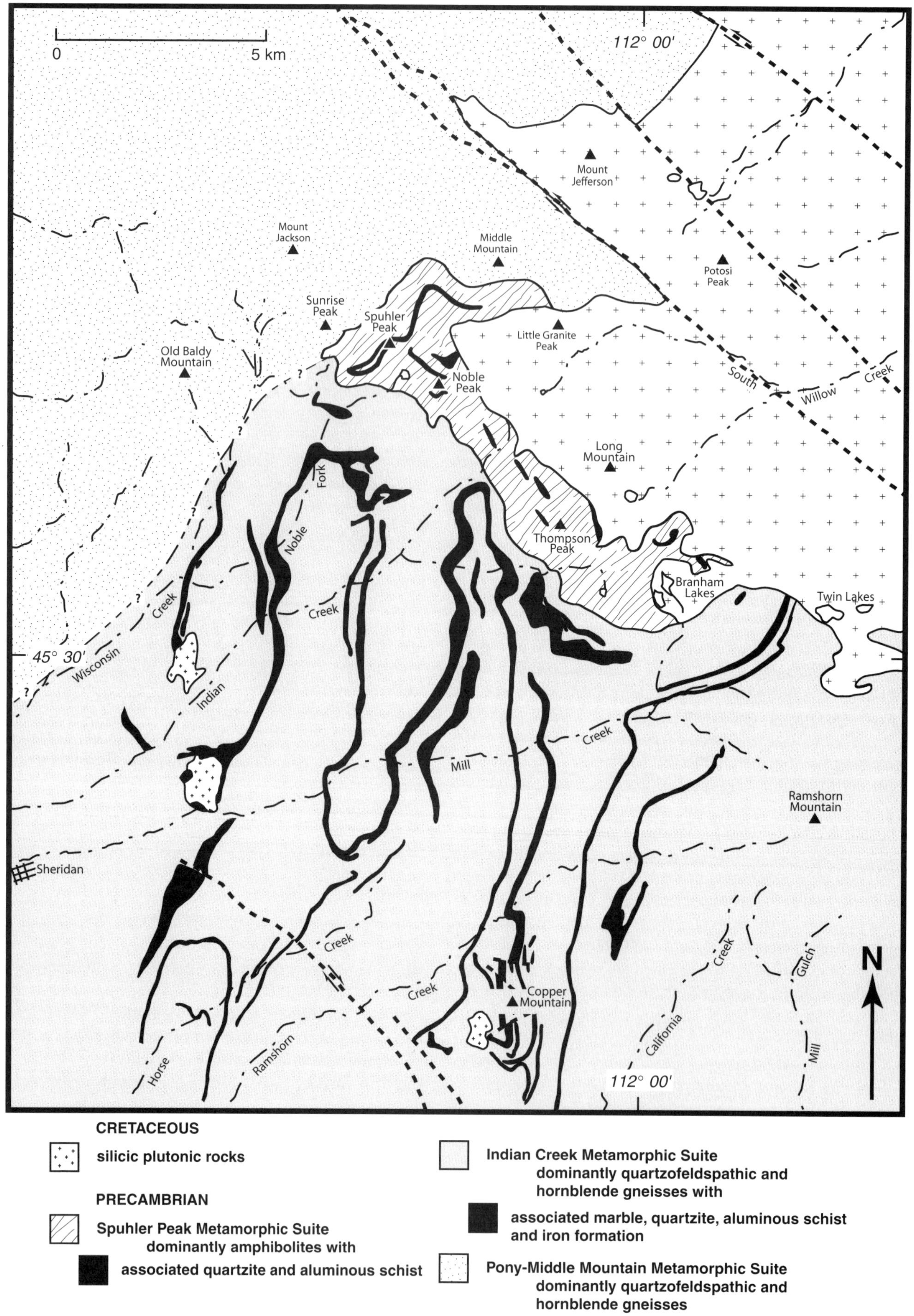

Figure 3. Schematic geologic map (after Vitaliano et al., 1979a) of the Tobacco Root Mountains area investigated for this volume. Boundary between rocks typically mapped as Indian Creek Metamorphic Suite and those mapped as Pony–Middle Mountain Metamorphic Suite is after O'Neill (1983).

(Cordua points out that much of the quartzofeldspathic gneiss in his area is blastomylonitic.) Lineations in metabasalt dikes and sills parallel lineations in adjacent, foliated rocks. These also parallel lineations in the north-plunging antiforms and synforms. Accordingly, Cordua postulated two major events: (1) an early granulite facies metamorphism and recumbent folding and (2) a later upper amphibolite facies (locally granulite facies) event associated with formation of major folds and their subsequent rotation. Using recently determined dates from nearby areas, Cordua correlated the older event with an age of 2830 Ma (Giletti, 1968) and the younger with the 1600 Ma age (Giletti, 1966).

Hanley (1975, 1976) undertook the final mapping project in Precambrian rocks in the Tobacco Root Mountains. His work was in the northwestern part of the range that contains primarily Pony metamorphics. Hanley separated three dominant map units: gray gneiss, hornblende gneiss, and hornblende-plagioclase gneiss. The gray gneiss varies in composition from granite to granodiorite to quartz diorite, and the potassium feldspar in these rocks commonly is present as augen. Hanley also noted the rare presence of thin beds of iron formation and quartzite, and, in a few locations, isolated lenses and thin beds of marble. Also, Hanley discusses the metabasalts (which he refers to as metabasites) in great detail and points out their importance in unraveling Precambrian events. He proposed two events, as did most other previous workers. The first, consisting of isoclinal folding and high-grade metamorphism, is followed by intrusion of metabasites, which in turn are followed by another period of isoclinal folding and high-grade metamorphism.

Other studies during these two decades, with the exception of geochronological results (summarized below), include a detailed petrologic study of Gillmeister's Spuhler Peak Formation (Friberg, 1976), additional work on the metamorphosed mafic dikes (Hanley and Vitaliano, 1983), studies of unmetamorphosed Precambrian diabase dikes (Koehler, 1972, 1976; Wooden et al., 1978), iron formation (Immega, 1976; Immega and Klein, 1976), and ultramafic intrusives (Tendall, 1978). Subsequent to the investigations of the 1960s and 1970s, little work has been done in the Tobacco Root Mountains other than unpublished master's theses (Angeloni, 1988; Duncan, 1978, 1982), limited geochronological work (summarized below), and checking and slight remapping of the Middle Mountain–Tobacco Root roadless area (O'Neill, 1983).

Keck Geology Consortium Research

The Keck Geology Consortium is a consortium of 12 liberal arts colleges whose undergraduate research is, in part, funded by the W.M. Keck Foundation. Financial support from the Keck Foundation is used primarily to fund collaborative geology research projects by groups of undergraduates. In the summer of 1993 a group of ten students and three faculty (John B. Brady, H. Robert Burger, and John T. Cheney) traveled to the Tobacco Root Mountains to study the Archean rocks in the west-central part of the range. This first summer's focus was to explore the nature of the contact between Burger's (1966) anthophyllite-gneiss assemblage (Gillmeister's [1971] Spuhler Peak Formation) and what had been referred to as Pony and Cherry Creek rocks. It soon became apparent that this original focus needed to be expanded to include structural, petrologic, geothermobarometric, and geochronologic studies of all metamorphic rock suites. Work conducted during the first field season included structural analysis, geochronologic, petrologic and geothermobarometric studies, and protolith determination (Archuleta, 1994; Cady, 1994; Fisher, 1994; Jacob, 1994; King, 1994, 1995; Lowell, 1994; Peck, 1994; Poulsen, 1994; Sincock, 1994; and Tierney, 1994). A second project took place during the summer of 1995 involving 13 students and two additional faculty (Tekla A. Harms and Kathleen E. Johnson). Projects from this summer covered many of the same general topics but expanded into the northern and southern parts of the range and added mineral fabrics and meta-ultramafic rocks (Cox, 1996; DeGraff, 1996; Hirst, 1996; Kovaric et al., 1996; Kranenburg, 1996; MacFarlane, 1996; Martin, 1996; Mohlman, 1996; Owen, 1996; Pufall, 1996; Thomas, 1996; Tuit, 1996; and Wegmann, 1996). The final project took place in the summer of 1997, involved nine students, and attempted to focus on unanswered questions from the previous two projects (Abeyta, 1998; Blednick, 1998; Carmichael, 1998; Frisch, 1998; Harris, 1998; Hatch, 1998; Monteleone, 1998; Picornell, 1998; and Steffen, 1998). This Special Paper is mainly the result of the work of these 32 students, their faculty advisers, and the project faculty who have guided and extended the students' contributions. Abstracts published from this work include Brady et al. (1994, 1998), Burger et al. (1996, 1998, 1999), Cheney et al. (1994, 1996, 1998, 1999), Harms et al. (1996), and Johnson et al. (1996).

At the beginning of this research effort we adopted several changes in lithologic unit terminology. Many recent workers in the Precambrian of southwestern Montana have expressed concern about using past conventions when referring to lithologic packages such as the Cherry Creek and Pony (Vitaliano et al., 1979b). In the case of Cherry Creek rocks, such usage is problematic, as it is not at all clear that the type section of Cherry Creek rocks in the Gravelly Range correlates with somewhat similar (but different in detail) rocks in the Tobacco Root Mountains. Due to the complex history of these rocks and the likelihood of numerous tectonic juxtapositions within the Montana metasedimentary terrane, basing correlation primarily on the presence of carbonate units seems unwise. In the case of Pony rocks, past use implies certain age relationships and protolith, and essentially excludes any rock sequence containing marble, even if thin or locally present. Therefore, we adopted the following terminology to designate suites of rocks where well exposed and consisting of volumetrically significant and compositionally distinctive lithologies (Fig. 3): The Indian Creek Metamorphic Suite refers to rocks that represent part of those that in the past would be assigned to the Cherry Creek. This package is best exposed along Indian Creek in the western part of the range and is dominated by quartzofeldspathic gneisses but contains the distinctive metasupracrustal assemblage of marble, quartzite, pelitic schist, and iron formation. The Pony–Middle Mountain Metamorphic Suite is dominated by quartzofeldspathic gneisses with subordinate hornblende gneiss. The Pony–Middle Mountain Metamorphic Suite is best represented by rocks near Pony in the northeastern part of the Tobacco Root Mountains and by those exposed in and around Middle Mountain in the central part of the range. As noted previously, both Burger and Gillmeister mapped a distinctive sequence of rocks as a separate unit that prior workers had placed within the Cherry Creek. Gillmeister's term "Spuhler Peak Formation" seemed reasonable, but we changed his unit to Spuhler Peak Metamorphic Suite for consistency with our Indian Creek Metamorphic Suite and Pony–Middle Mountain Metamorphic Suite terminology. Finally, we adopted the acronym MMDS to refer to "metamorphosed mafic dikes and sills." These are the mafic intrusives that cut gneissic banding and, therefore, are unequivocally younger than such banding. The MMDS do, however, possess a faint foliation and high-grade mineral assemblage. These intrusives, as noted, have been recognized by virtually all previous workers in the Tobacco Root Mountains, but have been referred to by varying terminology. Because of their key importance in relating structures and mineralogy to specific thermotectonic events, we choose to coin a new acronym for these intrusive bodies.

Geochronology Summary

In preparing the reader for the chapters that follow, it seems useful to provide a synopsis of existing geochronological data for the region. In Table 1, we compile ages from Precambrian rocks in the Tobacco Root Mountains and for other mountain ranges within the Montana metasedimentary terrane that either flank the Tobacco Root Mountains or extend away from the range in a northeasterly direction (parallel to the trend of the Great Falls tectonic zone). This compilation emphasizes that all the metamorphic rocks are Archean in age. No known metamorphosed Early Proterozoic rocks are present in this part of the Montana metasedimentary terrane. It also is clear that Early Proterozoic ages abound and that these likely reflect a major event. The exact nature and significance of this event was not yet clear before our work, although the Highland Mountains result (O'Neill et al., 1988) suggested that these ages mark a major orogenic event. There also are ages that suggest thermotectonic events in the Archean and the very earliest Proterozoic. Although we recognize that thermotectonic events predating the Early Proterozoic occurred, exactly what these ages represent still is uncertain. For the most part, the new ages reported in this paper greatly clarify and expand our knowledge concerning the Early Proterozoic orogenic event and differentiate Early Proterozoic metamorphism and deformation from those that definitely are Archean.

New Findings in This Volume

The findings of three Keck Geology Consortium summer projects plus material contributed by Mogk, Mueller, and Wooden are important, and significantly expand our knowledge of the Precambrian history of the Tobacco Root Mountains as well as shed light on major regional relationships.

Our work has produced extensive new analytical data that include *(a)* ~280 new whole-rock analyses for major elements, a large percentage of which have minor and trace element data, *(b)* numerous geothermobarometric determinations from several similar compositional units in each of the three major rock suites (Indian Creek, Spuhler Peak, and Pony–Middle Mountain Metamorphic Suites) as well as the MMDS, *(c)* new age determinations using $^{40}Ar/^{39}Ar$, U/Pb, and Th/Pb, and *(d)* new structural analytical data that support a regional simple shear event.

Many of our results provide new insights into Archean and Proterozoic events that affected Tobacco Root rocks, and several substantially change conventional wisdom regarding the evolution of this part of the Wyoming province. Following is a sampling of the more important findings.

1. Structural data support a regional, intense, simple shear event (the last major structural event to affect these rocks) that produced a regional-scale sheath fold, as well as outcrop-scale sheath folds throughout the range.

2. Field and petrologic observations, coupled with thermobarometric determinations, support a prolonged metamorphic event that began with a high-pressure (>1.1 GPa), high-temperature (>700 °C) phase that involved partial melting and that ended with reequilibration at temperatures and pressures of the upper amphibolite facies.

3. Multiple lines of evidence suggest that the Spuhler Peak Metamorphic Suite, which is dominated by tholeiitic and komatiitic basalts, most likely is allochthonous and possibly is Proterozoic in age.

4. Although metasedimentary rocks are present in the Indian Creek, Pony–Middle Mountain, and Spuhler Peak Metamorphic Suites, geochemical studies suggest that the protolith for each of these suites had dominant igneous components of a volcanic and/or volcaniclastic nature.

5. Ages, major element compositions, and rare earth element patterns suggest that the MMDS represent a subalkaline, tholeiitic basalt that intruded in a continental rift setting at 2.06 Ga.

6. The protolith for the meta-ultramafic rocks likely was an ultramafic cumulate rich in orthopyroxene, which formed in a continental setting and could be the source body for the MMDS.

7. Ages, structural and petrologic relationships, and regional relationships suggest that most of the metamorphic minerals, fabrics, and dominant folds in the region are related to an Early Proterozoic orogenic event (1.78 to 1.72 Ga) and are not Archean (2.7 Ga) as previously believed. This orogenic event is designated as the Big Sky orogeny.

ACKNOWLEDGMENTS

Funding from the W.M. Keck Foundation supported students and faculty for three field seasons in the Tobacco Root Mountains. That support is directly responsible for this volume.

TABLE 1. AGE COMPILATION FOR TOBACCO ROOT MOUNTAINS AND ADJACENT RANGES

Age (Ma)	Published Interpretation	Location	Method	Reference
3930–3130	Maximum age of quartzite equals 3130 Ma	Tobacco Root Mountains	U/Pb, detrital zircons	(Mueller et al., 1998b)
3370	Protolith crystallization age, amphibolitic-tonalitic gneiss	Tobacco Root Mountains	U/Pb, zircons	(Krogh et al., 1997)
3300	Protolith crystallization age, ICMS gneiss	Tobacco Root Mountains	U/Pb, zircons	(Mueller et al., 1996a)
2750	Age of metamorphism	Ruby Range, Tobacco Root Mountains, and Gallatin	Rb/Sr, whole-rock (multiple samples)	(James and Hedge, 1980)
2667	Age of metamorphism	Tobacco Root Mountains	Rb/Sr, whole-rock	(Mueller and Cordua, 1976)
2475	Cryptic tectonothermal event	Tobacco Root and Highland Mountains, Ruby Range	^{207}Pb/^{206}Pb (step-leach) monazite-dominated garnet	Roberts, et al., 2002
2400	Tectonothermal event	Tobacco Root Mountains	U/Pb, zircons	(Krogh et al., 1997)
3000–2200	Presence of 2500–2200 Ma crust beneath area	Tobacco Root Mountains	U/Pb, pre-magmatic zircons from Tobacco Root batholith	(Mueller et al., 1996c)
1800–1700	Tectonothermal event	Tobacco Root Mountains, Ruby Range	^{40}Ar/^{39}Ar, hornblende, phlogopite	(Brady et al., 1994, Kovaric et al., 1996)
1770	Tectonothermal event	Tobacco Root Mountains	^{207}Pb/^{206}Pb, zircons and baddeleyite from MMDS	(Burger et al., 1999)
1770	Tectonothermal event	Tobacco Root Mountains	^{207}Pb/^{206}Pb, zircons from pelitic melt zone associated with boudinaged amphibolite	(Burger et al., 1999)
1775	High-grade metamorphism	Tobacco Root Mountains	Th/Pb, monazites	(Cheney et al., 1999)
1820–1785	Tectonothermal event	Tobacco Root Mountains	^{40}Ar/^{39}Ar, biotite	Roberts et al., 2002
1600	Regional metamorphism	Tobacco Root and Highland Mountains, Ruby, Blacktail, northern Gravelly, and northern Madison Ranges	K/Ar, mainly biotite	(Giletti 1966, 1968)
1822	Tectonothermal event	Ruby Range	^{207}Pb/^{206}Pb (step-leach), garnet, monazite	(Dahl et al., 1999a)
1820–1793	Cooling ages following high-grade metamorphism	Highland Mountains	^{40}Ar/^{39}Ar, biotite and hornblende	(Harlan et al., 1996)
1800	Tectonothermal event	Highland Mountains	U/Pb, zircons	(O'Neill et al., 1988)
3300	Protolith crystallization age, tonalitic and trondhjemitic gneisses	Southern Madison Range	Rb/Sr, whole-rock; U/Pb, zircons	(Mueller et al., 1993)
2760–2680	Protolith crystallization age, granitic gneisses	Southern Madison Range	^{207}Pb/^{206}Pb, whole-rock; U/Pb, zircons	(Mueller et al., 1993)
2680	Meta-andesite crystallization age	Southern Madison Range	^{207}Pb/^{206}Pb, zircons	(Mueller et al., 1993)
2600	Protolith crystallization age, granodioritic gneiss; intrudes metasupracrustals	Southern Madison Range	Rb/Sr, whole-rock; U/Pb, zircons	(Mueller et al., 1993)
1900–1800	Thrusting in Madison mylonite zone	Southern Madison Range	^{40}Ar/^{39}Ar, muscovite and hornblende	(Erslev and Sutter, 1990)
2770	Age of felsic gneisses	Little Rocky Mountains	Rb/Sr, whole-rock	(Peterman, 1981)
1750–1710	Thermal resetting	Little Rocky Mountains	K/Ar, biotite and hornblende	(Peterman, 1981)
1880	Gneissic protolith crystallization age	Little Belt Mountains	U/Pb, zircons	(Mueller et al., 1996b)
1880–1860	Tectonothermal event	Little Belt Mountains	U/Pb, zircons	(Mueller et al., 1999)
2050–1740	Gneissic protolith crystallization age	Little Belt Mountains	U/Pb, zircons	(Mueller et al., 1999)
1850	Tectonothermal event	Little Belt Mountains	U/Pb, zircons	(Mueller et al., 1996b)
1800–1700	Protolith crystallization age, granitic augen gneiss	Little Belt Mountains	U/Pb, zircon	(Catanzaro and Kulp, 1964)

Note: ICMS—Indian Creek Metamorphic Suite; MMDS—metamorphosed mafic dikes and sills.

REFERENCES CITED

Abeyta, R.L., 1998, A geochemical and petrographic analysis of Archean quartzofeldspathic gneisses in the Indian Creek Metamorphic Suite and Pony–Middle Mountain Metamorphic Suite, Tobacco Root Mountains, southwestern Montana [senior thesis]: Colorado Springs, Colorado College, 75 p.

Angeloni, L.M., 1988, The role of water in the formation of granulite and amphibolite facies rocks, Tobacco Root Mountains, Montana [M.S. thesis]: Missoula, University of Montana, 51 p.

Archuleta, L.L., 1994, The mineralogy and petrology of garnet-amphibole rocks from the Spuhler Peak Formation along Thompson Peak ridge, Tobacco Root Mountains, SW Montana [senior thesis]: Claremont, California, Pomona College, 84 p.

Blednick, D.G., 1998, Ductile deformation in the Tobacco Root Mountains of southwest Montana [senior thesis]: Amherst, Massachusetts, Amherst College, 58 p.

Boerner, D.E., Craven, J.A., Kurtz, R.D., Ross, G.M., and Jones, F.W., 1998, The Great Falls tectonic zone: Suture or intracontinental shear zone?: Canadian Journal of Earth Sciences, v. 35, p. 175–183.

Brady, J.B., Burger, H.R., Cheney, J.T., King, J.T., Tierney, K.A., Peck, W.H., Poulsen, C.J., Cady, P., Lowell, J., Sincock, M.J., Archuleta, L.L., Fisher, R., and Jacob, L.J., 1994, Geochemical and $^{40}Ar/^{39}Ar$ evidence for terrane assembly in the Archean of southwestern Montana: Geological Society of America Abstracts with Programs, v. 26, no. 7, p. A-232.

Brady, J.B., Carmichael, S., Burger, H.R., Harris, C., Cheney, J.T., and Harms, T.A., 1998, Thermobarometry and geochemistry of metamorphosed mafic dikes and sills (MMDS) of the Tobacco Root Mountains, Montana: Geological Society of America Abstracts with Programs, v. 30, no. 7, p. A-97.

Burger, H.R., 1966, Structure, petrology, and economic geology of the Sheridan district, Madison County, Montana [Ph.D. thesis]: Bloomington, Indiana University, 156 p.

Burger, H.R., 1967, Bedrock geology of the Sheridan district, Madison County, Montana: Montana Bureau of Mines and Geology Memoir 41, 22 p.

Burger, H.R., 1969, Structural evolution of the southwestern Tobacco Root Mountains, Montana: Geological Society of America Bulletin, v. 80, p. 1329–1342.

Burger, H.R., Brady, J.B., Cheney, J.T., Harms, T.A., Johnson, K.E., DeGraff, K.A., Mohlman, H.K., and Owen, D.R., 1996, Geochemical evidence for a major reinterpretation of Archean rock protoliths, Tobacco Root Mountains, southwestern Montana: Geological Society of America Abstracts with Programs, v. 28, no. 7, p. A-492.

Burger, H.R., Brady, J.B., Steffen, K.J., Abeyta, R., Frisch, J.D., Cheney, J.T., and Harms, T.A., 1998, Geochemical and geothermobarometric appraisal of the Archean Pony–Middle Mountain Metamorphic Suite (PMMMS), Tobacco Root Mountains, Montana: Geological Society of America Abstracts with Programs, v. 30, no. 7, p. A-97.

Burger, R., Brady, J., Cheney, J., Harms, T., Mueller, P., Heatherington, A., and Wooden, J., 1999, Evidence for a major Early Proterozoic orogenic event in the Tobacco Root Mountains of southwestern Montana: Geological Society of America Abstracts with Programs, v. 31, no. 7, p. A177–178.

Cady, P., 1994, Petrographic and geochemical analysis of amphibolites from the Tobacco Root Mountains, Montana [senior thesis]: Colorado Springs, Colorado College, 86 p.

Carmichael, S., 1998, Geothermobarometry of metamorphosed mafic dikes and sills, Tobacco Root Mountains, southwestern Montana [senior thesis]: Northampton, Massachusetts, Smith College, 51 p.

Catanzaro, E.J., and Kulp, J.L., 1964, Discordant zircons from the Little Belt (Montana), Beartooth (Montana), and Santa Catalina (Arizona) Mountains: Geochimica et Cosmochimica Acta, v. 28, p. 87–124.

Cheney, J.T., Brady, J.B., Burger, H.R., King, J.T., Tierney, K.A., Peck, W.H., Poulsen, C.J., Cady, P., Lowell, J., Sincock, M.J., Archuleta, L.L., Fisher, R., and Jacob, L., 1994, Metamorphic evolution of Archean rocks in the Tobacco Root Mountains of SW Montana: Geological Society of America Abstracts with Programs, v. 26, no. 7, p. A-226.

Cheney, J.T., Brady, J.B., Burger, H.R., Harms, T.A., Johnson, K.E., DeGraff, K.A., King, J.F., Mohlman, H.K., Tuit, C.B., Hirst, J., Wegmann, K.W., and Thomas, R.B., 1996, Tectonometamorphic evolution of Archean amphibolites from the Tobacco Root Mountains of Montana: Geological Society of America Abstracts with Programs, v. 28, no. 7, p. A-45.

Cheney, J.T., Hatch, C.E., Harms, T.A., Brady, J.B., Burger, H.R., and Monteleone, B.D., 1998, Sapphirine symplectite bearing kyanite-orthopyroxene orthoamphibolite from the Archean Indian Creek Metamorphic Suite (ICMS), Tobacco Root Mountains, Montana: Geological Society of America Abstracts with Programs, v. 30, no. 7, p. A-97.

Cheney, J.T., Harms, T.A., Brady, J.B., and Burger, H.R., 1999, Early Proterozoic $^{208}Pb/^{232}Th$ in situ ion probe dating of monazite from Archean metamorphic suites, Tobacco Root Mountains, Montana: Geological Society of America Abstracts with Programs, v. 31, no. 7, p. A-178.

Cordua, W.S., 1973, Precambrian geology of the southern Tobacco Root Mountains, Madison County, Montana [Ph.D. thesis]: Bloomington, Indiana University, 248 p.

Cox, M.J., 1996, Geochemistry and petrography of amphibolites from the Spuhler Peak Metamorphic Suite near the Branham Lakes area, Tobacco Root Mountains, southwestern Montana [senior thesis]: Lexington, Virginia, Washington and Lee University, 64 p.

Dahl, P.S., Holm, D.K., Gardner, E.T., Hubacher, F.A., and Foland, K.A., 1999a, New constraints on the timing of Early Proterozoic tectonism in the Black Hills (South Dakota), with implications for docking of the Wyoming province with Laurentia: Geological Society of America Bulletin, v. 111, p. 1335–1349.

Dahl, P.S., Dorais, M.J., Roberts, H.J., Kelley, S.P., and Frei, R., 1999b, Electron microprobe geochronometry of age-zoned monazite crystals in Archean metapelites from the Wyoming province: The nature of Pb rejuvenation and implications for regional tectonism: Geological Society of America Abstracts with Programs, v. 31, no. 7, p. A39.

Davis, W.J., Berman, R., and Kyarsgaard, B., 1995, U-Pb geochronology and isotopic studies of crustal xenoliths from the Archean Medicine Hat Block, northern Montana and southern Alberta: Paleoproterozoic reworking of Archean crust, *in* Ross, G.M., ed., 1995 Alberta Basement Transects Workshop: Lithoprobe Secretariat, University of British Columbia, Lithoprobe Report 47, p. 330–335.

DeGraff, K.A., 1996, Petrology of amphibolites from the Archean Spuhler Peak Metamorphic Suite of the Tobacco Root Mountains, southwestern Montana [senior thesis]: Amherst, Massachusetts, Amherst College, 145 p.

Duncan, D.W., 1978, Structural analysis of the central Tobacco Root Mountains, southwest Montana [M.S. thesis]: University Park, Pennsylvania State University, 79 p.

Duncan, D.W., 1982, Kinematic and structural analysis of pre-Beltian rocks, west-central Tobacco Root Mountains, southwest Montana: Geological Society of America Abstracts with Programs, v. 14, no. 6, p. 310.

Erslev, E.A., and Sutter, J.F., 1990, Evidence for Proterozoic mylonitization in the northwestern Wyoming province: Geological Society of America Bulletin, v. 102, p. 1681–1694.

Fisher, R.G.M., 1994, Comparative metamorphic histories of three high-grade gneiss terrains in the Tobacco Root Mountains of southwest Montana [senior thesis]: Northampton, Massachusetts, Smith College, 71 p.

Friberg, N., 1976, Petrology of a metamorphic sequence of upper-amphibolite facies in the central Tobacco Root Mountains, southwestern Montana [Ph.D. thesis]: Bloomington, Indiana University, 146 p.

Frisch, J.D., 1998, Geothermobarometry of garnet amphibolites from the Indian Creek and Pony–Middle Mountain Metamorphic Suites, Tobacco Root Mountains, Montana [senior thesis]: Amherst, Massachusetts, Amherst College, 86 p.

Giletti, B.J., 1966, Isotopic ages from southwestern Montana: Journal of Geophysical Research, v. 71, p. 4029–4036.

Giletti, B.J., 1968, Isotopic geochronology of Montana and Wyoming, *in* Hamilton, E.I., and Farquhar, R.M., eds., Radiometric dating for geologists: London, Interscience Publishers, p. 111–146.

Gillmeister, N.M., 1971, Petrology of Precambrian rocks in the central Tobacco Root Mountains, Madison Country, Montana [Ph.D. thesis]: Cambridge, Massachusetts, Harvard University, 210 p.

Gillmeister, N.M., 1972, Cherry Creek Group–Pony Group relationship in the central Tobacco Root Mountains, Madison County, Montana: Northwest Geology, v. 1, p. 21–24.

Hanley, T.B., 1975, Structure and petrology of the northwestern Tobacco Root Mountains, Madison County, Montana [Ph.D. thesis]: Bloomington, Indiana University, 289 p.

Hanley, T.B., 1976, Stratigraphy and structure of the central fault block, northwestern Tobacco Root Mountains, Madison County, Montana, *in* Tobacco Root Geological Society, 1976 Field Conference Guidebook: Montana Bureau of Mines and Geology Special Publication 73, p. 7–14.

Hanley, T.B., and Vitaliano, C.J., 1983, Petrography of Archean mafic dikes of the Tobacco Root Mountains, Madison County, Montana: Northwest Geology, v. 12, p. 43–55.

Harlan, S.S., Geissman, J.W., Snee, L.W., and Reynolds, R.L., 1996, Late Cretaceous remagnetization of Proterozoic mafic dikes, southern Highland Mountains, southwestern Montana: A paleomagnetic and $^{40}Ar/^{39}Ar$ study: Geological Society of America Bulletin, v. 108, p. 653–668.

Harms, T.A., Burger, H.R., Brady, J.B., Cheney, J.T., Johnson, K.E., Owen, D.R., King, J.T., Kranenburg, S., Sincock, M.J., and Lowell, J., 1996, Structural setting of Archean rock suites in the Tobacco Root Mountains, southwestern Montana: Geological Society of America Abstracts with Programs, v. 28, no. 7, p. A-245.

Harris, C., 1998, Geochemistry of metamorphosed mafic dikes, Tobacco Root Mountains, southwestern Montana [senior thesis]: Claremont, California, Pomona College, 58 p.

Hatch, C.E., 1998, Sapphirine symplectites in orthoamphibolites of the Archean Indian Creek Metamorphic Suite (ICMS), Tobacco Root Mountains, Montana [senior thesis]: Amherst, Massachusetts, Amherst College, 110 p.

Hayden, F.V., 1872, Preliminary report of the U.S. Geological Survey of Montana and portions of adjacent territories: U.S. Geological Survey of Territories, 5th Annual Report, p. 33–43.

Hess, D.F., 1967, Geology of pre-Beltian rocks in the central and southern Tobacco Root Mountains with reference to superposed effects of the Laramide-age Tobacco Root batholith [Ph.D. thesis]: Bloomington, Indiana University, 333 p.

Hirst, J., 1996, Mineralogy, petrology, and geothermobarometry of aluminous Archean rocks in the Spuhler Peak Metamorphic Suite of the Tobacco Root Mountains, southwestern Montana [senior thesis]: Claremont, California, Pomona College, 64 p.

Hoffman, P.E., 1989, Precambrian geology and tectonic history of North America, *in* Bally, A.W., and Palmer, A.R., eds., The geology of North America—An overview: Boulder, Colorado, Geological Society of America, Geology of North America, v. A, p. 447–512.

Houston, R.S., Erslev, E.A., Frost, C.D., Karlstrom, K.E., Page, N.J., Zientek, M.L., Reed, J.C., Jr., Snyder, G.L., Worl, R.G., Bryant, B., Reynolds, M.W., and Peterman, Z.E., 1993, The Wyoming province, *in* Reed, J.C., et al., eds., Precambrian: Conterminous U.S.: Boulder, Colorado, Geological Society of America, Geology of North America, v. C-2, p. 121–170.

Immega, I.P., 1976, Mineralogy and petrology of some Pre-Cambrian iron formations in southwestern Montana [Ph.D. thesis]: Bloomington, Indiana University, 97 p.

Immega, I.P., and Klein, C., 1976, Mineralogy and petrology of some metamorphic Precambrian iron-formations of southwestern Montana: American Mineralogist, v. 61, p. 1117–1144.

Jacob, L.J., 1994, A geochemical and geochronological characterization of the Indian Creek Metamorphic Suite in the Noble Lake area, south-central Tobacco Root Mountains, southwestern Montana [senior thesis]: Northampton, Massachusetts, Smith College, 73 p.

James, H.L., and Hedge, C.E., 1980, Age of the basement rocks of southwest Montana: Geological Society of America Bulletin, v. 91, p. 11–15.

Johnson, K.E., Burger, H.R., Brady, J.B., Cheney, J.T., Harms, T.A., Thomas, R.B., and MacFarlane, W.A., 1996, Archean meta-ultramafics in the Tobacco Root Mountains, SW Montana: An evaluation of the metamorphic history and protolith: Geological Society of America Abstracts with Programs, v. 28, no. 7, p. A-493.

King, J.F., 1995, The metamorphism of the Spuhler Peak Formation along Noble Lake, Thompson Peak ridge, and Branham Lake, southwestern Montana [senior thesis]: Amherst, Massachusetts, Amherst College, 77 p.

King, J.T., 1994, A structural analysis of Archean gneisses in the Tobacco Root Mountains, southwestern Montana [senior thesis]: Amherst, Massachusetts, Amherst College, 44 p.

Koehler, S.W., 1972, Petrology and petrogeny of the diabase dikes of the Tobacco Root Mountains, southwestern Montana [M.S. thesis]: Bloomington, Indiana University, 60 p.

Koehler, S.W., 1976, Petrology of the diabase dikes of the Tobacco Root Mountains, Montana, *in* Tobacco Root Geological Society, 1976 Field Conference Guidebook: Montana Bureau of Mines and Geology Special Publication 73, p. 27–36.

Kovaric, D.N., Brady, J.B., Cheney, J.T., Grove, M., Jacob, L.J., and King, J.T., 1996, $^{40}Ar/^{39}Ar$ evidence for reheating events affecting basement rocks in the Tobacco Root, Ruby, and Highland Mountains, SW Montana: Geological Society of America Abstracts with Programs, v. 28, no. 7, p. A-493.

Kranenburg, S.R., 1996, Structural relationships among metamorphic suites in the Tobacco Root Mountains of southwestern Montana [senior thesis]: Huntsville, Texas, Sam Houston State University, 41 p.

Krogh, T.E., Kamo, S., and Hess, D.F., 1997, Wyoming province 3300+ gneiss with 2400 Ma metamorphism, northwestern Tobacco Root Mountains, Madison County, Montana: Geological Society of America Abstracts with Programs, v. 29, no. 6, p. A-408.

Levandowski, D.W., 1956, Geology and mineral deposits of the Sheridan-Alder area, Madison County, Montana [Ph.D. thesis]: Ann Arbor, University of Michigan, 318 p.

Lowell, J., 1994, A metamorphic and deformational history of the Spuhler Peak Formation, Tobacco Root Mountains, Montana [senior thesis]: Colorado Springs, Colorado College, 100 p.

MacFarlane, W.A., 1996, A geochemical and petrographic analysis of the meta-ultramafic rocks in the Tobacco Root Mountains, southwest Montana [senior thesis]: Colorado Springs, Colorado College, 132 p.

Martin, J., 1996, Petrology of mafic and felsic gneisses of the Indian Creek Metamorphic Suite, Tobacco Root Mountains, southwestern Montana [senior thesis]: Lancaster, Pennsylvania, Franklin and Marshall College, 53 p.

Mogk, D.W., and Henry, D.J., 1988, Metamorphic petrology of the northern Archean Wyoming province, southwestern Montana: Evidence for Archean collisional tectonics, *in* Ernst, W.G., ed., Metamorphism and crustal evolution in the western U.S.: Englewood Cliffs, New Jersey, Prentice-Hall, p. 363–382.

Mogk, D.W., Mueller, P.A., Wooden, J.L., and Bowes, D.R., 1992a, The northern Wyoming province: Contrasts in Archean crustal evolution, *in* Bartholomew, M.J., et al., eds., Characterization and comparison of ancient and Mesozoic continental margins—Proceedings of the 8th International Conference on Basement Tectonics: Dordrecht, Netherlands, Kluwer Academic Publishers, p. 283–297.

Mogk, D.W., Mueller, P.A., and Wooden, J.L., 1992b, The nature of Archean terrane boundaries: An example from the northern Wyoming province: Precambrian Research, v. 55, p. 155–168.

Mohlman, H.K., 1996, Tectonometamorphic evolution of metamorphosed mafic dikes in the Archean rocks of the Tobacco Root Mountains, Montana [senior thesis]: Amherst, Massachusetts, Amherst College, 82 p.

Monteleone, B., 1998, The petrology and geothermobarometry of the Indian Creek Metamorphic Suite and the Spuhler Peak Metamorphic Suite, Tobacco Root Mountains, southwestern Montana [senior thesis]: Wooster, Ohio, College of Wooster, 78 p.

Mueller, P.A., and Cordua, W.S., 1976, Rb-Sr whole rock age of gneisses from the Horse Creek area, Tobacco Root Mountains, Montana: Isochron/West, v. 16, p. 33–36.

Mueller, P.A., and D'Arcy, K.A., 1990, Sm-Nd systematics of the Late Archean Wyoming Greenstone Terrane: Eos (Transactions, American Geophysical Union), v. 71, p. 662.

Mueller, P.A., Shuster, R.D., Wooden, J.L., Erslev, E.A., and Bowes, D.R., 1993, Age and composition of Archean crystalline rocks from the southern Madison Range, Montana: Geological Society of America Bulletin, v. 105, p. 437–446.

Mueller, P.A., Wooden, J.L., Mogk, D.W., Nutman, A.P., and Williams, I.S., 1996a, Extended history of a 3.5 Ga trondhjemitic gneiss, Wyoming province, USA: Evidence from U-Pb systematics in zircon: Precambrian Research, v. 78, p. 41–52.

Mueller, P., Heatherington, A., Wooden, J., Mogk, D., and Nutman, A., 1996b, Proterozoic evolution of the northwestern Wyoming craton: Geological Society of America Abstracts with Programs, v. 28, no. 7, p. A-314.

Mueller, P.A., Heatherington, A.L., D'Arcy, K.A., Wooden, J.L., and Nutman, A.P., 1996c, Contrasts between Sm-Nd whole-rock and U-Pb zircon systematics in the Tobacco Root batholith, Montana: Implications for the determination of crustal age provinces: Tectonophysics, v. 265, p. 169–179.

Mueller, P., Heatherington, A., Wooden, J., Mogk, D., and Bowes, D., 1998a, The Wyoming province and its place in the evolution of Precambrian North America: Geological Society of America Abstracts with Programs, v. 30, no. 7, p. A-46.

Mueller, P.A., Wooden, J.L., Nutman, A.P., and Mogk, D.W., 1998b, Early Archean crust in the northern Wyoming province: Evidence from U-Pb ages of detrital zircons: Precambrian Research, v. 91, p. 295–307.

Mueller, P., Heatherington, A., Mogk, D., and Wooden, J., 1999, Paleoproterozoic tectonothermal activity along the NW margin of the Archean Wyoming province: Geological Society of America Abstracts with Programs, v. 31, no. 7, p. A107.

O'Neill, J.M., 1983, Geologic map of the Middle Mountain–Tobacco Root roadless area, Madison County, Montana: U.S. Geological Survey Miscellaneous Field Studies Map MF-1590-A, 1:50,000.

O'Neill, J.M., 1998, The Great Falls tectonic zone, Montana-Idaho: An Early Proterozoic collisional orogen beneath and south of the Belt Basin, *in* Berg, R.B., ed., Belt Symposium III—1993: Montana Bureau of Mines and Geology Special Publication 112, p. 222–228.
O'Neill, J.M., and Lopez, D.A., 1985, Character and significance of Great Falls tectonic zone, east-central Idaho and west-central Montana: American Association of Petroleum Geologists Bulletin, v. 69, p. 437–447.
O'Neill, J.M., Duncan, M.S., and Zartman, R.E., 1988, An Early Proterozoic gneiss dome in the Highland Mountains, southwestern Montana, *in* Lewis, S.E., and Berg, R.B., eds., Precambrian and Mesozoic plate margins; Montana, Idaho, and Wyoming, with field guides for the 8th International Conference on Basement Tectonics: Montana Bureau of Mines and Geology Special Publication 96, p. 81–88.
Owen, D.R., 1996, Archean deformation of the Tobacco Root Mountains, southwestern Montana [senior thesis]: Amherst, Massachusetts, Amherst College, 70 p.
Pardee, J.T., 1950, Late Cenozoic block faulting in western Montana: Geological Society of America Bulletin, v. 61, p. 359–406.
Peale, A.C., 1896, Geological atlas of the United States, No. 24, Three Forks Folio: U.S. Geological Survey, 7 p., 3 maps, 2 sections, 1:250,000.
Peck, W.H., 1994, Geochemistry of ortho-amphibole-bearing Archean rocks, Central Tobacco Root Mountains, southwestern Montana, [senior thesis]: Beloit, Wisconsin, Beloit College, 43 p.
Peterman, A.E., 1981, Archean gneisses in the Little Rocky Mountains, Montana: U.S. Geological Survey Professional Paper 1199A, p. 1–14.
Picornell, C.M., 1998, Structure of Archean marbles of the Indian Creek Metamorphic Suite (ICMS), Tobacco Root Mountains, Montana, *in* Mendelson, C.V., and Mankiewicz, C., eds., Eleventh Keck research symposium in Geology: Amherst, Massachusetts, Amherst College, p. 188–190.
Poulsen, C.J., 1994, Origin, metamorphic history, and tectonic setting of Archean rocks from the Spuhler Peak Assemblage, Tobacco Root Mountains, Montana [senior thesis]: Northfield, Minnesota, Carleton College, 54 p.
Pufall, A., 1996, Geochemistry of the Spuhler Peak and Pony–Middle Mountain Metamorphic Suites, Tobacco Root Mountains, Montana, *in* Mendelson, C.V., and Mankiewicz, C., eds., Ninth Keck research symposium in geology: Williamstown, Massachusetts, Williams College, p. 118–121.
Reid, R.R., 1957, Bedrock geology of the north end of the Tobacco Root Mountains, Madison County, Montana: Montana Bureau of Mines and Geology Memoir 36, 25 p.
Reid, R.R., 1963, Metamorphic rocks of the northern Tobacco Root Mountains, Madison County: Geological Society of America Bulletin, v. 74, p. 293–306.
Roberts, H., Dahl, P., Kelley, S., and Frei, R., 2002, New ^{207}Pb-^{206}Pb and ^{40}Ar-^{39}Ar ages from SW Montana, USA: Constraints on the Proterozoic and Archean tectonic and depositional history of the Wyoming province: Precambrian Research, v. 117, p. 119–143.
Root, F.K., 1965, Structure, petrology, and mineralogy of pre-Beltian metamorphic rocks of the Pony-Sappington area, Madison County, Montana [Ph.D. thesis]: Bloomington, Indiana University, 184 p.
Ross, G.M., Milkereit, B., Eaton, D., White, D., Kanasewich, E.R., and Burianyk, M.J.A., 1995, Paleoproterozoic collisional orogen beneath the western Canada sedimentary basin imaged by Lithoprobe crustal seismic-reflection data: Geology, v. 23, p. 195–199.
Schmidt, C.J., 1975, An analysis of folding and faulting in the northern Tobacco Root Mountains, southwestern Montana [Ph.D. thesis]: Bloomington, Indiana University, 480 p.
Sincock, M.J., 1994, Structural and metamorphic history of an intensely deformed zone, lower Branham Lake, southern Tobacco Root Mountains, Montana [senior thesis]: Lancaster, Pennsylvania, Franklin and Marshall College, 31 p.
Steffen, K.J., 1998, Geothermobarometry of Archean garnet-amphibolites: Pony–Middle Mountain and Spuhler Peak Metamorphic Suites, Tobacco Root Mountains, southwest Montana [senior thesis]: Northfield, Minnesota, Carleton College, 101 p.
Tansley, W., Schafer, P.A., and Hart, L.H., 1933, A geological reconnaissance of the Tobacco Root Mountains, Madison County, Montana: Montana Bureau of Mines and Geology Memoir 9, 57 p.
Tendall, B.A., 1978, Mineralogy and petrology of Precambrian ultramafic bodies from the Tobacco Root Mountains, Madison County, Montana [M.S. thesis]: Bloomington, Indiana University, 126 p.
Thomas, R.B., 1996, Metamorphic history of Archean ultramafic pods in the Tobacco Root Mountains, southwestern Montana [senior thesis]: Williamstown, Massachusetts, Williams College, 125 p.
Tierney, K.A., 1994, The origin and evolution of the Spuhler Peak Formation along the western ridge of Thompson Peak, Tobacco Root Mountains, Montana [senior thesis]: Amherst, Massachusetts, Amherst College, 87 p.
Tuit, C.B., 1996, Geothermobarometry of the Indian Creek Metamorphic Suite, Tobacco Root Mountains, Montana [senior thesis]: Beloit, Wisconsin, Beloit College, 43 p.
Vitaliano, C.J., Burger, H.R., Cordua, W.S., Hanley, T.B., Hess, D.F., and Root, F.K., 1979a, Geologic map of southern Tobacco Root Mountains, Madison County, Montana: Geological Society of America Map and Chart Series MC31, scale 1:62,500, 1 sheet, 8 p. text.
Vitaliano, C.J., Burger, H.R., Cordua, W.S., Hanley, T.B., Hess, D.F., and Root, F.K., 1979b, Explanatory text to accompany geologic map of the southern Tobacco Root Mountains, Madison County, Montana: Geological Society of America Map and Chart Series MC31, scale 1:62,500, 1 sheet, 8 p. text.
Vitaliano, C.J., Kish, S.A., and Towell, D.G., 1980, K-Ar dates and Sr isotopic values for rocks of the Tobacco Root batholith: Isochron/West, v. 28, p. 13–15.
Wegmann, K.W., 1996, Metamorphic evolution of the Archean Pony Middle Mountain Metamorphic Suite, Tobacco Root Mountains of southwestern Montana [senior thesis]: Walla Walla, Washington, Whitman College, 74 p.
Winchell, A.N., 1914, Mining districts of the Dillon quadrangle, Montana, and adjacent areas: U.S. Geological Survey Bulletin 574, 191 p.
Wooden, J.L., Vitaliano, C.J., Koehler, S.W., and Ragland, P.C., 1978, The late Precambrian mafic dikes of the southern Tobacco Root Mountains, Montana: Geochemistry, Rb-Sr geochronology, and relationship to Belt tectonics: Canadian Journal of Earth Sciences, v. 15, p. 467–479.

Manuscript Accepted by the Society September 12, 2003

Printed in the USA

Geological Society of America
Special Paper 377
2004

Geochemistry of quartzofeldspathic gneisses and metamorphic mafic rocks of the Indian Creek and Pony–Middle Mountain Metamorphic Suites, Tobacco Root Mountains, Montana

David W. Mogk
Department of Earth Sciences, Montana State University, Bozeman, Montana 59717, USA

H. Robert Burger
Department of Geology, Smith College, Northampton, Massachusetts 01063, USA

Paul A. Mueller
Kimberly D'Arcy
Ann L. Heatherington
Department of Geological Sciences, University of Florida, Gainesville, Florida 32611, USA

Joseph L. Wooden
U.S. Geological Survey, 345 Middlefield Road, Menlo Park, California 94025, USA

Reyna L. Abeyta
Colorado College, Colorado Springs, Colorado 80903, USA

Jennifer Martin
Franklin & Marshall College, Lancaster, Pennsylvania 17604-3003, USA

Lisa J. Jacob
Department of Geology, Smith College, Northampton, Massachusetts 01063, USA

ABSTRACT

Quartzofeldspathic gneisses and metamorphic mafic rocks are the dominant lithologies in the Indian Creek and Pony–Middle Mountain Metamorphic Suites of the Tobacco Root Mountains. Field relations, geochemical discriminant analysis, and isotopic systematics indicate that these rocks derive from a bimodal volcanic suite ca. 3.3 Ga. The quartzofeldspathic gneisses contain sodic rocks of the tonalite-trondhjemite-granodiorite suite as well as potassic varieties. This suite of rocks most likely contains some lithologies derived from sedimentary or volcaniclastic sources, and there is evidence that alkali metasomatism occurred prior to or during subsequent major tectonothermal events. The entire suite of gneisses and metamorphic mafic rocks has geochemical characteristics that are indicative of an active continental arc setting, with deposition most likely in an extensional, backarc setting similar to the Mesozoic through Tertiary rocks of the eastern Sierra Nevada Mountains or Mojave Desert. The formation of these rocks represents an early, distinct stage of crustal evolution that preceded the (unconformable?) deposition of one or more platform-type

Mogk, D.W., Burger, H.R., Mueller, P.A., D'Arcy, K., Heatherington, A.L., Wooden, J.L., Abeyta, R.L., Martin, J., and Jacob, L.J., 2004, Geochemistry of quartzofeldspathic gneisses and metamorphic mafic rocks of the Indian Creek and Pony–Middle Mountain Metamorphic Suites, Tobacco Root Mountains, Montana, *in* Brady, J.B., Burger, H.R., Cheney, J.T., and Harms, T.A., eds., Precambrian geology of the Tobacco Root Mountains, Montana: Boulder, Colorado, Geological Society of America Special Paper 377, p. 15–46. For permission to copy, contact editing@geosociety.org.

sedimentary sequences (e.g., marbles, pelitic schists, quartzites, banded iron formations). All primary lithologic contacts and textures or structures indicative of possible protoliths have been largely obliterated due to transposition during Archean and Paleoproterozoic (ca. 2.4 and ca. 1.8 Ga) deformation and metamorphism.

Keywords: quartzofeldspathic gneiss, metamorphosed mafic rocks, geochemistry, protolith, tectonic setting.

INTRODUCTION

Archean rocks are extensively exposed in the Tobacco Root Mountains, which lie along the western margin of the Wyoming province (see Burger, this volume, Chapter 1, his Figure 1). Previous work recognized three distinct domains of Archean rocks in the Wyoming province: the Beartooth-Bighorn magmatic zone in the east, the Wyoming greenstone province in the south, and the Montana metasedimentary terrane in the west (Mogk et al., 1992). Crystalline rocks in the Beartooth and Bighorn Mountains are well established as metaplutonic based on field relations, geochemical properties, and isotopic systematics. The crystalline rocks of the Beartooth-Bighorn magmatic terrane largely belong to the tonalite-trondhjemite-granodiorite (TTG) suite and formed ca. 2800 Ma (Wooden et al., 1988; Mueller et al., 1988; Wooden and Mueller, 1988). In contrast, the Montana metasedimentary terrane is distinguished by prominent sequences of metasedimentary rocks, including marble, quartzite, iron formation, and pelitic schist (Mogk et al., 1992). Although distinctive, these metasupracrustal associations are volumetrically subordinate to a variety of quartzofeldspathic gneisses. In some parts of the Montana metasedimentary terrane, such as the Spanish Peaks area of the northern Madison Range, the quartzofeldspathic gneisses belong largely to the TTG plutonic suite and formed at ca. 3.3–3.0 Ga (Mogk et al., 1987; Salt, 1987). Gneisses of the southern Madison Range also contain 3.3 Ga tonalitic gneisses as well as a younger suite of 2.7 Ga granitic rocks that clearly intrude the older gneisses and a metasupracrustal suite (Mueller et al., 1993).

In contrast, the quartzofeldspathic gneisses of the Tobacco Root Mountains include significant volumes of granitic gneisses interlayered with TTG gneisses. Metasedimentary rocks are spatially associated with both the TTG and granitic quartzofeldspathic gneisses and metamorphic mafic rocks, particularly in the western and southern parts of the Tobacco Root Mountains (Vitaliano et al., 1979; see reprinted map accompanying this volume; Cheney et al., this volume, Chapter 6). Numerous workers have described the metasedimentary sequences in the Gravelly Range (Peale, 1896; Heinrich and Rabbit, 1960), Madison Range (Spencer and Kozak, 1975; Erslev, 1983; Salt, 1987), Ruby Range (Garihan, 1979; Garihan and Williams, 1976), and Tobacco Root Mountains (Tansley et al., 1933; Reid, 1957, 1963; Root, 1965; Hess, 1967; Burger, 1969; Gillmeister, 1971; Cordua, 1973; Hanley, 1975; Friberg, 1976), but to date, consensus has not been achieved on the origin of these rocks, or on their relationship to the quartzofeldspathic gneisses and metamorphic mafic rocks in the Tobacco Root Mountains. In this contribution, we present new geochemical data to address the protolith(s) and tectonic setting of the quartzofeldspathic gneisses and metamorphic mafic rocks that are associated with the metasedimentary sequences in the Tobacco Root Mountains.

Four distinct lithologic associations are defined in this study (Burger, this volume, Chapter 1): Indian Creek Metamorphic Suite, Pony–Middle Mountain Metamorphic Suite, Spuhler Peak Metamorphic Suite, and metamorphosed mafic dikes and sills.

This contribution is limited to a discussion of the two most extensive units, the Indian Creek Metamorphic Suite and the Pony–Middle Mountain Metamorphic Suite. The Indian Creek Metamorphic Suite occurs in the west-central part of the Tobacco Root Mountains and includes the diagnostic marble-quartzite-pelitic schist-banded iron formation–bearing supracrustal sequences, in addition to the dominant quartzofeldspathic gneisses and metamorphosed mafic rocks. The Pony–Middle Mountain Metamorphic Suite occurs in the northeastern part of the range and contains mostly quartzofeldspathic gneisses and metamorphic mafic rocks with few demonstrable metasupracrustal units. Earlier workers assigned the metasupracrustal sequences to the "Cherry Creek Series" (e.g., Tansley et al., 1933), distinguished in the Gravelly Range (Peale, 1896) by extensive occurrences of marbles. This correlation is not followed in the present study because there is no unequivocal evidence for establishing temporal or geospatial relations among these occurrences. Similarly, the term "Pony gneiss" was coined by Tansley et al. (1933) for quartzofeldspathic and mafic gneisses in the northeastern part of the range. Tansley et al. (1933) interpreted the "Cherry Creek" rocks as being younger than the "Pony gneisses," whereas Reid (1957) presented just the opposite interpretation. Unfortunately, these designations of rock units ultimately evolved into genetic interpretations.

The question of the relationship between the supracrustal rocks, quartzofeldspathic gneisses, and various metamorphic mafic rock units of the Indian Creek Metamorphic Suite and Pony–Middle Mountain Metamorphic Suite persists today. The Tobacco Root Mountains have experienced at least two major metamorphic events at the granulite to upper amphibolite facies at ca. 2.4 and ca. 1.8 Ga, respectively (Mueller et al., this volume, Chapter 9; Cheney et al., this volume, Chapter 6). The last major structural event at 1800 Ma involved penetrative deformation, including regional simple shear and formation of sheath folds (Harms et al., this volume, Chapter 10). Consequently, primary lithologic field relations have been largely obliterated by these tectonothermal events (e.g., Passchier et al., 1990).

Multiple hypotheses exist for the origin and evolution of the following lithologic associations.

- The quartzofeldspathic gneisses (of plutonic or supracrustal origin) are basement to unconformable supracrustal rocks prior to Late Archean and Paleoproterozoic metamorphism/deformation.
- The quartzofeldspathic gneisses are derived from arkosic, volcaniclastic, or volcanic supracrustal sequences (i.e., paragneisses) that were originally interlayered with the unequivocally supracrustal marbles, quartzites, iron formation, and pelitic schists.
- The supracrustal rocks were intruded by felsic magmas (i.e., orthogneisses) prior to metamorphism/deformation.
- The gneisses, mafic rocks, and supracrustal rocks may have been tectonically juxtaposed prior to the last major tectonothermal event.

In a similar setting in the adjacent Ruby Range, the Dillon (quartzofeldspathic) gneiss has alternately been interpreted as both a paragneiss and orthogneiss (Garihan and Williams, 1976; Garihan, 1979). Correlative occurrences of Dillon gneiss in the Blacktail Mountains have been interpreted as paragneisses from volcanic or volcaniclastic protoliths (Clark, 1987). We present a summary of lithologic associations, geochemical discriminants, and isotope systematics to interpret the origins of the quartzofeldspathic gneisses and related metamorphic mafic rocks in the Tobacco Root Mountains.

LITHOLOGIC ASSOCIATIONS IN THE INDIAN CREEK METAMORPHIC SUITE AND PONY–MIDDLE MOUNTAIN METAMORPHIC SUITE

Rocks in the Indian Creek Metamorphic Suite are dominantly quartzofeldspathic gneisses, with lesser abundances of amphibolites and hornblende gneisses, and the characteristic metasupracrustal suite. Rocks in the Pony–Middle Mountain Metamorphic Suite are fundamentally the same, although the metasupracrustal suite is not as prevalent. Compositional layering occurs throughout these lithologies on scales ranging from centimeters to hundreds of meters (Harms et al., this volume, Chapter 10, their Figure 3). Map units (e.g., Vitaliano et al., 1979) have been defined by the relative proportions of rock types within them, recognizing that in detail compositional variation persists down to the centimeter scale. Lithologic contacts in these units are largely concordant as a result of transposition during D1 isoclinal folding, and further reorientation during tight to isoclinal folding and simple shearing during D2 (Harms et al., this volume, Chapter 10). In both the Indian Creek and Pony–Middle Mountain Metamorphic Suites, quartzofeldspathic and mafic gneisses are intimately interlayered.

The gneisses are typically tonalitic to granitic in composition based on their modal mineralogy (International Union of Geological Sciences [IUGS] classification), although some samples are unusually rich in potassium feldspar and are classified as alkali feldspar granites. Biotite and hornblende vary from 0 to 25 mod%, and in some cases the quartzofeldspathic gneisses grade into semipelitic schists or mafic gneisses. Some of the quartzofeldspathic gneisses are quite aluminous, with biotite defining foliation, sillimanite (fibrolite) growing in micaceous layers, and garnets (with cordierite coronas) growing in discrete layers. Multiple generations of migmatite and pegmatite occur throughout the quartzofeldspathic gneisses. Migmatitic leucosomes are locally conformable, folded, or crosscutting, attesting to a long and varied petrogenetic history. Pegmatitic pods, up to meters across, typically cut the foliation and dominantly contain microcline, quartz, and cm-scale books of muscovite and biotite. There are also layers of an unusual garnet-gedrite unit that are conformable in the quartzofeldspathic gneisses, as well as dismembered pods of meta-ultramafic rocks (Vitaliano et al., 1979; Johnson et al., this volume, Chapter 4).

Metamorphic mafic rocks also occur in layers tens of centimeters to hundreds of meters thick. In most cases, the metamorphic mafic rocks occur as well-foliated and lineated amphibolite or hornblende gneiss. Some mafic layers preserve an early granulite mineralogy with granoblastic texture that may be variably overprinted by amphibolite facies assemblages. In rare cases, subeuhedral plagioclase is present, which may represent relict subophitic texture.

GEOCHEMISTRY

Given the polymetamorphic history of the Tobacco Root Mountains, it is likely that there are many crustal components of different ages and origins within the lithologic units recognized as the Indian Creek Metamorphic Suite and Pony–Middle Mountain Metamorphic Suite map units, and few with unaltered compositions relative to their protoliths. For example, a "gray gneiss" (trondhjemitic composition, IUGS classification) unit from the North Snowy Block, Beartooth Mountains, has yielded at least three distinct populations of zircons, each representing distinct stages of magmatism and/or metamorphism (Mueller et al., 1996). The present-day geochemical signature may be a composite of multiple igneous intrusions or extrusions, hydrothermal alteration and/or weathering, and if supracrustal in origin, mixing of different sources. Nonetheless, geochemical discrimination diagrams are useful as guides to the dominant petrologic affinities, petrogenetic evolution, protoliths, and tectonic setting. This section discusses a composite geochemical database derived from contributions from the Keck Geology Consortium (Abeyta, Martin, Jacob), University of Florida (Mueller, Heatherington, D'Arcy), and Montana State University (Mogk). The composite database and analytical methods are presented in Tables 1 and 2. The terms "granitic" and "tonalitic" are used in a strictly descriptive sense to distinguish potassic versus sodic felsic gneiss, and do not imply any genetic interpretations.

Major and Trace Element Whole-Rock Analyses

Rock types are represented on a series of chemical variation diagrams in groups according to SiO_2 content: metamorphic mafic rocks <52 wt% (open squares), intermediate gneisses

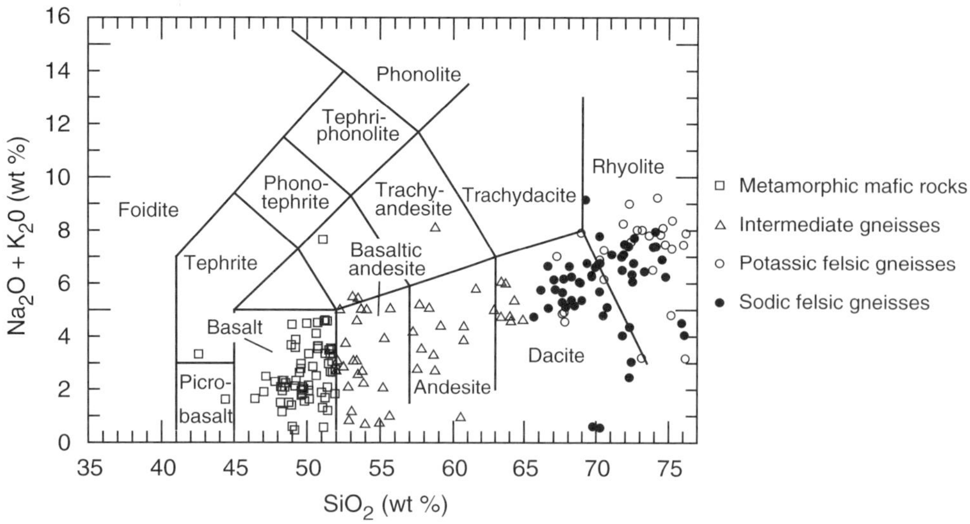

Figure 1. Geochemical classification of the metamorphic mafic rocks and quartzofeldspathic gneisses of the Tobacco Root Mountains (after LeBas et al., 1986).

52–65 wt% (open triangles), and felsic gneisses >65 wt% (Hyndman, 1985). Granitic gneisses (potassic; open circles) are further distinguished from tonalitic gneisses (sodic; closed circles) for rocks where K_2O wt% > Na_2O wt%. Overall, the quartzofeldspathic gneisses, mafic gneisses, and related metamorphic mafic rocks span a range of compositions from picritic basalt to rhyolite (Fig. 1, after LeBas et al., 1986). There is generally a bimodal distribution of rock types with basalts and basaltic andesites on the mafic end and dacites and rhyolites on the felsic end; although present, gneisses of intermediate composition are subordinate (Fig. 2). This bimodal distribution of rock types can also be seen in a series of Harker variation diagrams (Figs. 3A–L). Most of the rocks have compositions that fall within the expected range of the basalt-andesite-dacite-rhyolite series (e.g., Hyndman, 1985). The majority of the rocks are subalkaline as defined by total alkali versus silica (Fig. 4, after Irvine and Baragar, 1971). Most of the mafic rocks follow a tholeiitic trend on an AFM diagram (Fig. 5), although a few mafic rocks and intermediate gneisses plot as calc-alkaline (Fig. 6). Most of the metamorphic mafic rocks are quartz-normative, although a few samples are olivine-normative (Table 1). For the felsic gneisses (i.e., SiO_2 > 65 wt%), potassic versus sodic varieties can be distinguished on a CaO versus K_2O/Na_2O diagram, with higher CaO content and K_2O/Na_2O ratio <1 for the TTG suite (Fig. 7). The felsic and intermediate gneisses mostly follow a calc-alkaline trend (Fig. 5), and most are metaluminous (as indicated by normative diopside), although about a quarter of the samples are peraluminous (as indicated by normative corundum, Figure 8).

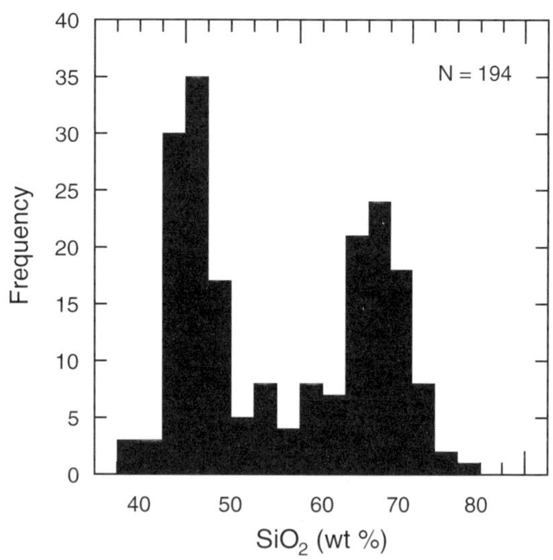

Figure 2. Histogram of weight percent SiO_2 showing the bimodal distribution of the metamorphosed mafic rocks and quartzofeldspathic gneisses of the Tobacco Root Mountains.

Given the likelihood of alteration and mobilization of major elements, and particularly of the alkali and alkali earth elements, in the Tobacco Root rocks during their long geologic histories, elements recognized as being immobile during hydrothermal alteration and metamorphism should provide the most useful indicators of magmatic affinities and precursor rock types (MacLean and Barrett, 1993). Using their lithogeochemical variation diagrams, a plot of Y versus Zr (Fig. 9) shows that the mafic rocks lie in the tholeiitic field, the intermediate rocks lie in the transitional field, and the felsic rocks mostly lie in the calc-alkaline field. However, the regression lines through these data points have slopes (indicated by lines of constant Zr/Y ratios, of 6–7 for rocks of mafic and intermediate compositions, respectively) indicating a transitional (i.e., mildly calc-alkaline) affinity, and the felsic gneisses in aggregate are represented by a regressed line with Zr/Y ratio of 9–10, indicating a calc-alkaline affinity.

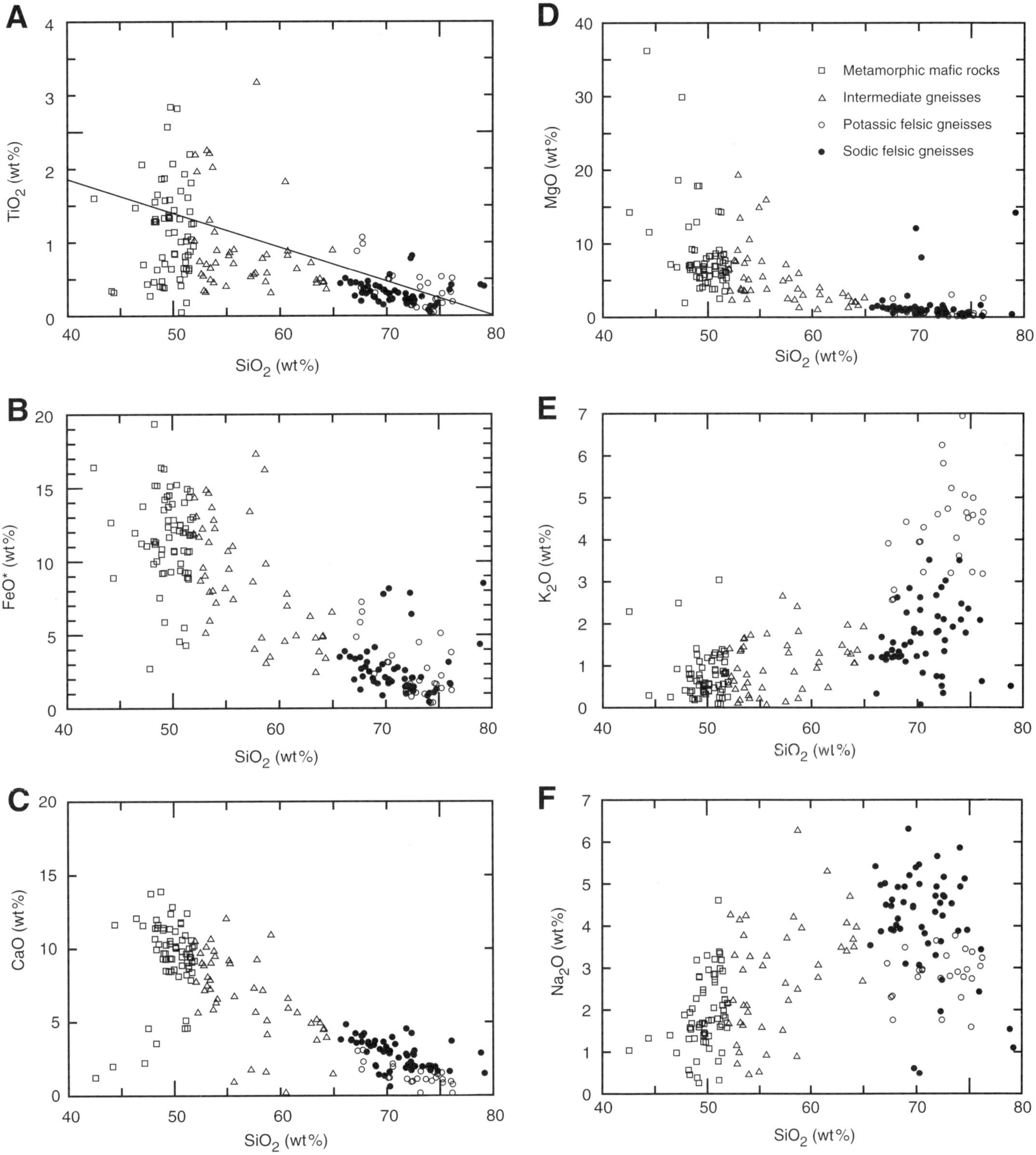

Figure 3 (*on this and following page*). Harker chemical variation diagrams for all metamorphic mafic rocks and quartzofeldspathic gneisses for selected major and trace elements. The reference line in A delineates rocks that are likely to have an igneous origin (below the line) and those that may have a sedimentary origin (above the line; after Tarney, 1976).

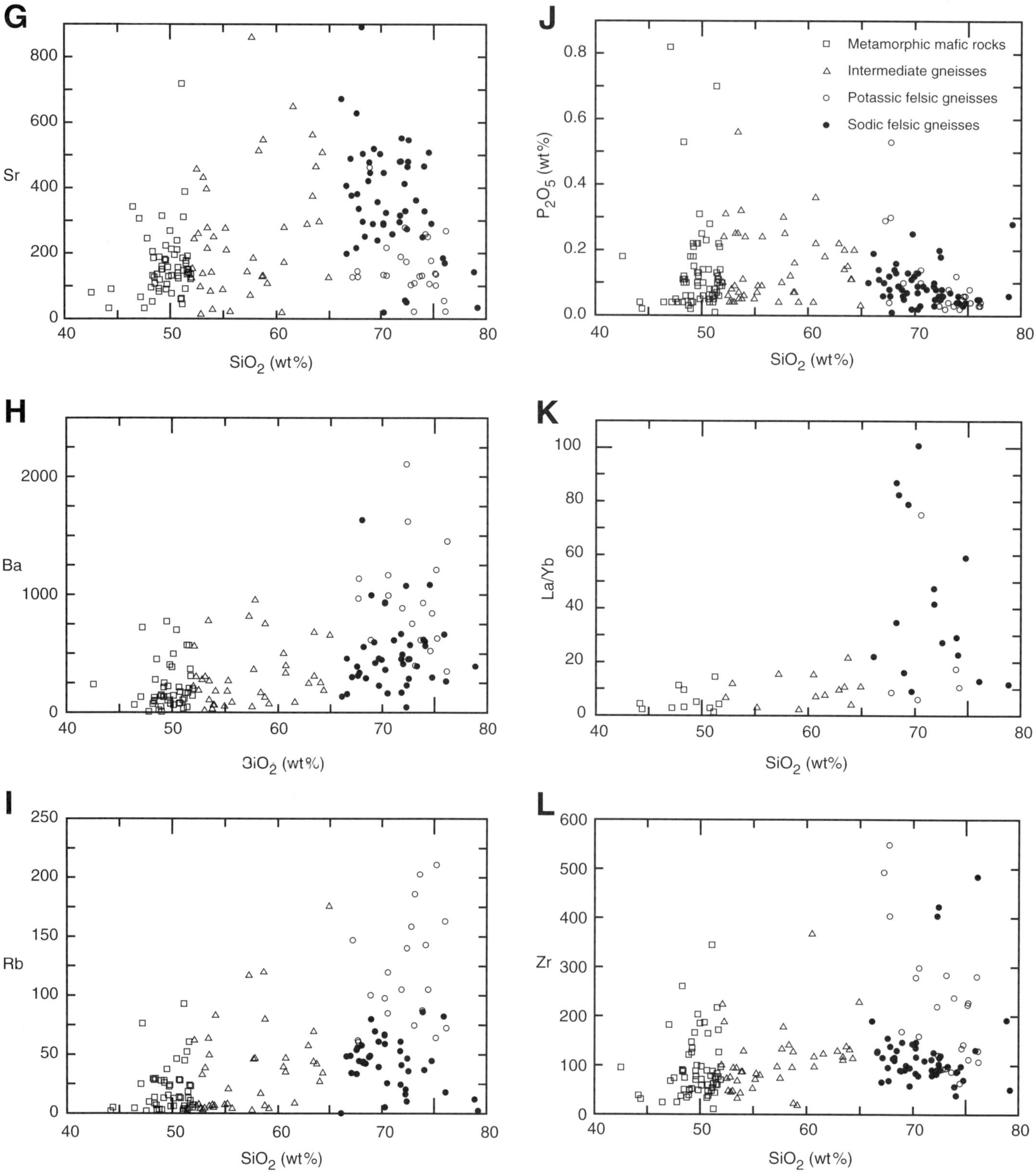

Figure 3 (*continued*).

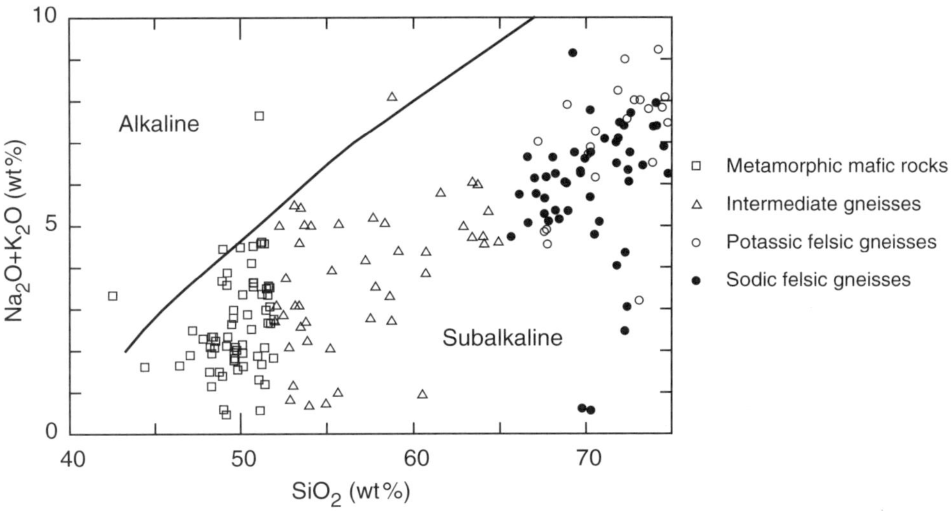

Figure 4. Total alkalis (weight percent $Na_2O + K_2O$) vs. SiO_2 used as a discriminant for alkaline vs. subalkaline rocks (after Irvine and Baragar, 1971).

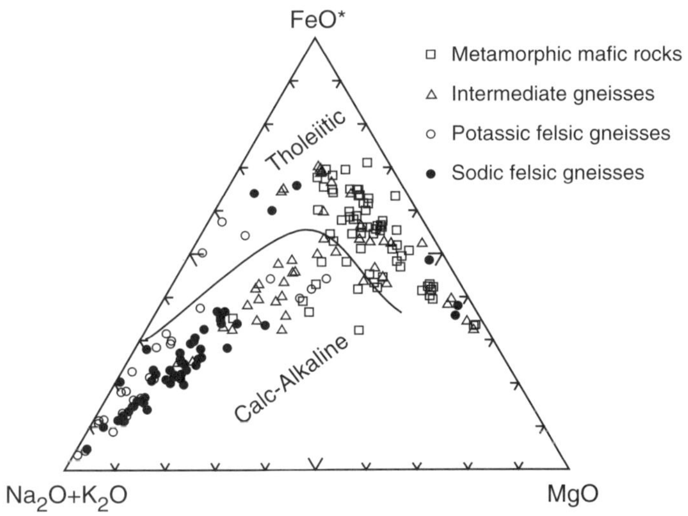

Figure 5. AFM diagram (after Irvine and Baragar, 1971) for all metamorphic mafic rocks and quartzofeldspathic gneisses. The mafic rocks generally follow a tholeiitic trend, and the quartzofeldspathic gneisses follow a calc-alkaline trend.

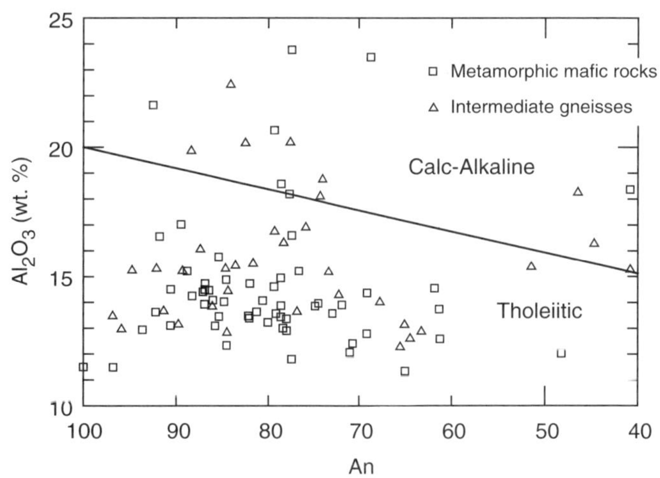

Figure 6. Al_2O_3 vs. normative anorthite (An) diagram for metamorphic mafic rock and gneisses of intermediate composition showing the overall tholeiitic compositions of the metamorphic mafic rocks with subordinate calc-alkaline samples (after Irvine and Baragar, 1971).

A plot of TiO_2 versus Zr (Fig. 10, after MacLean and Barrett, 1993) shows an idealized fractionation curve for immobile elements for a basalt-andesite-dacite-rhyolite sequence. Intersection of the regression lines for each suite of metamorphic rocks with the fractionation curve identifies the original (pre-metamorphic) mafic rocks as basaltic andesites, the intermediate gneisses as andesites (many are high-silica varieties), and the felsic gneisses as midway between dacite and rhyolite in composition. The regression lines for the potassic and sodic gneisses are nearly colinear, indicating that the precursor rocks for these two types of gneisses may have been identical in composition.

Plots of immobile compatible-incompatible element pairs (e.g., Al_2O_3 vs. Zr; after MacLean and Barrett, 1993) may be used to identify one or more homogeneous precursor rocks. For each distinct precursor rock, it is expected that a distinct line should be defined from the origin through the precursor composition, with altered equivalents lying along this line. Figure 11 shows the regression lines for the metamorphic mafic rocks and intermediate, sodic, and potassic gneisses. There is poor correlation of the data points in each of these suites of rocks ($r < 0.47$ in all cases), and the regression lines do not pass through the origin. The premise of these plots is that they

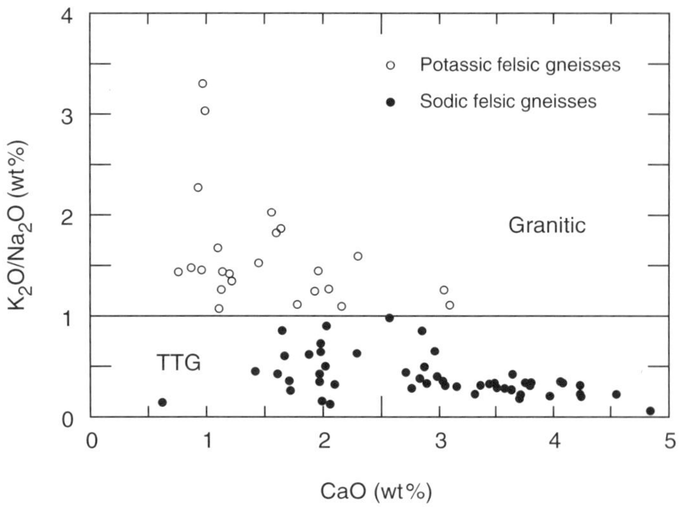

Figure 7. K_2O/Na_2O vs. CaO used to distinguish quartzofeldspathic gneisses of the TTG suite and granitic varieties.

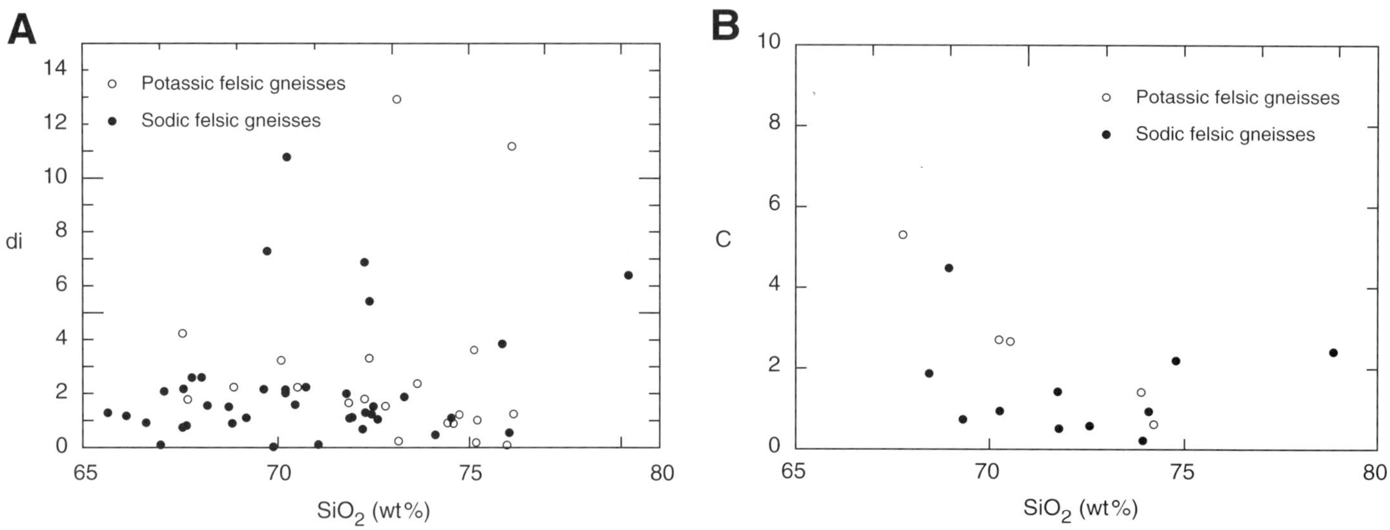

Figure 8. Normative diopside (di) and normative corundum (C) vs. SiO_2 for the potassic gneisses (open circles) and sodic gneisses (closed circles). Most of the gneisses are metaluminous, although ~25% of the gneisses have peraluminous compositions.

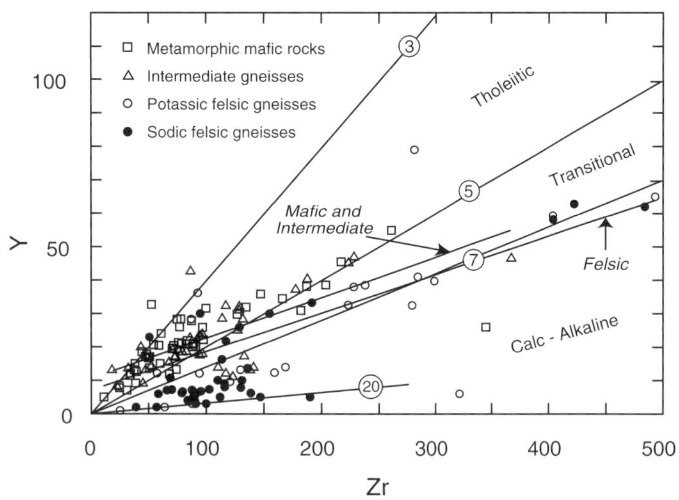

Figure 9. Y-Zr plot used for discrimination of magmatic affinity (after MacLean and Barrett, 1993). Reference lines with numbers indicate constant Zr/Y ratios, which delineate fields of magmatic affinities. Regression lines for mafic and intermediate rocks have Zr/Y ratios of 6–7, indicating an affinity with transitional magmas, and for felsic rocks have Zr/Y ratios of 9–10, indicating a calc-alkaline affinity.

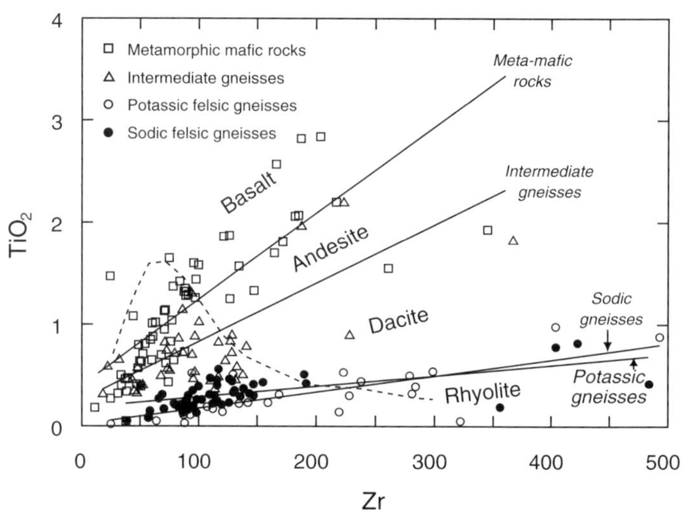

Figure 10. TiO_2-Zr plot showing an idealized fractionation curve (dashed line; after MacLean and Barrett, 1993). The intersection of the regression lines through the data points for each suite of rocks identifies the precursor of each.

TABLE 1. WHOLE-ROCK ANALYSES FOR METAMORPHOSED MAFIC ROCKS AND QUARTZOFELDSPATHIC GNEISSES OF THE TOBACCO ROOT MOUNTAINS

Sample	OO1	OO2	OO3	OO4	OO5b	OO6	OO9	O1O	O11	O12	O13	O14	O15	O17	O18	018 II	O19	019 II	O2O
Data Source	1	1	1	1	1	1	1	1	1	1	1	1	1	1	1	1	1	1	1
Major Elements (wt%)																			
SiO_2	53.49	44.17	49.84	58.76	70.24	68.08	48.32	70.54	53.88	75.91	47.05	67.77	53.42	53.79	n.a.	67.59	66.64	65.65	51.00
TiO_2	0.49	0.34	0.64	0.58	0.29	0.30	1.32	0.23	1.14	0.35	2.06	0.98	1.30	0.45	n.a.	0.38	0.44	0.44	0.65
Al_2O_3	16.05	5.00	15.76	22.42	14.63	16.79	13.74	14.29	13.66	12.48	14.55	13.08	15.17	15.31	n.a.	13.18	15.09	14.88	11.49
Fe_2O_3T	8.78	14.07	10.34	3.40	1.99	3.15	16.87	2.02	14.22	3.49	12.48	7.30	12.56	8.87	n.a.	6.50	3.90	3.89	13.32
Fe_2O_3	n.a.	n.a.	n.a.	n.a.	n.a.	n.a.	n.a.	n.a.	n.a.	n.a.	n.a.	n.a.	n.a.	n.a.	n.a.	n.a.	n.a.	n.a.	n.a.
FeO	n.a.	n.a.	n.a.	n.a.	n.a.	n.a.	n.a.	n.a.	n.a.	n.a.	n.a.	n.a.	n.a.	n.a.	n.a.	n.a.	n.a.	n.a.	n.a.
MgO	7.29	36.20	7.97	1.27	0.57	0.99	3.43	0.63	5.05	1.62	6.78	0.31	5.31	7.91	n.a.	2.51	1.33	1.30	14.39
MnO	0.28	0.15	0.27	0.02	0.03	0.08	0.23	0.02	0.21	0.03	0.19	0.12	0.15	0.15	n.a.	0.10	0.05	0.06	0.52
CaO	10.64	1.99	12.84	4.14	2.85	2.96	9.95	2.16	9.49	1.65	11.58	2.30	7.25	9.84	n.a.	1.78	3.80	3.81	4.58
Na_2O	2.10	n.d.	1.23	6.27	3.07	4.02	1.94	2.94	1.94	2.43	0.98	1.76	2.95	2.10	n.a.	2.30	3.87	3.54	1.79
K_2O	0.46	n.d.	0.32	1.81	2.62	2.62	0.40	3.22	0.28	2.08	0.92	2.80	1.63	0.58	n.a.	2.56	1.20	1.20	0.09
P_2O_5	0.07	0.04	0.05	0.07	0.07	0.14	0.10	0.10	0.11	0.05	0.82	0.53	0.56	0.05	n.a.	0.04	0.11	0.10	0.06
Total	99.63	102.00	99.26	98.73	96.36	99.13	99.29	96.15	99.97	100.10	97.41	96.95	100.30	99.03		96.94	96.40	94.87	97.86
LOI	−0.52	n.a.	−0.45	−1.13	−0.49	−0.54	−0.26	−0.39	−0.25	−0.51	−0.46	−0.21	−1.47	−0.68	−1.06	−1.06	−0.38	−0.41	−0.33
Rare earth and trace elements (ppm)																			
Sc	30.7	24.1	28.0	n.d.	3.0	4.7	50.3	1.1	39.2	124.6	15.7	n.d.	16.2	42.4	13.7	n.a.	0.4	n.a.	48.7
V	167.1	104.8	192.4	53.1	24.4	38.3	362.2	42.6	292.9	49.3	269.1	9.7	190.3	173.6	86.5	n.a.	53.5	n.a.	208.4
Cr	156.9	5751.0	564.5	47.4	66.3	55.2	125.8	51.9	104.4	244.9	223.3	63.5	131.5	377.8	440.5	n.a.	124.3	n.a.	1135.0
Co	36.5	148.1	43.9	11.5	5.3	4.9	72.9	4.7	54.6	9.0	58.0	7.9	50.7	33.4	17.3	n.a.	8.8	n.a.	68.1
Ni	90.7	1897.0	94.8	13.8	1.9	4.7	n.d.	4.0	18.6	78.2	137.5	10.5	122.7	125.4	100.6	n.a.	35.1	n.a.	236.8
Cu	37.7	6.7	37.3	35.6	8.3	71.9	30.4	5.0	55.6	18.7	123.1	26.1	156.2	17.9	20.6	n.a.	10.2	n.a.	89.7
Zn	52.1	49.4	85.4	48.8	24.7	29.7	138.5	30.7	107.4	45.0	112.9	56.4	119.8	50.2	58.1	n.a.	40.7	n.a.	88.9
Ga	16.0	4.5	16.6	25.2	16.8	17.3	23.5	17.0	20.7	14.7	21.0	18.1	20.2	15.0	17.1	n.a.	17.9	n.a.	12.5
Rb	6.0	2.0	6.0	79.7	66.4	57.8	8.6	85.0	4.9	82.4	24.9	60.5	63.4	7.0	101.2	n.a.	48.5	n.a.	2.3
Sr	213.6	32.2	129.0	545.5	292.0	891.7	131.7	216.8	140.9	185.1	306.0	127.9	395.7	83.0	200.0	n.a.	198.4	n.a.	59.4
Y	15.3	9.1	16.7	8.2	13.6	16.2	27.8	12.3	22.7	10.0	31.0	59.3	23.3	19.8	16.3	n.a.	26.0	n.a.	20.9
Zr	67.9	38.6	49.7	22.1	135.8	113.6	87.4	159.2	86.2	131.0	182.1	403.8	94.0	43.3	110.2	n.a.	128.8	n.a.	78.8
Nb	2.3	1.7	2.7	7.9	3.6	19.8	5.9	2.8	4.2	7.4	10.9	21.4	11.4	4.0	7.5	n.a.	7.3	n.a.	4.0
Ba	104.0	n.d.	74.2	750.9	927.6	1633.0	100.7	1166.0	61.2	663.5	131.2	1136.0	776.2	27.3	783.4	n.a.	156.2	n.a.	104.5
La	7.7	5.8	6.7	3.8	n.d.	26.8	2.9	42.8	6.1	27.8	15.6	43.0	24.3	7.5	23.9	n.a.	20.8	n.a.	7.3
Ce	n.a.	n.d.	n.a.	n.a.	n.a.	n.a.	18.8	87.7	n.a.	n.a.	n.a.	n.a.	n.a.	n.a.	n.a.	n.a.	n.a.	n.a.	2.1
Hf	1.6	0.7	1.2	n.d.	4.7	2.3	1.6	5.6	2.0	4.7	3.7	12.2	2.0	1.2	3.4	n.a.	4.2	n.a.	1.6
Th	0.8	1.8	0.6	n.d.	n.d.	7.3	0.5	32.0	n.d.	13.3	2.1	9.2	1.5	1.4	10.3	n.a.	2.6	n.a.	n.d.
U	n.a.	0.0	n.a.	n.a.	n.a.	n.a.	0.1	n.d.	n.a.	n.a.	n.a.	n.a.	n.a.	n.a.	n.a.	n.a.	n.a.	n.a.	0.6

(continued)

TABLE 1. WHOLE-ROCK ANALYSES FOR METAMORPHOSED MAFIC ROCKS AND QUARTZOFELDSPATHIC GNEISSES OF THE TOBACCO ROOT MOUNTAINS (*continued*)

Sample	020 II	O21	O22	O23	O24	O25	O26	O27	O29	O30	O31	O32	032 II	O34	O35	035 II	O36	O37	O38
Data Source	1	1	1	1	1	1	1	1	1	1	1	1	1	1	1	1	1	1	1
Major Elements (wt%)																			
SiO_2	51.24	51.07	62.89	67.83	51.39	51.42	72.46	53.12	67.72	57.24	70.55	48.32	49.68	81.32	49.19	49.62	73.89	52.83	46.44
TiO_2	0.64	0.57	0.72	0.37	1.08	0.83	0.26	0.70	1.07	0.53	0.54	1.29	1.33	0.19	1.33	1.34	0.44	0.54	1.47
Al_2O_3	11.49	14.36	15.26	17.68	18.19	14.40	15.43	20.15	12.99	15.79	14.70	13.92	14.08	12.03	13.38	13.36	13.13	12.27	17.02
Fe_2O_3T	13.34	11.96	6.93	2.41	9.92	11.84	2.31	5.69	8.04	14.87	3.99	12.44	12.95	1.32	15.03	15.25	3.04	13.57	13.29
Fe_2O_3	n.a.	n.a.	n.a.	n.a.	n.a.	n.a.	n.a.	n.a.	n.a.	n.a.	n.a.	n.a.	n.a.	n.a.	n.a.	n.a.	n.a.	n.a.	n.a.
FeO	n.a.	n.a.	n.a.	n.a.	n.a.	n.a.	n.a.	n.a.	n.a.	n.a.	n.a.	n.a.	n.a.	n.a.	n.a.	n.a.	n.a.	n.a.	n.a.
MgO	14.28	9.15	3.19	1.17	5.05	7.37	0.55	3.80	1.19	4.90	0.51	7.11	7.29	0.05	6.29	6.37	0.40	9.05	7.17
MnO	0.52	0.19	0.11	0.01	0.14	0.19	0.02	0.07	0.17	0.25	0.08	0.19	0.19	0.01	0.19	0.19	0.06	0.23	0.14
CaO	4.60	10.92	4.90	4.23	8.84	10.02	2.87	7.48	3.09	1.75	1.96	11.98	12.38	1.99	9.29	9.35	1.93	8.86	12.07
Na_2O	1.60	0.78	3.49	3.89	3.19	0.98	4.24	4.14	2.33	1.51	2.97	1.58	1.46	3.96	1.59	1.24	2.90	1.15	1.40
K_2O	0.08	0.53	1.48	1.22	1.39	0.22	2.10	1.34	2.58	2.65	4.29	0.36	0.36	0.62	0.53	0.53	3.61	0.92	0.25
P_2O_5	0.06	0.04	0.22	0.01	0.70	0.07	0.06	0.09	0.30	0.04	0.14	0.12	0.12	0.01	0.22	0.22	0.12	0.08	0.04
Total	97.85	99.56	99.17	98.80	99.87	97.34	100.30	96.58	99.46	99.52	99.71	97.28	99.82	101.48	97.02	97.47	99.53	99.50	99.30
LOI	–0.33	–0.68	–0.70	–0.49	–0.68	–0.59	–0.45	–0.56	n.a.	–0.57	–0.56	–0.25	n.a.	–0.40	–0.91	–0.91	–0.40	0.00	–0.52
Rare earth and trace elements (ppm)																			
Sc	n.a.	56.3	7.8	n.d.	8.6	45.5	n.d.	6.3	16.3	19.9	10.3	38.2	n.a.	n.d.	47.0	n.a.	7.2	104.1	30.6
V	n.a.	243.1	68.4	29.2	214.1	272.9	24.9	144.0	60.9	128.4	22.1	297.7	n.a.	n.d.	252.2	n.a.	20.0	223.0	436.5
Cr	n.a.	418.8	140.9	58.6	302.5	261.7	44.1	180.2	62.7	472.7	38.5	212.3	n.a.	62.8	159.3	n.a.	45.6	401.5	252.3
Co	n.a.	55.2	19.9	5.3	35.2	46.5	4.8	19.8	13.1	35.5	4.4	65.0	n.a.	2.3	65.9	n.a.	4.4	67.4	71.2
Ni	n.a.	63.3	73.9	11.4	118.2	62.8	0.9	86.0	17.1	210.1	3.4	97.8	n.a.	5.4	50.7	n.a.	1.3	127.3	135.6
Cu	n.a.	53.7	39.5	11.1	21.6	66.8	8.3	46.5	25.4	40.6	7.7	146.8	n.a.	n.d.	13.1	n.a.	6.1	21.4	88.4
Zn	n.a.	91.3	73.7	21.5	102.8	83.6	31.9	69.9	118.1	106.9	57.3	92.6	n.a.	25.6	104.1	n.a.	45.4	148.5	68.2
Ga	n.a.	15.4	21.5	19.8	23.5	17.5	18.2	26.1	19.9	22.4	18.5	20.0	n.a.	26.0	20.7	n.a.	16.6	19.8	18.8
Rb	n.a.	4.0	57.4	44.6	52.4	3.4	46.7	38.6	62.0	116.8	119.8	13.6	n.a.	14.3	13.1	n.a.	87.5	32.6	4.4
Sr	n.a.	62.3	288.4	336.3	387.8	91.5	274.7	430.5	145.4	142.4	131.8	204.6	n.a.	110.9	150.0	n.a.	131.7	96.6	342.2
Y	n.a.	17.8	31.0	6.2	18.5	21.3	7.3	17.5	52.0	23.8	39.7	20.1	n.a.	133.2	35.8	n.a.	38.5	19.1	8.3
Zr	n.a.	48.3	128.6	138.6	44.2	76.1	102.7	94.4	548.9	96.8	298.7	91.0	n.a.	356.0	147.4	n.a.	237.9	75.5	24.3
Nb	n.a.	2.4	10.0	3.3	3.3	4.8	3.6	6.2	23.3	8.8	15.5	5.8	n.a.	82.0	11.9	n.a.	13.8	4.2	1.4
Ba	n.a.	34.6	245.1	336.1	572.9	45.4	454.4	301.2	967.1	813.8	993.7	126.0	n.a.	21.0	132.4	n.a.	930.8	275.8	64.9
La	n.a.	0.7	27.6	6.1	20.2	4.8	6.5	5.1	31.2	23.5	47.8	7.5	n.a.	77.4	19.5	n.a.	54.1	6.3	1.7
Ce	n.a.	n.a.	42.1	n.a.	n.a.	n.a.	n.a.	22.7	77.8	50.9	n.a.	n.a.	n.a.	n.a.	n.a.	n.a.	132.7	n.a.	n.a.
Hf	n.a.	1.3	3.8	4.8	0.7	2.0	3.4	3.2	13.9	2.2	10.0	2.1	n.a.	15.0	3.5	n.a.	7.5	1.9	0.3
Th	n.a.	n.d.	6.3	1.4	1.0	0.9	0.3	n.d.	1.1	8.4	5.2	n.d.	n.a.	47.0	0.2	n.a.	6.4	0.0	0.2
U	n.a.	n.a.	n.a.	n.a.	n.a.	n.a.	n.a.	n.d.	n.d.	n.d.	n.a.	n.a.	n.a.	n.a.	n.a.	n.a.	n.d.	n.a.	n.a.

(*continued*)

TABLE 1. WHOLE-ROCK ANALYSES FOR METAMORPHOSED MAFIC ROCKS AND QUARTZOFELDSPATHIC GNEISSES OF THE TOBACCO ROOT MOUNTAINS (*continued*)

Sample	O45	RC-1	RC-2	RC-3	RC-4	RC-4 II	RC-5	RC-5 II	RC-6	RC-6 II	RC-7	RC-8	RC-9	RC-9 II	RC-10	RC-10 II	RC-11	RC-12	RC-13
Data Source	1	1	1	1	1	1	1	1	1	1	1	1	1	1	1	1	1	1	1
Major Elements (wt%)																			
SiO_2	49.26	53.06	55.20	52.50	50.13	50.17	54.09	55.69	49.01	49.17	51.94	67.69	64.03	64.09	68.45	70.77	74.78	67.11	60.71
TiO_2	1.57	0.32	0.82	0.74	0.85	0.84	0.89	0.90	0.39	0.38	1.26	0.31	0.45	0.46	0.33	0.34	0.17	0.45	0.82
Al_2O_3	13.00	13.14	14.00	12.82	13.87	13.85	18.10	18.75	9.56	9.42	14.87	17.28	15.41	15.43	15.29	15.77	13.85	16.87	15.37
Fe_2O_3T	15.80	10.02	11.88	12.95	11.93	11.86	7.96	8.22	10.23	10.26	14.47	2.99	5.38	5.45	2.78	2.96	1.43	3.74	8.62
Fe_2O_3	n.a.	n.a.	n.a.	n.a.	n.a.	n.a.	n.a.	n.a.	n.a.	n.a.	n.a.	n.a.	n.a.	n.a.	n.a.	n.a.	n.a.	n.a.	n.a.
FeO	n.a.	n.a.	n.a.	n.a.	n.a.	n.a.	n.a.	n.a.	n.a.	n.a.	n.a.	n.a.	n.a.	n.a.	n.a.	n.a.	n.a.	n.a.	n.a.
MgO	5.34	13.44	7.54	7.63	8.33	8.25	3.58	3.75	17.83	17.85	6.39	1.48	1.99	1.99	0.90	0.93	0.52	1.19	3.97
MnO	0.22	0.19	0.18	0.24	0.20	0.20	0.12	0.12	0.17	0.17	0.21	0.03	0.08	0.08	0.04	0.04	0.02	0.05	0.11
CaO	9.30	10.12	9.19	9.70	10.12	10.05	6.57	6.75	9.70	9.69	9.18	4.08	4.47	4.52	3.36	3.49	1.67	3.58	6.60
Na_2O	1.65	0.98	1.58	2.22	1.38	1.08	3.27	3.28	0.39	0.26	2.16	4.62	3.68	3.50	3.93	3.82	3.90	4.50	2.77
K_2O	0.69	0.17	0.46	0.62	0.57	0.55	1.72	1.75	0.21	0.21	0.60	1.55	1.05	1.05	1.23	1.28	2.35	1.27	1.07
P_2O_5	0.15	0.04	0.07	0.06	0.06	0.06	0.24	0.24	0.04	0.04	0.10	0.06	0.11	0.11	0.09	0.09	0.04	0.10	0.36
Total	96.98	101.47	100.89	99.48	97.42	96.91	96.52	99.45	97.51	97.45	101.16	100.08	96.64	96.68	96.37	99.46	98.73	98.86	100.39
LOI	–0.29	–0.51	–0.51	–0.49	–0.42	–0.44	–0 62	–0.69	–0.60	–0.64	–0.57	–0.57	–0.54	–0.58	–1.28	–1.35	–0.77	–1.16	–0.59
Rare earth and trace elements (ppm)																			
Sc	46.9	36.6	45.7	18.1	54.3	n.a.	7.6	n.a.	22.9	n.a.	46.3	n.d.	13.0	n.a.	n.d.	n.a.	n.d.	1.7	13.5
V	369.2	134.8	267.9	163.0	288.4	n.a.	184.4	n.a.	163.0	n.a.	298.9	54.2	76.3	n.a.	39.3	n.a.	14.1	64.4	210.0
Cr	182.4	920.0	231.9	207.7	393.8	n.a.	56.5	n.a.	2078.8	n.a.	231.3	136.2	86.9	n.a.	15.9	n.a.	50.3	47.8	203.7
Co	69.8	60.9	57.5	66.9	57.2	n.a.	23.5	n.a.	80.4	n.a.	60.4	7.5	11.2	n.a.	6.7	n.a.	4.1	9.0	27.1
Ni	1.0	356.6	117.6	179.3	112.7	n.a.	53.4	n.a.	644.0	n.a.	93.0	21.7	18.8	n.a.	2.8	n.a.	3.9	5.5	62.0
Cu	153.5	55.0	113.5	171.1	81.7	n.a.	19.2	n.a.	58.0	n.a.	37.0	13.1	19.1	n.a.	13.2	n.a.	6.8	15.8	25.0
Zn	132.1	65.0	78.8	91.2	90.5	n.a.	92.6	n.a.	64.7	n.a.	106.7	41.6	75.7	n.a.	46.2	n.a.	24.8	56.3	81.0
Ga	20.8	11.0	16.9	14.0	15.6	n.a.	25 4	n.a.	8.1	n.a.	21.4	20.9	18.9	n.a.	18.6	n.a.	17.3	20.3	23.4
Rb	25.9	4.3	7.9	7.5	6.2	n.a.	82.9	n.a.	3.1	n.a.	7.3	55.9	26.8	n.a.	42.4	n.a.	44.7	34.4	35.1
Sr	150.4	136.6	275.8	455.8	77.0	n.a.	248.9	n.a.	118.9	n.a.	134.9	381.4	296.3	n.a.	251.6	n.a.	290.8	376.6	171.3
Y	32.0	13.0	19.0	16.8	20.4	n.a.	32.1	n.a.	13.7	n.a.	25.9	10.7	28.3	n.a.	7.8	n.a.	7.2	9.0	32.3
Zr	134.7	47.5	71.1	72.7	59.1	n.a.	128.7	n.a.	50.7	n.a.	97.0	68.9	131.9	n.a.	130.0	n.a.	70.3	115.8	117.1
Nb	8.4	2.3	4.1	3.6	3.0	n.a.	8.5	n.a.	2.4	n.a.	4.9	5.0	5.3	n.a.	4.1	n.a.	2.5	3.8	14.3
Ba	249.8	15.2	177.4	185.5	142.4	n.a.	179.2	n.a.	7.0	n.a.	200.6	312.0	250.8	n.a.	290.3	n.a.	297.0	299.7	332.7
La	15.6	2.2	3.4	5.4	5.0	n.a.	14.6	n.a.	1.1	n.a.	9.2	15.9	12.2	n.a.	14.2	n.a.	4.9	11.3	20.7
Ce	n.a.	13.5	10.3	31.7	6.6	n.a.	37.5	n.a.	4.5	n.a.	23.2	38.8	39.6	n.a.	29.2	n.a.	10.4	27.9	54.1
Hf	3.0	1.1	1.4	1.9	1.2	n.a.	3.5	n.a.	1.1	n.a.	2.0	2.8	4.2	n.a.	4.6	n.a.	3.0	4.4	3.1
Th	2.1	n.d.	n.d.	n.d.	n.d.	n.a.	2.1	n.a.	0.4	n.a.	0.9	2.5	n.d.	n.a.	5.8	n.a.	n.d.	n.d.	3.9
U	n.a.	n.d.	n.d.	n.d.	1.3	n.a.	n.d.	n.a.	1.5	n.a.	n.d.	n.d.	n.d.	n.a.	n.d.	n.a.	n.d.	n.d.	0.9

(*continued*)

TABLE 1. WHOLE-ROCK ANALYSES FOR METAMORPHOSED MAFIC ROCKS AND QUARTZOFELDSPATHIC GNEISSES OF THE TOBACCO ROOT MOUNTAINS (*continued*)

Sample	RC-14	RC-15	RC-17	RC-19	RC-20	RC-21	RC-22	RC-22b	RC-23	RC-24	RC-25	RC-26	RC-27	RC-28	RC-29	RC-30	RC-31	RC-33	RC-33 II
Data Source	1	1	1	1	1	1	1	1	1	1	1	1	1	1	1	1	1	1	1
Major Elements (wt%)																			
SiO_2	57.68	49.23	58.35	51.60	63.41	64.33	47.81	67.61	50.41	49.49	71.81	70.49	57.82	60.72	53.50	72.30	50.64	48.42	48.22
TiO_2	0.58	0.71	0.78	1.42	0.44	0.36	0.27	0.20	2.82	2.57	0.21	0.22	3.17	0.88	0.65	0.20	0.50	1.28	1.28
Al_2O_3	16.74	20.66	19.85	12.90	16.90	15.50	28.09	16.96	12.41	12.59	16.20	15.96	14.28	16.30	20.18	15.50	23.77	13.96	13.90
Fe_2O_3T	4.47	6.53	5.32	11.90	2.70	3.76	3.02	1.40	16.92	16.03	1.62	2.05	19.22	7.72	6.61	1.44	5.07	12.58	12.67
Fe_2O_3	n.a.	n.a.	n.a.	n.a.	n.a.	n.a.	n.a.	n.a.	n.a.	n.a.	n.a.	n.a.	n.a.	n.a.	n.a.	n.a.	n.a.	n.a.	n.a.
FeO	n.a.	n.a.	n.a.	n.a.	n.a.	n.a.	n.a.	n.a.	n.a.	n.a.	n.a.	n.a.	n.a.	n.a.	n.a.	n.a.	n.a.	n.a.	n.a.
MgO	2.47	3.92	2.32	6.21	1.27	1.56	1.95	0.57	3.80	4.10	0.80	0.86	5.19	3.04	3.47	0.91	3.81	7.09	6.99
MnO	0.06	0.11	0.05	0.23	0.03	0.05	0.04	0.01	0.23	0.21	0.02	0.02	0.25	0.10	0.10	0.02	0.07	0.20	0.20
CaO	5.66	10.26	7.16	9.42	3.76	3.94	13.76	3.64	8.13	8.46	4.55	3.97	9.25	5.93	9.03	4.24	11.80	11.43	11.40
Na_2O	3.71	2.78	4.21	1.87	4.07	3.97	1.88	4.47	1.76	1.69	3.30	3.97	2.22	3.07	3.77	3.63	2.85	1.54	1.31
K_2O	1.47	1.10	0.84	0.80	1.96	1.36	0.41	1.19	1.11	0.95	0.74	0.82	1.30	1.29	1.65	0.73	1.26	0.80	0.78
P_2O_5	0.30	0.21	0.12	0.11	0.20	0.20	0.04	0.08	0.23	0.22	0.03	0.03	0.25	0.22	0.05	0.05	0.05	0.11	0.11
Total	93.14	95.51	98.98	96.46	94.74	95.03	97.27	96.13	97.82	96.31	99.25	98.38	112.95	99.26	98.99	99.01	99.79	97.38	96.86
LOI	–0.73	–0.84	–0.40	–0.32	–0.77	–0.38	–1.15	–1.07	–1.30	–1.64	–1.03	–1.37	–1.15	–0.36	–1.62	–0.40	–1.17	–0.59	–0.59
Rare earth and trace elements (ppm)																			
Sc	n.d.	9.3	n.d.	40.9	n.d.	5.2	n.d.	10.7	35.3	44.1	n.d.	0.5	41.4	7.2	9.2	n.d.	n.d.	37.7	n.a.
V	86.6	124.7	139.4	295.6	45.5	60.1	51.5	17.6	504.7	530.9	23.0	18.1	582.0	110.6	144.8	14.2	101	306.3	n.a.
Cr	69.7	102.4	127.8	214.8	78.6	92.0	194.7	72.8	178.8	230.8	105.5	71.0	247.2	113.8	47.5	87.0	190.9	280.5	n.a.
Co	12.9	21.1	15.1	76.0	7.4	8.2	11.6	3.6	59.5	59.7	6.3	6.0	66.2	19.9	18.4	4.8	15.5	68.6	n.a.
Ni	41.8	99.5	55.1	44.6	21.4	22.0	76.2	n.d.	n.d.	2.4	11.6	10.1	n.d.	39.8	28.9	18.1	38.4	117.2	n.a.
Cu	7.5	42.0	37.5	51.4	6.1	23.5	32.6	5.8	243.2	206.0	6.7	7.8	245.4	29.0	23.2	11.6	10.8	169.2	n.a.
Zn	54.4	58.0	45.1	160.4	46.0	58.3	30.4	24.7	132.2	139.3	25.6	27.0	137.8	79.8	49.3	21.9	36.8	103.5	n.a.
Ga	22.5	22.8	21.9	19.0	23.3	20.5	19.8	18.2	26.9	25.0	14.4	15.1	25.5	18.1	20.5	14.9	19.8	21.2	n.a.
Rb	45.9	27.9	16.8	4.9	69.4	34.3	9.7	33.6	46.4	37.5	24.4	25.7	46.6	46.9	49.4	20.3	28.5	29.5	n.a.
Sr	858.6	314.5	512.1	135.6	561.2	507.2	244.7	628.2	194.9	181.3	316.0	324.0	184.1	278.9	277.3	329.5	238.4	185.1	n.a.
Y	17.2	32.7	13.8	21.9	13.8	28.5	7.7	6.7	38.1	34.5	6.6	6.2	37.2	17.8	13.5	6.1	11.4	21.0	n.a.
Zr	133.0	52.5	141.0	83.8	110.3	114.0	24.6	95.4	187.2	166.2	79.6	78.9	177.7	96.8	32.2	91.5	33.7	88.9	n.a.
Nb	5.6	6.0	4.5	5.1	6.4	7.2	1.9	3.5	12.2	11.5	2.5	2.5	11.6	8.6	2.7	2.5	3.2	7.1	n.a.
Ba	361.9	131.2	166.2	204.8	307.1	185.3	7.1	389.6	701.2	772.5	169.5	162.7	951.7	395.5	263.9	229.9	74.4	277.5	n.a.
La	n.a.	20.4	10.8	6.9	37.0	37.0	5.4	11.9	20.2	14.6	16.2	12.3	16.0	20.4	8.3	12.3	n.d.	9.2	n.a.
Ce	n.a.	n.a.	n.a.	n.a.	n.a.	n.a.	n.a.	n.a.	n.a.	n.a.	n.a.	n.a.	n.a.	n.a.	n.a.	n.a.	n.d.	n.a.	n.a.
Hf	n.a.	1.0	3.9	2.0	2.9	2.9	0.3	2.4	3.9	3.5	2.3	2.4	3.6	2.5	0.6	3.1	1.8	2.0	n.a.
Th	n.a.	0.6	2.3	n.d.	14.6	14.6	n.d.	2.2	2.7	2.8	2.5	2.1	2.2	4.2	2.0	0.1	0.7	n.d.	n.a.
U	n.a.	n.a.	n.a.	n.a.	n.a.	n.a.	n.a.	n.a.	n.a.	n.a.	n.a.	n.a.	n.a.	n.a.	n.a.	n.a.	n.d.	n.a.	n.a.

(*continued*)

TABLE 1. WHOLE-ROCK ANALYSES FOR METAMORPHOSED MAFIC ROCKS AND QUARTZOFELDSPATHIC GNEISSES OF THE TOBACCO ROOT MOUNTAINS (*continued*)

Sample	RP-21	RP-25	RP-25 II	RP-26	RP-26 II	RP-29	RP-31	RP-31 II	RP-32	RP-33	RP-33 II	RP-35	RP-36	RP-36 II	RP-37	RP-38	TR-1	TR-2	TR-4
Data Source	1	1	1	1	1	1	1	1	1	1	1	1	1	1	1	1	2	2	2
Major Elements (wt%)																			
SiO_2	20.29	52.26	53.41	70.25	70.13	23.93	49.70	51.40	58.64	51.59	53.18	47.19	52.09	53.71	50.64	50.12	96.60	98.75	63.72
TiO_2	0.05	1.96	2.21	0.50	0.50	0.01	1.58	1.61	0.46	2.20	2.25	0.70	2.19	2.02	1.32	1.44	0.05	0.02	0.50
Al_2O_3	2.75	14.81	12.87	13.31	13.43	2.32	13.09	13.45	15.01	12.79	13.13	11.50	12.60	15.22	14.72	14.49	1.32	0.36	16.74
Fe_2O_3T	2.32	14.50	16.27	3.48	3.50	1.67	16.12	16.60	18.04	15.96	16.49	15.29	15.93	15.18	13.44	14.26	0.09	0.08	4.26
Fe_2O_3	n.a.	n.a.	n.a.	n.a.	n.a.	n.a.	n.a.	n.a.	n.a.	n.a.	n.a.	n.a.	n.a.	n.a.	n.a.	n.a.	0.09	0.08	1.16
FeO	n.a.	n.a.	n.a.	n.a.	n.a.	n.a.	n.a.	n.a.	n.a.	n.a.	n.a.	n.a.	n.a.	n.a.	n.a.	n.a.	n.d.	n.d.	2.79
MgO	31.37	2.27	3.58	0.70	0.67	30.57	5.78	6.01	3.79	3.52	3.62	18.61	3.50	2.37	6.94	6.41	n.d.	n.d.	2.05
MnO	0.98	0.21	0.22	0.07	0.07	0.90	0.27	0.28	0.19	0.21	0.22	0.38	0.22	0.22	0.21	0.22	n.d.	n.d.	0.07
CaO	49.04	5.64	7.87	1.22	1.20	46.32	8.48	8.68	1.60	7.87	8.07	2.23	7.74	5.84	11.71	11.19	0.09	0.08	4.93
Na_2O	n.d.	4.26	1.63	2.94	2.78	n.d.	1.67	1.65	0.89	1.99	1.69	n.d.	1.67	4.24	2.05	1.75	0.01	n.d.	4.70
K_2O	n.d.	0.73	1.44	3.95	3.94	n.d.	0.42	0.43	2.40	1.35	1.38	2.49	1.40	0.77	0.47	0.40	0.54	0.16	1.27
P_2O_5	0.02	0.31	0.25	0.13	0.13	0.04	0.13	0.13	0.07	0.22	0.25	0.04	0.24	0.32	0.14	0.14	0.03	n.d.	0.15
Total	106.80	96.94	99.73	96.53	96.35	105.75	97.21	100.21	101.07	97.69	100.26	98.44	97.56	99.88	101.63	100.41	98.82	99.53	102.34
LOI	−35.27	−0.42	−0.42	−0.60	−0.66	−31.84	−1.03	−1.07	−0.09	−0.02		−0.74			−0.30	−0.14	0.80	0.38	1.34
Rare earth and trace elements (ppm)																			
Sc	n.d.	14.4	n.a.	12.7	n.a.	n.d.	n.a	n.a.	n.a.	35.3	n.a.	64.6	40.7	n.a.	39.0	39.8	n.a.	n.a.	n.a.
V	n.d.	50.1	n.a.	21.4	n.a.	n.d.	n.a.	n.a.	n.a.	392.4	n.a.	290.5	380.7	n.a.	290.3	291.9	n.a.	n.a.	n.a.
Cr	27.9	36.8	n.a.	64.4	n.a.	24.6	n.a.	n.a.	n.a.	130.2	n.a.	1346.9	140.7	n.a.	221.8	165.8	n.a.	n.a.	n.a.
Co	1.2	27.3	n.a.	5.3	n.a.	1.6	65.0	n.a.	28.8	60.2	n.a.	86.3	59.8	n.a.	56.2	58.0	80.0	162.0	36.0
Ni	n.d.	7.8	n.a.	6.2	n.a.	n.d.	57.0	n.a.	126.2	48.7	n.a.	321.8	48.0	n.a.	85.1	69.5	31.0	4.0	31.0
Cu	8.0	23.0	n.a.	8.0	n.a.	4.2	113.7	n.a.	32.1	97.2	n.a.	37.0	96.8	n.a.	138.9	155.3	344.0	21.0	26.0
Zn	5.0	175.0	n.a.	38.3	n.a.	2.4	152.7	n.a.	90.6	156.7	n.a.	240.1	157.0	n.a.	106.5	111.2	8.0	4.0	63.0
Ga	n.d.	28.9	n.a.	18.4	n.a.	n.d.	22.9	n.a.	20.9	25.6	n.a.	14.0	26.8	n.a.	19.7	19.9	1.0	1.0	21.0
Rb	1.7	6.4	n.a.	97.7	n.a.	1.7	22.7	n.a.	119.7	60.7	n.a.	76.4	61.9	n.a.	13.9	12.7	9.0	3.0	42.0
Sr	38.3	242.6	n.a.	133.9	n.a.	39.9	104.5	n.a.	127.6	149.1	n.a.	65.7	152.1	n.a.	210.1	212.1	10.0	7.0	464.0
Y	6.5	40.3	n.a.	32.5	n.a.	6.1	31.6	n.a.	23.3	45.5	n.a.	11.9	45.1	n.a.	20.7	22.1	6.0	1.0	13.0
Zr	27.5	187.8	n.a.	279.0	n.a.	23.1	99.8	n.a.	95.5	217.0	n.a.	68.2	223.4	n.a.	89.4	97.7	322.0	25.0	138.0
Nb	1.9	11.2	n.a.	12.1	n.a.	1.8	4.6	n.a.	8.1	15.9	n.a.	2.5	14.7	n.a.	8.5	9.4	n.a.	n.a.	n.a.
Ba	n.d.	300.5	n.a.	935.0	n.a.	n.d.	n.d.	n.a.	n.d.	570.1	n.a.	721.4	561.1	n.a.	213.3	496.5	n.a.	n.a.	n.a.
La	1.1	22.8	n.a.	n.a.	n.a.	5.4	6.7	n.a.	27.8	34.5	n.a.	4.1	39.0	n.a.	14.2	9.2	24.0	5.0	14.0
Ce	n.a.	49.6	n.a.	n.a.	n.a.	n.a.	n.a.	n.a.	n.a.	n.a.	n.a.	15.3	n.a.	n.a.	29.4	n.a.	45.0	11.0	37.0
Hf	0.8	4.0	n.a.	n.a.	n.a.	0.7	2.3	n.a.	2.4	4.8	n.a.	1.3	4.9	n.a.	1.9	2.2	n.a.	n.a.	n.a.
Th	0.6	0.1	n.a.	n.a.	n.a.	n.d.	2.0	n.a.	9.6	7.3	n.a.	n.d.	8.3	n.a.	n.d.	1.1	12.0	2.0	2.0
U	n.a.	n.d.	n.a.	n.a.	n.a.	n.a.	n.a.	n.a.	n.a.	n.a.	n.a.	n.d.	n.a.	n.a.	0	n.a.	2	2	1

(*continued*)

TABLE 1. WHOLE-ROCK ANALYSES FOR METAMORPHOSED MAFIC ROCKS AND QUARTZOFELDSPATHIC GNEISSES OF THE TOBACCO ROOT MOUNTAINS (*continued*)

Sample	TR-5	TR-6	TR-7	TR-8A	TR-8B	TR-9	TR-10	TR-11	TR-12	TR-13	TR-14	TR-14A	TR-15	TR-16	TR-17	TR-18	TR-20	TR-20DUP	TR-21
Data Source	2	2	2	2	2	2	2	2	2	2	2	2	2	2	2	2	1	1	1
Major Elements (wt%)																			
SiO_2	68.87	67.02	52.64	74.21	74.44	67.59	68.78	48.49	44.39	52.88	76.03	75.22	67.22	73.16	51.10	51.18	n.a.	n.a.	n.a.
TiO_2	0.25	0.27	0.57	0.03	0.05	0.43	0.26	0.63	0.32	0.34	0.32	0.30	0.88	0.39	1.93	0.73	n.a.	n.a.	n.a.
Al_2O_3	16.72	17.10	15.31	13.58	13.65	15.69	16.54	14.46	18.58	6.39	11.54	11.53	13.53	13.26	23.49	13.63	n.a.	n.a.	n.a.
Fe_2O_3T	2.02	2.18	9.64	0.38	0.42	3.86	2.27	11.15	9.89	10.63	1.88	2.47	5.40	2.08	6.11	10.30	n.a.	n.a.	n.a.
Fe_2O_3	0.82	0.62	2.13	0.18	0.15	1.40	1.07	3.18	1.92	1.57	1.01	1.49	2.09	0.92	1.48	2.91	n.a.	n.a.	n.a.
FeO	1.08	1.40	6.76	0.18	0.24	2.21	1.08	7.17	7.17	8.15	0.78	0.88	2.98	1.04	4.17	6.65	n.a.	n.a.	n.a.
MgO	0.85	1.16	7.53	0.08	0.16	1.46	1.12	9.26	11.57	19.28	0.21	0.20	1.08	0.32	2.51	8.31	n.a.	n.a.	n.a.
MnO	0.01	0.02	0.17	n.d.	0.01	0.06	0.02	0.31	0.22	0.27	0.03	0.02	0.07	0.04	0.10	0.13	n.a.	n.a.	n.a.
CaO	3.72	4.23	9.02	0.99	1.60	4.06	3.44	11.63	11.63	7.17	0.96	1.10	3.04	1.64	5.12	8.23	n.a.	n.a.	n.a.
Na_2O	4.93	5.01	3.30	2.29	2.78	3.92	4.56	1.68	1.33	0.71	3.04	2.74	3.11	2.80	4.61	3.39	n.a.	n.a.	n.a.
K_2O	1.09	1.13	0.43	6.94	5.06	1.37	1.49	0.40	0.29	0.10	4.42	4.58	3.91	5.22	3.04	1.19	n.a.	n.a.	n.a.
P_2O_5	0.06	0.06	0.04	0.02	0.05	0.12	0.08	0.06	0.02	0.04	0.03	0.04	0.29	0.08	0.05	0.08	n.a.	n.a.	n.a.
Total	100.42	100.20	107.54	98.88	98.61	102.17	100.71	108.42	107.33	107.53	100.25	100.57	103.60	100.95	103.71	106.73			
LOI	0.90	0.90	1.71	0.96	1.40	1.42	1.30	2.15	2.33	2.54	0.70	0.80	0.88	0.72	1.58	1.72	n.a.	n.a.	n.a.
Rare earth and trace elements (ppm)																			
Sc	n.a.	n.a.	n.a.	n.a.	n.a.	n.a.	n.a.	n.a.	n.a.	n.a.	n.a.	n.a.	n.a.	n.a.	n.a.	n.a.	n.d.	n.a.	n.d.
V	n.a.	n.a.	n.a.	n.a.	n.a.	n.a.	n.a.	n.a.	n.a.	n.a.	n.a.	n.a.	n.a.	n.a.	n.a.	n.a.	27.6	n.a.	20.5
Cr	n.a.	n.a.	n.a.	n.a.	n.a.	n.a.	n.a.	n.a.	n.a.	n.a.	n.a.	n.a.	n.a.	n.a.	n.a.	n.a.	122.9	n.a.	56.6
Co	27.0	22.0	61.0	35.0	42.0	39.0	31.0	67.0	70.0	70.0	46.0	57.0	35.0	40.0	66.0	56.0	13.9	13.5	8.1
Ni	7.0	18.0	192.0	6.0	n.d.	21.0	15.0	126.0	324.0	291.0	5.0	5.0	11.0	5.0	150.0	230.0	30.1	25.2	18.2
Cu	11.0	16.0	14.0	10.0	5.0	4.0	15.0	49.0	14.0	10.0	5.0	4.0	19.0	5.0	52.0	46.0	13.0	11.6	5.8
Zn	31.0	50.0	92.0	7.0	11.0	52.0	43.0	128.0	82.0	91.0	18.0	37.0	82.0	50.0	115.0	86.0	10.8	9.6	23.4
Ga	20.0	21.0	20.0	16.0	16.0	20.0	20.0	15.0	15.0	9.0	16.0	18.0	19.0	17.0	29.0	17.0	2.8	3.2	1.2
Rb	49.0	49.0	4.0	143.0	105.0	54.0	48.0	7.0	5.0	3.0	163.0	211.0	147.0	186.0	93.0	45.0	40.4	39.4	22.3
Sr	478.0	489.0	259.0	259.0	251.0	216.0	421.0	127.0	91.0	13.0	55.0	102.0	127.0	109.0	718.0	185.0	42.4	42.3	53.2
Y	5.0	7.0	17.0	3.0	2.0	30.0	4.0	17.0	7.0	9.0	79.0	38.0	65.0	41.0	26.0	20.0	19.3	20.6	15.4
Zr	89.0	66.0	48.0	88.0	64.0	155.0	90.0	51.0	31.0	45.0	281.0	228.0	493.0	284.0	345.0	88.0	104.1	108.5	55.0
Nb	n.a.	n.a.	n.a.	n.a.	n.a.	n.a.	n.a.	n.a.	n.a.	n.a.	n.a.	n.a.	n.a.	n.a.	n.a.	n.a.	4.7	3.7	3.5
Ba	n.a.	n.a.	n.a.	n.a.	n.a.	n.a.	n.a.	n.a.	n.a.	n.a.	n.a.	n.a.	n.a.	n.a.	n.a.	n.a.	224.2	n.a.	10.2
La	11.0	8.0	12.0	4.0	7.0	18.0	11.0	4.0	3.0	13.0	71.0	70.0	77.0	68.0	63.0	10.0	13.9	n.a.	7.9
Ce	21.0	14.0	24.0	13.0	11.0	45.0	25.0	4.0	4.0	3.0	142.0	130.0	153.0	126.0	110.0	16.0	n.a.	n.a.	n.a.
Hf	n.a.	n.a.	n.a.	n.a.	n.a.	n.a.	n.a.	n.a.	n.a.	n.a.	n.a.	n.a.	n.a.	n.a.	n.a.	n.a.	3.3	n.a.	1.6
Th	4.0	2.0	n.d.	n.d.	n.d.	5.0	1.0	n.d.	n.d.	1.0	23.0	18.0	17.0	14.0	21.0	1.0	2.9	n.a.	2.0
U	2	2	1	2	2	1	1	4	1	1	2	3	4	2	1	2	n.a.	n.a.	n.a.

(continued)

TABLE 1. WHOLE-ROCK ANALYSES FOR METAMORPHOSED MAFIC ROCKS AND QUARTZOFELDSPATHIC GNEISSES OF THE TOBACCO ROOT MOUNTAINS (*continued*)

Sample	TR-21DUP	TR-22	TR-23	TR-30	TR-31	TR-32	TR-33	TR-34	TR-35	TR-36	TR-37	TR-37 II	TR-38	TR-39A	TR-39B	TR-40	TR-41	TR-42	TR-43
Data Source	1	1	1	1	1	1	1	1	1	1	1	1	1	1	1	1	1	1	1
Major Elements (wt%)																			
SiO_2	n.a.	n.a.	n.a.	48.30	51.75	58.74	53.98	60.51	70.28	57.53	79.19	69.77	55.64	72.41	72.28	51.14	76.16	73.12	95.11
TiO_2	n.a.	n.a.	n.a.	1.55	1.25	0.83	0.72	1.82	0.56	0.56	0.40	0.35	0.71	0.82	0.78	0.81	0.51	0.52	0.09
Al_2O_3	n.a.	n.a.	n.a.	16.59	15.21	15.24	13.63	28.71	10.58	13.46	10.22	9.15	12.96	12.65	12.02	13.62	13.19	14.64	4.18
Fe_2O_3T	n.a.	n.a.	n.a.	21.51	13.13	10.92	13.60	5.06	9.05	9.57	9.44	8.63	12.25	7.12	8.71	15.58	4.23	4.99	0.90
Fe_2O_3	n.a.	n.a.	n.a.	n.a.	n.a.	n.a.	n.a.	n.a.	n.a.	n.a.	n.a.	n.a.	n.a.	n.a.	n.a.	n.a.	n.a.	n.a.	n.a.
FeO	n.a.	n.a.	n.a.	n.a.	n.a.	n.a.	n.a.	n.a.	n.a.	n.a.	n.a.	n.a.	n.a.	n.a.	n.a.	n.a.	n.a.	n.a.	n.a.
MgO	n.a.	n.a.	n.a.	6.73	5.31	5.92	10.52	1.01	8.09	7.09	14.19	12.08	15.91	1.22	1.60	7.56	2.59	3.04	0.55
MnO	n.a.	n.a.	n.a.	0.28	0.19	0.13	0.32	0.07	0.08	0.16	0.14	0.12	0.12	0.08	0.08	0.40	0.04	0.06	0.04
CaO	n.a.	n.a.	n.a.	3.56	10.11	5.09	6.31	0.19	0.62	7.32	1.51	1.35	0.93	2.06	1.72	9.66	0.17	0.19	0.17
Na_2O	n.a.	n.a.	n.a.	0.46	1.82	2.49	0.46	n.d.	0.49	2.63	1.10	0.61	0.92	2.71	1.96	0.33	n.d.	n.d.	n.d.
K_2O	n.a.	n.a.	n.a.	0.69	0.84	0.21	0.21	0.93	0.07	0.13	n.d.	n.d.	0.06	0.34	0.51	0.24	3.18	3.20	0.47
P_2O_5	n.a.	n.a.	n.a.	0.53	0.14	0.16	0.06	0.04	0.13	0.10	0.28	0.25	0.09	0.18	0.20	0.05	0.03	0.04	0.01
Total				100.21	99.75	99.71	99.79	98.33	99.93	98.54	116.50	102.30	99.57	99.59	99.84	99.37	100.10	99.79	101.50
LOI	n.a.	n.a.	n.a.	–0.22	–0.78	–0.46	–0.63	–1.44	–1.64	–0.54	–0.76	n.a.	–1.21	–0.32	–0.39	–0.88	–2.33	–2.73	–0.71
Rare earth and trace elements (ppm)																			
Sc	n.a.	n.d.	42.1	67.2	35.1	62.2	58.9	59.5	39.4	31.2	43.1	n.a.	63.8	14.7	6.0	52.2	23.0	29.9	2.9
V	n.a.	14.7	178.8	229.0	256.9	243.9	229.8	400.9	143.4	174.9	138.9	n.a.	229.4	12.9	11.3	268.8	91.6	116.3	2.6
Cr	n.a.	37.4	648.3	116.0	136.6	169.1	769.5	422.7	253.3	588.9	411.4	n.a.	786.4	48.1	40.7	169.1	307.1	341.9	120.7
Co	8.3	4.4	47.6	46.8	53.8	42.9	59.9	19.6	33.8	46.1	41.4	n.a.	58.2	7.5	8.4	67.5	13.4	17.3	7.2
Ni	19.7	11.0	161.0	15.9	69.2	60.1	129.9	24.8	44.7	113.5	75.0	n.a.	129.4	n.d.	1.1	67.8	61.8	82.9	18.7
Cu	5.5	5.1	10.4	12.9	89.9	8.8	9.3	52.7	84.0	35.6	479.5	n.a.	4.7	7.0	8.0	14.0	20.3	20.6	25.0
Zn	23.6	21.8	61.7	51.7	143.3	55.1	57.5	42.6	34.7	61.4	46.9	n.a.	63.6	18.4	20.2	100.9	21.3	22.7	4.1
Ga	1.2	0.4	11.1	9.9	19.2	16.7	13.6	23.9	8.9	12.5	8.4	n.a.	12.7	18.1	18.5	15.3	14.2	16.3	2.0
Rb	23.3	13.7	26.2	28.3	23.1	6.5	6.5	39.4	5.3	2.5	2.4	n.a.	4.0	10.2	16.3	6.2	72.5	74.7	19.6
Sr	54.2	55.4	68.3	51.9	140.0	130.1	27.4	18.4	19.4	71.0	34.3	n.a.	21.1	50.6	56.3	44.5	22.7	22.2	18.1
Y	15.0	13.4	16.7	54.9	29.8	25.0	42.6	46.5	21.8	17.1	23.0	n.a.	18.4	62.8	58.1	20.6	13.2	12.1	12.1
Zr	54.9	51.1	61.2	260.9	126.8	127.4	86.7	367.1	117.1	73.4	50.7	n.a.	79.5	422.7	404.0	54.3	129.7	94.2	57.4
Nb	4.1	2.2	2.3	31.7	7.4	6.1	5.7	24.5	5.3	4.2	2.4	n.a.	3.9	21.7	19.5	2.8	5.4	5.4	1.7
Ba	n.a.	n.d.	123.0	92.4	308.4	127.8	59.4	501.4	n.d.	45.7	n.d.	n.a.	83.8	n.d.	45.3	117.0	348.3	398.6	n.d.
La	n.a.	9.6	8.8	50.9	13.5	20.4	0.5	n.a.	10.8	11.2	3.4	n.a.	8.1	59.7	51.3	0.2	12.5	15.0	0.4
Ce	n.a.	n.a.	n.a.	109.2	21.1	n.a.	n.a.	n.a.	n.a.	20.6	20.4	n.a.	n.a.	128.9	115.9	12.4	31.6	n.a.	0.6
Hf	n.a.	1.4	1.7	4.6	2.8	3.6	2.5	n.a.	4.0	1.7	1.4	n.a.	2.6	13.3	11.0	0.9	4.2	3.7	1.8
Th	n.a.	n.d.	0.3	2.8	2.4	5.7	0.5	n.a.	2.7	2.8	0.7	n.a.	1.2	13.2	10.4	0.3	6.7	4.5	n.d.
U	n.a.	n.a.	n.a.	0	0	n.a.	n.a.	n.a.	n.a.	n.d.	2	n.a.	n.a.	3	0	n.d.	n.d.	n.a.	0

(continued)

TABLE 1. WHOLE-ROCK ANALYSES FOR METAMORPHOSED MAFIC ROCKS AND QUARTZOFELDSPATHIC GNEISSES OF THE TOBACCO ROOT MOUNTAINS (*continued*)

Sample	TR-44	TR-45	TR-46	TR-47	TR-48	TR-49	TR-50	TR-51	TR-52	TR-53	TR-54	TR-55	TR-56	TR-57	TR-59	TR-60	WC-1	WC-2	WC-3
Data Source	1	1	1	1	1	1	1	1	1	1	1	1	1	1	1	1	1	1	1
Major Elements (wt%)																			
SiO_2	48.55	49.62	49.78	47.55	54.95	48.76	72.32	64.92	n.a.	55.29	71.87	75.16	78.86	73.65	68.91	71.89	69.68	48.19	72.40
TiO_2	1.65	1.37	2.84	0.43	0.40	0.40	0.11	0.89	n.a.	0.86	0.18	0.53	0.42	0.11	0.31	0.16	0.15	0.46	0.14
Al_2O_3	12.94	13.56	13.43	6.35	7.76	16.55	13.90	16.88	n.a.	14.43	14.37	11.78	10.22	14.63	15.98	16.25	16.57	12.03	13.67
Fe_2O_3T	16.85	15.24	16.81	12.29	9.06	8.36	1.17	7.25	n.a.	10.53	1.24	5.68	4.84	1.07	2.61	1.09	0.96	10.95	0.87
Fe_2O_3	n.a.	n.a.	n.a.	n.a.	n.a.	n.a.	n.a.	n.a.	n.a.	n.a.	n.a.	n.a.	n.a.	n.a.	n.a.	n.a.	n.a.	n.a.	n.a.
FeO	n.a.	n.a.	n.a.	n.a.	n.a.	n.a.	n.a.	n.a.	n.a.	n.a.	n.a.	n.a.	n.a.	n.a.	n.a.	n.a.	n.a.	n.a.	n.a.
MgO	6.44	6.60	4.62	29.90	14.89	9.13	0.10	2.59	n.a.	5.45	0.19	0.24	0.40	0.12	0.68	0.41	0.47	12.30	0.46
MnO	0.32	0.32	0.26	0.38	0.34	0.23	0.02	0.08	n.a.	0.15	0.02	0.09	0.09	0.04	0.02	0.02	0.01	0.19	0.01
CaO	10.32	10.53	9.32	4.58	12.04	13.90	1.14	1.46	n.a.	8.99	1.13	1.56	2.89	1.11	2.05	2.71	2.98	10.67	0.97
Na_2O	2.03	1.40	1.43	n.d.	0.52	1.32	3.29	2.68	n.a.	3.05	3.65	1.59	1.54	3.77	3.49	4.93	4.47	0.57	1.76
K_2O	0.21	0.56	0.59	n.d.	0.20	0.18	4.73	1.92	n.a.	0.86	4.60	3.22	0.51	4.04	4.42	2.17	1.79	0.93	5.81
P_2O_5	0.10	0.09	0.31	0.05	0.05	0.03	0.02	0.03	n.a.	0.09	0.05	0.08	0.06	0.03	0.10	0.05	0.02	0.04	0.05
Total	99.40	99.27	99.36	101.50	100.20	98.86	97.31	98.68	n.a.	99.69	97.30	99.91	99.82	98.54	98.58	99.65	97.10	96.33	96.14
LOI	−1.13	−0.63	−0.45	−1.84	−1.06	−0.57	−0.09	−0.98		−0.48	−0.45	−0.22	−0.03	−0.15	−0.54	−0.42	−0.58	−1.22	−0.70
Rare earth and trace elements (ppm)																			
Sc	58.7	53.1	34.8	26.4	60.2	26.8	6.0	35.5	n.a.	28.3	5.0	13.3	9.3	6.4	6.4	n.d.	n.d.	41.0	n.d.
V	448.7	333.7	348.7	107.4	175.0	157.6	n.d.	201.2	n.a.	207.3	3.3	5.8	1.7	1.4	1.4	9.6	11.6	199.0	12.7
Cr	198.6	136.9	113.5	3168.0	1862.0	873.3	35.1	331.7	n.a.	21.8	20.3	21.5	39.3	16.9	16.9	51.8	41.1	1313.0	45.7
Co	78.0	64.0	60.0	99.9	48.3	39.2	1.5	27.2	n.a.	49.5	25.0	4.8	6.5	2.4	5.2	3.3	3.4	67.0	3.7
Ni	77.8	64.3	39.9	1384.0	197.8	138.4	1.4	69.0	n.a.	25.1	12.4	0.0	n.d.	0.8	2.2	0.1	n.d.	313.2	n.d.
Cu	142.2	147.8	107.8	26.9	38.1	23.5	2.0	79.8	n.a.	37.3	20.3	13.4	10.6	2.7	24.9	4.9	4.0	27.5	4.5
Zn	112.8	113.4	132.6	73.7	65.4	62.7	21.1	108.9	n.a.	80.3	60.2	52.3	36.1	34.1	38.5	25.1	22.3	71.2	11.9
Ga	23.4	21.7	28.6	3.8	8.5	14.2	21.4	27.4	n.a.	19.2	21.7	16.4	14.7	25.9	17.9	19.5	17.6	9.2	12.6
Rb	3.1	15.4	24.0	2.0	7.1	3.5	158.5	175.5	n.a.	20.6	105.0	64.3	12.1	202.8	100.3	41.1	39.6	29.3	140.2
Sr	107.3	89.9	157.1	31.9	88.1	135.9	103.6	125.2	n.a.	208.6	189.3	138.7	142.9	130.2	462.7	480.8	357.0	95.6	174.1
Y	28.3	28.4	38.6	13.3	14.2	12.0	36.2	46.9	n.a.	18.5	17.1	32.6	33.3	28.3	14.0	7.1	5.9	13.1	9.6
Zr	74.8	78.2	203.6	74.2	52.5	34.2	92.8	228.2	n.a.	82.8	n.d.	223.1	191.7	87.1	168.7	87.8	58.5	37.5	120.6
Nb	3.6	3.2	15.0	2.8	3.3	1.3	11.6	17.7	n.a.	4.6	7.4	9.6	8.8	9.7	2.6	2.5	2.6	1.4	2.6
Ba	450.5	221.5	403.5	n.d.	51.4	19.4	755.1	655.5	n.a.	265.8	886.2	1211.0	392.9	616.1	616.1	494.3	228.0	72.2	1621.0
La	4.5	5.1	23.9	1.5	7.2	n.a.	22.2	33.3	n.a.	11.0	34.9	43.6	42.9	21.2	77.0	4.8	n.a.	n.a.	n.a.
Ce	13.3	n.d.	n.a.	n.a.	18.5	n.a.	n.a.	71.3	n.a.	n.a.	n.a.	n.a.	84.4	31.6	n.a.	n.a.	n.a.	n.a.	n.a.
Hf	1.4	1.4	4.4	2.1	1.0	n.a.	3.5	6.4	n.a.	1.9	5.4	7.2	6.0	3.0	5.1	2.4	n.a.	n.a.	n.a.
Th	n.d.	n.d.	3.1	0.5	1.5	n.a.	14.9	14.3	n.a.	0.8	23.2	20.6	23.9	12.2	28.8	1.8	n.a.	n.a.	n.a.
U	n.d.	n.d.	n.a.	n.a.	n.d.	n.a.	n.a.	n.d.	n.a.	n.a.	n.a.	n.a.	1	n.d.	n.a.	n.a.	n.a.	n.a.	n.a.

(continued)

TABLE 1. WHOLE-ROCK ANALYSES FOR METAMORPHOSED MAFIC ROCKS AND QUARTZOFELDSPATHIC GNEISSES OF THE TOBACCO ROOT MOUNTAINS (*continued*)

Sample	WC-4	WC-5	WC-8	16b	29b	25b	3	28b	33a	26b	15a	38-p	70b-p	72-p	28-p	66-p	37-p	31-p	26-p
Data Source	1	1	1	3	3	3	3	3	3	3	3	4	4	4	4	4	4	4	4
Major Elements (wt%)																			
SiO_2	51.48	51.92	48.95	50.74	50.77	51.25	51.51	51.63	51.71	68.23	70.24	63.40	71.79	68.96	66.13	68.23	71.76	61.57	69.33
TiO_2	0.62	0.95	0.47	0.41	0.05	0.18	0.65	0.88	1.04	0.47	0.21	0.48	0.30	0.41	0.51	0.40	0.22	0.64	0.39
Al_2O_3	14.94	14.50	11.35	14.72	14.02	21.64	15.22	11.8	14.06	15.51	16.13	16.27	15.05	14.91	16.52	15.77	14.47	16.72	16.32
Fe_2O_3T	10.26	13.15	11.67	10.43	10.98	4.78	9.80	13.10	14.30	4.28	1.87	5.30	2.88	4.62	4.32	3.49	2.22	5.48	2.29
Fe_2O_3	n.a.	n.a.	n.a.	n.a.	n.a.	n.a.	n.a.	n.a.	n.a.	n.a.	n.a.	n.a.	n.a.	n.a.	n.a.	n.a.	n.a.	n.a.	n.a.
FeO	n.a.	n.a.	n.a.	n.a.	n.a.	n.a.	n.a.	n.a.	n.a.	n.a.	n.a.	n.a.	n.a.	n.a.	n.a.	n.a.	n.a.	n.a.	n.a.
MgO	8.50	7.24	12.94	8.66	8.31	5.65	8.60	7.70	6.07	1.27	0.76	2.79	1.04	2.89	1.53	1.15	0.68	2.27	1.13
MnO	0.20	0.23	0.21	0.21	0.19	0.09	0.16	0.23	0.21	0.07	0.03	0.08	0.05	0.07	0.05	0.05	0.02	0.06	0.03
CaO	9.67	10.21	11.28	9.46	10.61	12.41	9.46	10.58	9.10	3.51	1.61	5.16	2.83	1.98	4.48	3.64	1.88	5.62	3.15
Na_2O	2.11	1.58	0.77	2.72	2.86	2.99	2.93	2.47	2.24	4.17	5.46	3.94	4.71	3.10	5.42	4.92	4.33	5.30	5.20
K_2O	0.87	0.25	0.63	0.91	0.78	0.37	0.36	1.09	0.82	1.20	2.32	1.31	1.79	2.26	0.33	1.33	2.67	0.47	1.56
P_2O_5	0.10	0.09	0.02	0.04	0.05	0.01	0.12	0.06	0.09	0.16	0.09	0.14	0.10	0.05	0.19	0.13	0.09	0.18	0.13
Total	98.73	100.10	98.25	100.10	99.13	100.00	99.55	100.35	100.50	99.55	100.03	100.55	100.89	99.71	100.26	99.53	98.95	98.47	100.60
LOI	–0.73	–0.41	–0.78	1.82	0.52	0.62	0.54	0.81	0.87	0.69	1.31	1.69	0.37	0.47	0.40	0.42	0.62	0.42	0.43
Rare earth and trace elements (ppm)																			
Sc	34.0	53.0	38.6	n.a.	n.a.	n.a.	n.a.	n.a.	n.a.	n.a.	n.a.	11.0	0.0	3.0	8.0	5.0	3.0	11.0	4.0
V	204.9	264.0	184.3	157.0	252.0	62.0	170.0	237.0	281.0	51.0	27.0	75.0	15.0	24.0	52.0	39.0	15.0	88.0	39.0
Cr	408.9	102.9	1354.0	212.0	372.0	157.0	409.0	429.0	26.0	14.0	13.0	111.0	23.0	257.0	15.0	12.0	22.0	57.0	14.0
Co	46.6	57.0	71.0	n.a.	n.a.	n.a.	n.a.	n.a.	n.a.	n.a.	n.a.	17.4	6.7	18.7	10.5	7.6	4.2	18.0	7.8
Ni	182.5	7.5	339.3	97.0	144.0	58.0	179.0	120.0	62.0	11.0	6.0	77.0	23.0	139.0	25.0	18.0	n.d.	n.d.	17.0
Cu	37.0	44.3	56.9	9.0	41.0	90.0	12.0	43.0	85.0	19.0	7.0	n.a.	n.a.	n.a.	n.a.	n.a.	n.a.	n.a.	n.a.
Zn	81.2	91.9	73.1	77.0	83.0	34.0	72.0	114.0	84.0	66.0	34.0	58.0	68.0	87.0	63.0	64.0	47.0	88.0	63.0
Ga	16.7	15.3	10.2	12.8	14.4	13.8	15.2	14.7	19.0	18.6	18.1	n.a.	n.a.	n.a.	n.a.	n.a.	n.a.	n.a.	n.a.
Rb	18.5	5.0	27.7	28.0	13.0	4.0	13.0	24.0	11.0	43.0	67.0	44.0	61.0	80.0	0.2	44.0	53.0	8.6	69.7
Sr	168.5	121.1	82.4	140.0	123.0	311.0	177.0	191.0	142.0	297.0	288.0	374.0	480.0	446.0	672.0	504.0	296.0	648.0	519.0
Y	20.3	19.5	12.6	17.0	15.0	5.0	15.0	24.0	26.0	10.0	4.0	12.0	3.0	5.0	5.0	8.0	5.0	11.0	3.0
Zr	72.0	70.6	39.3	47.0	38.0	11.0	60.0	61.0	77.0	110.0	84.0	111.0	96.0	141.0	112.0	102.0	132.0	114.0	105.0
Nb	5.9	3.7	1.7	1.9	0.6	n.d.	2.4	2.4	2.4	4.2	2.2	n.a.	n.a.	n.a.	n.a.	n.a.	n.a.	n.a.	n.a.
Ba	167.3	141.9	48.9	n.a.	n.a.	n.a.	n.a.	n.a.	n.a.	n.a.	n.a.	697.0	369.0	898.0	221.0	506.0	615.0	84.0	357.0
La	n.a.	n.a.	n.a.	n.a.	3.6	2.9	n.a.	7.7	n.a.	27.8	n.a.	10.4	22.1	29.0	13.9	76.5	26.1	9.0	34.7
Ce	n.a.	n.a.	n.a.	n.a.	9.0	4.0	n.a.	18.0	n.a.	49.0	n.a.	22.9	39.0	52.6	20.3	111.0	42.4	22.3	55.7
Hf	n.a.	n.a.	n.a.	n.a.	n.a.	n.a.	n.a.	n.a.	n.a.	n.a.	n.a.	3.2	3.0	3.6	3.8	2.9	5.1	3.4	3.2
Th	n.a.	n.a.	n.a.	1.5	n.d.	4.5	0.8	3.2	0.0	2.3		0.3	5.3	12.1	0.6	n.d.	n.d.	0.3	9.8
U	n.a.	n.a.	n.a.	n.d.	0.9	0.8	1.0	1.0	0.6	0.3	1.5	0.4	0.7	1.7	0.5	0.8	0.7	n.d.	n.d.

(continued)

TABLE 1. WHOLE-ROCK ANALYSES FOR METAMORPHOSED MAFIC ROCKS AND QUARTZOFELDSPATHIC GNEISSES OF THE TOBACCO ROOT MOUNTAINS (*continued*)

Sample	75a-p	40a-p	59-I	50-I	81-I	82-I	52-I	LJ 5	LJ 8	LJ 47	LJ 53	LJ 73	LJ 79-2	LJ 80	LJ 81	LJ82-1	LJ 82-2	LJ84-1
Data Source	4	4	4	4	4	4	4	5	5	5	5	5	5	5	5	5	5	5
Major Elements (wt%)																		
SiO_2	69.67	72.57	59.13	70.27	76.09	73.93	74.08	74.12	74.60	72.51	73.31	72.62	72.23	74.46	71.09	74.54	71.96	69.93
TiO_2	0.35	0.16	0.31	0.30	0.42	0.08	0.05	0.13	0.22	0.25	0.26	0.21	0.19	0.23	0.31	0.13	0.23	0.32
Al_2O_3	15.46	16.15	18.25	16.44	12.21	13.96	15.73	14.79	13.57	15.41	14.63	15.39	15.2	13.33	14.27	14.86	15.93	15.94
Fe_2O_3T	3.33	1.27	3.88	2.19	1.81	0.94	0.47	1.13	1.87	2.31	2.35	1.61	1.68	1.81	3.10	1.81	1.73	2.35
Fe_2O_3	n.a.	n.a.	n.a.	n.a.	n.a.	n.a.	n.a.	n.a.	n.a.	n.a.	n.a.	n.a.	n.a.	n.a.	n.a.	n.a.	n.a.	n.a.
FeO	n.a.	n.a.	n.a.	n.a.	n.a.	n.a.	n.a.	n.a.	n.a.	n.a.	n.a.	n.a.	n.a.	n.a.	n.a.	n.a.	n.a.	n.a.
MgO	1.41	0.50	2.95	1.08	0.14	0.23	0.17	0.53	0.58	0.85	1.04	0.62	0.77	0.67	1.62	0.45	0.70	1.08
MnO	0.06	0.01	0.17	0.02	0.02	0.02	0.01	0.03	0.04	0.06	0.04	0.03	0.03	0.03	0.05	0.03	0.03	0.03
CaO	3.65	3.05	10.92	3.03	3.71	2.03	1.71	2.02	1.22	2.76	1.97	1.98	2.29	1.45	2.57	1.97	2.10	3.31
Na_2O	4.44	5.16	3.95	4.99	3.44	3.88	5.86	4.93	3.45	4.72	4.53	4.69	4.54	2.96	3.58	5.12	5.66	5.39
K_2O	1.87	1.60	0.43	1.77	0.62	3.50	2.09	2.47	4.64	1.34	1.92	3.02	2.86	4.51	3.51	1.78	1.82	1.22
P_2O_5	0.11	0.06	0.04	0.02	0.05	0.04	0.06	0.04	0.06	0.08	0.07	0.06	0.06	0.06	0.08	0.03	0.08	0.12
Total	100.91	100.89	100.36	100.68	98.90	98.86	100.58	100.18	100.23	100.26	100.11	100.21	99.87	99.80	100.17	100.09	100.24	99.69
LOI	0.57	0.37	0.34	0.57	0.04	0.26	0.35	n.a.	n.a.	n.a.	n.a.	n.a.	n.a.	n.a.	n.a.	n.a.	n.a.	n.a.
Rare earth and trace elements (ppm)																		
Sc	7.0	0.0	50.0	3.0	4.0	1.0	0.0	n.a.	n.a.	n.a.	n.a.	n.a.	n.a.	n.a.	n.a.	n.a.	n.a.	n.a.
V	55.0	15.0	230.0	24.0	0.0	9.0	0.0	14.9	23.2	25.5	25.6	17.1	26.5	26.7	42.8	11.4	20.0	30.6
Cr	39.0	4.0	144.0	10.0	4.0	3.0	1.0	4.3	9.4	7.7	9.6	4.5	8.1	8.3	76.4	1.2	1.4	8.3
Co	n.a.	2.9	32.6	5.8	1.4	2.6	1.2	n.a.	n.a.	n.a.	n.a.	n.a.	n.a.	n.a.	n.a.	n.a.	n.a.	n.a.
Ni	28.0	9.0	131.0	19.0	11.0	n.d.	n.d.	11.4	8.7	8.5	11.4	10.4	8.9	8.7	35.1	6.8	6.4	11.2
Cu	n.a.	n.a.	n.a.	n.a.	n.a.	n.a.	n.a.	n.a.	n.a.	n.a.	n.a.	n.a.	n.a.	n.a.	n.a.	n.a.	n.a.	n.a.
Zn	35.0	31.0	77.0	37.0	18.0	41.0	20.0	22.5	31.8	53.8	59.1	29.4	23.9	19.3	48.5	15.5	35.0	40.5
Ga	n.a.	n.a.	n.a.	n.a.	n.a.	n.a.	n.a.	n.a.	n.a.	n.a.	n.a.	n.a.	n.a.	n.a.	n.a.	n.a.	n.a.	n.a.
Rb	61.0	36.0	4.0	59.0	18.0	86.0	37.0	n.a.	n.a.	n.a.	n.a.	n.a.	n.a.	n.a.	n.a.	n.a.	n.a.	n.a.
Sr	240.0	480.0	107.0	446.0	170.0	250.0	466.0	329.0	106.7	464.5	362.4	546.0	412.4	177.4	259.0	507.8	551.6	503.6
Y	39.0	3.0	13.0	5.0	62.0	2.0	2.0	n.a.	n.a.	n.a.	n.a.	n.a.	n.a.	n.a.	n.a.	n.a.	n.a.	n.a.
Zr	88.0	64.0	13.0	121.0	399.0	90.0	48.0	86.6	134.8	115.8	98.5	119.9	82.7	142.3	109.4	98.1	126.2	143.6
Nb	n.a.	n.a.	n.a.	n.a.	n.a.	n.a.	n.a.	n.a.	n.a.	n.a.	n.a.	n.a.	n.a.	n.a.	n.a.	n.a.	n.a.	n.a.
Ba	426.0	447.0	169.0	417.0	259.0	629.0	742.0	566.8	524.1	287.2	392.7	574.4	1075.1	842.7	614.4	1083.6	411.5	447.8
La	29.0	10.4	3.0	49.4	90.9	11.7	5.7	13.1	30.1	13.5	13.7	19.6	7.4	31.8	24.6	5.7	21.6	13.0
Ce	52.0	16.6	4.3	84.4	144.0	20.4	10.1	29.2	59.9	30.3	29.3	37.7	19.5	67.9	53.6	17.5	45.0	29.6
Hf	3.0	2.5	0.7	4.4	14.1	2.5	2.0	n.a.	n.a.	n.a.	n.a.	n.a.	n.a.	n.a.	n.a.	n.a.	n.a.	n.a.
Th	2.2	3.6	1.6	30.4	27.7	2.3	0.1	n.a.	n.a.	n.a.	n.a.	n.a.	n.a.	n.a.	n.a.	n.a.	n.a.	n.a.
U	1.7	n.d.	n.d.	n.d.	6	n.d.	0.64	n.a.	n.a.	n.a.	n.a.	n.a.	n.a.	n.a.	n.a.	n.a.	n.a.	n.a.

(*continued*)

TABLE 1. WHOLE-ROCK ANALYSES FOR METAMORPHOSED MAFIC ROCKS AND QUARTZOFELDSPATHIC GNEISSES OF THE TOBACCO ROOT MOUNTAINS (*continued*)

Sample	RB-1	RB-2	RB-3	RB-4	RB-6
Data Source	5	5	5	5	5
Major Elements (wt%)					
SiO_2	66.61	69.23	76.20	75.25	72.28
TiO_2	0.46	0.19	0.19	0.17	0.14
Al_2O_3	15.89	16.88	12.76	13.36	14.51
Fe_2O_3T	3.90	1.86	1.35	1.50	1.85
Fe_2O_3	n.a.	n.a.	n.a.	n.a.	n.a.
FeO	n.a.	n.a.	n.a.	n.a.	n.a.
MgO	2.28	1.13	0.59	0.50	0.70
MnO	0.07	0.03	0.02	0.03	0.03
CaO	3.76	1.42	0.76	0.87	0.93
Na_2O	4.97	6.31	3.24	3.38	2.75
K_2O	1.68	2.84	4.65	4.99	6.25
P_2O_5	0.14	0.07	0.04	0.04	0.03
Total	99.76	99.97	99.97	100.09	99.46
LOI	n.a.	n.a.	n.a.	n.a.	n.a.
Rare earth and trace elements (ppm)					
Sc	n.a.	n.a.	n.a.	n.a.	n.a.
V	67.3	30.1	22.9	16.6	14.0
Cr	107.5	3.0	8.4	6.1	22.6
Co	n.a.	n.a.	n.a.	n.a.	n.a.
Ni	40.8	11.3	9.4	4.4	13.3
Cu	n.a.	n.a.	n.a.	n.a.	n.a.
Zn	45.8	22.8	15.4	18.5	19.9
Ga	n.a.	n.a.	n.a.	n.a.	n.a.
Rb	n.a.	n.a.	n.a.	n.a.	n.a.
Sr	406.4	290.0	268.8	135.8	278.9
Y	n.a.	n.a.	n.a.	n.a.	n.a.
Zr	126.6	90.9	107.3	112.2	219.6
Nb	n.a.	n.a.	n.a.	n.a.	n.a.
Ba	457.9	597.5	1453.9	630.2	2106.1
La	14.5	5.9	36.8	33.4	118.9
Ce	37.6	14.7	79.0	77.4	203.8
Hf	n.a.	n.a.	n.a.	n.a.	n.a.
Th	n.a.	n.a.	n.a.	n.a.	n.a.
U	n.a.	n.a.	n.a.	n.a.	n.a.

Note: Data sources: 1. P.A. Mueller, K. D'Arcy, A. Heatherington, University of Florida (unpubl.); 2. Mogk et al., 1992; 3. Martin, 1996; 4. Abeyta, 1998; 5. Jacob, 1994. All analyses were done by X-ray fluorescence (XRF), except for set 5, which was analyzed using inductively-coupled plasma (ICP) spectrometry.
n.a.—not analyzed; n.d.—not detected.

can "see through" imposed alteration events. Interpretation of these plots is enabled in circumstances where unaltered or "least-altered" rocks can be identified. This is clearly not possible in high-grade gneiss terrains. Although rocks from each of these suites share similar geochemical affinities and their precursors may be broadly characterized compositionally as basalts, andesites, dacites, and rhyolites, in detail, each suite is most likely a composite of numerous discrete magmatic events that have aggregated over long spans of geologic time. Thus, it is impossible to discern individual alteration lines for discrete rock suites. Interpretations of these plots may be further confounded if these suites are from supracrustal sources (i.e., volcanic) and there is significant mixing with other supracrustal rocks (see next section).

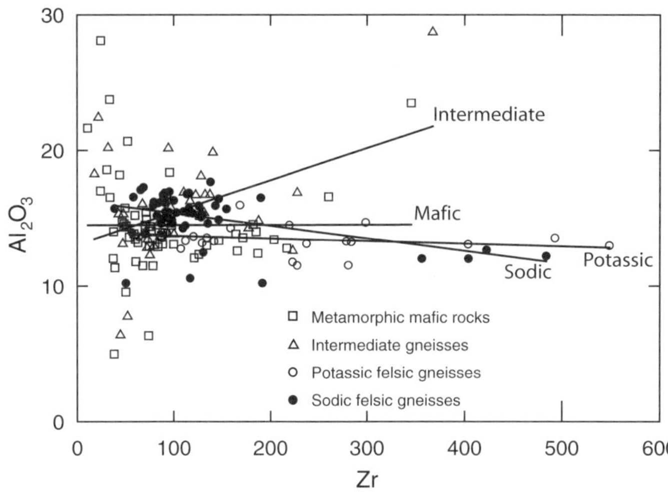

Figure 11. Al_2O_3-Zr plot used to distinguish alteration trends from precursor rocks (after MacLean and Barrett, 1993). Alteration lines should pass through the origin and the precursor composition for homogeneous rocks that have undergone varying degrees of alteration. The orientation of the regression lines, and scatter of data for each of the rock suites, most likely represents primary compositional variations of numerous precursor rocks in each suite rather than progressive alteration of single precursors.

Multi-element variation diagrams (so-called "spider diagrams") that are normalized to primitive mantle (Sun and McDonough, 1989) are also widely used to interpret magmatic affinities. Figures 12A–C show spider diagrams for the metamorphosed mafic rocks, intermediate gneisses, and granitic and tonalitic subsets of felsic gneisses, respectively. Large ion lithophile elements (LILEs) show erratic patterns in all rock suites, reflecting their high mobility. In general, the LILEs for all three rock suites are enriched with respect to the rare earth elements (REEs), and there is pronounced depletion of Nb and P in the intermediate and felsic gneisses (and a subset of the mafic rocks). This pattern has been interpreted as the signature for rocks from volcanic arcs (e.g., Davidson, 1996), and approaches that expected for the bulk continental crust (e.g., Kerrich and Wyman, 1997).

One subset of the mafic rocks has compositions suggesting komatiitic affinities, with MgO > 18 wt%, negative Nb and P anomalies (Fig. 12A), negative correlation of MgO with Al_2O_3 and CaO, and positive correlation of MgO, Cr, and Ni, indicating magmatic fractionation or accumulation of olivine (Fig. 13; Kerrich and Wyman, 1997).

TABLE 2. TOBACCO ROOT MOUNTAINS: INSTRUMENTATION NEUTRON ACTIVATION ANALYSIS DATA

Sample	002	009	010	014	022	030	036	RC-2	RC-9	RC-10	RC-11	RC-13	RC-22	RC-24
							wt%							
Na_2O	0.12	2.78	3.85	2.55	4.29	2.05	3.54	1.96	4.32	4.52	4.25	3.19	2.59	2.53
K_2O	n.d.	0.42	3.44	2.82	1.51	2.59	3.78	0.51	1.32	1.23	2.42	0.91	0.53	0.93
CaO	1.50	9.90	2.11	2.40	4.60	1.62	1.89	9.00	4.20	3.33	1.68	6.56	13.40	8.20
FeO	10.18	15.40	1.80	6.52	5.83	12.19	2.62	10.47	4.68	2.42	1.31	7.38	2.83	14.86
							ppm							
Sc	12.85	41.70	2.53	10.20	13.49	16.19	6.64	39.10	13.56	3.62	1.67	15.98	4.94	34.10
Cr	4960.00	33.30	6.16	1.60	71.20	389.00	5.50	180.40	61.70	6.94	8.45	183.10	123.90	28.10
Co	136.70	57.30	5.55	8.43	21.49	31.80	5.82	41.90	13.19	8.53	5.20	28.20	14.01	51.30
Ni	1860.00	59.00	n.d.	n.d.	63.00	187.00	n.d.	47.00	33.00	n.d.	n.d.	62.00	58.00	45.00
Zn	n.d.	n.d.	n.d.	n.d.	n.d.	n.d.	n.d.	n.d.	n.d.	n.d.	n.d.	n.d.	n.d.	n.d.
As	0.58	n.d.	n.d.	n.d.	n.d.	n.d.	n.d.	n.d.	n.d.	n.d.	n.d.	n.d.	2.28	1.40
Br	n.d.	n.d.	0.58	n.d.	0.78	0.52	0.70	0.46	n.d.	1.29	1.75	0.62	1.23	4.80
Rb	n.d.	n.d.	89.00	56.40	63.80	119.00	90.00	7.30	28.80	43.60	44.90	36.00	9.90	35.30
Sr	n.d.	136.00	246.00	121.00	330.00	169.00	143.00	80.00	355.00	312.00	325.00	206.00	272.00	202.00
Zr	n.d.	110.00	155.00	410.00	159.00	131.00	248.00	n.d.	141.00	135.00	49.00	128.00	n.d.	180.00
Sb	n.d.	n.d.	n.d.	n.d.	n.d.	n.d.	n.d.	n.d.	n.d.	n.d.	n.d.	n.d.	0.03	0.22
Cs	1.48	n.d.	1.21	0.32	2.33	1.71	0.45	n.d.	0.33	0.39	0.15	0.35	0.28	1.03
Ba	n.d.	102.00	1224.00	1089.00	247.00	696.00	955.00	66.00	298.00	332.00	344.00	322.00	76.00	199.00
La	2.62	7.93	49.50	65.10	30.10	32.50	69.40	6.20	11.18	18.96	6.49	21.39	3.47	17.27
Ce	6.90	20.40	93.50	133.40	70.10	57.20	130.60	16.00	28.30	32.70	10.97	49.40	6.52	42.20
Nd	n.d.	11.30	37.00	62.00	31.00	21.90	54.00	7.20	18.80	12.00	4.00	25.50	2.30	25.00
Sm	0.77	3.59	5.45	14.98	6.74	3.80	8.21	2.58	5.21	1.76	0.69	6.59	0.61	6.53
Eu	0.25	1.18	0.92	3.15	1.26	0.90	1.59	0.82	0.90	0.68	0.39	1.83	0.39	1.98
Tb	0.18	0.77	n.d.	2.24	0.87	0.60	1.16	0.59	0.91	0.14	0.07	0.99	0.09	1.18
Yb	0.60	2.54	0.66	7.55	3.17	2.11	3.97	2.16	2.84	0.23	0.11	3.06	0.31	3.40
Lu	0.08	0.38	0.11	1.12	0.47	0.31	0.55	0.33	0.39	0.03	0.02	0.45	0.05	0.50
Hf	0.66	2.45	5.48	11.83	4.51	3.15	7.67	2.02	3.80	3.81	1.79	3.81	0.17	4.96
Ta	0.08	0.29	0.10	1.66	1.01	0.70	0.86	0.24	0.28	0.13	0.08	2.09	0.12	0.77
W	1.10	0.90	9.80	16.30	10.00	7.65	13.20	4.34	3.90	14.78	12.50	6.60	9.60	7.30
Th	0.44	0.53	29.20	11.93	6.76	11.19	7.09	0.55	0.12	5.07	0.13	3.66	0.33	4.00
U	n.d.	0.21	0.51	2.15	1.41	1.07	0.46	0.23	n.d.	0.13	0.13	1.30	0.18	1.06

Sample	RP-25	RP-26	RP-35	TR-2	TR-4	TR-8A	TR-12	TR-13	TR-30	TR-34	TR-40	TR-43	TR-51	TR-56
							wt%							
Na_2O	5.14	3.52	0.30	0.09	4.48	2.60	1.45	0.67	0.99	0.05	0.82	0.14	3.26	1.92
K_2O	0.71	4.20	2.68	0.20	1.31	7.50	0.25	n.d.	0.75	0.95	n.d.	0.55	1.94	n.d.
CaO	5.10	n.d.	1.86	n.d.	4.75	0.91	11.10	6.90	3.50	n.d.	9.08	0.11	1.40	2.70
FeO	13.05	2.89	13.34	0.08	3.76	0.21	8.72	9.85	18.56	4.30	14.18	0.85	5.89	3.99
							ppm							
Sc	15.16	7.26	38.40	0.07	6.58	0.24	15.03	31.10	41.30	39.00	44.20	2.63	22.48	8.76
Cr	1.75	5.50	1105.00	9.25	32.90	0.97	223.20	1579.00	74.80	286.00	126.40	41.40	287.00	2.18
Co	23.14	6.53	73.70	173.60	30.70	46.10	63.20	69.10	32.80	24.52	59.70	10.33	28.10	7.51
Ni	n.d.	n.d.	329.00	n.d.	24.00	n.d.	326.00	289.00	n.d.	n.d.	59.00	26.00	77.00	n.d.
Zn	n.d.	n.d.	n.d.	n.d.	n.d.	n.d.	n.d.	n.d.	n.d.	n.d.	n.d.	n.d.	n.d.	n.d.
As	1.40	n.d.	n.d.	n.d.	n.d.	n.d.	n.d.	n.d.	n.d.	18.70	n.d.	0.33	n.d.	n.d.
Br	n.d.	1.48	n.d.	n.d.	0.59	n.d.	0.21	n.d.	n.d.	n.d.	n.d.	0.16	n.d.	n.d.
Rb	n.d.	93.00	76.00	4.20	42.10	156.00	6.30	n.d.	25.80	40.60	n.d.	20.40	172.00	12.40
Sr	265.00	156.00	59.00		480.00	314.00	108.00	n.d.	n.d.	n.d.	54.00	n.d.	152.00	137.00
Zr	230.00	370.00	n.d.	29.00	147.00	64.00	n.d.	n.d.	240.00	420.00	n.d.	30.00	190.00	193.00
Sb	n.d.	n.d.	n.d.	0.53	n.d.	n.d.	n.d.	n.d.	n.d.	0.29	n.d.	0.02	n.d.	n.d.
Cs	0.73	0.48	2.28	0.08	0.50	1.15	0.53	0.36	1.13	0.79	0.25	0.39	4.36	0.06
Ba	256.00	910.00	475.00	101.00	380.00	2100.00	69.00	n.d.	94.00	160.00	37.00	22.00	521.00	414.00
La	25.39	20.18	3.76	1.71	21.94	3.71	1.54	9.69	53.90	94.10	2.84	2.75	54.30	44.60
Ce	60.10	44.40	7.80	3.04	43.30	5.04	4.13	18.50	111.60	183.80	6.65	5.50	107.40	84.90
Nd	36.00	24.00	n.d.	n.d.	19.80	n.d.	3.80	9.00	54.00	69.00	n.d.	2.00	48.00	29.70
Sm	8.88	5.73	0.75	0.17	3.33	0.40	0.79	1.42	11.31	13.39	1.59	0.44	9.30	6.22
Eu	2.79	1.43	0.45	0.06	0.94	0.82	0.38	0.38	2.08	2.01	0.49	0.12	1.35	1.22
Tb	1.36	0.99	0.17	n.d.	0.39	0.07	0.18	0.21	1.90	1.75	0.46	0.10	1.40	0.97
Yb	3.86	3.34	1.35	0.06	1.02	0.35	0.68	0.82	5.65	6.15	2.19	1.51	5.08	3.79
Lu	0.55	0.49	0.20	n.d.	0.15	0.07	0.10	0.12	0.84	0.87	0.31	0.21	0.72	0.53
Hf	5.85	11.27	1.66	0.47	3.36	3.03	0.58	1.07	7.16	11.75	1.12	1.23	5.99	5.68
Ta	0.82	0.78	0.11	7.68	1.39	4.69	0.34	0.41	2.20	2.37	0.13	0.18	1.33	0.76
W	5.70	15.00	3.90	2650.00	302.00	1583.00	73.30	143.40	13.30	17.60	9.50	30.80	12.60	22.30
Th	1.07	0.70	0.53	0.76	2.39	0.88	0.09	1.73	6.94	26.30	0.45	0.75	20.71	19.93
U	0.91	0.42	n.d.	n.d.	0.18	1.08	n.d.	0.39	2.02	5.60	n.d.	0.87	3.85	3.18

Note: Data set from P. Mueller, K. D'Arcy, A. Heatherington, University of Florida (2003, personal commun.).
n.d.—not detected.

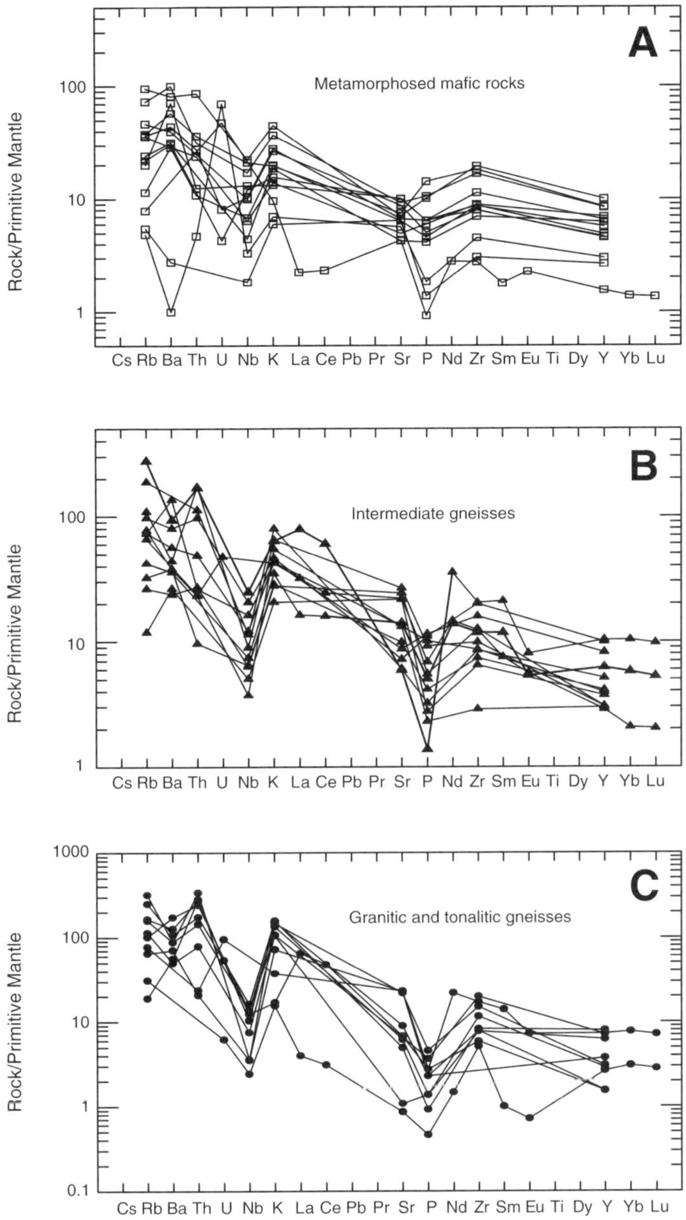

Figure 12. Primitive mantle–normalized "spider" diagrams (Sun and McDonough, 1989) for metamorphosed mafic rocks (A), intermediate gneisses (B), and granitic and tonalitic gneisses (C). Relative enrichment of LILEs to REEs and depletion of Nb and P suggest that these rocks formed in an arc environment.

Chondrite-normalized rare earth element plots are shown in Figures 14A–F. The metamorphic mafic rocks exhibit a range of REE patterns (Fig. 14A). Two of the samples have a relatively flat, mid-oceanic-ridge basalt–type pattern, with total REEs in the range of 2–20 times chondritic values (Sun and McDonough, 1989). Six other samples show LREE (light rare earth element) enrichment with La_N as high as 200 and La_N/Yb_N ratios ranging up to 10. The sample with the highest LREE enrichment also has a negative Eu anomaly and the sample with the lowest total REE content has a slight positive Eu anomaly; otherwise, most of these rocks do not exhibit Eu anomalies. The intermediate gneisses typically exhibit moderate LREE enrichment with La_N values of 25–100 and La_N/Yb_N ratios up to 8, with no significant Eu anomalies (Fig. 14B). However, three samples with among the highest total REE content show greater LREE enrichment with La_N of 150–400, and with slight negative Eu anomalies.

The sodic and potassic gneisses show a number of distinct REE patterns (Figs. 14C–F). Subsets of the tonalitic gneiss samples have been distinguished by the presence of negative, positive, or lack of Eu anomalies (Figs. 14C–E). One set of tonalitic gneiss samples with small negative Eu anomalies exhibits LREE enrichment with La_N of 100–400 and La_N/Yb_N ratios of ~10, similar to the three intermediate gneiss samples with the highest REE content (Fig. 14C). In contrast, other tonalitic samples with positive Eu anomalies (Fig. 14D) have generally lower total REE content, lower La_N (~20–100), and higher La_N/Yb_N ratios (~30). Many of the tonalitic samples exhibit no Eu anomaly (Fig. 14E), with a range of La_N from 20 to 400 and La_N/Yb_N ratios ranging from ~20 to 100. Granitic gneisses typically have REE patterns similar to the tonalitic gneisses with negative Eu anomalies (La_N of 100–300 and La_N/Yb_N ratios of ~10), although one sample has no discernable Eu anomaly and La_N/Yb_N ratio of ~25, similar to tonalitic gneiss counterparts (Fig. 14F). These varied REE patterns indicate that the aggregate rocks now present in the felsic gneisses were derived from different sources and/or evolved via different petrogenetic processes.

In addition to the suites of metamorphic mafic rocks and quartzofeldspathic gneisses that are compositionally equivalent to modern basalts, andesites, dacites, and rhyolites, there are a number of compositional variants that may indicate substantial pre- or synmetamorphic alteration (e.g., weathering and/or hydrothermal alteration). For mafic rocks, unusually high enrichment of Mg and Fe may reflect formation of Mg-rich alteration minerals (e.g., smectites) and Fe oxides during low-temperature hydrothermal alteration or weathering. Similarly, the unusually high contents and scatter of alkali and alkali earth elements (e.g., K, Na, Rb, Sr, Ba) in the quartzofeldspathic gneisses and some of the rocks of mafic and intermediate composition may indicate substantial mobilization during pre- or synmetamorphic alteration (e.g., potassic alteration, quartz-sericite alteration).

Protolith(s) of the Quartzofeldspathic Gneisses

Chemical variation diagrams provide some criteria for distinguishing ortho- from paragneisses, and if of supracrustal origin, distinguishing immature sediments (e.g., arkoses, "wackes," or volcaniclastic rocks) from more mature sediments.

Harker variation diagrams may be used to observe enrichment of elements related to detrital minerals that may have been concentrated through weathering or sedimentary processes. For the SiO_2 versus TiO_2 plot (Fig. 3A), most of the granitic and tonalitic gneisses fall in the magmatic field below the reference

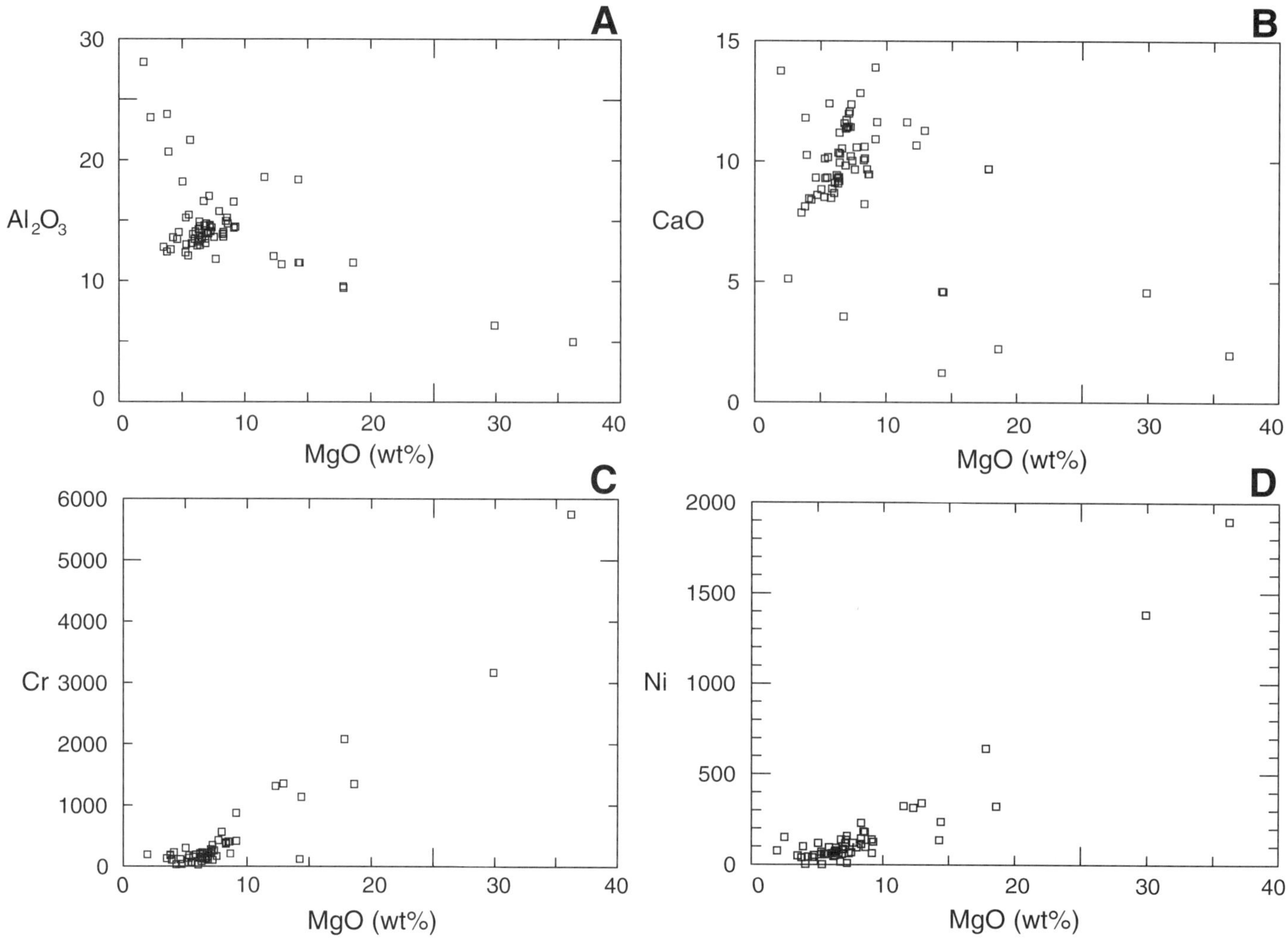

Figure 13. Elemental variation diagrams of Al_2O_3 vs. MgO (A), CaO vs. MgO (B), Cr vs. MgO (C), and Ni vs. MgO (D). Samples with high MgO (>18 wt%) have negative correlation with Al_2O_3 and CaO and positive correlation with Cr and Ni, which may be related to olivine fractionation or accumulation. These rocks may have a komatiitic affinity.

line proposed by Tarney (1976), although a significant number of samples lie above the line in the sedimentary field, presumably due to enrichment of detrital ilmenite. Similarly, there is a wide scatter and enrichment of Zr (related to zircon, Figure 3L) and P (related to apatite, Figure 3J) in the felsic gneisses.

Niggli numbers (i.e., molecular norms; see Barth, 1962) have been used to distinguish between quartzofeldspathic metasedimentary and igneous rocks by representing chemical variation trends indicative of sedimentary and igneous processes (Leake and Singh, 1986). Figure 15A is a plot of Niggli al-alk versus c values for all felsic gneisses. Rocks that have evolved from igneous fractionation would be expected to follow the trend indicated by the line with a positive slope, which represents feldspar crystallization. Sedimentary rocks would be expected to exhibit a negative correlation of these parameters because clays are rich in al-alk, but are typically devoid of Niggli c, arkoses straddle the igneous trend, and carbonates (off the graph to the lower right) are rich in c but deficient in al-alk. Most of the felsic gneisses generally fall in the igneous field suggested by Leake and Singh (1986), although a number of samples appear to be enriched in al-alk, possibly indicating enrichment in clays, and many other samples fall below the igneous trend in the field indicative of "sands." Niggli si, al-alk, and P_2O_5 wt% have also been plotted against Niggli mg in Figures 15B–D. In all cases most of the samples fall well within the compositional field of southern California batholithic rocks, although numerous other samples fall within the field of Californian arkoses.

In Figure 16, meta-igneous rocks and metasedimentary rocks have been distinguished by plotting data in the ternary CaO-MgO-Al_2O_3 system (Leyreloup et al., 1977). Primary igneous rocks should have relatively higher Ca in feldspar and Mg in ferromagnesian minerals, whereas weathered sediments should

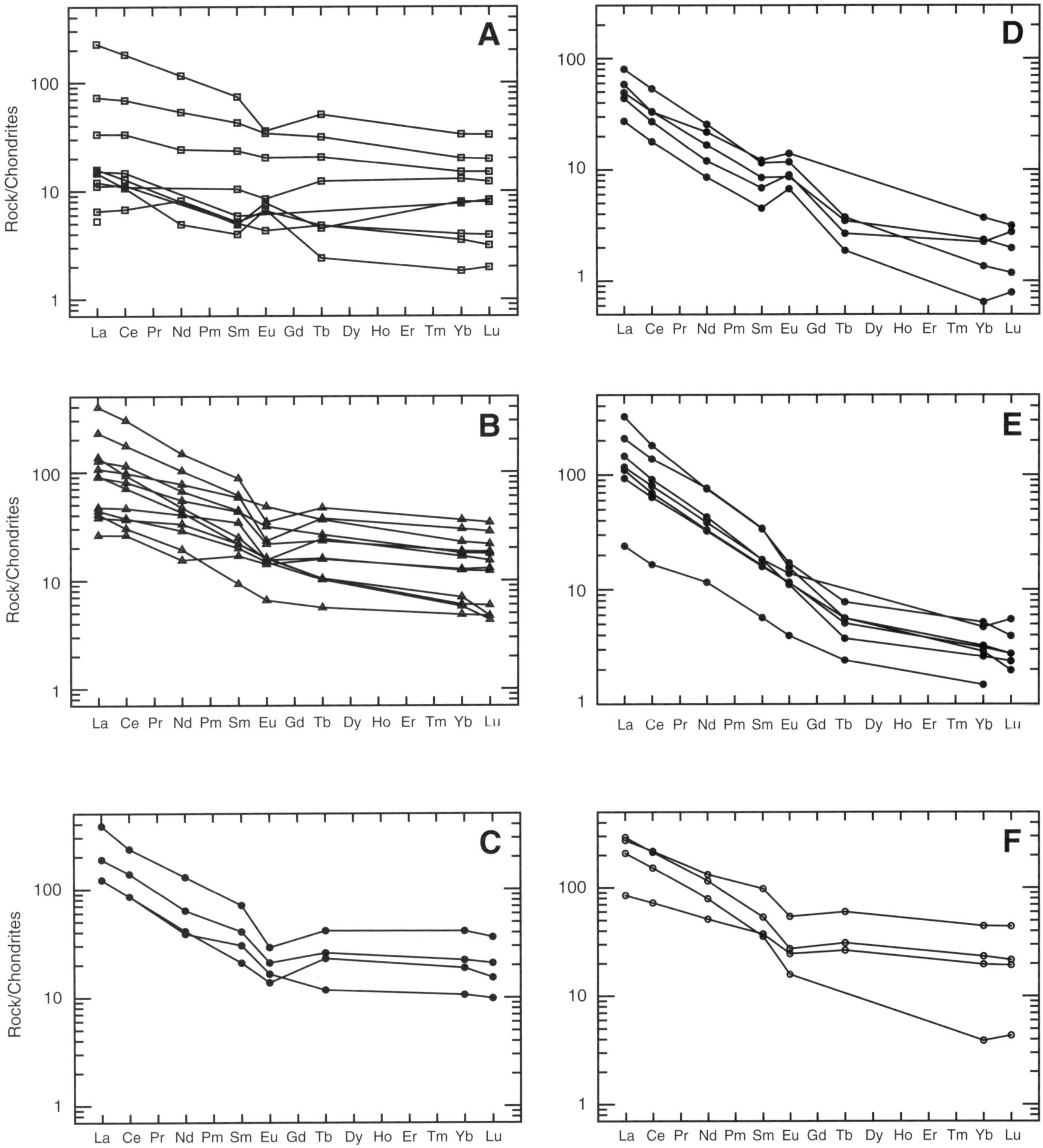

Figure 14. Plots of rare earth element normalized to chondrites (Sun and McDonough, 1989) for mafic rocks (A), intermediate gneisses (B), sodic gneisses with negative Eu anomaly (C), sodic gneisses with positive Eu anomaly (D), sodic gneisses with little or no Eu anomaly (E), and potassic gneisses (F).

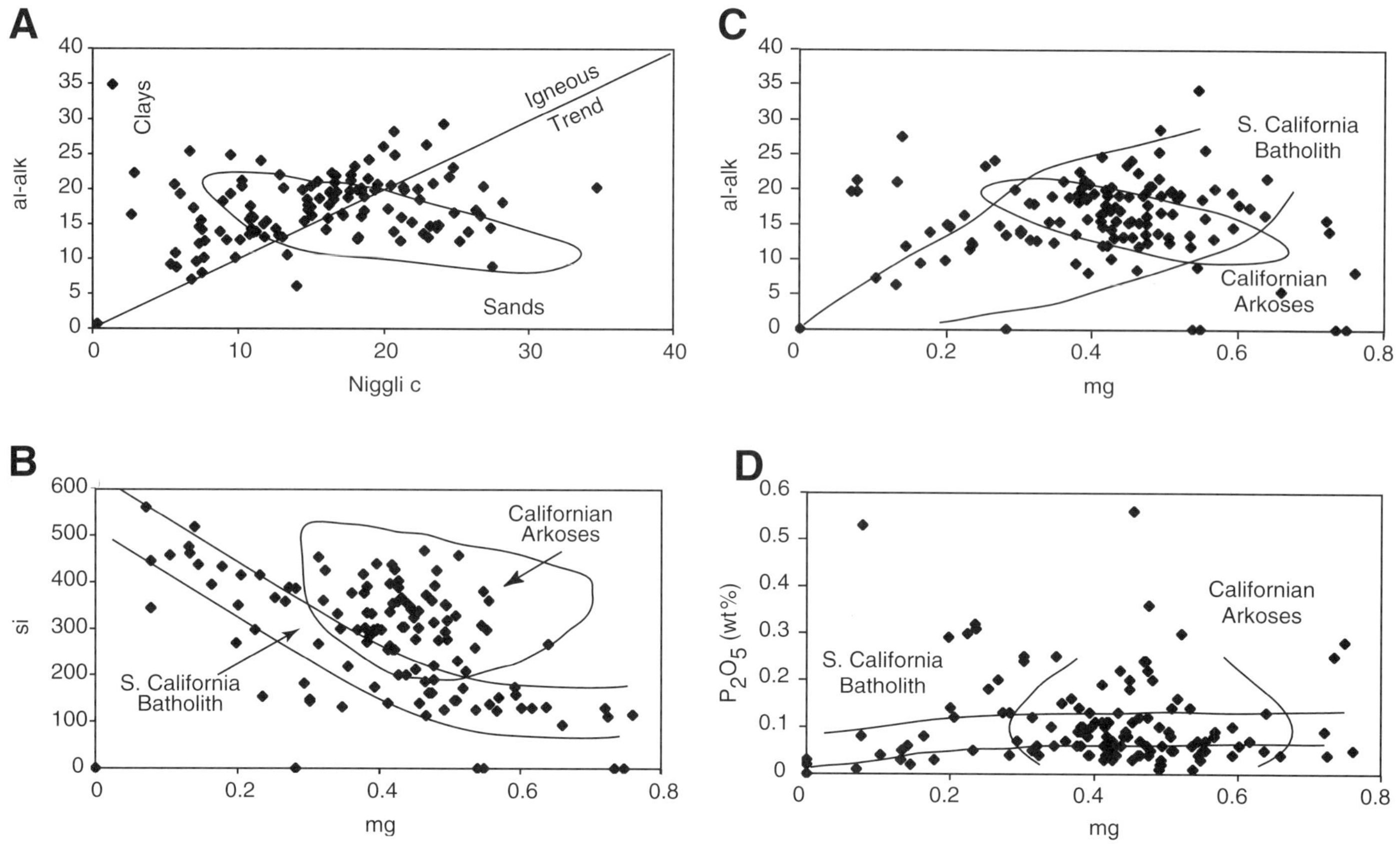

Figure 15. Variation diagrams for all felsic gneisses using Niggli values for c vs. al-alk (A), mg vs. si (B), mg vs. al-alk (C), and mg vs. P_2O_5 weight percent (D). In Figure 15A, the reference line shows the igneous trend that represents feldspar crystallization. A negative trend indicates probably sedimentary protoliths for the gneisses, with enrichment of clays represented in the upper right, arkosic sands in the middle, and carbonate rocks in the lower left. In Figures 15B–D, the fields of southern California batholith and Californian arkoses are outlined. (after Leake and Singh, 1986).

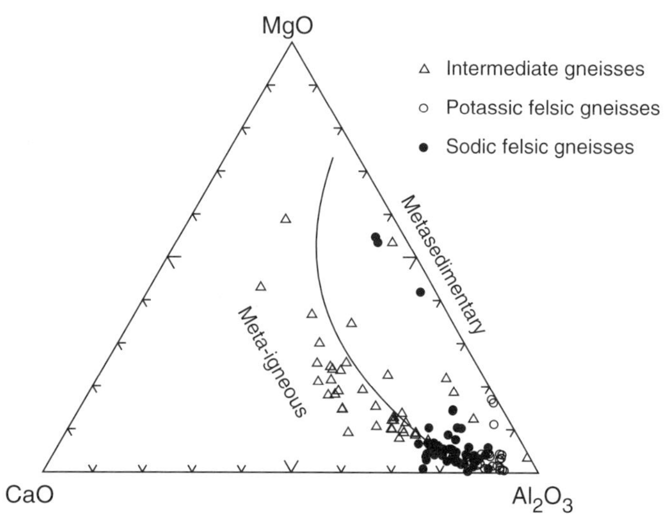

Figure 16. Ternary CaO-MgO-Al_2O_3 variation diagram showing the expected fields of meta-igneous and metasedimentary rocks for all intermediate and felsic gneisses (after Leyreloup et al., 1977).

show a loss of Ca due to weathering and enrichment in Al. Werner (1987) used P_2O_5/TiO_2 versus MgO/CaO plots to distinguish magmatic from sedimentary rocks (Fig. 17). And Shaw (1972) used discriminant functions of the form

$$DF3 = 10.44 - 0.21SiO_2 - 0.32Fe_2O_3(t) - 0.98MgO + 0.55CaO + 1.46Na_2O + 0.54K_2O$$

to distinguish igneous versus sedimentary protoliths; igneous rocks have a positive DF3 value and sedimentary rocks have a negative value. On the Shaw diagram (Fig. 18), the felsic and intermediate gneisses appear to be largely of igneous origin, although a significant number of gneisses also have compositional characteristics that could indicate a sedimentary protolith. The sedimentary components of these gneisses may derive from weathering of the igneous rocks (e.g., formation of clays, Figure 15A; leaching of alkali and alkali earth elements, Figures 16 and 18), or mixing with sediments that have undergone at least one cycle of erosion, transportation, and deposition (e.g., Figs. 3A, 3J, 3L, 15B–D).

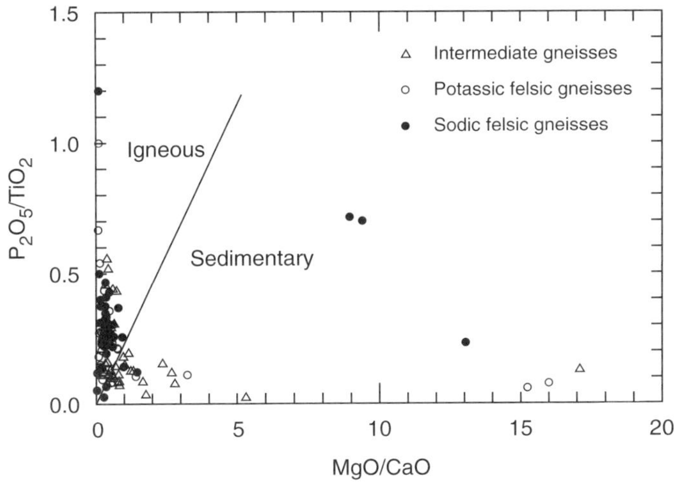

Figure 17. Variation diagram for P_2O_5/TiO_2 vs. MgO/CaO indicating igneous vs. sedimentary protoliths for all intermediate and felsic gneisses (after Werner, 1987).

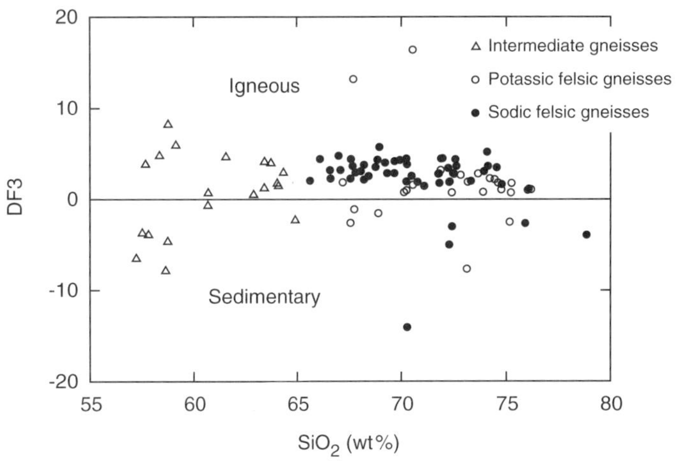

Figure 18. Compositional discriminant diagram showing the discriminant function 3 (DF3) vs. SiO_2 for all felsic gneisses (after Shaw, 1972). Rocks with positive DF3 values are interpreted as having an igneous origin, and rocks with a negative value are interpreted as having a sedimentary origin.

The Potential Influence of Alteration

Interpretation of the protoliths of the quartzofeldspathic gneisses is further complicated due to the potential influence of pre- or synmetamorphic hydrothermal alteration. As noted above, the large ion lithophile elements (e.g., K, Na, Rb, Ba, Sr) are widely scattered for all compositions of gneisses (and metamorphic mafic rocks). This suggests that the entire sequence, mafic and felsic rocks alike, may have experienced potassic alteration (and mobilization of other incompatible elements) to varying degrees. There are also rocks of both mafic and felsic composition that have unusually high iron content, which may be the result of pre-metamorphic weathering or hydrothermal alteration. Garnet-gedrite rocks interlayered in the quartzofeldspathic gneisses are similar to those described at Orijarvi, Finland, which have been interpreted as metamorphosed rocks of mafic composition that have undergone pre-metamorphic hydrothermal alteration (e.g., to smectite, as opposed to alternative explanations that suggest that these rocks are restites after a partial melting event; Smith et al., 1992).

Potassic alteration of volcanic sequences has been well documented in contemporary settings (e.g., Creede, Colorado; Socorro, New Mexico; Harcuvar Mountains, Arizona; Rougvie and Sorensen, 2002; eastern Sierra Nevada, Sorensen et al., 1998; and the Mojave Desert, California, Glazner, 1988). Sorensen et al. (1998) used a series of variation diagrams using elemental ratios of incompatible elements (Rb, Sr, K, Na) and "immobile" elements (Zr, Ti, Al) to determine whether or not the highly potassic rocks of eastern California are calc-alkaline or peralkaline volcanic rocks, or their altered equivalents. They concluded that the alkali systematics of these rocks reflects potassic metasomatism, based on comparisons with analyses of calc-alkaline arc rocks and altered Tertiary rocks from California and Arizona. A comparison of the quartzofeldspathic gneisses of the Indian Creek and Pony–Middle Mountain Metamorphic Suites with these modern analogues indicates that the sodic gneisses largely fall in the fields of continental arc rocks, whereas the potassic gneisses have closer affinities with the altered volcanic rocks (Figs. 19A–D). Further, REE patterns of the potassic gneisses (Fig. 14F) are similar to those of the sodic gneisses (Figs. 14C and 14E). If the REEs are relatively immobile during hydrothermal alteration and metamorphism, it could be the case that these gneisses derive from the same igneous protoliths, but the higher K content of the granitic gneisses may have been introduced as potassic alteration prior to or during metamorphism. Such potassic alteration (e.g., quartz-sericite) may also have imparted peraluminous characteristics to some of the gneisses (Fig. 7B). Sorensen et al. (1998) also note that alkaline and peralkaline rhyolites tend to display very large negative Eu anomalies, a characteristic that is not developed in any of the quartzofeldspathic gneisses. It is most likely that the quartzofeldspathic gneisses of the Tobacco Root Mountains had protoliths that were originally volcanic rocks of a calc-alkaline affinity that were subjected to potassic alteration to varying degrees. Although some of these rocks may have originated as true rhyolites, many of the rocks that are presently highly potassic may also have been part of a TTG suite that was subsequently metasomatized.

Tectonic Setting

Chemical variation diagrams for mafic to felsic rocks have been used to compare ancient rocks with those from known contemporary tectonic settings. For the metamorphic mafic rocks, a plot of Zr-Ti-Sr (Pearce and Cann, 1973) shows most samples falling in the ocean floor field; outliers in the calc-alkali and

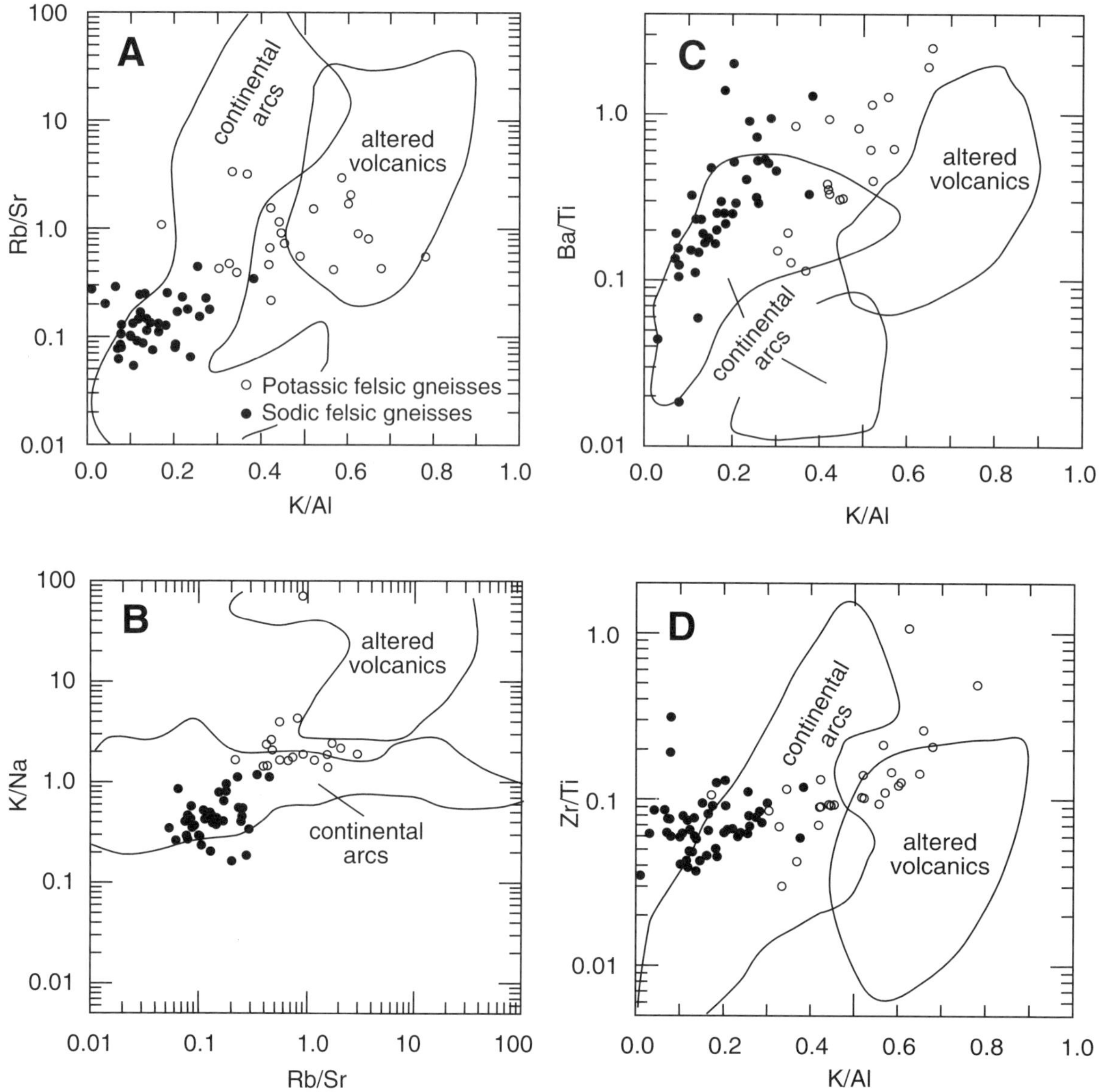

Figure 19. Rb/Sr vs. K/Al (A), K/Na vs. Rb/Sr (B), Ba/Ti vs. K/Al (C), and Zr/Ti vs. K/Al (D) used to distinguish unaltered continental arc volcanic rocks from altered volcanic rocks. Fields for the continental arc rocks and altered volcanic rocks are based on compilations of Sorensen et al. (1998).

island-arc basalt fields may be the result of Sr mobility in some samples (Fig. 20A). A plot of Zr-Ti-Y (Pearce and Cann, 1973) indicates a range of possible tectonic settings, with most samples in the calc-alkali or ocean floor fields (Fig. 20B). The Mn-Ti-P plot developed by Mullen (1983) shows most samples falling in the calc-alkali basalt and island-arc tholeiite fields (Fig. 20C). Given the association with large volumes of felsic gneisses, it is unlikely that the metamorphic mafic rocks derive from ocean floor basalts, and island-arc or calc-alkali provinces are the most likely settings for these associations.

Trace element (Nb, Y, Rb, Ta) variation diagrams for felsic gneisses are also presented in Figures 21A–D (Pearce et al., 1984). These diagrams assume that the rocks are of igneous origin, and we recognize that some of the protoliths may have a sedimentary component, or may have been significantly altered. Nonetheless, two of the diagrams (Figs. 21A and 21D) involve elements that are generally recognized as immobile, and all of the diagrams strongly indicate a source as either volcanic-arc or syncollisional granites (labels of these compositional fields using the terminology of Pearce et al., 1984).

A series of variation diagrams that represent the fields of Paleozoic sediments from known tectonic settings, as developed by Bhatia (1983), is presented in Figures 22A–C. Samples plotted on these diagrams are restricted to felsic gneisses interpreted as being of sedimentary origin (e.g., arkosic or volcaniclastic) based on DF3 values (Shaw, 1972) less than zero. These samples generally fall in the fields of active continental margins or continental arcs. Bhatia (1983) defines active continental margins as basins adjacent to Andean-type magmatic arcs on thick continental margins or adjacent to transverse plate boundaries, whereas continental arcs are defined as backarc basins on the continental side of island arcs or as basins adjacent to arcs formed on thin continental margins (e.g., Cascades, western United States).

GEOCHRONOLOGY AND ISOTOPE SYSTEMATICS

Detailed geochronologic studies are presented in the accompanying chapters by Mueller et al. (this volume, Chapter 9), and Cheney et al. (this volume, Chapter 6). However, a summary of some of the results of isotopic analysis is presented here because of their importance in interpreting the age, petrogenesis, and evolution of the Indian Creek and Pony–Middle Mountain Metamorphic Suite rocks. Because primary relations are obscured by deformation and metamorphism, the age relations among the many rocks of the Indian Creek and Pony–Middle Mountain Metamorphic Suites can only be resolved through studies of mineralogic chronometers. Limiting age relations are summarized as follows.

- U-Pb analyses of zircons extracted from the quartzofeldspathic gneisses yield ages in the range of 3.3–3.5 Ga (Mueller et al., this volume, Chapter 9; Mueller et al., 1996; Krogh et al., 1997). This time interval represents a major stage of crustal growth in the Wyoming province, as similar ages have been demonstrated in the surrounding exposures of Archean basement (Mueller et al., 1993, 1996, 1998).
- T_{dm}s (Sm-Nd depleted mantle model ages) of the quartzofeldspathic gneisses and metamorphic mafic rocks range from 3.07 to 3.94 Ga, although most of these rocks are in the range of 3.3–3.5 Ga. The depleted mantle model age is very close to the crystallization age of these rocks based on the U/Pb zircon analyses of the gneisses, and indicates that limited crustal recycling has occurred (Mueller et al., this volume, Chapter 9). Epsilon Nd values range from –7.94 (T_{dm} = 3.22 Ga) to –47.6 (T_{dm} = 3.13 Ga) for metamorphic mafic rocks to felsic gneisses, respectively (Fig. 23).
- Detrital zircons recovered from quartzites in the Tobacco Root Mountains yield ages that range from ca. 4.0 Ga to 2.9 Ga, with a majority of ages of 3.3–3.2 Ga (Mueller et al., 1998, this volume, Chapter 9). Consequently, the quartzites, and by association the other stable platform metasedimentary rocks (e.g., marbles, pelitic schists, iron formations), must be significantly younger than the surrounding quartzofeldspathic gneisses to have received zircons up to 300 Ma younger.

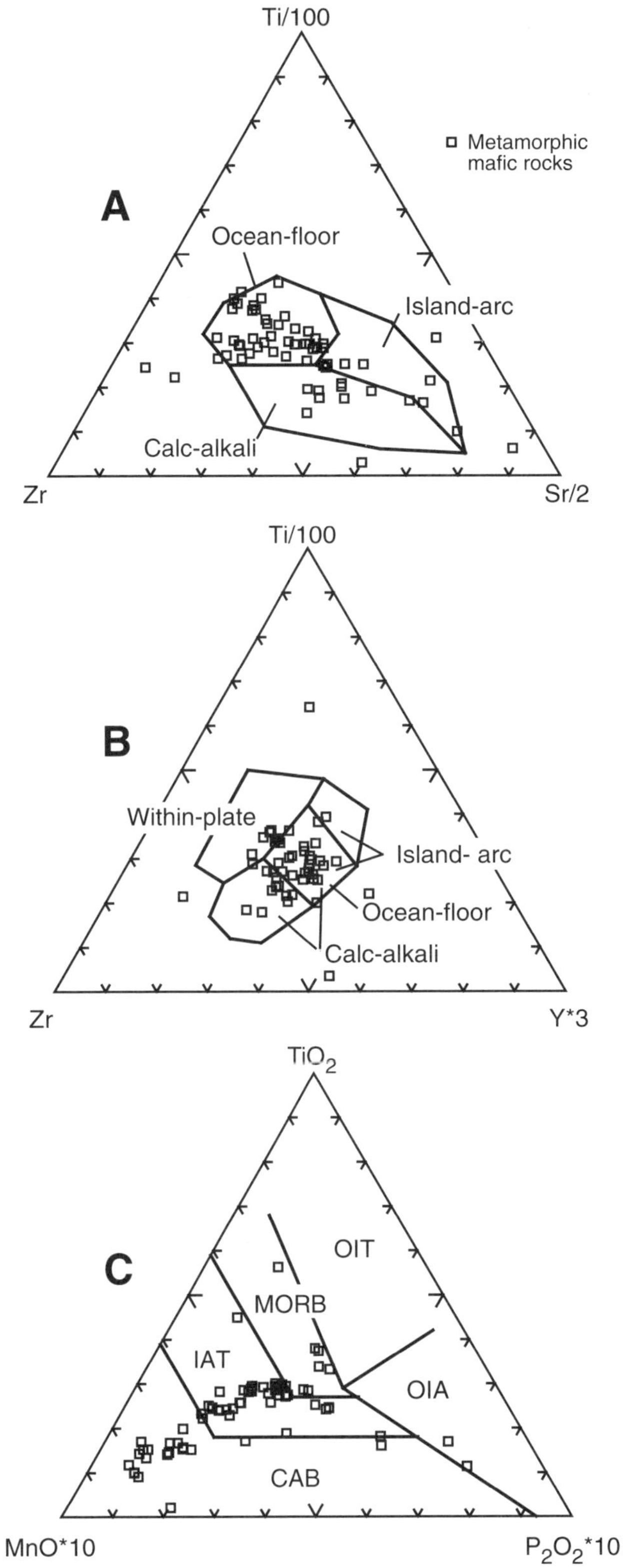

Figure 20. Trace element variation diagrams for metamorphic mafic rocks. A and B: After Pearce and Cann (1973). C: After Mullen (1983). Most metamorphic mafic rocks fall in the fields of calc-alkaline or island-arc tholeiite basalts. OIT—ocean island tholeiites; MORB—mid-oceanic-ridge basalt; IAT—island-arc tholeiites; OIA—ocean island alkalic basalts; CAB—calc-alkaline basalt.

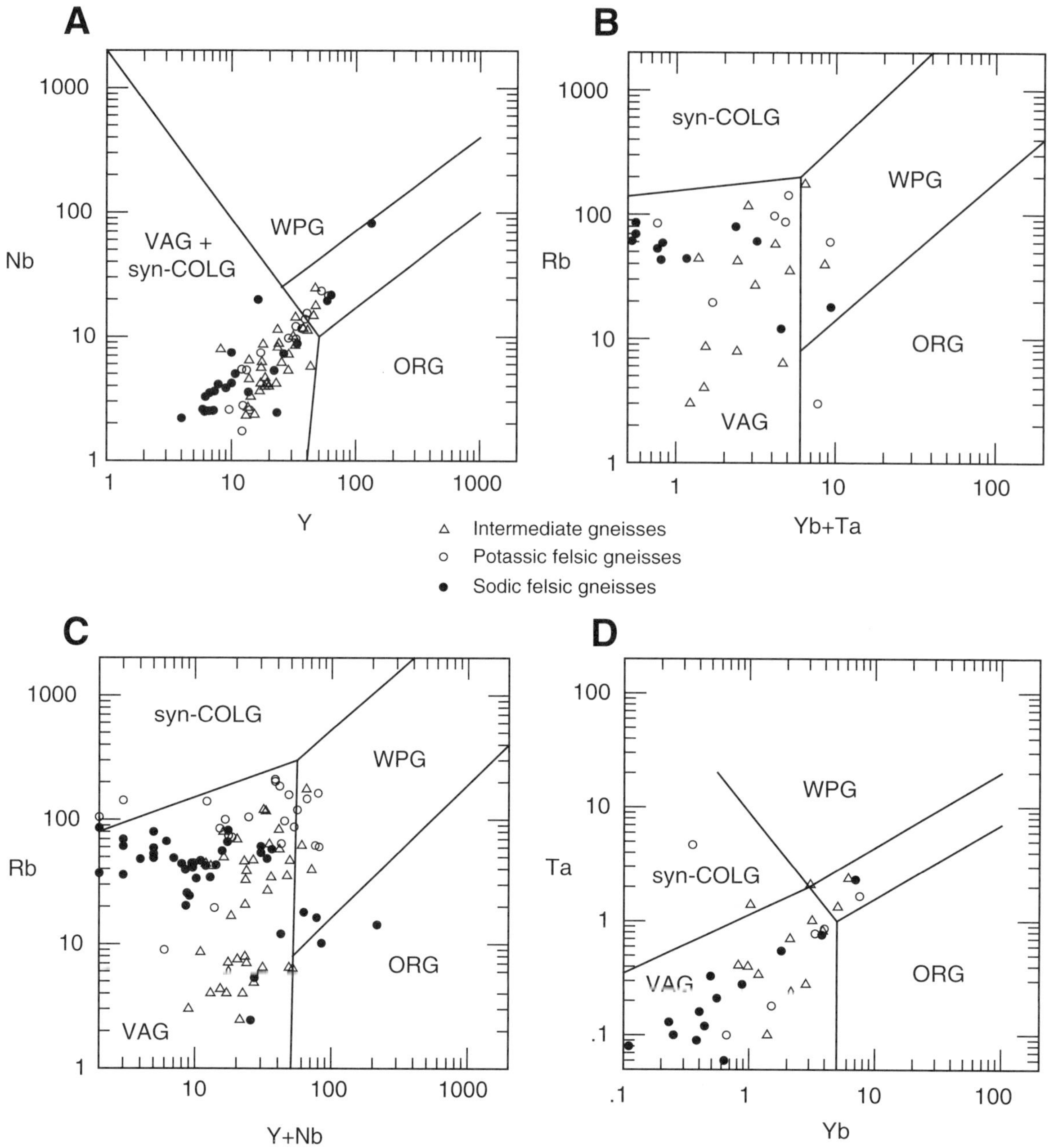

Figure 21. Trace element variation diagrams for intermediate and felsic gneisses (after Pearce et al., 1984). Most samples fall in the field of syncollisional or volcanic-arc granites. VAG—volcanic-arc granite; syn-COLG—syncollisional granite; WPG—within-plate granites; ORG—oceanic ridge granite.

- Post-Archean metamorphism of the supracrustal rocks occurred at ca. 2450 Ma, based on Th/Pb dating of monazite inclusions in garnets (Cheney et al., this volume, Chapter 6) and U/Pb dating of zircons from migmatitic gneisses (Krogh et al., 1997; Mueller et al., this volume, Chapter 9). A recent interpretation that the metasedimentary rocks are of Paleoproterozoic age presented by Roberts et al. (2002) is clearly untenable in light of these metamorphic ages.
- A second major tectonothermal event occurred in the Paleoproterozoic (ca. 1780–1720 Ma; Giletti, 1966, 1971; Erslev and Sutter, 1990; Burger et al., 1999; Cheney et al., 1999, this volume, Chapter 6; Mueller et al., this volume, Chapter 9). Although this event is not germane to the genesis of the crystalline rocks, it is possible that the high-grade metamorphism and deformation may have resulted in local to regional-scale elemental mobility.

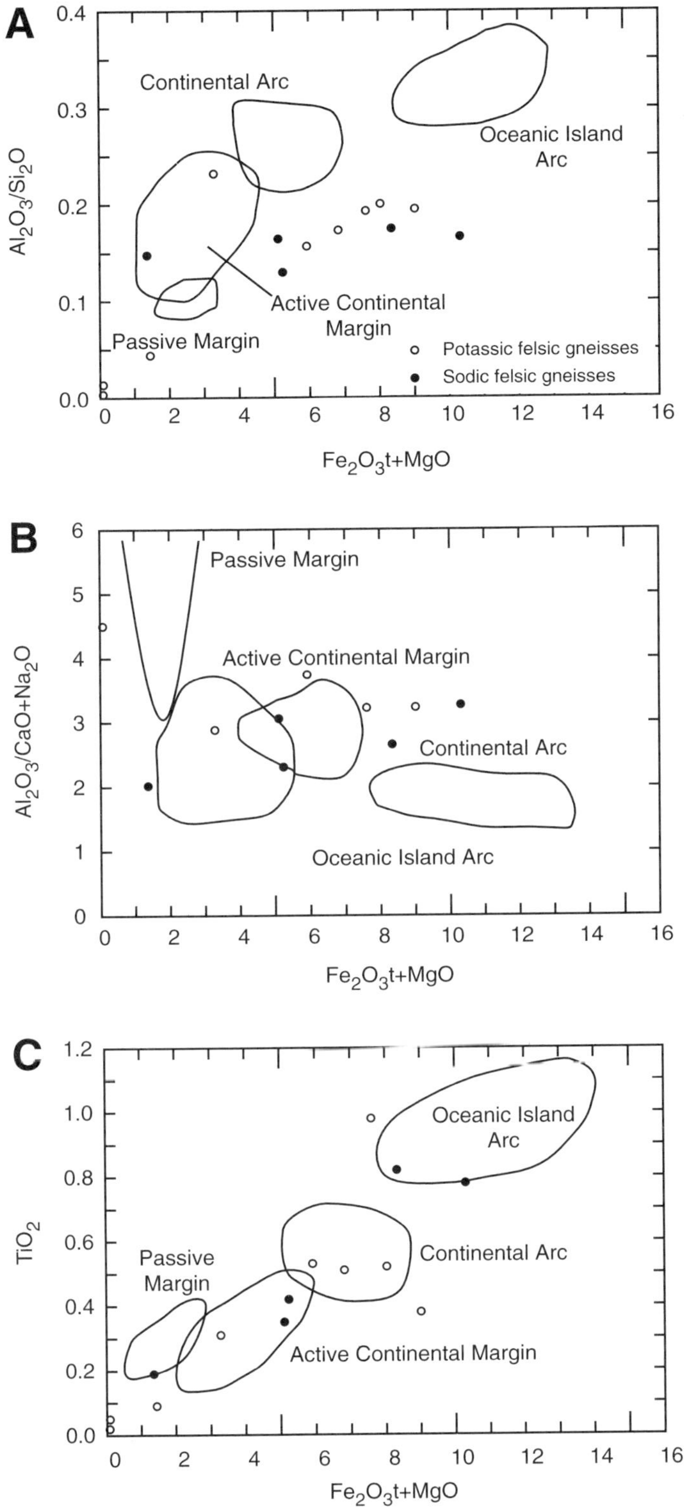

Figure 22. Elemental variation diagrams delineating compositional fields of sediments from known tectonic settings. Samples represented are felsic gneisses with Shaw DF3 values less than 1 (i.e., interpreted as sedimentary). These samples generally have compositions indicative of deposition in a continental arc or active continental margin setting (after Bhatia, 1983).

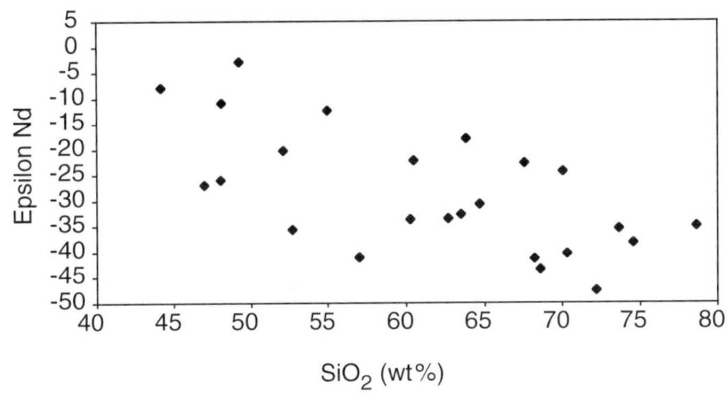

Figure 23. Epsilon Nd values for metamorphic mafic rocks and quartzofeldspathic gneisses plotted against SiO_2.

- These two Paleoproterozoic tectonothermal events have largely obscured any primary relations between the metamorphic mafic rock–quartzofeldspathic gneiss sequence and the platform-type supracrustal rocks. The bimodal volcanic suite was originally deposited in the Early Archean (3.5–3.3 Ga) in an active continental margin (i.e., basins adjacent to Andean-type magmatic arcs on thick continental margins, as defined by Bhatia [1983]). The bimodal volcanic suites most likely did not occur along an arc axis (e.g., Wilson, 1989, p. 191–225), but rather in a backarc setting associated with crustal extension (e.g., Hazlett, 1990; Glazner, 1990). In contrast, the quartzites of the metasedimentary sequence were receiving detrital zircons at least until 2900 Ma in a stable platform setting. There is presently no way to determine if the original contacts between these sequences formed at an unconformity or at a fault contact. However, these data conclusively demonstrate that the quartzofeldspathic gneisses predate the deposition of the metasupracrustal rocks by several hundred million years. Given the intensity of metamorphism and deformation of the 2450 and the 1780–1720 Ma orogenic events, it is not possible to determine whether the bimodal suite had experienced yet an earlier orogenic cycle prior to emplacement (structural or depositional) of the protoliths of the metasedimentary rocks.

DISCUSSION

Multiple lines of evidence indicate that the metamorphic mafic rock–quartzofeldspathic gneiss suites formed as one or more bimodal volcanic associations. Field relations show that the metamorphic mafic rocks and quartzofeldspathic gneisses are interlayered on a meter to hundreds of meters scale. This compositional variation is in sharp contrast to the large volumes of granitoids that occur across the northern Wyoming province, where plutonic origins are clearly evident in km-scale homogeneous rock units, preserved plutonic textures, and clear crosscutting field relations (e.g., Beartooth and Bighorn Mountains, northern and southern parts of the Madison Range). Although there are scattered examples of Late Archean granites, ca. 2.55 and 2.7–2.8 Ga (e.g., Mogk et al., 1988; Mueller et al., 1993), a major crust-forming event occurred in the northern Wyoming province (Montana metasedimentary terrane) that resulted in extensive formation of rocks of the TTG suite at a time of ca. 3.3–3.2 Ga. In contrast, the quartzofeldspathic gneisses of the Tobacco Root Mountains that date to 3.3–3.2 Ga contain rocks of both the TTG suite and a relatively high abundance of interlayered potassic rocks.

Chemical variation diagrams of the quartzofeldspathic gneisses suggest that most of these rocks have igneous protoliths. It is difficult to differentiate magmatic rocks that may be of plutonic origin from their volcanic, volcaniclastic, or arkosic equivalents. In addition to the general trends that indicate a magmatic origin for most of the gneisses, many samples have chemical characteristics that indicate that weathering and other sedimentary processes may have contributed to their present geochemical signature. A small but significant number of samples show enrichment of trace elements (Zr, Ti, P) that may be the result of concentration of detrital minerals during weathering and sedimentary processes. Other samples show relative enrichment of aluminum (corundum-normative rocks, high Niggli al-alk numbers) that may be the result of weathering. The intimate interlayering of rocks with both magmatic and sedimentary geochemical signatures indicates that these rocks are a supracrustal sequence that may include volcanic or hypabyssal igneous rocks and associated volcaniclastic or immature sediments (e.g., arkoses) that derive from the igneous suite.

In addition, both the metamorphic mafic rock and quartzofeldspathic gneisses exhibit geochemical signatures that indicate some degree of metasomatic alteration. All rock types include samples with unusually high potassium, and there is corresponding enrichment and scatter of the trace elements Rb, Sr, and Ba. It's interesting to note that potassic alteration associated with detachment zones is commonly associated with hematitic alteration and mineral deposits of Cu, Fe, Au, Ag, Mn, and barite (e.g., Harcuvar Mountains, Arizona, Roddy et al., 1988). The Tobacco Root Mountains are a historic mining district (e.g., Tansley et al., 1933) for quartz-vein gold, and throughout the range there are scattered "glory holes" where small manganese deposits have been explored. The original source of the gold in this district may have been epithermally altered felsic volcanic sequences.

The tectonic setting of these rocks is interpreted as an active continental margin or continental arc, based on the trends observed in numerous trace element variation diagrams. The metamorphic mafic rocks are not very diagnostic with respect to tectonic setting, and are best represented as calc-alkaline basalts or island-arc tholeiites. However, numerous variation diagrams of the magmatic rocks of felsic and intermediate composition clearly indicate an origin in a continental volcanic-arc or syncollisional tectonic setting. Sedimentary protoliths of the quartzofeldspathic gneisses associated with or derived from these volcanic sequences most likely were deposited in basins adjacent to continental arcs built on thick continental crust, e.g., in retroarc or pull-apart basins (Bhatia, 1983).

These settings are identical to those described by Sorensen et al. (1998) for the eastern Sierra Nevada Mountains, where continental arc rocks have been host to extensive potassic alteration. In the case of the Mojave Desert (Glazner, 1988), volcanism and metasomatism occurred at the same time the tectonic regime changed from subduction to transform-fault tectonics off the coast of California. In the Sleeping Beauty area of the central Mojave Desert, the volcanic stratigraphy consists of five units: *(a)* andesitic to dacitic flows, pyroclastic rocks, shallow intrusions, and volcaniclastic sediments of the Sleeping Beauty Ridge formation; *(b)* basalt to andesite flows, basalt scoria and silicic tuffs, and clastic sedimentary rocks of the formation of Argos Station formation; *(c)* the dacite of Cady Mountains; *(d)* the Peach Springs Tuff, a rhyolite ignimbrite sheet; and *(e)* fluvial, alluvial, and lacustrine sediments of the Hector Formation (Glazner, 1988). These field, petrographic, and geochemical studies from the Mojave Desert have provided evidence for wholesale potassic metasomatism of these units during at least two stages of extensional faulting. The Tertiary volcanic and volcaniclastic rocks of the North American Cordillera, therefore, provide a good modern analog, in terms of the lithologic associations and tectonic setting, for the protoliths of the Archean metamorphic mafic rock–quartzofeldspathic rocks of the Tobacco Root Mountains.

Subsequently, if such a bimodal volcanic sequence were buried to 30 km, underwent polyphase metamorphism in the granulite and amphibolite facies, and experienced polyphase deformation (e.g., isoclinal folding, development of sheath folds), we believe that the resulting metamorphic equivalents, metamorphic mafic rocks, and quartzofeldspathic gneisses would look very similar to the Indian Creek and Pony–Middle Mountain Metamorphic Suites. The metasupracrustal rocks (quartzites, marbles, pelitic schists, iron formations) were most likely unconformably deposited or tectonically emplaced at a much younger time (ca. 300 Ma) as a stable platform sequence, either directly on the bimodal volcanic rocks or after these rocks had undergone at least one cycle of orogenesis (i.e., they were already gneisses and metamorphic mafic rocks). Both sequences were metamorphosed and deformed in Paleoproterozoic events at 2450 Ma and ca. 1800 Ma (Mueller et al., this volume, Chapter 9; Cheney et al., this volume, Chapter 6; Harms et al., this volume, Chapter 10).

ACKNOWLEDGMENTS

Special thanks to Robert Burger, John Brady, Jack Cheney, and Tekla Harms for their many years of supervision of students on this Keck Geology Consortium research project in the Tobacco Root Mountains, and for their efforts in compiling and editing this GSA Special Paper. Reviews by Andy Barth and Klaus Schulz greatly improved the overall quality of this manuscript. The interpretations, however, remain the responsibility of the authors.

REFERENCES CITED

Abeyta, R.L., 1998, A geochemical and petrographic analysis of Archean quartzofeldspathic gneisses in the Indian Creek Metamorphic Suite and Pony–Middle Mountain Metamorphic Suite, Tobacco Root Mountains, southwestern Montana [senior thesis]: Colorado Springs, Colorado College, 75 p.

Barth, T.F.W., 1962, Theoretical petrology (2nd edition): New York, Wiley, p. 416.

Bhatia, M.R., 1983, Plate tectonics and geochemical composition of sandstones: Journal of Geology, v. 91, p. 611–627.

Burger, H.R., 1969, Structural evolution of the southwestern Tobacco Root Mountains, Montana: Geological Society of America Bulletin, v. 80, p. 1329–1342.

Burger, H.R., 2004, General geology and tectonic setting of the Tobacco Root Mountains, Montana, *in* Brady, J.B., et al., eds., Precambrian geology of the Tobacco Root Mountains, Montana: Boulder, Colorado, Geological Society of America Special Paper 377, p. 1–14 (this volume).

Burger, H.R., Brady, J.B., Cheney, J.T., Harms, T.A., Mueller, P., Heatherington, A., and Wooden, J., 1999, Evidence for a major, Early Proterozoic orogenic event in the Tobacco Root Mountains of southwestern Montana: Geological Society of America Abstracts with Programs, v. 31, no. 7, p. 177–178.

Cheney, J.T., Harms, T.A., Brady, J.B., and Burger, H.R., 1999, Early Proterozoic ^{208}Pb/^{232}Th in situ ion probe dating of monazite from Archean metamorphic suites, Tobacco Root Mountains, Montana: Geological Society of America Abstracts with Programs, v. 31, no. 7, p. A-178.

Cheney, J.T., Brady, J.B., Tierney, K.A., DeGraff, K.A., Mohlman, H.K., Frisch, J.D., Hatch, C.E., Steiner, M.L., Carmichael, S.K., Fisher, R.G.M., Tuit, C.B., Steffen, K.J., Cady, P., Lowell, J., Archuleta, L.L., Hirst, J., Wegmann, K.W., and Monteleone, B., 2004, Proterozoic metamorphism of the Tobacco Root Mountains, Montana, *in* Brady, J.B., et al., eds., Precambrian geology of the Tobacco Root Mountains, Montana: Boulder, Colorado, Geological Society of America Special Paper 377, p. 105–129 (this volume).

Clark, M., 1987, Protolith and tectonic setting of an Archean quartzofeldspathic gneiss sequence in the Blacktail Mountains, Beaverhead County, Montana [M.S. thesis]: Bozeman, Montana State University, 86 p.

Cordua, 1973, Precambrian geology of the southern Tobacco Root Mountains, Madison County, Montana [Ph.D. thesis]: Bloomington, Indiana University, 248 p.

Davidson, J.P., 1996, Deciphering mantle and crustal signatures in subduction zone magmatism, *in* Bebout, G.E., et al., eds., Subduction top to bottom: American Geophysical Union Monograph 96, p. 251–262.

Erslev, E.A., 1983, Pre-Beltian geology of the southern Madison Range, southwestern Montana: Montana Bureau of Mines and Geology Memoir 55, 26 p.

Erslev, E.A., and Sutter, J., 1990, Evidence for Proterozoic mylonitization in the northwestern Wyoming province: Geological Society of America Bulletin, v. 102, p. 1681–1694.

Friberg, N., 1976, Petrology of a metamorphic sequence of upper-amphibolite facies in the central Tobacco Root Mountains, southwestern Montana [Ph.D. thesis]: Bloomington, Indiana University, 146 p.

Garihan, J.M., 1979, Geology and structure of the central Ruby Range, Madison County, Montana: Geological Society of America Bulletin, v. 90, p. 323–326 (Part I), p. 695–788 (Part II).

Garihan, J.M., and Williams, K., 1976, Petrography, modal analyses and origin of Dillon quartzofeldspathic and pre–Cherry Creek gneisses, Ruby Range, southwestern Montana: Northwest Geology, v. 5, p. 42–49.

Giletti, B., 1966, Isotopic ages from southwestern Montana: Journal of Geophysical Research, v. 71, p. 4029–4036.

Giletti, B., 1971, Discordant isotopic ages and excess argon in biotites: Earth and Planetary Science Letters, v. 110, p. 157–164.

Gillmeister, N.M., 1971, Petrology of Precambrian rocks in the central Tobacco Root Mountains, Madison country, Montana [Ph.D. thesis]: Cambridge, Massachusetts, Harvard University, 201 p.

Glazner, A.F., 1988, Stratigraphy, structure, and potassic alteration of Miocene volcanic rocks in the Sleeping Beauty area, central Mojave Desert, California: Geological Society of America Bulletin, v. 100, p. 424–435.

Glazner, A.F., 1990, Recycling of continental crust in Miocene volcanic rocks from the Mojave block, southern California, *in* Anderson, J.L., ed., The nature and origin of Cordilleran magmatism: Geological Society of America Memoir 174, p. 147–168.

Hanley, T.B., 1975, Structure and petrology of the northwestern Tobacco Root Mountains, Madison County, Montana [Ph.D. thesis]: Bloomington, Indiana University, 289 p.

Harms, T.A., Burger, H.R., Blednick, D.G., Cooper, J.M., King, J.T., Owen, D.R., Lowell, J., Sincock, M.J., Kranenburg, S.R., Pufall, A., and Picornell, C.M., 2004, Character and origin of Precambrian fabrics and structures in the Tobacco Root Mountains, Montana, *in* Brady, J.B., et al., eds., Precambrian geology of the Tobacco Root Mountains, Montana: Boulder, Colorado, Geological Society of America Special Paper 377, p. 203–226 (this volume).

Hazlett, R.W., 1990, Extension-related Miocene volcanism in the Mopah Range volcanic field southeastern California, *in* Anderson, J.L., ed., The nature and origin of Cordilleran magmatism: Geological Society of America Memoir 174, p. 133–145.

Heinrich, E.W., and Rabbit, J.E., 1960, Pre-Beltian geology of the Cherry Creek and Ruby Mountains areas, southwestern Montana: Montana Bureau of Mines and Geology Memoir 38, 40 p.

Hess, D.F., 1967, Geology of pre-Beltian rocks in the central and southern Tobacco Root Mountains with reference to superposed effects of the Laramide-age Tobacco Root batholith [Ph.D. thesis]: Bloomington, Indiana University, 333 p.

Hyndman, D.W., 1985, Petrology of igneous and metamorphic rocks (2nd edition): New York, McGraw-Hill, 786 p.

Irvine, T.N., and Baragar, W.R.A., 1971, A guide to the chemical classification of the common volcanic rocks: Canadian Journal of Earth Sciences, v. 8, p. 523–548.

Jacob, L.J., 1994, A geochemical and geochronological characterization of the Indian Creek Metamorphic Suite in the Noble Lake area, south-central Tobacco Root Mountains, southwestern Montana [senior thesis]: Northampton, Massachusetts, Smith College, 73 p.

Johnson, K.E., Brady, J.B., MacFarlane, W.A., Thomas, R.B., Poulsen, C., and Sincock, M.J., 2004, Precambrian meta-ultramafic rocks from the Tobacco Root Mountains, Montana, *in* Brady, J.B., et al., eds., Precambrian geology of the Tobacco Root Mountains, Montana: Boulder, Colorado, Geological Society of America Special Paper 377, p. 71–87 (this volume).

Kerrich, R., and Wyman, D.A., 1997, Review of developments in trace-element fingerprinting of geodynamic settings and their implications for mineral exploration: Australian Journal of Earth Sciences, v. 44, p. 465–487.

Krogh, T.E., Kamo, S., and Hess, D.F., 1997, Wyoming province 3300+ gneiss with 2400 Ma metamorphism, northwestern Tobacco Root Mountains, Madison, County, Montana: Geological Society of America Abstracts with Programs, v. 29, no. 6, p. A-408.

Leake, B.E., and Singh, D., 1986, The Delaney Dome Formation, Connemara, West Ireland, and the geochemical distinction of ortho- and para-quartzofeldspathic rocks: Mineralogical Magazine, v. 50, p. 205–215.

LeBas, M.J., Le Maitre, R.W., Streckeisen, A., and Zanettin, B., 1986, A chemical classification of volcanic rocks based on the total alkali silica diagram: Journal of Petrology, v. 27, p. 745–750.

Leyreloup, A., Dupuy, C., and Andriambololona, R., 1977, Catazonal xenoliths in French Neogene volcanic rocks: Constitution of the lower crust: Contributions to Mineralogy and Petrology, v. 62, p. 283–300.

MacLean, W.H., and Barrett, T.J., 1993, Lithogeochemical techniques using immobile elements: Journal of Geochemical Exploration, v. 48, p. 109–133.

Martin, J., 1996, Petrology of mafic and felsic gneisses of the Indian Creek Metamorphic Suite, Tobacco Root Mountains, southwestern Montana [senior thesis]: Lancaster, Pennsylvania, Franklin and Marshall College, 53 p.

Mogk, D.W., Mueller, P.A., Wooden, J.L., and Bowes, D.R., 1987, Granitoids from an Archean collision zone, northern Madison Range, SW Montana: Eos (Transactions, American Geophysical Union), v. 68, p. 1515.

Mogk, D.W., Mueller, P.A., and Wooden, J.L., 1988, Archean tectonics of the North Snowy Block, Beartooth Mountains, Montana: Journal of Geology, v. 96, p. 125–141.

Mogk, D.W., Mueller, P.A., and Wooden, J.L., 1992, The nature of Archean terrane boundaries; an example from the northern Wyoming province: Precambrian Research, v. 55, p. 155–168.

Mueller, P.A., Shuster, R.D., Graves, M.A., Wooden, J.L., and Bowes, D.R., 1988, Age and composition of a late Archean magmatic complex, Beartooth Mountains, Montana-Wyoming, *in* Lewis, S.E., and Berg, R.B., eds., Precambrian and Mesozoic plate margins, Montana, Idaho, and Wyoming: Montana Bureau of Mines and Geology Special Publication 96, p. 7–22.

Mueller, P.A., Shuster, R.D., Wooden, J.L., Erslev, E.A., and Bowes, D.R., 1993, Age and composition of Archean crystalline rocks from the southern Madison Range, Montana: Implications for crustal evolution in the Wyoming craton: Geological Society of America Bulletin, v. 105, p. 437–446.

Mueller, P.A., Wooden, J.L., Mogk, D.W., Nutman, S.P., and Williams, I.S., 1996, Extended history of a 3.5 Ga trondhjemitic gneiss, Wyoming province, USA; evidence from U-Pb systematics in zircon: Precambrian Research 78, p. 41–52.

Mueller, P.A., Wooden, J.L., Nutman, A.P., and Mogk, D.W., 1998, Early Archean crust in the northern Wyoming province; evidence from U-Pb ages of detrital zircons: Precambrian Research, v. 91, p. 295–307.

Mueller, P.A., Burger, H.R., Wooden, J.L., Heatherington, A.L., Mogk, D.W., and D'Arcy, K., 2004, Age and evolution of the Precambrian crust of the Tobacco Root Mountains, Montana, *in* Brady, J.B., et al., eds., Precambrian geology of the Tobacco Root Mountains, Montana: Boulder, Colorado, Geological Society of America Special Paper 377, p. 181–202 (this volume).

Mullen, E.D., 1983, $MnO/TiO_2/P_2O_5$: A minor element discriminant for basaltic rocks of oceanic environments and its implications for petrogenesis: Earth and Planetary Science Letters, v. 62, p. 53–62.

Passchier, C.W., Myers, J.S., and Kroner, A., 1990, Field geology of high-grade gneiss terrains: Berlin, Springer Verlag, 150 p.

Peale, A.C., 1896, Geological atlas of the United States, No. 24, Three Forks Folio: U.S. Geological Survey, 7 p., 3 maps, 2 sections.

Pearce, J.A., and Cann, J.R., 1973, Tectonic setting of basic volcanic rocks determined using trace element analysis: Earth and Planetary Science Letters, v. 19, p. 290–300.

Pearce, J.A., Harris, N.B.W., and Tindle, A.G., 1984, Trace element discrimination diagrams for the tectonic interpretation of granitic rocks: Journal of Petrology, v. 25, p. 959–978.

Reid, R.R., 1957, Bedrock geology of the north end of the Tobacco Root Mountains, Madison County, Montana: Montana Bureau of Mines and Geology Memoir 36, 25 p.

Reid, R.R., 1963, Metamorphic rocks of the northern Tobacco Root Mountains, Madison County, Montana: Geological Society of America Bulletin, v. 74, p. 293–306.

Roberts, H., Dahl, P., Kelley, S., and Frei, R., 2002, New ^{207}Pb-^{206}Pb and ^{40}Ar-^{39}Ar ages from SW Montana, USA: Constraints on the Proterozoic and Archean tectonic and depositional history of the Wyoming province: Precambrian Research, v. 117, p. 119–143.

Roddy, M.S., Reynolds, S.J., Smith, B.M., and Ruiz, J., 1988, K-metasomatism and detachment-related mineralization, Haracuvar Mountains, Arizona: Geological Society of America Bulletin, v. 100, p. 1627–1639.

Root, F.K., 1965, Structure, petrology, and mineralogy of pre-Beltian metamorphic rocks of the Pony-Sappington area, Madison County, Montana [Ph.D. thesis]: Bloomington, Indiana University, 184 p.

Rougvie, J.R., and Sorensen, S.S., 2002, Cathodoluminescence record of K-metasomatism in ash-flow tuffs: Grain-scale mechanisms and large-scale geochemical implications: Geology, v. 30, p. 307–310.

Salt, K., 1987, Archean geology of the Spanish Peaks area, southwestern Montana [M.S. thesis]: Bozeman, Montana State University, 80 p.

Shaw, D.M., 1972, The origin of the Apsley Gneiss, Ontario: Canadian Journal of Earth Sciences, v. 9, p. 18–35.

Smith, M.S., Dymek, R.F., and Schneiderman, J.S., 1992, Implications of trace element geochemistry for the origin of cordierite-orthoamphibole rocks from Orijarvi, SW Finland: Journal of Geology, v. 100, p. 545–559.

Sorensen, S.S., Dunne, G.C., Hanson, R.B., Barton, M.D., Becker, M., Tobisch, O.T., and Fiske, R.S., 1998, From Jurassic shores to Cretaceous plutons: Geochemical evidence for paleoalteration environments of metavolcanic rocks, eastern California: Geological Society of America Bulletin, v. 110, p. 326–343.

Spencer, E.W., and Kozak, S.J., 1975, Precambrian evolution of the Spanish Peaks area, Montana: Geological Society of America Bulletin, v. 86, p. 785–792.

Sun, S.-s., and McDonough, W.F., 1989, Chemical and isotopic systematics of oceanic basalts: Implications for mantle compositions and processes, *in* Saunders, A.D., and Norry, M.J., eds., Magmatism in the ocean basins: Geological Society [London] Special Publication 42, p. 313–345.

Tansley, W., Schafer, F.A., and Hart, L.A., 1933, A geological reconnaissance of the Tobacco Root Mountains, Madison County, Montana: Montana Bureau of Mines and Geology Memoir 9, 57 p.

Tarney, J., 1976, Geochemistry of Archean high-grade gneisses with implications as to the origin and evolution of the Precambrian crust, *in* Windley, B.F., ed., The early history of the Earth: London, Wiley, p. 405–417.

Vitaliano, C.J., Burger, H.R., III, Cordua, W.S., Hanley, T.B., Hess, D.F., and Root, F.K., 1979, Geologic map of southern Tobacco root Mountains, Madison County, Montana: Geological Society of America Map and Chart Series MC31, scale 1:62,500, 1 sheet, 8 p. text.

Werner, C.D., 1987, Saxonian granulites—Igneous or lithogenous: A contribution to the geochemical diagnosis of the original rocks in high grade metamorphic complexes, *in* Gerstenberger, H., ed., Contributions to the geology of the Saxonian granulite massif (Sachsisches Granulitgebirge) Zfl-Mitteilungen Nr. 133, p. 221–250.

Wilson, M., 1989, Igneous petrogenesis: A global tectonic approach: London, Unwin Hyman, 466 p.

Wooden, J.L., and Mueller, P.A., 1988, Pb, Sr, and Nd isotopic compositions of a suite of Late Archean, igneous rocks, eastern Beartooth Mountains: Implications for crust-mantle evolution: Earth and Planetary Science Letters, v. 87, p. 59–72.

Wooden, J.L., Mueller, P.A., Mogk, D.W., and Bowes, D.R., 1988, A review of the geochemistry and geochronology of Archean rocks of the Beartooth Mountains, Montana and Wyoming, *in* Lewis, S.E., and Berg, R.B., eds., Precambrian and Mesozoic plate margins, Montana, Idaho, and Wyoming: Montana Bureau of Mines and Geology Special Publication 96, p. 7–22.

Manuscript Accepted by the Society September 12, 2003

Printed in the USA

Geological Society of America
Special Paper 377
2004

Geology and geochemistry of the Spuhler Peak Metamorphic Suite

H. Robert Burger
Department of Geology, Smith College, Northampton, Massachusetts 01063, USA

William H. Peck
Department of Geology, Colgate University, Hamilton, New York 13346, USA

Kathleen E. Johnson*
Department of Geology and Geophysics, University of New Orleans, New Orleans, Louisiana 70148, USA

Kara A. Tierney
Department of Geology, Amherst College, Amherst, Massachusetts 01002, USA

Chris J. Poulsen
Carleton College, Northfield, Minnesota 55057, USA

Pamela Cady
Josh Lowell
William A. MacFarlane
Colorado College, Colorado Springs, Colorado 80903, USA

M. Jennifer Sincock
Franklin & Marshall College, Lancaster, Pennsylvania 17604-3003, USA

LeAndra L. Archuleta
Pomona College, Claremont, California 91711-6339, USA

Ann Pufall
Department of Geology, Smith College, Northampton, Massachusetts 01063, USA

M. Jason Cox
Washington & Lee University, Lexington, Virginia 24450, USA

ABSTRACT

The Spuhler Peak Metamorphic Suite consists of a thick sequence of Archean metamorphic rocks, dominated by amphibolite, orthoamphibole-garnet gneiss, and hornblende gneiss and containing minor quantities of quartzite, aluminous schist, quartzofeldspathic gneiss, and meta-ultramafic rocks. Although the Spuhler Peak Metamorphic Suite represents less than 3% of Precambrian exposures in the Tobacco Root Mountains, it has been recognized as a unique suite in the region, due in large part to the unusually abundant amphibolite and the spectacular appearance of the orthoamphibole-garnet gneisses.

*Present address: P.O. Box 8352, Monterey, California 93940, USA, volcan@mac.com.

Burger, H.R., Peck, W.H., Johnson, K.E., Tierney, K.A., Poulsen, C.J., Cady, P., Lowell, J., MacFarlane, W.A., Sincock, M.J., Archuleta, L.L., Pufall, A., and Cox, M.J., 2004, Geology and geochemistry of the Spuhler Peak Metamorphic Suite, *in* Brady, J.B., Burger, H.R., Cheney, J.T., and Harms, T.A., eds., Precambrian geology of the Tobacco Root Mountains, Montana: Boulder, Colorado, Geological Society of America Special Paper 377, p. 47–70. For permission to copy, contact editing@geosociety.org.

An analysis of mafic rocks within the Spuhler Peak Metamorphic Suite, using classification diagrams, tectonic discrimination diagrams, and rare earth element plots, provides results consistent with primary igneous processes and pre-metamorphic hydrothermal alteration. Spuhler Peak Metamorphic Suite amphibolites compare well to typical Archean tholeiitic basalts with arc-related affinities. The major element geochemistry of the Spuhler Peak Metamorphic Suite orthoamphibole-garnet gneisses is within the range of the suite's amphibolites, excepting CaO and MgO contents. This supports the hypothesis that the amphibolites represent metamorphosed equivalents to the unaltered orthoamphibole-garnet gneiss protolith. When the effects of hydrothermal alteration are taken into account, the geochemistry of the Spuhler Peak Metamorphic Suite orthoamphibole-garnet gneisses is consistent with a volcanic protolith, genetically related to the protolith of its amphibolites. Outcropping within the suite are a number of small bodies of meta-ultramafic rocks, but these rocks are chemically distinct from the suite. We believe these bodies possibly are tectonically emplaced and not directly related to the Spuhler Peak Metamorphic Suite protolith. This protolith, therefore, is interpreted as largely mafic volcanic rocks (90% of the total package) deposited in a marine environment and intercalated with minor amounts of sedimentary material.

The Spuhler Peak Metamorphic Suite is distinct from the other Archean suites in the Tobacco Root Mountains—the Indian Creek Metamorphic Suite, which consists of quartzofeldspathic gneisses and metasedimentary rocks including marble, and the Pony–Middle Mountain Metamorphic Suite, which is composed primarily of quartzofeldspathic gneisses with subsidiary hornblende gneiss. Field relationships and radiometric dating demonstrate that the Spuhler Peak Metamorphic Suite was brought into fault contact with the Indian Creek and Pony–Middle Mountain Metamorphic Suites during the Big Sky orogeny, a major tectonothermal event at 1780–1720 Ma.

Keywords: Tobacco Root Mountains, Archean, amphibolite, oxygen isotopes, hydrothermal alteration.

INTRODUCTION

The Spuhler Peak Metamorphic Suite consists of a thick sequence of diverse rock types dominated by amphibolite, orthoamphibole-garnet gneiss, and hornblende gneiss. Also present, but in substantially lesser quantities, are quartzite, aluminous schist, quartzofeldspathic gneiss, and meta-ultramafic rocks. This lithologic package is exposed over an area of ~30 km^2 in the west-central Tobacco Root Mountains—an area representing <3% of Precambrian exposures in these mountains. Spuhler Peak Metamorphic Suite exposures begin at Branham Lakes (Fig. 2 *in* Burger, 2004, this volume, Chapter 1) and extend as an ~2-km-wide strip to the northwest, terminating at Sunrise Lake. A short, additional strip of the suite extends east from Sunrise Lake to just north of Little Granite Peak and south of Middle Mountain. Examination of this exposure pattern suggests that as much as 50% of the original extent of Spuhler rocks have been obscured by intrusion of the Tobacco Root batholith.

Spuhler Peak Metamorphic Suite rocks overlie the Indian Creek Metamorphic Suite, which is exposed south of the Spuhler Peak Metamorphic Suite, and lie under the Pony–Middle Mountain Metamorphic Suite, which is present to the north. Detailed mapping around a major isoclinal fold in the vicinity of Spuhler Peak (Fig. 2 *in* Burger, 2004, this volume, Chapter 1) demonstrates that the Spuhler Peak Metamorphic Suite did originally overlay both the Indian Creek and Pony–Middle Mountain Metamorphic Suites. As discussed in Burger (2004, this volume, Chapter 1), earlier mapping (Tansley et al., 1933; Reid, 1957; Burger, 1966; and Gillmeister, 1971) noted the distinctive nature of the Spuhler Peak Metamorphic Suite. Burger (1966) was the first to separate these rocks as a distinct entity. Gillmeister (1971) concurred when extending Burger's (1966) mapping, and although their interpretations differ, both agree that Spuhler Peak Metamorphic Suite rocks were placed on Indian Creek and Pony–Middle Mountain Metamorphic Suite rocks. In addition to the work of Burger (1966) and Gillmeister (1971), the only detailed study prior to the 1990s to focus exclusively on Spuhler Peak Metamorphic Suite rocks was that of Friberg (1976), who undertook a detailed mineralogical and petrologic investigation of Spuhler Peak Metamorphic Suite rocks exposed in the vicinity of Noble Lake and Spuhler Peak.

Because one of the main objectives of the Keck Geology Consortium project in the Tobacco Root Mountains was to understand more fully the origin and evolution of the Spuhler Peak Metamorphic Suite and its relationship to the Indian Creek and Pony–Middle Mountain Metamorphic Suites, several detailed projects were designed to address these questions. The studies include work on Spuhler Peak Metamorphic Suite exposures in four critical locations: a study by Poulsen (1994) in the vicinity of Branham Lakes and Leggat Mountain (Fig. 2 *in* Burger, 2004, this volume, Chapter 1), a study by Archuleta (1994) and Tierney (1994) in the Thompson Peak area, work by Lowell (1994) in the Spuhler Peak area, and an investigation by Sincock (1994) in a shear zone along the contact of the Spuhler Peak and Indian Creek Metamorphic Suites. In addition, two detailed geochemical studies of Spuhler Peak Metamorphic Suite rocks were undertaken by Peck (1994), who investigated the orthoamphibole-garnet gneisses, and Cox (1996), who focused on amphibolites and hornblende gneisses. Finally, MacFarlane (1996) undertook a geochemical study of meta-ultramafic rocks, including those in the Spuhler Peak Metamorphic Suite. Here we synthesize these data and offer an interpretation of the protolith and petrogenesis of the Spuhler Peak Metamorphic Suite.

LITHOLOGY OF THE SPUHLER PEAK METAMORPHIC SUITE

Thompson Peak Ridge

One of the longest and best-exposed sections of the Spuhler Peak Metamorphic Suite is along the ridge extending west from Thompson Peak (Fig. 2 *in* Burger, 2004, this volume, Chapter 1) for ~1700 m to the contact of the Spuhler Peak Metamorphic Suite with the Indian Creek Metamorphic Suite. Archuleta (1994) and Tierney (1994) compiled a detailed description of the best-exposed portion of this traverse that yields an ~750-m-thick detailed section (Fig. 1) of the Spuhler Peak Metamorphic Suite. Rock types present include, in order of abundance, amphibolite and hornblende gneiss, orthoamphibole-garnet gneiss, quartzite and aluminous gneiss, and quartzofeldspathic gneiss. Mafic rocks constitute more than 80% of these exposures, which is typical for the Spuhler Peak Metamorphic Suite in other areas where exposures are relatively complete. Although no large-scale isoclinal folds or shear zones were identified along this traverse, folds and shear zones are abundant within the Spuhler Peak Metamorphic Suite both to the east and west of the ridge. Lateral continuity is difficult to establish due to the steep slopes and the spotty exposures on these slopes, so it is likely that the sequence along the ridge contains portions repeated due to folding and shearing and portions omitted due to faulting. Nevertheless, this section provides an excellent representation of the varieties and relative proportions of rock types present in the Spuhler Peak Metamorphic Suite. The most abundant lithologies are amphibolite and hornblende gneiss. We use amphibolite as a field term for rocks that are dark gray to black, lack even a crude gneissic banding, are generally equigranular and composed primarily of amphibole and plagioclase, but typically contain visible garnet. These rocks may contain up to 10% quartz and minor amounts of biotite. In many samples, hornblende, garnet, and biotite represent 70% of the minerals. The dominant amphibole is either hornblende or magnesiohornblende in most cases, but cummingtonite also is common. In a few samples, cummingtonite is the dominant amphibole. Garnets are typically unzoned. A representative garnet composition is $Alm_{62}Py_{24}Gr_{12}Sp_2$. See Cheney et al. (2004a, this volume, Chapter 6) for full analyses. Hornblende gneisses are similar in composition, except that they contain more quartz and less amphibole, are generally light gray in color, and typically have a crude gneissic banding.

Orthoamphibole-garnet gneisses are the second most abundant rock type. These are present both as spectacular specimens (Fig. 2) with almandine-rich garnets up to 10 cm in diameter and orthoamphiboles as long as 15 cm, and as finer-grained varieties with grains averaging 0.1 mm. A representative mode for typical gneiss is orthoamphibole (gedrite) (35%), garnet (30%), plagioclase (15%), quartz (10%), cordierite (5%), and biotite (5%). Orthoamphibole is generally gedrite ($Mg/(Mg + Ca) \approx 0.60$) and contains anthophyllite-rich orthoamphibole exsolution lamellae, indicating initial growth above the orthoamphibole miscibility gap (>600 °C) at the peak of metamorphism (Spear, 1993; Fig. 15 *in* Johnson et al., 2004, this volume, Chapter 4). Cummingtonite may be present replacing gedrite. Typically, finer-grained varieties of the gedrite-gneiss are well foliated, whereas coarser-grained rocks are nonfoliated. These unique rocks range in thickness from <1 m to as much as 30 m. Many layers appear laterally continuous for the extent of the Thompson Peak ridge exposure, but others occur as laterally discontinuous, lens-like bodies. The orthoamphibole-garnet gneisses are intimately intercalated with amphibolites, but some of the thinner-bedded and finer-grained varieties are located within sequences of aluminous schist and gneiss (Fig. 1). Orthoamphibole-garnet gneisses also are present in the Indian Creek Metamorphic Suite and within Indian Creek Metamorphic Suite–like rocks in the Ruby Range, but in significantly reduced amounts as compared to the Spuhler Peak Metamorphic Suite. In addition, these occurrences are not associated with the thick sequences of amphibolite and hornblende gneiss that are so typical of the Spuhler Peak Metamorphic Suite.

Quartzites, although a volumetrically relatively minor component of the Spuhler Peak Metamorphic Suite, are interspersed throughout the Thompson Peak ridge section. A thin (generally <1 m), relatively pure, white quartzite is present at the contact of Spuhler Peak Metamorphic Suite rocks with the Indian Creek Metamorphic Suite. This layer contains only small amounts of biotite and plagioclase, in distinct contrast to the other quartzites in the Spuhler Peak Metamorphic Suite. Most quartzites within this traverse have a rusty, orange-weathering color and contain varying proportions of biotite, Cr-mica, almandine garnet, and sillimanite as well as occasional grains of detrital zircon. Sillimanite-rich horizons occur within the quartzites. These horizons can contain up to 60% sillimanite with individual crystals in splays of coarse sillimanite having lengths as great as 6 cm. The thickest, relatively pure quartzite layer in this sequence is 15 m thick; others range from <1 m to 9 m thick.

The aluminous gneisses are quartz-rich rocks with abundant garnet, sillimanite, and biotite. A representative mode is quartz (25%), sillimanite (25%), garnet (25%), biotite (15%), plagioclase (5%), and cordierite (5%). Although quartz content typically ranges from 25% to 60%, these other major constituents are invariably present. The gneisses have a quite distinctive field presence with

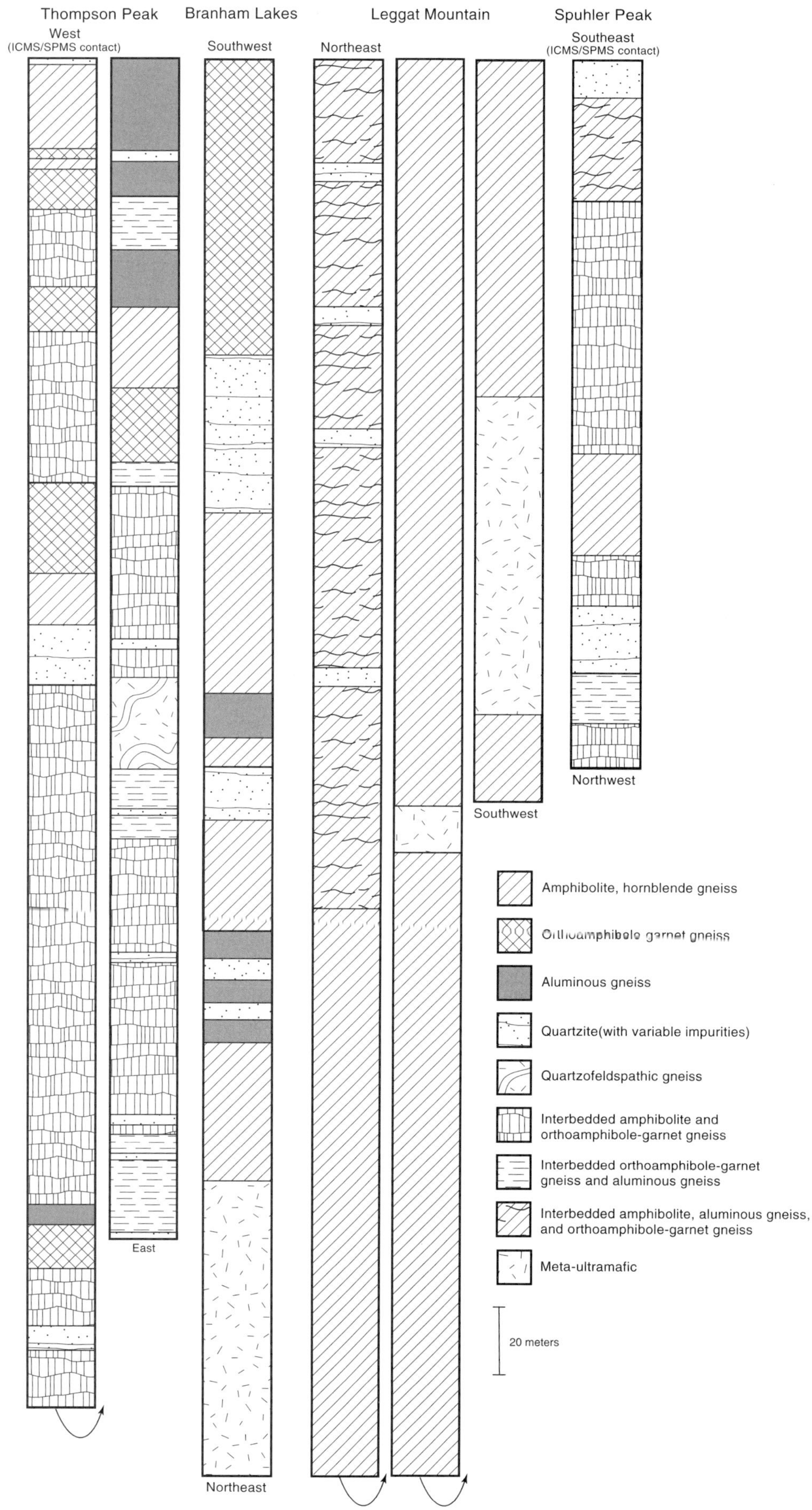

Figure 1. Measured sections of Spuhler Peak Metamorphic Suite exposures from Thompson Peak, Branham Lakes, Leggat Mountain, and Spuhler Peak.

their well-developed foliation, abundant biotite and sillimanite, and their dark orange-red weathering color. Where biotite and sillimanite are particularly abundant, this unit becomes more schistose than gneissic. Unlike the other units in this traverse, the aluminous gneiss (and especially the thicker layers) are concentrated predominantly in the middle of the ridge section. This particular lithology, especially when present with substantial quantities of amphibolite, is diagnostic of the Spuhler Peak Metamorphic Suite.

Spuhler Peak–Noble Lake Area

Although the Spuhler Peak Metamorphic Suite exposures at Spuhler Peak (Fig. 2 *in* Burger, 2004, this volume, Chapter 1) do not offer as thick a section as at Thompson Peak ridge, the exposures are spectacular and offer additional opportunity for determining lateral continuity. A total thickness of ~250 m is present from the Spuhler Peak Metamorphic Suite/Indian Creek Metamorphic Suite contact to the axial surface of a large isoclinal fold that repeats the sequence. This section was first described by Gillmeister (1971) and subsequently studied in great detail by Friberg (1976). Lowell (1994) also measured and described this section, focusing on a detailed description of Spuhler Peak Metamorphic Suite rocks at Spuhler Peak, in order to compare this section to sections at Thompson Peak ridge and at Branham Lakes, and to provide a detailed base for geochemical studies.

It is clear from these studies that the lateral continuity of individual rock types varies considerably. Many of the units pinch and swell and vary greatly in thickness, but some quartzites can be traced for at least 2 km. A particularly distinctive lithology (90% pink almandine garnet and 10% quartz) occurs in bands only a few centimeters thick but can be traced for hundreds of meters (Friberg, 1976, p. 21). Conversely, many of the orthoamphibole-garnet gneisses are demonstrably lensoid.

Virtually all major rock types present on the Thompson Peak ridge are also present in the Spuhler Peak–Noble Lake area. However, correlation of the sections has not been possible. This most likely is due to a combination of complex folding and repetition and/or omission by faulting, as well as the lack of lateral continuity of many units. No unequivocal shear zones were identified on Thompson Peak ridge, but Lowell (1994) mapped a several-meter-thick shear zone in a thick sequence of amphibolite that is present in the same stratigraphic position on the lower and upper limbs of the large isoclinal fold.

Two small outcrops of meta-ultramafic rock occur within the area. One is just southwest of Spuhler Peak, and the larger body (10 m by 15 m) is farther down the ridge to the southwest. Such bodies are not present within the Spuhler Peak Metamorphic Suite along Thompson Peak ridge but are relatively common in the Branham Lakes area.

Branham Lakes–Leggat Mountain Area

In addition to Spuhler Peak Metamorphic Suite exposures at Thompson Peak ridge and the Spuhler Peak–Noble Lake area, we measured two additional sections just west of Branham Lakes. One traverse is along a ridge located 1.6 km northwest of Branham Lakes. The other is along a ridge that begins west of Leggat Mountain (Fig. 2 *in* Burger, 2004, this volume, Chapter 1) and extends along that ridge to a point northwest of Gneiss Lake. These sections (Fig. 1) document a Spuhler Peak Metamorphic Suite sequence that is dominated by amphibolite and contains more meta-ultramafic rocks than other localities. Again, typical Spuhler Peak Metamorphic Suite quartzite, aluminous gneiss, and orthoamphibole-garnet gneiss are interspersed throughout the exposed sequence, but as noted by Cummings and McCulloch (1992, p. 333), the mass of amphibolite in this area is the thickest amphibolitic sequence anywhere in the Tobacco Roots. However, it is not possible to correlate directly the Spuhler Peak Metamorphic Suite sequences in this area with those at Thompson Peak ridge or Spuhler Peak–Noble Lake, either by attempting to match portions of measured sections or by tracing individual units in the field. Based on the thickness of

Figure 2. Photograph of coarse-grained variety of orthoamphibole-garnet gneiss. Lighter gray, equidimensional garnets and dark gray laths of gedrite dominate this view. Diameter of coin is 18 mm.

the amphibolites and the presence of the meta-ultramafic rocks, it is probable that these rocks represent a portion of the Spuhler Peak Metamorphic Suite that is not present, at least in its entirety, elsewhere in the suite's exposures.

Although the meta-ultramafic rocks are discussed in detail in Johnson et al. (2004, this volume, Chapter 4) and by Cummings and McCulloch (1992), a few comments are relevant here. The meta-ultramafic rocks in the Branham Lakes area occur both as somewhat equidimensional bodies no more than a few tens of meters wide, and as more sill-like objects that can be traced for a few hundred meters. Contact relationships are obscured by a hornblende-rich "blackwall" that is generally present between the meta-ultramafic bodies and amphibolite. Nevertheless, there are no field relationships that support a crosscutting, intrusive relationship for the meta-ultramafic rocks. The tabular bodies may represent sill-like intrusions or lava flows. The more equidimensional bodies suggest intrusion but can just as easily be explained as a boudinaged sill, lava flow, or perhaps a tectonically emplaced sliver. Indeed, one series of three meta-ultramafic bodies on the slopes just northwest of upper Branham Lake define a pattern that mimics a large-scale fold that is identified by change in orientation of a weakly developed foliation in amphibolite. Given the tight isoclinal folding present throughout the region and the likely competency of the meta-ultramafic rocks, it is logical to expect boudinaged meta-ultramafic layers. Alternatively, we cannot disprove with field relations either intrusion or tectonic emplacement. However, both the detailed study of Cummings and McCulloch (1992) and our own field work find an intimate association between the meta-ultramafic rocks and the amphibolites present at Branham Lakes. See Johnson et al. (2004, this volume, Chapter 4) for more details and additional hypotheses.

GEOCHEMISTRY OF THE SPUHLER PEAK METAMORPHIC SUITE

Our primary goal is to provide information concerning the protolith of the Spuhler Peak Metamorphic Suite, as this suite is a distinctive assemblage in the field and provides such a marked contrast with Indian Creek and Pony–Middle Mountain Metamorphic Suite rocks. Because of this unique character, any insights into the environment in which the Spuhler Peak Metamorphic Suite formed are important in fully understanding the Archean of southwestern Montana. We approach this characterization in two ways: (1) analysis of Spuhler Peak Metamorphic Suite mafic rocks using tectonic discriminant diagrams, and (2) comparison of geochemical characteristics of the Spuhler Peak Metamorphic Suite to those of other Precambrian mafic rocks of known origin.

As the Spuhler Peak Metamorphic Suite was subjected to at least two high-grade metamorphic events (Cheney et al., 2004a, this volume, Chapter 6), we acknowledge that the chemistry of these rocks could be sufficiently changed so as to make this exercise problematic. However, a variety of tectonic discriminant diagrams provide reasonably consistent results. Combining these results with analysis of immobile trace elements and oxygen isotope ratios, the observed geochemical variations are consistent with primary, igneous processes and premetamorphic hydrothermal alteration. We believe the results of our analysis provide specific constraints on any Spuhler Peak Metamorphic Suite protolith reconstruction.

In order to characterize more fully the mafic rocks of the Spuhler Peak Metamorphic Suite, 58 samples of amphibolite, 29 samples of orthoamphibole-garnet gneiss, and 33 samples of meta-ultramafic rocks were collected for chemical analysis. Major oxides were determined for all 120 samples, and trace and/or rare earth elements (REEs) were determined for 110 of the 120 samples. The results are presented in Tables 1 and 2 and also in Table 2 of Johnson et al. (2004, this volume, Chapter 4). Oxygen isotope ratios were determined for 25 samples of orthoamphibole-garnet gneiss. These data are presented in Table 3.

In most cases, major and trace element analyses were performed by X-ray fluorescence spectrometry (XRF) either at the Department of Geosciences, University of Massachusetts, Amherst, or at Franklin & Marshall College, and rare earth analyses were performed by INAA (instrumental neutron activation analysis) at the Radiation Center, Oregon State University (Archuleta, 1994; Cady, 1994; Lowell, 1994; Peck, 1994; Sincock, 1994; Tierney, 1994; MacFarlane, 1996; Pufall, 1996). Analytical procedures for these analyses are described in Rhodes (1996), Mertzman (2000), and at http://ne.oregonstate.edu/facilities/radiation_center/naa.html, respectively. The remaining analyses (Poulsen, 1994; Cox, 1996) were obtained mainly at XRAL Laboratories, Ontario, using XRF and inductively coupled plasma–mass spectrometry (ICP-MS), using analytical procedures described at http://www.sgs.ca/geochem/. In all cases, chemical analyses are reported on an anhydrous basis with total iron given as FeO. For the diagrams that follow we also show 20 analyses from Cummings and McCulloch (1992) and six analyses from Friberg (1976).

Whole-rock powders were analyzed for oxygen isotope ratio by laser fluorination at the University of Wisconsin, using the air-lock method of Spicuzza et al. (1998). Seventeen aliquots of powdered UWG-2 garnet standard were analyzed on the four days of analysis, $\delta^{18}O$ values averaged 5.55 ± 0.12‰, and daily precision averaged 0.08 ± 0.02‰. Analyses were adjusted by the amount that daily UWG-2 values deviated from 5.8‰ (the accepted value of UWG-2 based on a value of 9.59‰ for NBS-28; Valley et al., 1995).

Amphibolites

Workers in the Tobacco Roots have assumed that the amphibolites of the Spuhler Peak Metamorphic Suite are igneous in origin due to their composition, association with meta-ultramafic sequences, and the massive, poorly foliated to nonfoliated character of many, so we briefly consider this topic. A plot of Spuhler Peak Metamorphic Suite amphibolites (Fig. 3) on the classification diagram of LeBas et al. (1986) for volcanic rocks defines a

TABLE 1. WHOLE-ROCK MAJOR AND TRACE ELEMENT ANALYSES FOR SPUHLER PEAK METAMORPHIC SUITE AMPHIBOLITES

Sample	CP1	CP7	CP10	CP18	CP21	CP24	CP35	CP41	CP44	CP50	CP3	CP31	CP34	WP6a	MJS1	MJS141	PC2	PC15	PC20
Major Elements (wt%)																			
SiO_2	48.67	48.43	49.62	52.02	51.57	53.73	52.25	52.60	50.99	55.73	47.12	43.01	47.79	56.50	51.47	53.26	52.92	54.35	52.36
TiO_2	0.52	1.46	0.41	0.66	1.26	0.95	0.61	1.51	0.84	1.26	0.53	1.21	0.65	1.40	0.92	2.11	0.78	1.06	0.76
Al_2O_3	14.32	12.67	15.45	15.08	13.31	13.69	15.06	13.29	15.88	14.08	11.75	14.64	11.39	14.94	14.09	12.52	15.89	13.62	14.55
FeO*	9.55	15.08	7.58	10.16	13.36	10.61	8.62	15.54	11.17	10.81	11.28	15.99	12.07	10.36	12.73	12.68	11.19	14.11	10.18
MnO	0.28	0.29	0.23	0.31	0.28	0.15	0.31	0.42	0.22	0.16	0.35	0.45	0.46	0.21	0.20	0.24	0.21	0.34	0.19
MgO	10.19	6.76	8.80	7.58	6.47	6.34	7.54	5.64	7.63	6.17	13.96	8.92	12.73	7.05	7.82	6.31	7.95	6.04	8.07
CaO	11.69	11.27	13.23	9.96	11.01	10.31	11.51	8.48	9.77	5.57	10.85	12.41	11.42	5.90	11.23	10.02	7.93	6.47	9.97
Na_2O	2.05	1.65	2.52	2.27	0.80	2.26	2.00	1.15	1.07	2.14	1.40	1.27	1.38	2.91	1.00	1.95	2.42	1.30	2.09
K_2O	0.54	0.30	0.42	0.60	0.50	0.36	0.45	0.59	1.02	2.24	0.48	0.60	0.48	0.18	0.56	0.49	0.26	0.23	0.57
P_2O_5	0.02	0.12	0.02	0.06	0.10	0.08	0.06	0.12	0.05	0.15	0.02	0.13	0.08	0.37	0.09	0.24	0.14	0.62	0.04
Total	97.84	98.04	98.27	98.69	98.66	98.47	98.40	99.35	98.65	98.32	97.75	98.63	98.44	99.83	100.13	99.83	99.70	98.15	98.78
Rare earth elements (ppm)																			
La	9.00	6.80	N.D.[†]	N.D.	N.D.	7.30	N.D.	N.D.	4.30	27.70	N.D.	N.D.	N.D.	32.00	6.00	8.50	13.25	26.76	3.45
Ce	N.D.	18.70	N.D.	N.D.	N.D.	17.30	109.00	8.00	12.70	55.60	N.D.	N.D.	N.D.	76.00	15.70	24.30	30.89	55.14	7.73
Nd	N.D.	11.60	N.D.	N.D.	N.D.	9.50	N.D.	N.D.	8.80	23.90	N.D.	N.D.	N.D.	N.D.	10.10	18.10	N.D.	25.93	4.38
Sm	N.D.	3.37	N.D.	N.D.	N.D.	2.43	N.D.	N.D.	2.18	4.80	N.D.	N.D.	N.D.	N.D.	2.73	5.74	N.D.	5.85	1.86
Eu	N.D.	1.18	N.D.	N.D.	N.D.	0.87	N.D.	N.D.	0.86	1.60	N.D.	N.D.	N.D.	N.D.	0.89	1.99	N.D.	1.69	0.76
Tb	N.D.	0.75	N.D.	N.D.	N.D.	0.52	N.D.	N.D.	0.45	0.61	N.D.	N.D.	N.D.	N.D.	0.55	1.19	N.D.	0.88	0.30
Yb	N.D.	2.89	N.D.	N.D.	N.D.	1.27	N.D.	N.D.	1.76	2.13	N.D.	N.D.	N.D.	N.D.	2.23	5.23	N.D.	4.05	1.59
Lu	N.D.	0.39	N.D.	N.D.	N.D.	0.17	N.D.	N.D.	0.26	0.28	N.D.	N.D.	N.D.	N.D.	0.36	0.78	N.D.	0.63	0.25
Hf	N.D.	2.51	N.D.	N.D.	N.D.	1.96	N.D.	N.D.	1.27	4.11	N.D.	N.D.	N.D.	N.D.	1.86	4.54	N.D.	5.13	1.25
Ta	N.D.	0.31	N.D.	N.D.	N.D.	0.32	N.D.	N.D.	0.14	0.39	N.D.	N.D.	N.D.	N.D.	0.22	0.31	N.D.	0.90	N.D.
Th	N.D.	1.02	N.D.	N.D.	N.D.	0.82	N.D.	N.D.	1.00	2.02	N.D.	N.D.	N.D.	N.D.	1.04	1.04	2.94	4.63	0.08
U	N.D.	5.70	N.D.	N.D.	N.D.	6.00	N.D.	N.D.	4.80	6.00	N.D.	N.D.	N.D.	N.D.	6.88	8.10	1.56	3.05	0.59
Sc	N.D.	52.10	N.D.	N.D.	N.D.	42.80	N.D.	N.D.	36.20	27.60	N.D.	N.D.	N.D.	N.D.	44.00	46.50	N.D.	34.43	34.33
Cr	90.00	143.00	N.D.	N.D.	N.D.	38.30	109.00	8.00	268.00	86.60	N.D.	N.D.	N.D.	59.00	177.00	46.80	252.53	111.00	303.50
Co	N.D.	58.30	N.D.	N.D.	N.D.	54.20	N.D.	N.D.	57.80	36.00	N.D.	N.D.	N.D.	N.D.	58.10	38.50	N.D.	48.00	51.80
Rb	12.00	25.00	N.D.	N.D.	N.D.	25.00	3.00	9.00	33.30	55.70	N.D.	N.D.	N.D.	10.00	11.00	11.90	3.20	4.08	6.26
Cs	N.D.	1.10	N.D.	N.D.	N.D.	0.90	N.D.	N.D.	0.52	2.30	N.D.	N.D.	N.D.	N.D.	0.87	0.81	N.D.	N.D.	N.D.
Trace elements (ppm)																			
Ba	N.D.	179.00	N.D.	N.D.	N.D.	119.00	N.D.	N.D.	181.00	647.00	N.D.	N.D.	N.D.	86.00	189.00	356.00	61.88	88.74	160.76
Nb	6.00	6.00	N.D.	N.D.	N.D.	N.D.	5.00	4.00	N.D.	N.D.	N.D.	N.D.	N.D.	12.00	4.00	5.00	4.82	9.06	3.06
Ni	99.00	91.00	N.D.	N.D.	N.D.	59.00	84.00	N.D.	135.00	63.00	N.D.	N.D.	N.D.	64.00	98.00	170.00	90.32	30.50	148.54
Sr	121.00	360.00	N.D.	N.D.	N.D.	390.00	167.00	52.00	330.00	330.00	N.D.	N.D.	N.D.	60.00	360.00	360.00	115.87	32.51	152.38
V	N.D.	N.D.	N.D.	N.D.	N.D.	N.D.	N.D.	N.D.	N.D.	N.D.	N.D.	N.D.	N.D.	244.00	280.00	410.00	188.27	254.53	197.19
Y	10.00	26.00	N.D.	N.D.	N.D.	N.D.	15.00	25.00	N.D.	N.D.	N.D.	N.D.	N.D.	N.D.	20.00	48.00	22.09	32.79	12.63
Zn	N.D.	136.00	N.D.	N.D.	N.D.	102.00	N.D.	N.D.	99.00	84.00	N.D.	N.D.	N.D.	135.00	113.00	100.00	103.81	1003.81	94.84
Zr	34.00	78.00	N.D.	N.D.	N.D.	N.D.	71.00	85.00	N.D.	N.D.	N.D.	N.D.	N.D.	255.00	N.D.	N.D.	113.01	170.58	42.10

(continued)

TABLE 1. WHOLE-ROCK MAJOR AND TRACE ELEMENT ANALYSES FOR SPUHLER PEAK METAMORPHIC SUITE AMPHIBOLITES *(continued)*

Sample	PC26	PC36	PC70	JL1	JL7	JL9	JL11	JL38	JC1	JC3	JC6	JC10	JC12	JC16	JC21B	JC22A	JC24A	JC24D	JC25
Major Elements (wt%)																			
SiO_2	55.51	48.32	50.88	57.72	58.57	54.52	52.79	53.09	49.27	49.52	47.71	57.81	56.84	53.00	52.64	51.10	51.19	49.69	53.42
TiO_2	0.62	1.24	0.90	1.35	1.25	1.04	1.22	0.30	1.36	1.29	0.37	0.78	0.54	1.46	0.65	0.51	0.86	1.74	1.90
Al_2O_3	13.55	13.74	14.60	14.49	13.17	15.31	13.81	4.98	14.07	13.30	17.37	14.74	13.07	13.43	15.49	18.04	14.90	17.14	13.25
*FeO	9.97	16.89	12.21	10.17	11.10	10.45	13.44	8.48	14.94	13.81	7.79	10.15	8.86	15.50	8.40	9.25	12.01	15.70	15.41
MnO	0.24	0.26	0.20	0.23	0.22	0.22	0.28	0.14	0.43	0.27	0.26	0.17	0.17	0.31	0.24	0.24	0.30	1.10	0.33
MgO	9.59	6.55	7.53	6.96	5.27	6.37	5.57	24.90	6.90	7.34	10.15	5.33	7.94	6.04	8.54	6.62	6.86	4.19	5.06
CaO	7.18	9.90	11.58	5.83	5.92	6.13	8.82	5.85	10.28	10.44	10.83	8.38	7.83	9.70	12.80	12.11	9.57	6.58	7.21
Na_2O	2.33	1.16	0.70	1.80	3.59	3.91	2.60	0.91	2.09	2.26	2.02	2.86	2.46	0.63	1.24	2.50	3.26	1.80	1.37
K_2O	0.28	0.43	0.56	0.45	0.48	1.52	0.81	0.10	0.52	0.92	0.57	0.56	0.21	0.92	0.29	0.21	1.01	0.58	0.52
P_2O_5	0.08	0.02	0.07	0.32	0.11	0.11	0.10	0.04	0.12	0.09	0.00	0.14	0.07	0.10	0.03	0.03	0.09	0.26	0.15
Total	99.36	98.51	99.25	99.32	99.68	99.59	99.45	98.80	99.97	99.24	97.06	100.91	98.00	101.09	100.31	100.62	100.06	98.78	98.61
Rare earth elements (ppm)																			
La	15.72	4.26	5.63	28.20	23.00	12.00	1.74	4.09	14.20	3.80	1.70	19.50	14.10	2.50	3.10	3.30	7.30	10.50	13.60
Ce	26.55	9.30	13.76	66.10	49.52	20.74	10.31	1.73	29.50	12.70	4.90	37.80	24.00	7.30	8.50	7.00	14.70	22.20	29.70
Nd	N.D.	5.89	6.71	N.D.	N.D.	N.D.	N.D.	N.D.	12.00	6.30	1.00	17.00	11.60	4.80	1.30	4.70	8.50	13.40	13.20
Sm	N.D.	2.60	2.28	N.D.	N.D.	N.D.	N.D.	N.D.	4.70	3.50	0.82	4.00	2.60	2.10	1.90	1.40	2.60	4.40	4.80
Eu	N.D.	0.96	0.84	N.D.	N.D.	N.D.	N.D.	N.D.	N.D.	N.D.	N.D.	N.D.	N.D.	N.D.	N.D.	N.D.	N.D.	N.D.	N.D.
Tb	N.D.	0.51	0.50	N.D.	N.D.	N.D.	N.D.	N.D.	N.D.	N.D.	N.D.	N.D.	N.D.	N.D.	N.D.	N.D.	N.D.	N.D.	N.D.
Yb	N.D.	2.80	2.16	N.D.	N.D.	N.D.	N.D.	N.D.	3.6	1.1	0.72	2.7	1.8	2.2	1.5	1.3	1.9	4	3.4
Lu	N.D.	0.45	0.34	N.D.	N.D.	N.D.	N.D.	N.D.	0.52	0.2	0.09	0.32	0.25	0.31	0.21	0.19	0.27	0.57	0.42
Hf	N.D.	1.61	1.76	N.D.	N.D.	N.D.	N.D.	N.D.	N.D.	N.D.	N.D.	N.D.	N.D.	N.D.	N.D.	N.D.	N.D.	N.D.	N.D.
Ta	N.D.	N.D.	N.D.	N.D.	N.D.	N.D.	N.D.	N.D.	N.D.	N.D.	N.D.	N.D.	N.D.	N.D.	N.D.	N.D.	N.D.	N.D.	N.D.
Th	3.61	0.70	1.18	5.33	3.64	2.93	0.11	1.08	3.82	N.D.	N.D.	6.30	N.D.	2.24	2.10	1.30	2.16	N.D.	N.D.
U	2.18	N.D.	1.08	2.82	3.28	1.86	1.24	1.15	1.10	N.D.	N.D.	1.75	N.D.	0.14	0.47	0.45	1.36	N.D.	N.D.
Sc	N.D.	48.83	42.27	N.D.	N.D.	N.D.	N.D.	N.D.	N.D.	N.D.	N.D.	N.D.	N.D.	N.D.	N.D.	N.D.	N.D.	N.D.	N.D.
Cr	443.51	76.80	176.60	61.02	34.62	91.65	80.73	2867.75	123.37	<68.4	684.00	110.90	342.00	87.13	300.82	57.71	148.84	<68.4	<68.4
Co	N.D.	57.00	54.40	N.D.	N.D.	N.D.	N.D.	N.D.	N.D.	N.D.	N.D.	N.D.	N.D.	N.D.	N.D.	N.D.	N.D.	N.D.	N.D.
Rb	6.30	N.D.	6.48	13.17	9.49	38.03	12.38	1.96	17.49	42.00	9.00	7.64	<2	13.55	6.23	2.33	15.59	2.00	12.00
Cs	N.D.	N.D.	N.D.	N.D.	N.D.	N.D.	N.D.	N.D.	N.D.	N.D.	N.D.	N.D.	N.D.	N.D.	N.D.	N.D.	N.D.	N.D.	N.D.
Trace elements (ppm)																			
Ba	48.21	N.D.	52.86	196.09	145.03	871.93	116.49	8.76	N.D.	60.00	230.00	N.D.	80.00	N.D.	N.D.	N.D.	N.D.	130.00	100.00
Nb	5.71	6.00	40.10	16.41	5.63	4.84	4.73	1.76	5.25	6.00	<2	4.96	3.00	1.32	0.91	0.70	4.51	13.00	7.00
Ni	106.52	53.00	87.12	63.88	160.41	105.18	61.20	425.94	91.04	N.D.	N.D.	62.47	N.D.	58.08	94.62	44.96	97.22	N.D.	N.D.
Sr	51.45	N.D.	100.69	73.72	65.98	106.25	103.74	23.50	114.13	174.00	139.00	155.66	124.00	47.59	139.52	156.10	123.95	85.00	95.00
V	182.70	331.00	286.26	228.98	367.83	236.15	125.95	94.31	316.20	N.D.	N.D.	191.32	N.D.	383.47	210.19	169.19	263.20	N.D.	N.D.
Y	17.58	9.00	19.42	41.73	24.68	20.42	25.20	7.27	37.33	29.00	4.00	26.94	17.00	33.55	17.94	12.99	24.26	37.00	29.00
Zn	85.37	201.00	113.05	113.71	192.58	116.57	161.29	67.41	94.72	N.D.	N.D.	85.93	N.D.	113.89	70.00	82.67	98.61	N.D.	N.D.
Zr	100.93	53.00	54.78	245.78	136.34	104.47	73.62	33.81	113.82	79.00	31.00	129.83	78.00	92.34	38.31	31.76	56.03	130.00	120.00

(continued)

TABLE 1. WHOLE-ROCK MAJOR AND TRACE ELEMENT ANALYSES FOR SPUHLER PEAK METAMORPHIC SUITE AMPHIBOLITES *(continued)*

Sample	JC26	JC28B	JC29D	JC30E	JC31	JC31B	JC31C	JC32	JC33D	JC34E	JC35A	JC36B	JC37B	JC38A	JC39C	JC40C	JC40D	JC41	JC42
Major Elements (wt%)																			
SiO_2	52.54	50.48	47.14	49.65	49.48	55.96	53.18	52.59	45.67	50.72	50.01	59.70	53.90	50.32	52.61	50.99	51.87	51.75	50.76
TiO_2	1.18	0.83	0.40	0.93	1.55	0.78	0.70	0.52	0.30	1.16	1.08	0.73	0.66	1.97	0.55	0.84	1.52	1.09	1.16
Al_2O_3	14.20	16.22	17.78	14.24	13.80	14.22	15.76	8.49	20.03	14.08	13.88	14.43	14.98	13.17	13.73	15.88	14.88	15.57	14.26
*FeO	13.45	8.67	8.02	12.63	17.46	10.05	10.51	13.85	8.69	13.50	14.05	8.69	9.44	15.70	11.83	11.17	13.31	12.86	14.18
MnO	0.20	0.24	0.23	0.56	0.69	0.20	0.29	0.57	0.22	0.24	0.21	0.15	0.17	0.26	0.22	0.22	0.22	0.23	0.22
MgO	6.57	7.59	9.82	7.00	5.12	5.24	6.76	16.04	11.22	6.45	6.62	6.04	7.12	5.12	8.00	7.63	5.78	7.42	7.12
CaO	9.15	10.91	12.60	9.74	9.59	7.92	10.56	6.95	11.52	9.88	9.91	7.08	8.38	8.30	11.61	9.77	10.07	9.42	10.97
Na_2O	1.97	2.10	1.31	2.70	1.99	3.09	1.85	1.04	1.55	1.41	1.17	2.79	2.43	3.19	1.57	1.07	2.14	1.33	1.19
K_2O	0.74	0.49	0.27	0.59	0.58	0.66	0.49	0.14	0.42	0.71	0.99	0.47	1.49	0.47	0.25	1.02	0.46	0.36	0.26
P_2O_5	0.10	0.06	0.03	0.08	0.06	0.15	0.04	0.07	0.02	0.11	0.13	0.11	0.07	0.19	0.04	0.05	0.13	0.08	0.10
Total	100.11	97.59	97.60	98.12	100.33	98.27	100 15	100.26	99.63	98.27	98.07	100.20	98.64	98.69	100.40	98.65	100.39	100.10	100.22
Rare earth elements (ppm)																			
La	9.70	4.00	1.80	5.10	14.30	N.D.	N.D.	N.D.	N.D.	N.D.	N.D.	N.D.	N.D.	N.D.	N.D.	4.3	N.D.	N.D.	N.D.
Ce	19.40	8.30	5.40	12.60	22.80	N.D.	N.D.	N.D.	N.D.	N.D.	N.D.	N.D.	N.D.	N.D.	N.D.	12.7	N.D.	N.D.	N.D.
Nd	5.50	4.90	0.90	5.30	9.70	N.D.	N.D.	N.D.	N.D.	N.D.	N.D.	N.D.	N.D.	N.D.	N.D.	8.8	N.D.	N.D.	N.D.
Sm	3.20	1.70	0.92	2.60	2.40	N.D.	N.D.	N.D.	N.D.	N.D.	N.D.	N.D.	N.D.	N.D.	N.D.	2.18	N.D.	N.D.	N.D.
Eu	N.D.	N.D.	N.D.	N.D.	N.D.	N.D.	N.D.	N.D.	N.D.	N.D.	N.D.	N.D.	N.D.	N.D.	N.D.	0.86	N.D.	N.D.	N.D.
Tb	N.D.	N.D.	N.D.	N.D.	N.D.	N.D.	N.D.	N.D.	N.D.	N.D.	N.D.	N.D.	N.D.	N.D.	N.D.	0.45	N.D.	N.D.	N.D.
Yb	2.4	1.42	0.88	2.2	1.7	N.D.	N.D.	N.D.	N.D.	N.D.	N.D.	N.D.	N.D.	N.D.	N.D.	1.76	N.D.	N.D.	N.D.
Lu	0.35	0.21	0.11	0.29	0.26	N.D.	N.D.	N.D.	N.D.	N.D.	N.D.	N.D.	N.D.	N.D.	N.D.	0.26	N.D.	N.D.	N.D.
Hf	N.D.	N.D.	N.D.	N.D.	N.D.	N.D.	N.D.	N.D.	N.D.	N.D.	N.D.	N.D.	N.D.	N.D.	N.D.	1.27	N.D.	N.D.	N.D.
Ta	N.D.	N.D.	N.D.	N.D.	N.D.	N.D.	N.D.	N.D.	N.D.	N.D.	N.D.	N.D.	N.D.	N.D.	N.D.	0.14	N.D.	N.D.	N.D.
Th	4.60	N.D.	N.D.	N.D.	2.08	N.D.	3.54	2.02	1.47	N.D.	N.D.	12.64	N.D.	N.D.	2.95	1	2.32	1.23	1.78
U	0.39	N.D.	N.D.	N.D.	1.74	N.D.	-0.05	0.01	0.47	N.D.	N.D.	3.32	N.D.	N.D.	-0.22	4.8	-0.34	0.82	1.78
Sc	N.D.	N.D.	N.D.	N.D.	N.D.	N.D.	N.D.	N.D.	N.D.	N.D.	N.D.	N.D.	N.D.	N.D.	N.D.	36.2	N.D.	N.D.	N.D.
Cr	109.30	205.20	820.80	<68.4	44.46	<68.4	98.73	1208.21	236.72	<68.4	68.40	192.01	342.00	<68.4	158.62	268	86.57	256.41	150.77
Co	N.D.	N.D.	N.D.	N.D.	N.D.	N.D.	N.D.	N.D.	N.D.	N.D.	N.D.	N.D.	N.D.	N.D.	N.D.	57.8	N.D.	N.D.	N.D.
Rb	11.01	4.00	8.00	4.00	6.98	<2	6.75	2.38	12.83	4.00	8.00	11.71	41.00	12.00	6.57	33.3	9.13	6.58	6.34
Cs	N.D.	N.D.	N.D.	N.D.	N.D.	N.D.	N.D.	N.D.	N.D.	N.D.	N.D.	N.D.	N.D.	N.D.	N.D.	0.52	N.D.	N.D.	N.D.
Trace elements (ppm)																			
Ba	N.D.	140.00	50.00	120.00	N.D.	240.00	N.D.	N.D.	N.D.	70.00	140.00	N.D.	370.00	70.00	N.D.	50.00	N.D.	N.D.	N.D.
Nb	4.84	3.00	<2	12.00	1.57	6.00	1.27	0.58	-0.16	6.00	<2	7.58	2.00	7.00	0.83	9.00	5.90	1.77	4.35
Ni	82.61	N.D.	N.D.	N.D.	63.55	N.D.	80.34	215.68	324.78	N.D.	N.D.	152.34	N.D.	N.D.	97.42	N.D.	61.06	87.13	99.66
Sr	106.36	161.00	119.00	123.00	112.45	162.00	212.92	14.07	136.60	104.00	91.00	326.50	152.00	103.00	85.57	150.00	149.81	142.44	95.41
V	295.54	N.D.	N.D.	N.D.	1142.10	N.D.	389.27	156.34	81.49	N.D.	N.D.	147.68	N.D.	N.D.	219.33	N.D.	298.29	263.80	317.13
Y	29.56	11.00	5.00	16.00	23.26	22.00	14.06	13.22	8.16	25.00	17.00	36.01	23.00	39.00	15.61	27.00	29.59	25.02	30.00
Zn	110.72	N.D.	N.D.	N.D.	91.50	N.D.	74.76	103.88	85.21	N.D.	N.D.	78.37	N.D.	N.D.	88.77	N.D.	109.33	92.35	118.96
Zr	102.83	53.00	28.00	57.00	71.64	130.00	49.80	37.16	21.47	100.00	69.00	186.26	92.00	140.00	36.36	120.00	99.91	82.10	86.90

Note: Sources of analyses are detailed in the text.
*FeO: total iron as FeO.
[†]N.D.: no data.

TABLE 2. WHOLE-ROCK MAJOR AND TRACE ELEMENT ANALYSES FOR SPUHLER PEAK METAMORPHIC SUITE ORTHOAMPHIBOLE-GARNET GNEISSES

Sample	CP12	CP13	CP14	CP17	CP47	CP49	WP1	WP2	WP3a	WP11	WP12	WP13	WP14	WP20	WP21	WP4	WP5
Major Elements (wt%)																	
SiO_2	49.50	43.64	51.16	46.52	52.19	57.06	59.84	52.77	55.91	61.86	59.88	53.64	51.10	58.38	54.93	50.72	57.55
TiO_2	0.37	0.42	0.67	0.89	0.83	1.44	0.95	1.15	1.29	0.72	0.84	2.06	0.71	0.60	1.47	1.53	0.86
Al_2O_3	7.00	17.28	5.75	17.81	14.41	12.40	15.95	16.17	15.27	18.92	15.61	14.15	14.11	12.88	14.78	14.18	7.75
FeO*	9.26	16.48	10.98	15.74	15.43	15.92	8.08	12.52	14.16	8.36	10.03	14.93	12.06	8.65	10.78	15.32	11.66
MnO	0.39	1.13	0.45	0.41	0.24	0.22	0.16	0.24	0.35	0.22	0.21	0.25	0.14	0.14	0.20	0.11	0.14
MgO	21.45	11.82	26.03	11.32	11.11	9.45	5.43	9.75	9.00	5.92	6.08	6.34	19.50	10.20	6.10	13.29	18.10
CaO	10.79	5.45	4.61	2.37	2.31	2.03	3.35	1.89	2.34	0.37	4.31	7.01	1.18	6.72	8.16	1.92	1.97
Na_2O	0.76	0.75	0.50	2.13	2.00	0.90	6.09	3.53	1.42	0.52	4.04	1.05	1.04	1.91	3.27	2.35	0.72
K_2O	0.16	1.73	0.12	1.48	0.37	0.09	0.10	2.07	0.10	2.85	0.52	0.53	0.20	0.16	0.31	0.38	0.67
P_2O_5	0.04	0.02	0.06	0.04	0.08	0.20	0.10	0.10	0.26	0.04	0.18	0.21	0.09	0.10	0.37	0.24	0.08
Total	99.73	98.72	100.33	98.71	98.97	99.71	100.04	100.18	100.09	99.79	101.69	100.17	100.14	99.73	100.36	100.06	99.50
Rare earth elements (ppm)																	
La	5.50	N.D.[†]	N.D.	3.30	11.10	13.60	11.00	7.00	15.00	15.59	1.50	11.00	17.58	9.00	36.00	19.00	19.45
Ce	14.50	N.D.	N.D.	8.70	25.10	31.00	2.30	16.00	42.00	30.42	37.00	25.00	35.31	22.00	76.00	1.00	40.89
Nd	6.00	N.D.	N.D.	5.90	12.50	15.20	N.D.	N.D.	N.D.	12.30	N.D.	N.D.	14.39	N.D.	N.D.	N.D.	18.59
Sm	1.53	N.D.	N.D.	2.13	2.64	3.54	N.D.	N.D.	N.D.	2.53	N.D.	N.D.	2.74	N.D.	N.D.	N.D.	3.87
Eu	0.37	N.D.	N.D.	0.32	0.92	0.91	N.D.	N.D.	N.D.	0.49	N.D.	N.D.	0.55	N.D.	N.D.	N.D.	0.96
Tb	0.03	N.D.	N.D.	0.68	0.81	0.72	N.D.	N.D.	N.D.	0.42	N.D.	N.D.	0.38	N.D.	N.D.	N.D.	0.58
Yb	0.94	N.D.	N.D.	3.18	3.18	2.73	N.D.	N.D.	N.D.	1.02	N.D.	N.D.	1.79	N.D.	N.D.	N.D.	1.82
Lu	0.14	N.D.	N.D.	0.52	0.44	0.31	N.D.	N.D.	N.D.	0.17	N.D.	N.D.	0.29	N.D.	N.D.	N.D.	0.26
Hf	0.67	N.D.	N.D.	1.80	2.96	3.61	N.D.	N.D.	N.D.	4.05	N.D.	N.D.	2.95	N.D.	N.D.	N.D.	3.71
Ta	0.16	N.D.	N.D.	0.20	0.51	0.30	N.D.	N.D.	N.D.	0.71	N.D.	N.D.	0.40	N.D.	N.D.	N.D.	0.79
Th	1.58	N.D.	N.D.	1.20	3.03	2.52	N.D.	N.D.	N.D.	10.10	N.D.	N.D.	3.60	N.D.	N.D.	N.D.	4.10
U	4.20	N.D.	N.D.	6.60	6.00	5.10	N.D.	N.D.	N.D.	1.90	N.D.	N.D.	N.D.	N.D.	N.D.	N.D.	0.30
Sc	20.60	N.D.	N.D.	64.40	45.90	31.00	N.D.	N.D.	N.D.	18.96	N.D.	N.D.	35.67	N.D.	N.D.	N.D.	18.46
Cr	4060.00	N.D.	1530.00	50.00	280.00	36.00	101.00	152.00	90.00	388.90	110.00	203.00	659.70	620.00	56.00	13.00	769.20
Co	72.50	N.D.	N.D.	46.50	52.30	46.60	N.D.	N.D.	N.D.	35.90	N.D.	N.D.	67.00	N.D.	N.D.	N.D.	71.40
Rb	20.00	N.D.	6.00	83.90	12.50	22.00	N.D.	N.D.	N.D.	94.90	N.D.	N.D.	N.D.	N.D.	N.D.	N.D.	21.30
Cs	0.69	N.D.	N.D.	4.10	0.37	0.72	N.D.	N.D.	N.D.	1.80	N.D.	N.D.	0.50	N.D.	N.D.	N.D.	1.80
Trace elements (ppm)																	
Ba	188.00	N.D.	N.D.	235.00	132.00	95.00	64.00	578.00	36.00	504.00	143.00	168.00	15.00	34.00	282.00	131.00	76.00
Nb	N.D.	N.D.	5.00	N.D.	8.00	N.D.	3.00	3.00	10.00	6.00	6.00	9.00	5.00	3.00	17.00	12.00	8.00
Ni	1210.00	N.D.	916.00	160.00	86.00	94.00	72.00	158.00	33.00	157.00	66.00	88.00	179.00	138.00	70.00	34.00	756.00
Sr	180.00	N.D.	62.00	390.00	420.00	2330.00	72.00	54.00	34.00	10.00	114.00	40.00	7.00	39.00	110.00	47.00	32.00
V	N.D.	N.D.	N.D.	N.D.	N.D.	N.D.	187.00	301.00	270.00	146.00	192.00	396.00	203.00	178.00	244.00	333.00	125.00
Y	N.D.	N.D.	11.00	N.D.	33.00	N.D.	N.D.	N.D.	N.D.	N.D.	N.D.	N.D.	N.D.	N.D.	N.D.	N.D.	N.D.
Zn	106.00	N.D.	N.D.	127.00	137.00	70.00	86.00	93.00	57.00	40.00	58.00	103.00	95.00	82.00	112.00	160.00	122.00
Zr	N.D.	N.D.	97.00	N.D.	92.00	N.D.	90.00	102.00	186.00	138.00	135.00	143.00	107.00	76.00	275.00	253.00	128.00

(continued)

TABLE 2. WHOLE-ROCK MAJOR AND TRACE ELEMENT ANALYSES FOR SPUHLER PEAK METAMORPHIC SUITE ORTHOAMPHIBOLE-GARNET GNEISSES (*continued*)

Sample	WP6b	WP7	WP26	WP15	WP16	WP17	WP18	WP27	WP29a	WP29b	WP29c	WP30
Major Elements (wt%)												
SiO_2	49.25	48.71	53.70	55.33	51.60	67.22	57.56	52.05	52.37	55.52	57.43	54.64
TiO_2	1.92	2.05	1.15	0.38	0.34	0.91	1.58	0.69	1.10	1.21	1.48	1.42
Al_2O_3	16.38	15.23	14.61	14.88	14.16	10.64	15.66	14.31	14.83	15.46	15.36	14.25
FeO	16.17	16.77	17.18	13.72	10.93	7.12	12.31	11.84	13.00	12.08	10.59	12.75
MnO	0.16	0.24	0.40	0.20	0.27	0.04	0.24	0.26	0.23	0.19	0.18	0.12
MgO	11.58	13.35	7.71	8.77	12.01	11.39	6.15	12.47	7.13	6.42	5.99	13.50
CaO	1.58	1.81	3.11	2.70	8.23	0.68	3.37	4.86	8.74	4.79	5.10	1.39
Na_2O	1.79	1.59	1.81	1.64	1.18	0.79	2.68	2.53	2.39	4.19	3.48	1.85
K_2O	1.10	0.13	0.15	1.80	1.07	0.44	0.13	0.80	0.46	0.23	0.40	0.19
P_2O_5	0.21	0.36	0.15	0.05	0.03	0.22	0.34	0.04	0.11	0.12	0.12	0.29
Total	100.13	100.22	99.97	99.48	99.82	99.45	100.02	99.87	100.36	100.20	100.12	100.39
Rare earth elements (ppm)												
La	11.00	38.37	15.00	4.00	7.32	7.00	27.00	3.38	13.00	11	26	13
Ce	28.00	80.48	31.00	10.00	20.62	20.00	61.00	7.69	39.00	26.00	52.00	35.00
Nd	N.D.	35.42	N.D.	N.D.	12.51	N.D.	N.D.	5.82	N.D.	N.D.	N.D.	N.D.
Sm	N.D.	7.74	N.D.	N.D.	2.42	N.D.	N.D.	1.81	N.D.	N.D.	N.D.	N.D.
Eu	N.D.	1.32	N.D.	N.D.	0.82	N.D.	N.D.	0.47	N.D.	N.D.	N.D.	N.D.
Tb	N.D.	1.69	N.D.	N.D.	0.47	N.D.	N.D.	0.34	N.D.	N.D.	N.D.	N.D.
Yb	N.D.	8.95	N.D.	N.D.	1.85	N.D.	N.D.	1.40	N.D.	N.D.	N.D.	N.D.
Lu	N.D.	1.48	N.D.	N.D.	0.31	N.D.	N.D.	0.20	N.D.	N.D.	N.D.	N.D.
Hf	N.D.	6.26	N.D.	N.D.	1.22	N.D.	N.D.	1.21	N.D.	N.D.	N.D.	N.D.
Ta	N.D.	1.14	N.D.	N.D.	0.32	N.D.	N.D.	0.25	N.D.	N.D.	N.D.	N.D.
Th	N.D.	7.50	N.D.	N.D.	0.30	N.D.	N.D.	0.50	N.D.	N.D.	N.D.	N.D.
U	N.D.	N.D.	N.D.	N.D.	N.D.	N.D.	N.D.		N.D.	N.D.	N.D.	N.D.
Sc	N.D.	57.41	N.D.	N.D.	38.19	N.D.	N.D.	34.67	N.D.	N.D.	N.D.	N.D.
Cr	81.00	38.30	67.00	2664.00	1337.90	22.00	35.00	47.70	275.00	102.00	52.00	90.00
Co	N.D.	41.40	N.D.	N.D.	58.10	N.D.	N.D.	46.70	N.D.	N.D.	N.D.	N.D.
Rb	N.D.	N.D.	N.D.	N.D.	18.90	N.D.	N.D.	25.10	N.D.	N.D.	N.D.	N.D.
Cs	N.D.	0.40	N.D.	N.D.	0.90	N.D.	N.D.	1.50	N.D.	N.D.	N.D.	N.D.
Trace elements (ppm)												
Ba	269.00	31.00	53.00	260.00	114.00	15.00	39.00	62.00	125.00	81.00	132.00	17.00
Nb	8.00	13.00	4.00	2.00	1.00	7.00	13.00	2.00	19.00	3.00	13.00	11.00
Ni	128.00	35.00	94.00	829.00	365.00	21.00	27.00	49.00	147.00	110.00	29.00	36.00
Sr	37.00	18.00	28.00	75.00	37.00	8.00	89.00	48.00	406.00	108.00	115.00	25.00
V	268.00	291.00	210.00	183.00	176.00	177.00	292.00	216.00	283.00	264.00	272.00	284.00
Y	N.D.	N.D.	N.D.	N.D.	N.D.	N.D.	N.D.	N.D.	N.D.	N.D.	N.D.	N.D.
Zn	88.00	79.00	81.00	53.00	75.00	53.00	83.00	104.00	112.00	52.00	85.00	97.00
Zr	163.00	274.00	100.00	42.00	39.00	140.00	236.00	42.00	180.00	97.00	228.00	203.00

Note: Sources of analyses are detailed in the text.
*FeO—total iron as FeO
†N.D.—no data

TABLE 3. WHOLE-ROCK OXYGEN ISOTOPE RATIOS FOR SPUHLER PEAK METAMORPHIC SUITE ORTHOAMPHIBOLE-GARNET GNEISSES

Sample	$\delta^{18}O$ (‰)	Sample	$\delta^{18}O$ (‰)
WP1	8.71 ± 0.04	WP16	8.01
WP2	8.14 ± 0.05	WP17	7.57
WP3a	7.31	WP18	6.63
WP4	6.06	WP20	6.40
WP5	7.13	WP21	6.52
WP6a	7.27	WP26	7.55
WP6b	6.63	WP27	5.22
WP7	6.13	WP29a	5.98 ± 0.02
WP11	7.24 ± 0.03	WP29b	6.54
WP12	7.65	WP29c	6.23
WP13	6.32	WP30	6.48 ± 0.04
WP14	6.55		
WP15	8.71		

Note: ± indicates reproducibility of duplicated analyses.

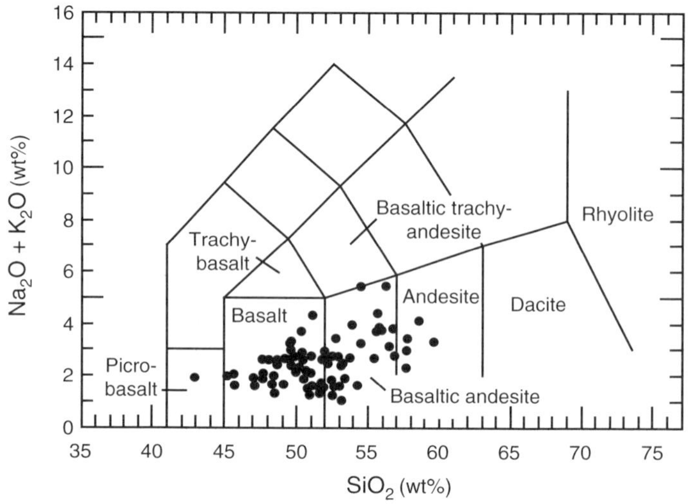

Figure 3. Classification diagram for volcanic rocks after LeBas et al. (1986). Points represent analyses for amphibolites (Table 1).

clear concentration in the basalt field (60% of all samples), with a substantial number of samples located in the basaltic andesite field (34%). Following the approach of Leake (1964), Niggli c and mg values were determined for 28 amphibolite samples that plot well within the basalt field of Figure 3, and Figure 4 illustrates the results. On both diagrams, Spuhler Peak Metamorphic Suite amphibolites follow a clear igneous trend with remarkably little scatter considering their complex metamorphic history. This evidence, taken together with field relationships and associations, strongly supports an igneous origin for Spuhler Peak Metamorphic Suite amphibolites. On an AFM diagram (Fig. 5) amphibolites show pronounced Fe enrichment, and 87% of the samples fall in the tholeiite field. This is consistent with a tholeiitic signature and the Al_2O_3 content of these rocks (Fig. 6).

To assess the effects of late-stage alteration common in Precambrian volcanic rocks, the Jensen cation plot (Jensen, 1976) was also used to evaluate the tholeiitic signature of the amphibolites. The Jensen cation plot uses Fe, Ti, Al, and Mg instead of alkali elements or Ca used by other classification schemes, in part to avoid the effects of late-stage hydrothermal alteration. Amphibolite samples

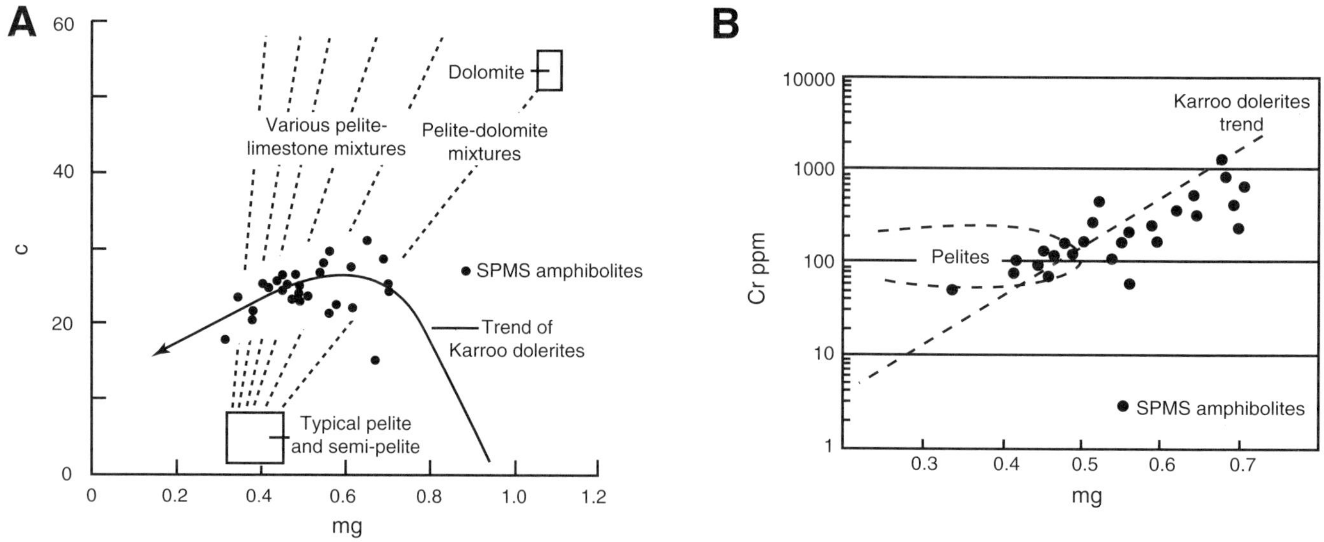

Figure 4. Plots after Leake (1964) for 28 Spuhler Peak Metamorphic Suite (SPMS) amphibolites. (A) Niggli mg and c. Dashed lines are mixing lines between pelitic and carbonate lithologies. (B) Niggli mg and Cr. Trend for the Karoo dolerites shown as an example of igneous differentiation.

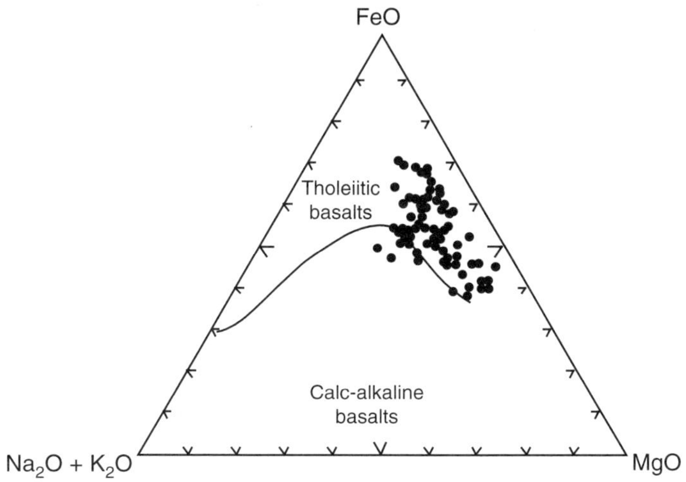

Figure 5. AFM diagram distinguishing tholeiitic basalts from calc-alkaline basalts. Line separating fields after Irvine and Baragar (1971, p. 528). Points represent analyses for amphibolites (Table 1). Total iron as FeO.

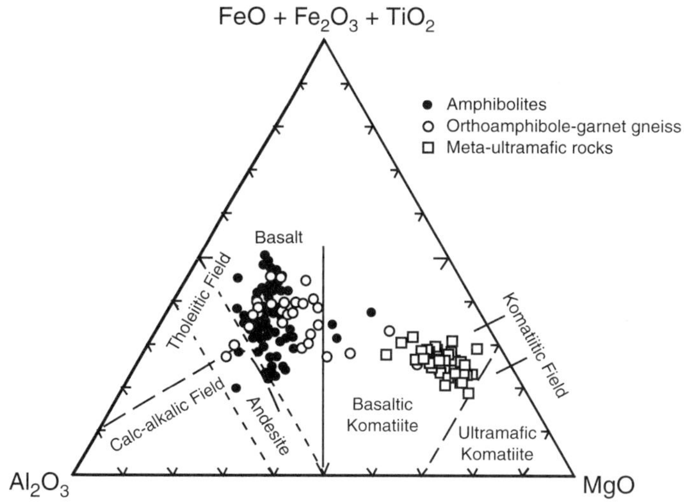

Figure 7. Jensen cation plot for classifying subalkalic volcanic rocks (Jensen, 1976). Analyses from Tables 1 and 2, and also from Table 2 of Johnson et al. (2004, this volume, Chapter 4).

plotted on a Jensen cation plot fall into a relatively tightly clustered group within the tholeiitic basalt field with only five samples plotting in the calc-alkaline field (Fig. 7). The persistence of tholeiitic geochemical signatures suggests limited element mobility for these elements after igneous crystallization of the rocks. Based on these three classification schemes, it appears that the majority of Spuhler Peak Metamorphic Suite amphibolites have compositions and differentiation trends similar to tholeiitic basalts, some are tholeiitic basaltic andesites, and a small minority is andesites.

Tectonic discriminant diagrams, taken together in conjunction with the major element classification diagrams and REE plots, provide important constraints on Spuhler Peak Metamorphic Suite protolith possibilities. A ternary diagram (Fig. 8) using MnO-TiO_2-P_2O_5 shows that almost all amphibolite samples plot within arc-related fields. On a Ti-Zr-Y diagram (Fig. 9) amphibolites predominantly occupy the calc-alkali basalt field and the calc-alkali/ocean-floor/island-arc overlap field (Pearce and Cann, 1973). The compositional space occupied by the amphibolites is

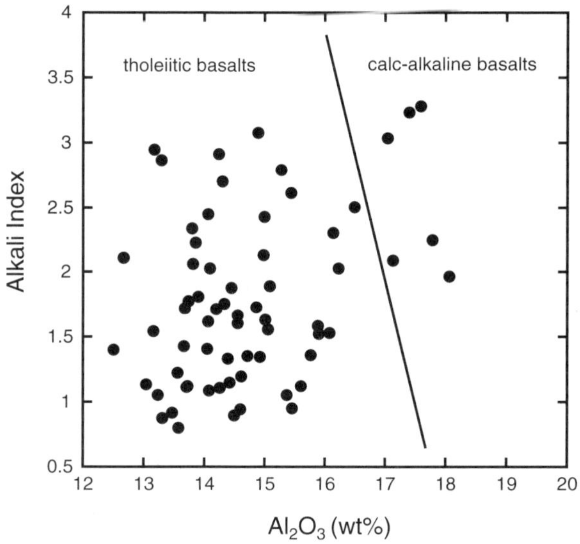

Figure 6. A plot of alkali index vs. Al_2O_3 used to distinguish tholeiitic basalts from calc-alkaline basalts. Line separating fields after Middlemost (1975). Alkali index = $(Na_2O + K_2O)/((SiO_2 - 43) \times 0.17)$. Points represent analyses for amphibolites (Table 1).

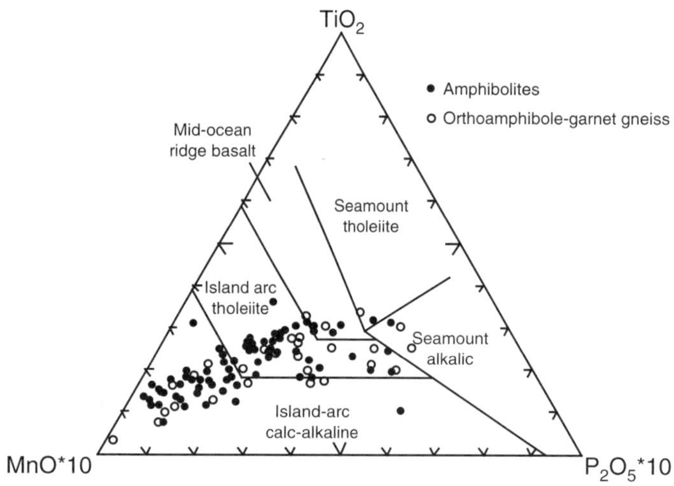

Figure 8. MnO-TiO_2-P_2O_5 after Mullen (1983) for discriminating among basaltic rocks of oceanic environments. Points represent analyses (wt%) from Tables 1 and 2.

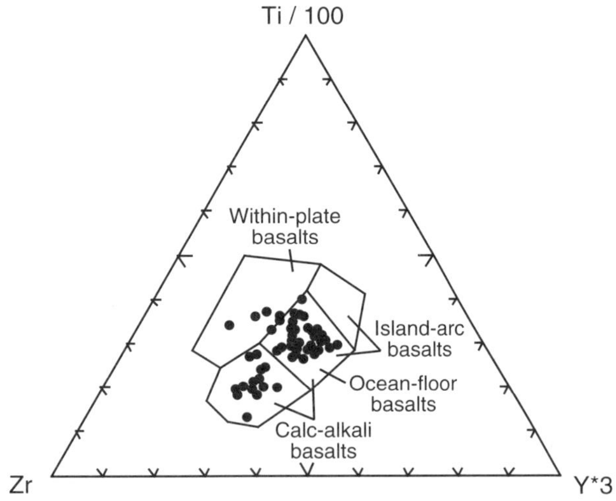

Figure 9. Ti-Zr-Y ternary plot after Pearce and Cann (1973) for distinguishing the tectonic setting of basaltic volcanic rocks. Points represent analyses (ppm) from Table 1.

similar to Phanerozoic basalts related to arc-continent collision zones, where variable continental crust, subducted sediment, and enriched mantle components cause an enrichment of Zr relative to Ti and Y (compared to arc-related basalts) on this discrimination diagram (Pearce, 1996). Condie (1990, p. 45) used Ta/Yb versus Th/Yb to compare Precambrian greenstone basalt distributions with those of modern basalt fields. Although our analytical data permit only six amphibolite samples to be plotted on this type of diagram (Fig. 10), these fall within modern arc-related fields and are located within the major concentrations of Archean samples. These three approaches for determining amphibolite protolith support an arc-related origin for the Spuhler Peak Metamorphic Suite.

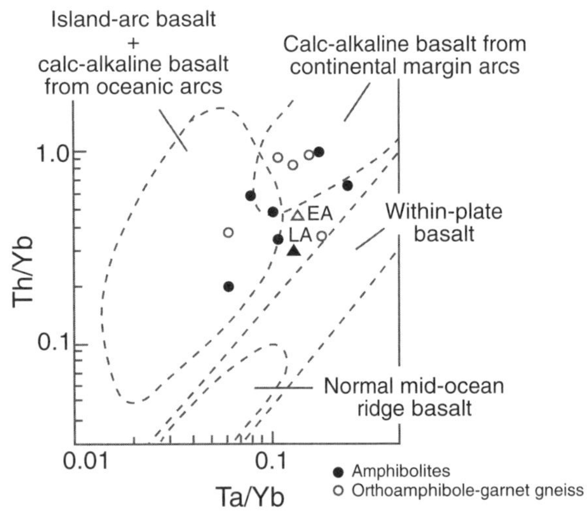

Figure 10. Comparison of Archean greenstone basalts with modern basalt fields by Condie (1990) using Pearce's (1983) plot of Ta/Yb versus Th/Yb. EA and LA designate the centers of concentration for Early Archean (older than 3.0 Ga) and Late Archean (3.0–2.5 Ga), respectively. Points represent analyses (weight ratios) from Tables 1 and 2.

In general, Spuhler Peak Metamorphic Suite amphibolites exhibit many similarities to typical Archean tholeiite basalts. Using Ti/Zr ratios to compare Spuhler Peak Metamorphic Suite amphibolites with Archean tholeiites (Fig. 11), 75% of the suite's amphibolites fall within the range of Archean tholeiites (younger than 3 Ga) presented by Cattell and Taylor (1990, p. 18). A few samples of Archean tholeiites (older than 3 Ga) in Cattell and Taylor's diagram have Ti/Zr values between 40 and 50, which is very similar to values for the remaining 25% of Spuhler Peak Metamorphic Suite amphibolites.

Chondrite-normalized REE plots for Spuhler Peak Metamorphic Suite amphibolites compare well with representative plots for Archean tholeiitic basalts (Fig. 12). Due to the clutter produced by plotting all Spuhler Peak Metamorphic Suite data, representative samples were chosen to illustrate all common patterns. In general, Archean tholeiitic basalts presented as representative by Cattell and Taylor (1990, p. 21) have flat REE patterns with a relatively narrow range of REE enrichment (10 to 30× chondrite). Spuhler Peak Metamorphic Suite amphibolites show some light rare earth element (LREE) enrichment (10 to 60× chondrite), with heavy rare earth element (HREE) and medium rare earth element (MREE) contents similar to average Archean tholeiitic basalts. REE patterns of Spuhler Peak Metamorphic Suite amphibolites roughly correlate with SiO_2 (consistent with fractional crystallization) and show some evidence for Ce and Eu mobility.

Based on the preceding analysis, Spuhler Peak Metamorphic Suite amphibolites are consistent with other Archean tholeiitic basalts with arc-like geochemical affinities.

Orthoamphibole-Garnet Gneiss

The rarity of orthoamphibole ± garnet ± cordierite rocks is due to their unusually Mg- and Al-rich and Ca-poor bulk compositions, which do not have equivalents in common sedimentary, igneous, or metamorphic rocks (Seki, 1957). Early investigators attributed these chemical characteristics to a number of metasomatic and magmatic processes, including removal of alkalis by fluids during metamorphism or by melt extraction. Vallance (1967) demonstrated that some cordierite-orthoamphibole rocks are equivalent to lower-grade quartz-chlorite lithologies, which are interpreted to be hydrothermally altered mafic volcanic rocks. This model explains the common association of cordierite-orthoamphibole rocks with the extensively altered stockwork zones of amphibolite-facies volcanogenic massive sulfide deposits (e.g., Bachinski, 1978). Field and geochemical studies have documented a hydrothermally altered volcanic protolith for many Early Archean to Phanerozoic cordierite-orthoamphibole rocks (e.g., Bachinski, 1978; Schumacher, 1988; Dymek and Smith, 1990; Smith et al., 1992a, 1992b; Pan and Fleet, 1995; Shandl et al., 1995; Zaleski and Peterson, 1995; Araujo et al., 1996; Peck

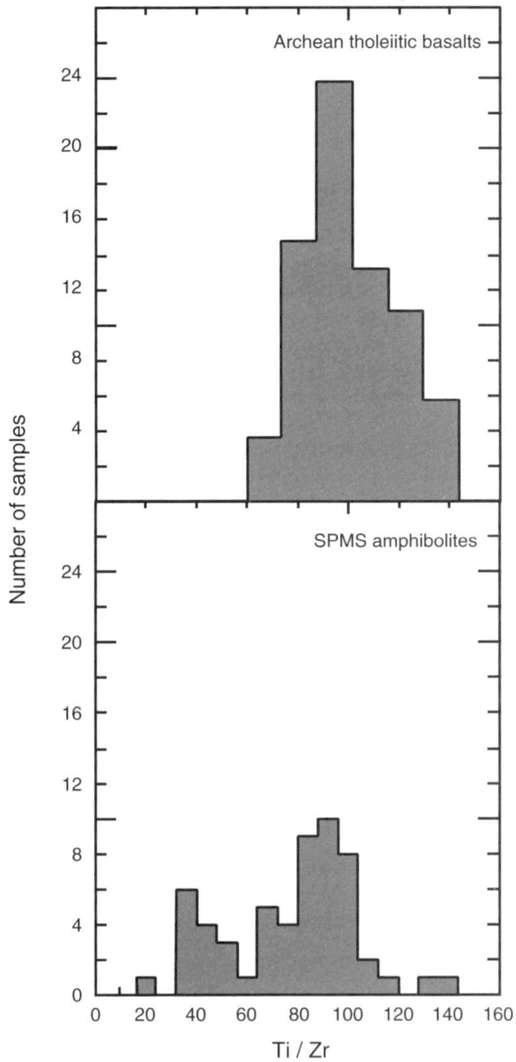

Figure 11. Ti/Zr histograms comparing Spuhler Peak Metamorphic Suite (SPMS) amphibolite analyses (weight ratios) from Table 1 with Archean tholeiitic basalts (Cattell and Taylor, 1990, p. 18).

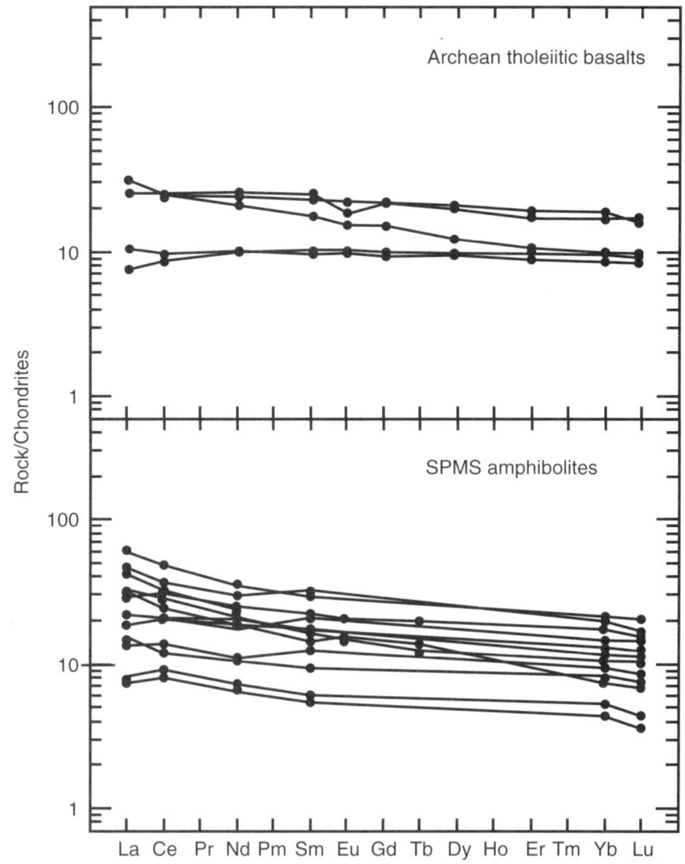

Figure 12. Chondrite-normalized rare earth element diagram comparing Spuhler Peak Metamorphic Suite (SPMS) amphibolites with representative Archean tholeiitic basalts as presented by Cattell and Taylor (1990, p. 21). SPMS amphibolite data from Table 1. Chondrite concentrations from Sun (1980).

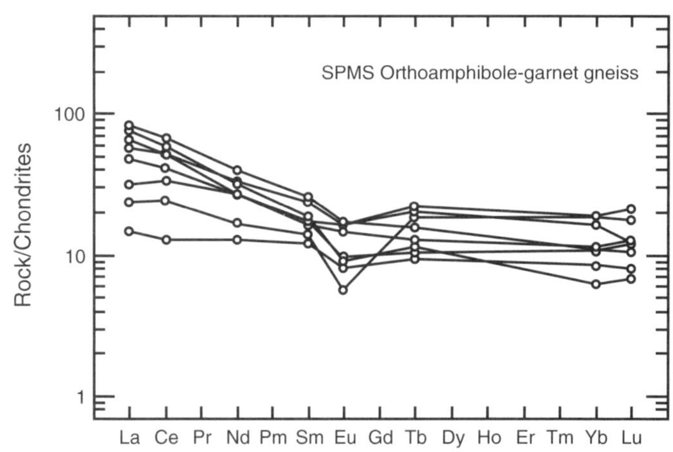

Figure 13. Chondrite-normalized rare earth element diagram for Spuhler Peak Metamorphic Suite (SPMS) orthoamphibole-garnet gneisses. SPMS orthoamphibole-garnet gneiss data from Table 2. Chondrite concentrations from Sun (1980).

and Valley, 2000), as well as for some sapphirine-bearing, granulite facies rocks (Warren, 1979; Wilson and Baksi, 1983; Peck and Valley, 1996). The geochemistry of these occurrences contrasts with Fe-poor cordierite-orthoamphibole rocks interpreted to be metamorphosed Mg-rich pelites (e.g., Reinhardt, 1987).

The chemistry and field relations of the Spuhler Peak Metamorphic Suite orthoamphibole-garnet (± cordierite) gneisses are related to the suite's amphibolites. On tectonic discrimination diagrams that use immobile elements (e.g., Figs. 8 and 10), the orthoamphibole-garnet gneiss and amphibolite analyses overlap and occupy similar fields. The orthoamphibole-garnet gneiss and amphibolite analyses also have similar LREE enrichment and flat HREE to MREE as amphibolite samples (Figs. 12 and 13). Additionally, many of the orthoamphibole-garnet gneiss samples display prominent negative Eu anomalies, similar to cordierite-orthoamphibole rocks worldwide and distinct from amphibolite samples. The main differences between the orthoamphibole-garnet gneisses and amphibolites are in the abundance of mobile elements and MgO, and the indication of plagioclase dissolution from the REE data. We interpret the amphibolites to be essentially isochemically metamorphosed representatives of the pre-alteration protoliths of the orthoamphibole-garnet gneisses.

The geochemical parameters that characterize the orthoamphibole-garnet gneiss analyses are low CaO and high MgO contents: 3.8 ± 2.8 wt% and 12.3 ± 6.3 wt% (1σ), respectively, for the entire data set. This strongly contrasts with 9.9 ± 2.0 wt% CaO and 7.8 ± 2.8 wt% MgO for the amphibolite data. Low CaO and high MgO relative to "normal" amphibolites has been noted in other orthoamphibole-garnet gneiss localities, and is also consistent with the common results of moderate- to high-temperature hydrothermal alteration of mafic volcanic rocks on the seafloor (e.g., Thompson, 1991). In addition, both field and laboratory studies document similar compositional changes for intermediate and felsic volcanic rocks (Hajash and Chandler, 1981; Shiraki et al., 1987; Thompson, 1991). Figure 14 shows orthoamphibole-garnet gneisses and amphibolites with Mg/((Al + Na + K)/2) plotted versus Ca/((Al + Na + K)/2), after Schumacher (1988). This figure shows the projected compositions of diopside, plagioclase, orthopyroxene, and olivine. Amphibolites plot as mixtures between plagioclase, diopside, and ferromagnesian minerals with no Ca. Orthoamphibole-garnet gneisses plot below the plagioclase-orthopyroxene+olivine mixing line, consistent with Ca depletion (by plagioclase dissolution) and precipitation of Al- and Mg-rich alteration minerals during hydrothermal alteration of the orthoamphibole-garnet gneiss protoliths.

Calcium depletion in orthoamphibole-garnet gneisses correlates with Sr content (Fig. 15), as amphibolites consistently have high CaO and Sr while orthoamphibole-garnet gneisses have lower values. Hydrothermal dissolution of a phase with a constant Ca/Sr ratio would cause protolith volcanic rocks to move in a line toward the origin of this plot. Orthoamphibole-garnet gneisses do not have Ca/Sr > 300, which we take to be the Ca/Sr ratio of plagioclase in the protolith, consistent with typical Ca/Sr ratios of intermediate plagioclase (see Smith, 1974). Figure 15 shows the effects of two apparent mechanisms of Ca depletion, one of which causes protolith volcanic rocks (represented by amphibolites) to lose Ca and Sr at Ca/Sr ratios >300 (probably representing dissolution of clinopyroxene, groundmass phases, and glass), and the other of which is plagioclase dissolution. In some rocks these other phases are exhausted before plagioclase dissolution is complete, and these rocks plot close to the Ca/Sr ≈ 300 line, where the position on the line away from the origin indicates the amount of plagioclase remaining. This Ca/Sr relationship between Ca-poor (orthoamphibole-bearing) and Ca-rich metavolcanic rocks (i.e., amphibolites) is observed between orthoamphibole-bearing rocks and amphibolites in the Ordovician Ammonoosuc volcanics, which also are interpreted as representing a variably hydrothermally-altered and metamorphosed volcanic sequence (Schumacher, 1988). Ca and Sr for the

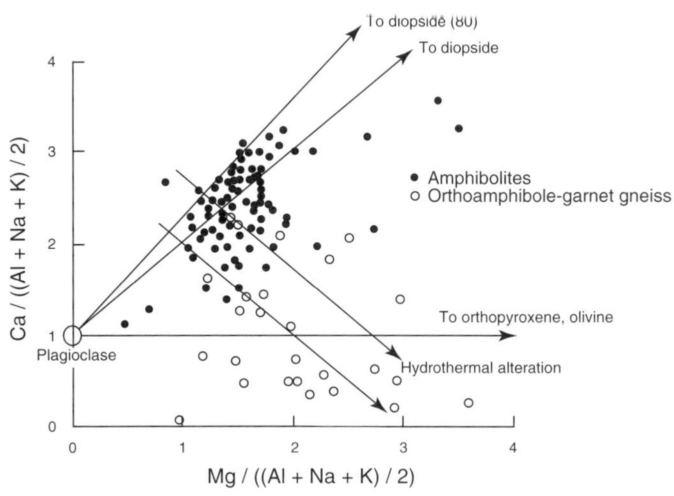

Figure 14. Major elements in Spuhler Peak Metamorphic Suite amphibolites and orthoamphibole-garnet gneisses. Amphibolites are consistent with mixtures of pyroxenes, plagioclase, and olivine. Orthoamphibole-garnet gneisses have lower Ca and higher Mg contents than amphibolites, consistent with hydrothermal alteration by seawater of the protolith of these rocks. After Schumacher (1988).

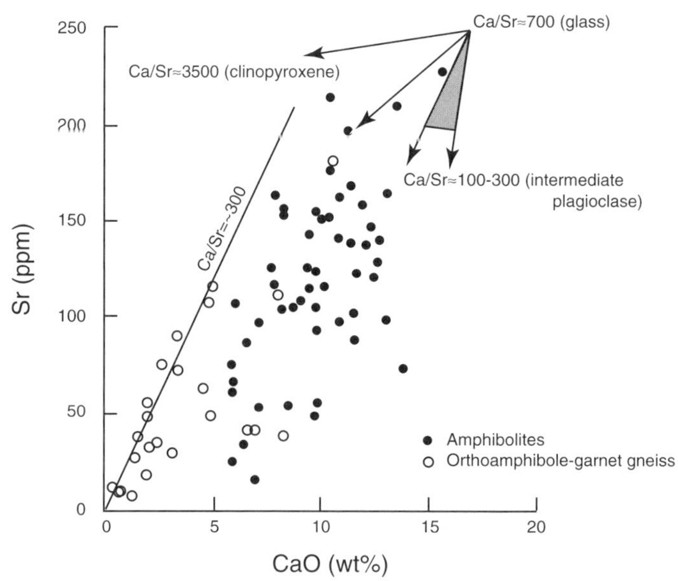

Figure 15. CaO (wt%) vs. Sr (ppm) for Spuhler Peak Metamorphic Suite amphibolites and orthoamphibole-garnet gneisses. Dissolution of plagioclase, glass, or other groundmass phases during hydrothermal alteration of the protolith of the orthoamphibole-garnet gneisses causes rocks to move towards the origin of this diagram. Note the limit of Ca/Sr ≈ 300 for orthoamphibole-garnet gneisses, which is taken to be the Ca/Sr ratio of plagioclase in the volcanic protolith.

Ammonoosuc amphibolites are higher and more variable than those for orthoamphibole-bearing rocks, which are bound at low Ca contents by a line of Ca/Sr ≈ 100, similar to the Spuhler Peak Metamorphic Suite data. In addition, CaO and Sr contents of cummingtonite-biotite-plagioclase and cordierite-orthoamphibole gneisses from Orijarvi, Finland (also with hydrothermally altered volcanic protoliths; Smith et al., 1992b) are bound at low CaO contents by a line of Ca/Sr ≈ 240.

Plagioclase dissolution is also supported by the negative Eu anomalies of orthoamphibole-garnet gneisses (Fig. 13). Samples with the lowest CaO contents also show the largest negative Eu anomalies. Plagioclase dissolution in the Spuhler Peak Metamorphic Suite protoliths is also consistent with low Ba contents in some samples. The geochemical signature of dissolution of minerals and glass in the protolith by seawater is also probably obscured by late-stage (lower-temperature) precipitation of phases in the waning hydrothermal system. This may account for the high Sr and Ba content of a few samples.

Oxygen isotope ratios also suggest a hydrothermally altered volcanic protolith for Spuhler Peak Metamorphic Suite orthoamphibole-garnet gneisses (Fig. 16). $\delta^{18}O$ values of Spuhler Peak Metamorphic Suite orthoamphibole-garnet whole rocks are restricted to 5.2 to 8.7‰, with an average of 7.0 ± 0.9‰ (1σ, n =24). Hydrothermal alteration by seawater can either raise or lower $\delta^{18}O$ values of volcanic rocks, depending on the protolith $\delta^{18}O$, the fluid history, and the temperature of alteration. Modern basalt that has been hydrothermally altered on the seafloor displays heterogeneous alteration on the millimeter to meter scale, dominated by crack-controlled disequilibrium exchange (e.g., Thompson, 1991). Gradients in chemistry and isotopic composition would be variably smoothed by oxygen exchange during high-grade metamorphism. Assuming a mantle-like protolith, $\delta^{18}O$ value of ~6‰, and hydrothermal alteration with seawater ($\delta^{18}O \approx 0$), temperatures of exchange below ~250 °C are required to produce $\delta^{18}O$ values >6‰ (Cole et al., 1987), such as those observed for the Spuhler Peak Metamorphic Suite orthoamphibole-garnet gneisses. Whole-rock $\delta^{18}O$ values of ~7‰ are common from the Mg-rich products of seafloor hydrothermal alteration of volcanic rocks (e.g., Muehlenbachs, 1986). Proterozoic cordierite-gedrite rocks from the Grenville province of Ontario have whole-rock $\delta^{18}O$ values of 6‰–8‰, similar to those of the Spuhler Peak Metamorphic Suite orthoamphibole-garnet gneisses, and these rocks are interpreted as having a hydrothermally altered volcanic protolith (Peck and Valley, 2000).

In the Green Tuff Belt of Japan, Ca-poor and Mg-rich sericite-chlorite rocks represent the most extreme results of hydrothermal alteration of volcanic rocks during formation of the Kuroko volcanogenic massive sulfide deposits (Green et al., 1983). These rocks have whole-rock geochemistries similar to the Spuhler Peak Metamorphic Suite orthoamphibole-garnet gneisses, and have statistically identical whole-rock $\delta^{18}O$ values (6.7 ± 1.3‰, n = 31) (Green et al., 1983). It is these sorts of lithologies (chlorite + clay + quartz products of seafloor alteration of volcanic rocks) that we infer to be the most likely protolith of the Spuhler Peak Metamorphic Suite orthoamphibole-garnet gneisses. The overall

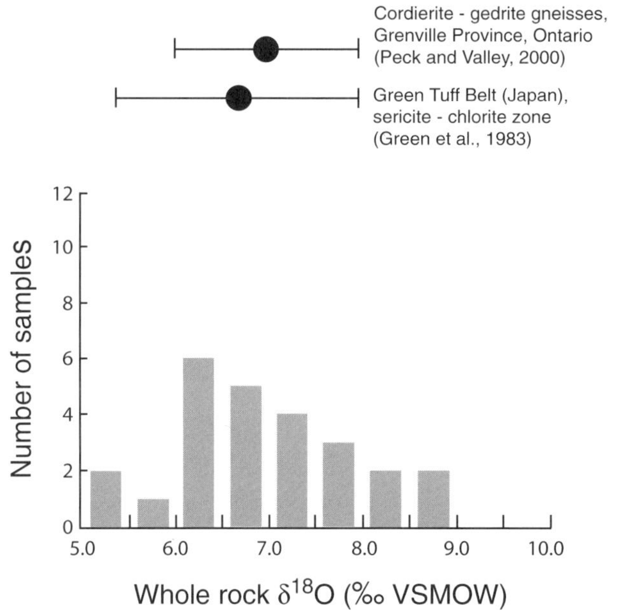

Figure 16. Oxygen isotope ratios for whole-rock samples of Spuhler Peak Metamorphic Suite orthoamphibole-garnet gneisses. These $\delta^{18}O$ values are consistent with low-temperature hydrothermal alteration of volcanic protoliths by seawater, in agreement with the Ca-poor and Mg-rich geochemistry of these rocks (see Table 3).

homogeneity in the $\delta^{18}O$ values of the Spuhler Peak Metamorphic Suite rocks suggests a common alteration history.

Although some orthoamphibole-bearing rocks have been assigned metasedimentary protoliths, this is not consistent with the major and trace element geochemistry or the oxygen isotope ratios of the Spuhler Peak Metamorphic Suite rocks. Lateritic soils typically have $\delta^{18}O$ values in near equilibrium with local meteoric water at the low temperatures of weathering, resulting in values of ~20‰ (e.g., Giral et al., 1993). Mudrocks also have high $\delta^{18}O$ values, ranging from ~10 to >20‰, although values lower than 10‰ are observed (e.g., Land and Lynch, 1996). Fe/Mg ratios are the best geochemical signature between Mg-rich metasedimentary rocks and hydrothermally altered volcanic rocks, as shown on Figure 17, a modified AFM diagram of Reinhardt (1987). In contrast to our data, Reinhardt (1987) describes low-Fe cordierite-gedrite rocks from Queensland that appear to have sedimentary protoliths, an interesting contrast to most other cordierite-orthoamphibole occurrences. On this diagram evaporitic clays plot at low Fe contents, as opposed to the higher Fe contents of seafloor alteration products and the Spuhler Peak Metamorphic Suite orthoamphibole-garnet gneisses.

The major and trace element geochemistry of the Spuhler Peak Metamorphic Suite orthoamphibole-garnet gneisses is within the range of the suite's amphibolites, excepting CaO and MgO contents. This supports the hypothesis that the amphibolites represent metamorphosed equivalents to the unaltered orthoamphibole-garnet gneiss protolith, consistent with the oxygen

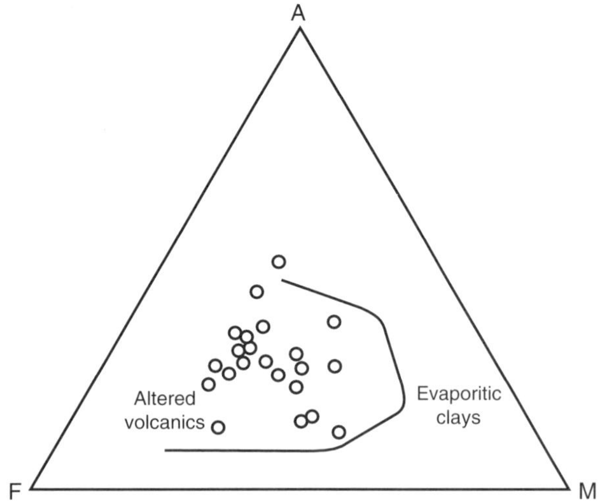

Figure 17. AFM diagram after Reinhardt (1987). All Spuhler Peak Metamorphic Suite orthoamphibole-garnet gneisses plot well within the field of altered volcanics rather than the field of evaporitic clays. Orthoamphibole-garnet gneiss data from Table 2.

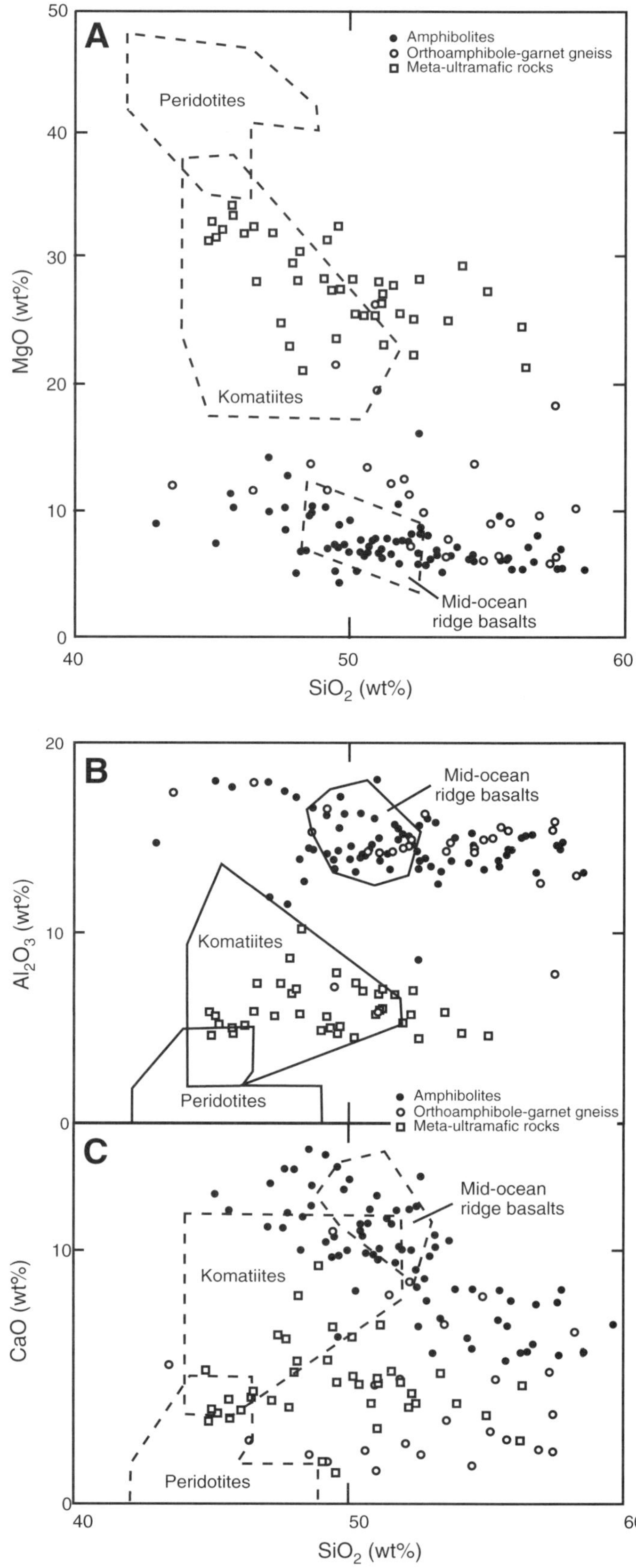

Figure 18. Discriminant diagram for CaO, MgO, and Al_2O_3 versus SiO_2. Fields from Herzberg (1993).

isotope data. As a result, these two lithologies plot in similar fields on major element discrimination diagrams that do not use CaO or MgO (e.g., Fig. 18B), but are slightly displaced on diagrams such as the Jensen Cation Plot (Fig. 7), which uses MgO as one of the coordinates (see also Figs. 18A and 18C). When the effects of hydrothermal alteration are taken into account, the geochemistry of the Spuhler Peak Metamorphic Suite orthoamphibole-garnet gneisses is consistent with a volcanic protolith, genetically related to the protolith of the suite's amphibolites.

Meta-ultramafic Rocks

Because meta-ultramafic rocks occur throughout the Precambrian of the Tobacco Root Mountains, they are discussed separately in Johnson et al. (2004, this volume, Chapter 4). However, due to their intimate association with Spuhler Peak Metamorphic Suite amphibolites, especially in the Branham Lakes region, and, therefore, their possible significance with regard to Spuhler Peak Metamorphic Suite protolith, the suite's meta-ultramafic rocks are discussed briefly here.

Cummings and McCulloch (1992) studied the mafic and meta-ultramafic rocks in the vicinity of Branham Lakes (Fig. 2 *in* Burger, 2004, this volume, Chapter 1). They conclude that those rocks we identify as Spuhler Peak Metamorphic Suite amphibolites represent oceanic basalts but do not specify exactly what the meta-ultramafic rocks represent. They note that there is no evidence for cold emplacement of these rocks during metamorphism and tectonism and reject the notion that they represent true komatiites. Although the meta-ultramafic rocks meet most of the compositional requirements for komatiites, Cummings and

McCulloch (1992, p. 338) argue that a $CaO:Al_2O_3$ ratio of less than one for all samples and the lack of spinifex textures in the rocks are inconsistent with komatiitic lava flows. The major element compositional characteristics of Spuhler Peak Metamorphic Suite meta-ultramafic rocks that bear on the questions of komatiitic parentage are presented in Figure 19. According to Arndt and Nisbet (1982), komatiites have $SiO_2 < 53\%$, $MgO > 9\%$, K_2O and $TiO_2 < 0.9\%$, and $CaO/Al_2O_3 \geq 1$. However, Cattell and Taylor (1990, p. 15) point out that most komatiites have CaO/Al_2O_3 between 0.8 and 1.0, which is the range illustrated by Spuhler Peak Metamorphic Suite meta-ultramafic rocks. In addition, Spuhler Peak Metamorphic Suite meta-ultramafic rocks meet the other major oxide compositional criteria. Cattell and Taylor (1990, p. 15) also note that more than half of all komatiite successions do not contain

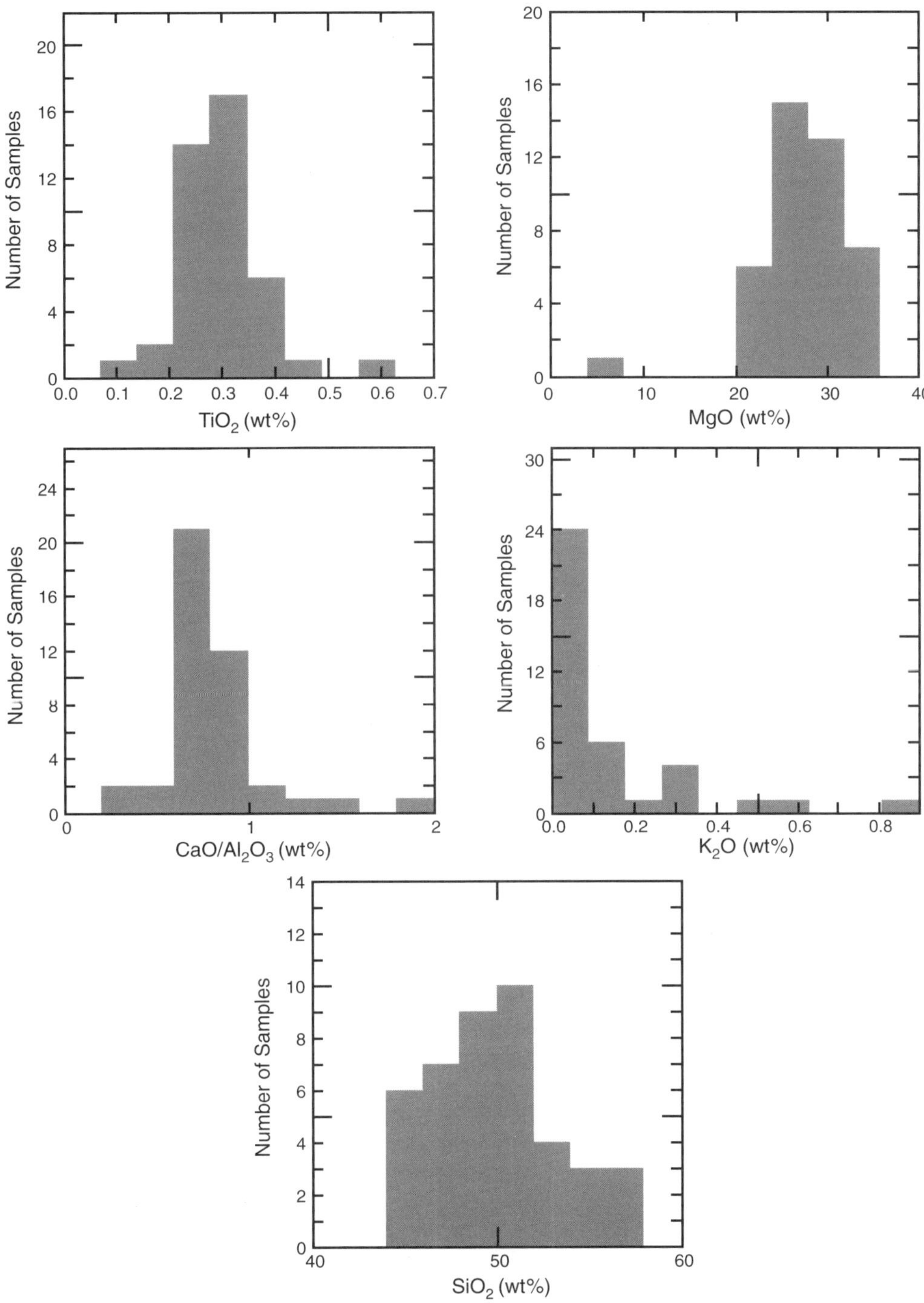

Figure 19. Histograms for selected major oxides from Spuhler Peak Metamorphic Suite meta-ultramafic samples. Data from Table 2 of Johnson et al. (2004, this volume, Chapter 4).

spinifex-textured rocks. It is unlikely that spinifex texture would survive the multiple periods of high-grade metamorphism and deformation that have affected Spuhler Peak Metamorphic Suite rocks, so the lack of such texture is equivocal.

The results from discriminant diagrams are somewhat mixed. In Figure 7, meta-ultramafic rocks form a tight cluster in the komatiitic field but are located in the basaltic komatiite portion of the field, not the ultramafic komatiite portion. In Figure 18B most points fall within the komatiite field. In Figure 18A a majority of points are within the field, and those that are not have higher proportions of metamorphic phases present (MacFarlane, 1996).

Comparison of REE for Spuhler Peak Metamorphic Suite meta-ultramafic rocks with Archean komatiitic basalts and komatiites (Fig. 20) illustrates that the suite's meta-ultramafic rocks are more similar to komatiitic basalt than komatiite. Spuhler Peak Metamorphic Suite meta-ultramafic rocks show LREE enrichment (10 to 100× chondrite), similar to Archean komatiitic basalts, with the addition of prominent negative Eu anomalies.

Similarly, comparison of Al_2O_3/TiO_2 and Ti/Zr ratios for Spuhler Peak Metamorphic Suite meta-ultramafic rocks with those for Archean komatiitic basalts and komatiites is contradictory. The Al_2O_3/TiO_2 values for Spuhler Peak Metamorphic Suite ultramafic rocks closely match those for Archean komatiites (Fig. 21). However, Ti/Zr ratios are quite similar to those for komatiitic basalts but not for komatiites (Fig. 22).

Thus, the protolith for the Spuhler Peak Metamorphic Suite meta-ultramafic rocks remains uncertain. Field relationships are not definitive, nor is the geochemistry. Spuhler Peak Metamorphic Suite meta-ultramafic rocks may represent komatiites and/or komatiitic basalts, but they almost certainly are not of peridotitic parentage. A komatiitic and/or komatiitic basalt protolith fits well with the tholeiitic basalt protolith for Spuhler Peak Metamorphic Suite amphibolites and orthoamphibole-garnet gneisses because of the close field associations of the suite's mafic and ultramafic rocks.

A Possible Protolith for the Spuhler Peak Metamorphic Suite

The preceding analysis demonstrates that the Spuhler Peak Metamorphic Suite is dominated by mafic volcanics with some

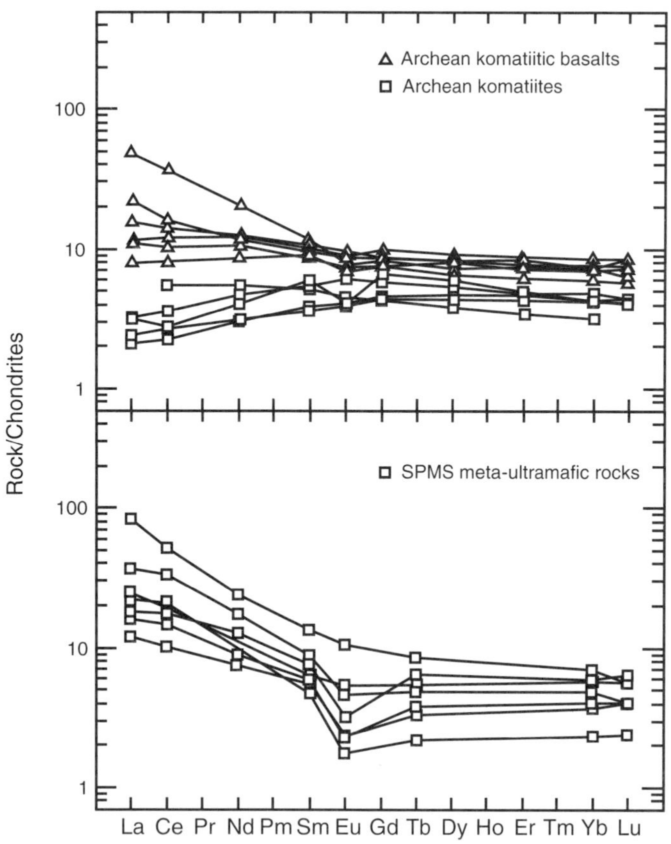

Figure 20. Chondrite-normalized rare earth element diagram for Spuhler Peak Metamorphic Suite (SPMS) meta-ultramafic rocks. Data from Table 2 of Johnson et al. (2004, this volume, Chapter 4). Chondrite concentrations from Sun (1980).

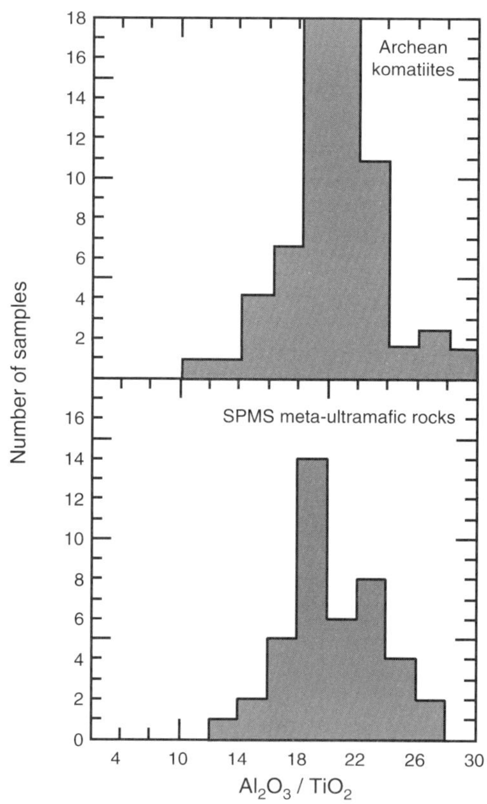

Figure 21. Histograms of Al_2O_3/TiO_2 values for Spuhler Peak Metamorphic Suite (SPMS) meta-ultramafic rocks and Archean komatiites. Archean komatiite data from Cattell and Taylor (1990, p. 17). SPMS meta-ultramafic data from Table 2 of Johnson et al. (2004, this volume, Chapter 4) (wt%).

ultramafic material present. The orthoamphibole-garnet gneisses are interpreted to represent mafic volcanics hydrothermally altered in a marine environment prior to metamorphism. If correct, this increases the dominance of mafic-ultramafic volcanics to close to 90% when taking all measured exposures into account (Fig. 1). Quartzite, impure quartzite, and aluminous gneiss, with a minor amount of quartzofeldspathic gneiss, dominate the remaining component of the Spuhler Peak Metamorphic Suite. As many, but not all, quartzites contain zircons with varying degrees of rounding, some amount of detrital component is present. Aluminous gneisses most likely began as muds. As no original textures survive in the Spuhler Peak Metamorphic Suite, it is impossible to define the exact sedimentary environment based on the rock types present.

Because the lithology of the Spuhler Peak Metamorphic Suite is so different from that of the Indian Creek and Pony–Middle Mountain Metamorphic Suites, it is natural to speculate on a depositional environment, but not possible to be definitive. The Spuhler Peak Metamorphic Suite appears to be a sequence dominated by metamorphosed tholeiitic basalts (and possibly some komatiitic basalts and komatiites) that were deposited in a marine environment. These are intercalated with volumetrically minor amounts of metamorphosed quartz sands and/or cherts, as well as muds. The quartz sands and/or cherts were sometimes relatively pure but often were mixed with varying amounts of mud. A very few, thin metamorphosed felsic volcanics complete the range of material present.

A common association in Archean greenstone belts consists of ultramafic-mafic metavolcanic sequences intercalated with thin, but laterally extensive, cherty metasedimentary rocks. Barley et al. (1979) describe a section of the Warrawoona Group, from the eastern Pilbara Block, Western Australia, that closely resembles the Spuhler Peak Metamorphic Suite as exposed in the Tobacco Root Mountains. A significant feature of these rocks is that, although metamorphosed to varying degrees, they were not subjected to high strain, and original structures and textures are commonly present. These features confirm a shallow-water origin for the metasedimentary rocks and, by association, the intercalated metavolcanic rocks. At North Pole in the eastern Pilbara Block, the cherty metasedimentary rocks (which often attain thicknesses >5 m) directly overlie ultramafic and mafic metavolcanic rocks. The cherty metasedimentary rocks represent silicified mud, sand, and conglomerate and contain abundant sedimentary features diagnostic of a shallow-water origin. Evaporite beds also are present, as are minor amounts of felsic metavolcanic rocks. This alternating sequence of dominantly mafic-ultramafic metavolcanic rocks with minor, but laterally extensive, cherty metasedimentary rocks is interpreted to have formed in an extensive shallow basin of subdued topography with local, more elevated felsic volcanoes and accompanying detritus (Barley et al., 1979, p. 81–82). The similarities between the sequences at North Pole and the Tobacco Roots provide a documented environment that represents a possible protolith for the Spuhler Peak Metamorphic Suite.

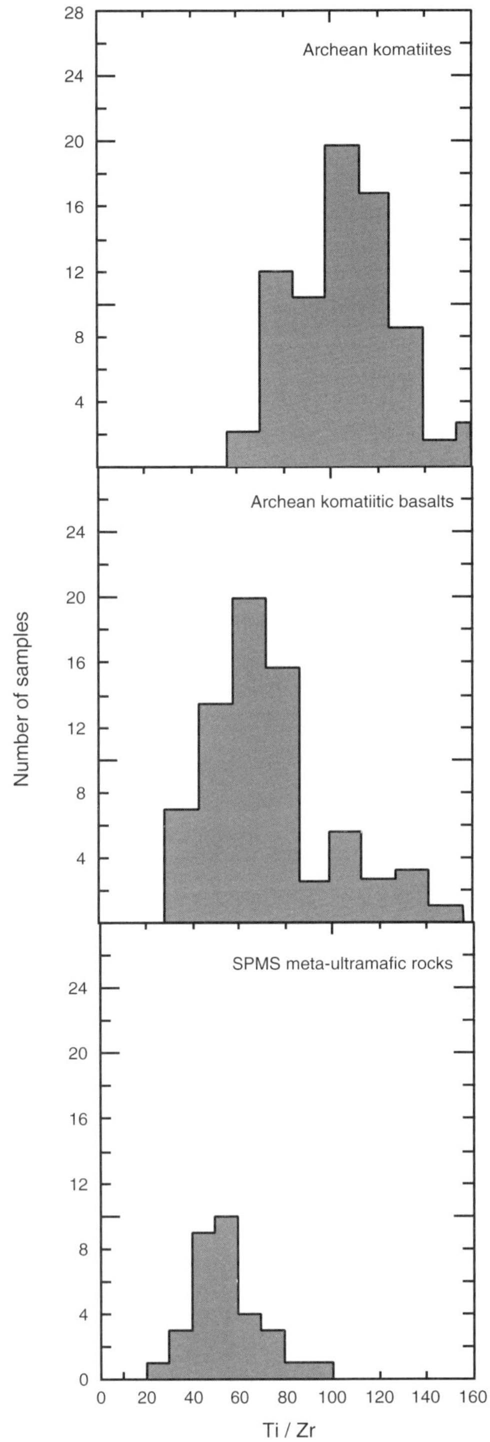

Figure 22. Histograms of Ti/Zr values (weight ratios) for Spuhler Peak Metamorphic Suite (SPMS) meta-ultramafic rocks and Archean komatiites and komatiitic basalts. Archean komatiite and komatiitic basalt data from Cattell and Taylor (1990, p. 18). SPMS meta-ultramafic data from Table 2 of Johnson et al. (2004, this volume, Chapter 4).

THE RELATIONSHIP OF THE SPUHLER PEAK METAMORPHIC SUITE TO OTHER ARCHEAN ROCKS IN THE TOBACCO ROOT MOUNTAINS

Although all workers in the Tobacco Root Mountains have recognized the distinctive character of the Spuhler Peak Metamorphic Suite, it has been interpreted to lie unconformably on the Indian Creek and Pony–Middle Mountain Metamorphic Suites (Gillmeister, 1971, 1972) or to be in fault contact with the Indian Creek and Pony–Middle Mountain Metamorphic Suites (Burger, 1966, 1967, 1969). Past and current workers agree that the Spuhler Peak Metamorphic Suite structurally overlies both the Indian Creek and the Pony–Middle Mountain Metamorphic Suites. Burger (1966, 1967, 1969) based his interpretation on the apparent truncation of foliation trends in the Indian Creek Metamorphic Suite by the Spuhler Peak Metamorphic Suite and what he perceived as different metamorphic histories between the two suites of rocks. He drew his fault contact along a zone of high strain separating the two units. Gillmeister (1971, 1972) based his view on the lateral persistence of quartzite that he observed to be everywhere present at the contact, and on the abundance of metamorphosed mafic dikes and sills that crosscut gneissic banding in the Indian Creek and Pony–Middle Mountain Metamorphic Suites and their absence in the Spuhler Peak Metamorphic Suite.

Our current study, and those of other workers, have both confirmed and questioned various details of this reasoning. As far as can be determined, the metamorphic histories of the Spuhler Peak, Indian Creek, and Pony–Middle Mountain Metamorphic Suites appear to be similar (see Cheney et al., 2004a, this volume, Chapter 6). Zones of high strain (intense folding) can be observed in many places at the contact between the Spuhler Peak Metamorphic Suite and the Indian Creek/Pony–Middle Mountain Metamorphic Suites. Shear zones paralleling the contact, but within the Spuhler Peak Metamorphic Suite, have been mapped by Cummings and McCulloch (1992). Cummings and McCulloch (1992, p. 328) also note foliation trends truncated by shear zones within the Spuhler Peak Metamorphic Suite. Metamorphosed mafic dikes and sills are abundant in the Indian Creek and Pony–Middle Mountain Metamorphic Suites within meters of the Spuhler Peak Metamorphic Suite, but never have been observed in it, despite concerted searches for their presence. Quartzite commonly is present at the contact between the Spuhler Peak Metamorphic Suite and Indian Creek/Pony–Middle Mountain Metamorphic Suites, but not invariably so, and when present, does possess compositional variations. Also, if the Spuhler Peak Metamorphic Suite lies unconformably on the Indian Creek and Pony–Middle Mountain Metamorphic Suites, one might reasonably expect to observe crosscutting rocks representing conduits for the thick mafic-ultramafic sequences in the Spuhler Peak Metamorphic Suite, but such rocks have never been recognized. Although we believe the sum total of these observations weigh in favor of emplacement of the Spuhler Peak Metamorphic Suite on the Indian Creek and Pony–Middle Mountain Metamorphic Suites, they do not conclusively disprove the presence of an unconformity. Indeed, there is no direct evidence that the Spuhler Peak Metamorphic Suite is Archean.

The best evidence supporting a fault origin for the Spuhler Peak/Indian Creek/Pony–Middle Mountain Metamorphic Suite contact comes from radiometric dating. Detrital zircon grains from Spuhler Peak Metamorphic Suite quartzites have an Archean provenance (Mueller et al., 2004, this volume, Chapter 9). Monazite ages reveal a widespread tectonothermal event at ca. 2450 Ma that affected the Indian Creek and Pony–Middle Mountain Metamorphic Suites. If the Spuhler Peak Metamorphic Suite is Archean in age and was originally deposited on the Indian Creek/Pony–Middle Mountain Metamorphic Suite, they should have shared the 2450 Ma tectonothermal event, but this event has not been recorded by monazite growth in the Spuhler Peak Metamorphic Suite (see Cheney et al., 2004b, this volume, Chapter 8). The petrographic, geochemical, and structural results of this study suggest that the Spuhler Peak Metamorphic Suite was emplaced following the 2450 Ma event and, therefore, most likely during the Big Sky orogeny (an intense tectonothermal event at 1780–1720 Ma (Brady et al., 2004, this volume, Chapter 7; Cheney et al., 2004b, this volume, Chapter 8; and Harms et al., 2004, this volume, Chapter 11).

Our interpretation is, therefore, that the Spuhler Peak Metamorphic Suite represents a shallow marine sequence of mafic (and possibly ultramafic) volcanics intercalated with cherty metasedimentary rocks that was brought into fault contact with the Indian Creek and Pony–Middle Mountain Metamorphic Suites during the Big Sky orogeny.

ACKNOWLEDGMENTS

Funding from the W.M. Keck Foundation supported the authors for three field seasons in the Tobacco Root Mountains. We are grateful for this support. We also thank John Valley for access to the University of Wisconsin Stable Isotope Laboratory, and Mike Spicuzza for laboratory assistance. Thoughtful reviews by Michael Smith and Darrell Henry are also acknowledged.

REFERENCES CITED

Araujo, S.M., Scott, S.D., and Longstaffe, F.J., 1996, Oxygen isotope composition of alteration zones of highly metamorphosed volcanogenic massive sulfide deposits, Geco, Canada, and Palmeirópolis, Brazil: Economic Geology, v. 91, p. 697–712.

Archuleta, L.L., 1994, The mineralogy and petrology of garnet-amphibole rocks from the Spuhler Peak Formation along Thompson Peak ridge, Tobacco Root Mountains, SW Montana [senior thesis]: Claremont, California, Pomona College, 84 p.

Arndt, N.T., and Nisbet, E.G., 1982, What is a komatiite?, *in* Arndt, N.T., and Nisbet, E.G., eds., Komatiites: London, George Allen & Unwin, p. 19–27.

Bachinski, D.J., 1978, Sulfur isotopic composition of thermally metamorphosed cupriferous iron sulfide ores associated with cordierite-anthophyllite rocks, Gull Pond, Newfoundland: Economic Geology, v. 73, p. 64–72.

Barley, M.E., Dunlop, J.S.R., Glover, J.E., and Groves, D.I., 1979, Sedimentary evidence for an Archaean shallow-water volcanic-sedimentary facies, eastern Pilbara block, western Australia: Earth and Planetary Science Letters, v. 43, p. 74–84.

Brady, J.B., Kovaric, D.N., Cheney, J.T., Jacob, L.J., and King, J.T., 2004, $^{40}Ar/^{39}Ar$ ages of metamorphic rocks from the Tobacco Root Mountains region, Montana, *in* Brady, J.B. et al, eds., Precambrian geology of the Tobacco Root Mountains, Montana: Boulder, Colorado, Geological Society of America Special Paper 377, p. 131–149 (this volume).

Burger, H.R., 1966, Structure, petrology, and economic geology of the Sheridan district, Madison County, Montana [Ph.D. thesis]: Bloomington, Indiana University, 156 p.

Burger, H.R., 1967, Bedrock geology of the Sheridan district, Madison County, Montana: Montana Bureau of Mines and Geology Memoir 41, 22 p.

Burger, H.R., 1969, Structural evolution of the southwestern Tobacco Root Mountains, Montana: Geological Society of America Bulletin, v. 80, p. 1329–1342.

Burger, H.R., 2004, General geology and tectonic setting of the Tobacco Root Mountains, Montana, *in* Brady, J.B., et al., eds., Precambrian geology of the Tobacco Root Mountains, Montana: Boulder, Colorado, Geological Society of America Special Paper 377, p. 1–14 (this volume).

Cady, P., 1994, Petrographic and geochemical analysis of amphibolites from the Tobacco Root Mountains, Montana [senior thesis]: Colorado Springs, Colorado College, 86 p.

Cattell, A.C., and Taylor, R.N., 1990, Archaean basic magmas, *in* Hall, R.P., and Hughes, D.J., eds., Early Precambrian basic magmatism: Glasgow and London, Blackie & Son, p. 36–37.

Cheney, J.T., Brady, J.B., Tierney, K.A., DeGraff, K.A., Mohlman, H.K., Frisch, J.D., Hatch, C.E., Steiner, M.L., Carmichael, S.K., Fisher, R.G.M., Tuit, C.B., Steffen, K.J., Cady, P., Lowell, J., Archuleta, L.L., Hirst, J., Wegmann, K.W, and Monteleone, B., 2004a, Proterozoic metamorphism of the Tobacco Root Mountains, Montana, *in* Brady, J.B. et al., eds. , Precambrian geology of the Tobacco Root Mountains, Montana: Boulder, Colorado, Geological Society of America Special Paper 377, p. 105–129 (this volume, Chapter 6).

Cheney, J.T., Webb, A.A.G., Coath, C.D., and McKeegan, K.D., 2004b, In situ ion microprobe $^{207}Pb/^{206}Pb$ dating of monazite from Precambrian metamorphic suites, Tobacco Root Mountains, Montana, *in* Brady, J.B., et al., eds., Precambrian geology of the Tobacco Root Mountains, Montana: Boulder, Colorado, Geological Society of America Special Paper 377, p. 151–179 (this volume).

Cole, D.R., Mottl, M.J., and Ohmoto, H., 1987, Isotope exchange in mineral-fluid systems, II: Oxygen and hydrogen isotopic investigation of the basalt-seawater system: Geochimica et Cosmochimica Acta, v. 51, p. 1523–1538.

Condie, K.C., 1990, Geochemical characteristics of Precambrian basaltic greenstones, *in* Hall, R.P., and Hughes, D.J., eds., Early Precambrian basic magmatism: Glasgow and London, Blackie & Son, p. 45.

Cox, M.J., 1996, Geochemistry and petrography of amphibolites from the Spuhler Peak Metamorphic Suite near the Branham Lakes area, Tobacco Root Mountains, southwestern Montana [senior thesis]: Lexington, Virginia, Washington and Lee University, 64 p.

Cummings, M.L., and McCulloch, W.R., 1992, Geochemistry and origin of amphibolite and ultramafic rocks, Branham Lakes area, Tobacco Root Mountains, southwestern Montana, *in* Bartholomew, M.J., et al., eds., Basement tectonics 8: Characterization and comparison of ancient and Mesozoic continental margins—Proceedings of the 8th International Conference on Basement Tectonics (Butte, Montana, 1988): Dordrecht, Netherlands, Kluwer Academic Publishers, p. 323–340.

Dymek, R.F., and Smith, M.S., 1990, Geochemistry and origin of Archaean quartz-cordierite gneisses from the Gothåbsfjord region, West Greenland: Contributions to Mineralogy and Petrology, v. 105, p. 715–730.

Friberg, N., 1976, Petrology of a metamorphic sequence of upper-amphibolite facies in the central Tobacco Root Mountains, southwestern Montana [Ph.D. thesis]: Bloomington, Indiana University, 146 p.

Gillmeister, N.M., 1971, Petrology of Precambrian rocks in the central Tobacco Root Mountains, Madison Country, Montana [Ph.D. thesis]: Cambridge, Massachusetts, Harvard University, 210 p.

Gillmeister, N.M., 1972, Cherry Creek Group–Pony Group relationship in the central Tobacco Root Mountains, Madison County, Montana: Northwest Geology, v. 1, p. 21–24.

Giral, S., Savin, S.M., Girard, J.P., and Nohon, D.B., 1993, The oxygen isotope geochemistry of kaolinites from lateritic profiles: Implications for pedology and paleoclimatology: Chemical Geology, v. 107, p. 237–240.

Green, G.R., Ohmoto, H., Date, J., and Takahashi, T., 1983, Whole-rock oxygen isotope distribution in the Fukazawa-Kosaka area, Hokuroku District, Japan, and its potential application to mineral exploration, *in* Ohmoto, H., and Skinner, B.J., eds., The Kuroko and related volcanogenic massive sulfide deposits: Economic Geology Monograph, v. 5, p. 395–411.

Hajash, A., and Chandler, G.W., 1981, An experimental investigation of high-temperature interactions between seawater and rhyolite, andesite, basalt and peridotite: Contributions to Mineralogy and Petrology, v. 78, p. 240–254.

Harms, T.A., Brady, J.B., Burger, H.R., and Cheney, J.T., 2004, Advances in the geology of the Tobacco Root Mountains, Montana, and their implications for the history of the northern Wyoming province, *in* Brady, J.B., et al., eds., Precambrian geology of the Tobacco Root Mountains, Montana: Boulder, Colorado, Geological Society of America Special Paper 377, p. 227–243 (this volume).

Herzberg, C.T., 1993, Lithosphere peridotites of the Kaapvaal craton: Earth and Planetary Science Letters, v. 120, p. 13–29.

Irvine, T.N., and Baragar, W.R.A., 1971, A guide to the chemical classification of the common volcanic rocks: Canadian Journal of Earth Sciences, v. 8, p. 523–548.

Jensen, L.S., 1976, A new cation plot for classifying subalkalic volcanic rocks: Ontario Division Mines Miscellaneous Paper 66, 22 p.

Johnson, K.E., Brady, J.B., MacFarlane, W.A., Thomas, R.B., Poulsen, C., and Sincock, M.J., 2004, Precambrian meta-ultramafic rocks from the Tobacco Root Mountains, Montana, *in* Brady, J.B., et al., eds., Precambrian geology of the Tobacco Root Mountains, Montana: Boulder, Colorado, Geological Society of America Special Paper 377, p. 71–87 (this volume).

Land, L.S., and Lynch, F.L., 1996, $\delta^{18}O$ values of mudrocks: More evidence for an ^{18}O-buffered ocean: Geochimica et Cosmochimica Acta, v. 60, p. 3347.

Leake, B.E., 1964, The chemical distinction between ortho- and para-amphibolites: Journal of Petrology, v. 5, p. 238Leake, B.E., 1964, The chemical distinction between ortho- and para-amphibolites: Journal of Petrology, v. 5, p. 238LeBas, M.J., LeMaitre, R.W., Streckeisen, A., and Zanettin, B., 1986, A chemical classification of volcanic rocks based on the total alkali silica diagram: Journal of Petrology, v. 27, p. 745–750.

Lowell, J., 1994, A metamorphic and deformational history of the Spuhler Peak Formation, Tobacco Root Mountains, Montana [senior thesis]: Colorado Springs, Colorado College, 100 p.

MacFarlane, W.A., 1996, A geochemical and petrographic analysis of the meta-ultramafic rocks in the Tobacco Root Mountains, southwest Montana [senior thesis]: Colorado Springs, Colorado College, 132 p.

Mertzman, S.A., 2000, K-Ar results from the southern Oregon–northern California Cascade Range: Oregon Geology, v. 62, no. 4, p. 99–122.

Middlemost, E.A.K., 1975, The basalt clan: Earth Science Reviews, v. 11, p. 337–364.

Muehlenbachs, K., 1986, Alteration of the ocean crust and the ^{18}O history of seawater, *in* Valley, J.W., et al., eds., Stable isotopes in high temperature geological processes: Reviews in Mineralogy, v. 16, p. 425–444.

Mueller, P.A., Burger, H.R., Wooden, J.L., Heatherington, A.L., Mogk, D.W., and D'Arcy, K., 2004, Age and evolution of the Precambrian crust of the Tobacco Root Mountains, Montana, *in* Brady, J.B., et al., eds., Precambrian geology of the Tobacco Root Mountains, Montana: Boulder, Colorado, Geological Society of America Special Paper 377, p. 181–202 (this volume).

Mullen, E.D., 1983, $MnO/TiO_2/P_2O_5$: A minor element discriminant for basaltic rocks of oceanic environments and its implications for petrogenesis: Earth and Planetary Science Letters, v. 62, p. 53–62.

Pan, Y., and Fleet, M.E., 1995, Geochemistry and origin of cordierite-orthoamphibole gneiss and associated rocks at an Archean volcanogenic massive sulphide camp, Manitouwadge, Ontario, Canada: Precambrian Research, v. 74, p. 73–89.

Pearce, J.A., 1983, Role of sub-continental lithosphere in magma genesis at active continental margins, *in* Hawkesworth, C.J., and Norry, M.J., eds., Continental basalts and mantle xenoliths: Nantwich, UK, Shiva, p. 230–249.

Pearce, J.A., 1996, A user's guide to basalt discrimination diagrams, *in* Wyman, D.A., ed., Trace element geochemistry of volcanic rocks: Applications for massive sulfide exploration: Geological Association of Canada Short Course Notes, v. 12, p. 79–114.

Pearce, J.A., and Cann, J.R., 1973, Tectonic setting of basic volcanic rocks determined using trace element analyses: Earth and Planetary Science Letters, v. 19, p. 290–300.

Peck, W.H., 1994, Geochemistry and petrology of Ca-poor amphibolites, Tobacco Root Mountains, southwestern Montana, [senior thesis]: Beloit, Wisconsin, Beloit College, 43 p.

Peck, W.H., and Valley, J.W., 1996, The Fiskenaesset Anorthosite Complex: Stable isotope evidence for shallow emplacement into Archean ocean crust: Geology, v. 24, p. 523–526.

Peck, W.H., and Valley, J.W., 2000, Genesis of cordierite-gedrite gneisses, Central Metasedimentary Belt boundary thrust zone, Grenville Province, Ontario, Canada: Canadian Mineralogist, v. 38, p. 511–524.

Poulsen, C.J., 1994, Origin, metamorphic history, and tectonic setting of Archean rocks from the Spuhler Peak Assemblage, Tobacco Root Mountains, Montana [senior thesis]: Northfield, Minnesota, Carleton College, 54 p.

Pufall, A., 1996, Geochemistry of the Spuhler Peak and Pony–Middle Mountain Metamorphic Suites, Tobacco Root Mountains, Montana, *in* Mendelson, C.V., and Mankiewicz, C., eds., Ninth Keck research symposium in geology: Williamstown, Massachusetts, Williams College, p. 118–121.

Reid, R.R., 1957, Bedrock geology of the north end of the Tobacco Root Mountains, Madison County, Montana: Montana Bureau of Mines and Geology Memoir 36, 25 p.

Reinhardt, J., 1987, Cordierite-anthophyllite rocks from northwest Queensland, Australia: metamorphosed magnesian pelites: Journal of Metamorphic Geology, v. 5, p. 451–472.

Rhodes, J.M., 1996, Geochemical stratigraphy of lava flows sampled by the Hawaiian Scientific Drilling Project: Journal of Geophysical Research, v. 101, p. 11,729–11,746.

Schumacher, J.C., 1988, Stratigraphy and geochemistry of the Ammonoosuc volcanics, central Massachusetts and southwestern New Hampshire: American Journal of Science, v. 288, p. 619–663.

Seki, Y., 1957, Petrological study of hornfelses in the central part of the median zone of the Kitakami Mountainland, Iwate Prefecture: Scientific Reports Saitama University, Series B, v. 2, p. 307–373.

Shandl, E., Gorton, M.P., and Wasteneys, H.A., 1995, Rare earth element geochemistry of the metamorphosed volcanogenic massive sulphide deposits of the Manitouwadge mining camp, Superior Province, Canada: A potential exploration tool?: Economic Geology, v. 90, p. 1217–1236.

Shiraki, R., Sakai, H., Endoh, M., and Kishima, N., 1987, Experimental studies on rhyolite- and andesite-seawater interactions at 300 °C and 1000 bars: Geochemical Journal, v. 21, p. 139–148.

Sincock, M.J., 1994, Structural and metamorphic history of an intensely deformed zone, lower Branham Lake, southern Tobacco Root Mountains, Montana [senior thesis]: Lancaster, Pennsylvania, Franklin and Marshall College, 31 p.

Smith, J.V., 1974, Feldspar minerals, Volume 2: Chemical and textural properties: New York, Springer Verlag, 690 p.

Smith, M.S., Dymek, R.F., and Chadwick, B., 1992a, Petrogenesis of Archaean Malene supracrustal rocks, NW Buksefjorden region, West Greenland: Geochemical evidence for highly evolved Archaean crust: Precambrian Research, v. 57, p. 49–90.

Smith, M.S., Dymek, R.F., and Schneiderman, J.S., 1992b, Implications of trace element geochemistry for the origin of cordierite-orthoamphibole rocks from Orijärvi, SW Finland: Journal of Geology, v. 100, p. 545–559.

Spicuzza, M.J., Valley, J.W., and McConnell, V.S., 1998, Oxygen isotope analysis of whole rock via laser fluorination: An air-lock approach: Geological Society of America Abstracts with Programs, v. 30, no. 7, p. 80.

Sun, S.S., 1980, Lead isotopic study of young volcanic rocks from mid-ocean ridges, ocean islands and island arcs: Royal Society of London Philosophical Transactions, ser. A: Mathematical and Physical Sciences, v. 297, p. 409–445.

Tansley, W., Schafer, P.A., and Hart, L.H., 1933, A geological reconnaissance of the Tobacco Root Mountains, Madison County, Montana: Montana Bureau of Mines and Geology Memoir 9, 57 p.

Thompson, G., 1991, Metamorphic and hydrothermal processes: basalt-seawater interaction, *in* Floyd, P.A., ed., Oceanic basalts: New York, Van Nostrand Reinhold, p. 149–171.

Tierney, K.A., 1994, The origin and evolution of the Spuhler Peak Formation along the western ridge of Thompson Peak, Tobacco Root Mountains, Montana [senior thesis]: Amherst, Massachusetts, Amherst College, 87 p.

Vallance, T.G., 1967, Mafic rock alteration and isochemical development of some cordierite-anthophyllite rocks: Journal of Petrology, v. 8, p. 84–96.

Valley, J.W., Kitchen, N., Kohn, M.J., Niendorf, C.R., and Spicuzza, M.J., 1995, UWG-2, a garnet standard for oxygen isotope ratios: Strategies for high precision and accuracy with laser heating: Geochimica et Cosmochimica Acta, v. 59, p. 5523–5531.

Warren, R.G., 1979, Sapphirine-bearing rocks with sedimentary and volcanogenic protoliths from the Arunta Block: Nature, v. 278, p. 159–161.

Wilson, A.F., and Baksi, A.K., 1983, Widespread ^{18}O depletion in some Precambrian granulites of Australia: Precambrian Research, v. 23, p. 33–56.

Zaleski, E., and Peterson, V.L., 1995, Depositional setting and deformation of massive sulphide deposits, iron-formation, and associated alteration in the Manitouwadge Greenstone Belt, Superior Province, Ontario: Economic Geology, v. 90, p. 2244–2261.

Manuscript Accepted by the Society September 12, 2003

Printed in the USA

Geological Society of America
Special Paper 377
2004

Precambrian meta-ultramafic rocks from the Tobacco Root Mountains, Montana

Kathleen E. Johnson*
Department of Geology and Geophysics, University of New Orleans, New Orleans, Louisiana 70148, USA

John B. Brady
Department of Geology, Smith College, Northampton, Massachusetts 01063, USA

William A. MacFarlane
Colorado College, Colorado Springs, Colorado 80903, USA

Rebecca B. Thomas
Williams College, Williamstown, Massachusetts 01267, USA

Chris J. Poulsen
Carleton College, Northfield, Minnesota 55057, USA

M. Jennifer Sincock
Franklin & Marshall College, Lancaster, Pennsylvania 17604-3003, USA

ABSTRACT

Meta-ultramafic rocks occur as small (2 to 100 m long), podiform bodies in all three major Precambrian rock suites of the Tobacco Root Mountains of southwest Montana. Most samples consist of a randomly oriented, coarse-grained assemblage of orthopyroxene, olivine, and magnesiohornblende ± spinel, partially replaced by a fine-grained assemblage that may include anthophyllite, talc, cummingtonite, magnesiohornblende, chlorite, serpentine, and/or magnetite. Blackwall reaction zones of anthophyllite, actinolite, chlorite, and/or biotite surround several of the meta-ultramafic bodies. The absence of clinopyroxene with orthopyroxene limits the metamorphic pressure-temperature history of these rocks to temperatures below ~800 °C. The presence of anthophyllite with olivine indicates that these rocks passed through 650–700 °C at pressures below 0.6 GPa. These constraints are consistent with the detailed pressure-temperature path determined for the surrounding upper-amphibolite to granulite facies metamorphic rocks. Whole-rock chemical analyses of the meta-ultramafic rocks show them to be rich in SiO_2 (44–54 wt%) and poor in MgO (21–34 wt%) relative to mantle peridotites and typical Alpine-type ultramafic rocks. Rare earth element concentrations are all enriched relative to chondritic values and the light rare earth elements are especially enriched (10–30 times), inconsistent with either an upper-mantle or a komatiitic origin. All samples have similar TiO_2/Zr ratios, suggesting that they have a common or related origin, and that TiO_2 and Zr were conserved in the processes that led to the observed chemical variations. Together, the chemical data point to a protolith that was an ultramafic cumulate rich in orthopyroxene, and therefore probably formed in a continental setting from a basaltic magma enriched in silica. One possible time of origin is a magmatic event during the continental rifting at 2.06 Ga that led to the intrusion of a suite of mafic dikes.

Keywords: Big Sky orogeny, orthopyroxene, cumulate.

*Present address: P.O. Box 8352, Monterey, California 93940, USA, volcan@mac.com.

Johnson, K.E., Brady, J.B., MacFarlane, W.A., Thomas, R.B., Poulsen, C.J., and Sincock, M.J., 2004, Precambrian meta-ultramafic rocks from the Tobacco Root Mountains, Montana, *in* Brady, J.B., Burger, H.R., Cheney, J.T., and Harms, T.A., eds., Precambrian geology of the Tobacco Root Mountains, Montana: Boulder, Colorado, Geological Society of America Special Paper 377, p. 71–87. For permission to copy, contact editing@geosociety.org.

INTRODUCTION

Lenses of deformed and metamorphosed ultramafic rocks occur within many Archean and Proterozoic supracrustal terranes (e.g., Paktunc, 1984; Srikantappa et al., 1984; Wilson and Hyndman, 1990; Occhipinti et al., 1998; Kalsbeek and Manatschal, 1999). The original character and origin of these meta-ultramafic rocks is largely obscured by the metamorphism to which they have been subjected, their tectonic disruption, and their isolation within more siliceous rocks. In some cases, relict grains from the original mineral assemblages suggest a parent material of mantle peridotite composition. In other cases, spinifex textures point to a komatiitic origin. Precambrian rocks of the Tobacco Root Mountains and the Ruby Range of the Wyoming province include a number of such ultramafic bodies (e.g., Merrill, 1895; Heinrich, 1963; Root, 1965; Burger, 1966, 1967; Hess, 1967; Gillmeister, 1972, 1976; Cordua, 1973; Hanley, 1975; Friberg, 1976; Tendall, 1978; Garihan, 1979; Desmarais, 1981; Cummings and McCulloch, 1992; MacFarlane, 1996a, 1996b; Thomas, 1996a, 1996b) whose tectonic and petrologic significance have been elusive.

A variety of hypotheses have been proposed for the origin of the meta-ultramafic pods in the Tobacco Root Mountains and adjacent Ruby Range. They have been interpreted as (1) forcibly emplaced hydrous peridotitic magmas (Heinrich, 1963); (2) "Alpine-type ultramafic intrusives" (Cordua, 1973); (3) "cumulates of a differentiating basaltic magma" emplaced in the solid state (Tendall, 1978); (4) tectonically emplaced, cold ultramafic bodies that were serpentinized prior to metamorphism (Desmarais, 1981); (5) "a sliver of oceanic crust generated during ensialic basin development" (Cummings and McCulloch, 1992); and (6) metamorphosed komatiite lavas or intrusions (MacFarlane, 1996b; Johnson et al., 1996). In this paper we present new geochemical and mineralogic data that place important constraints on the origin and history of the Tobacco Root meta-ultramafic rocks. We use these new data to argue for an igneous cumulate origin and, therefore, tectonic emplacement of these rocks, similar to the preferred interpretation of Tendall (1978) based on modal data.

FIELD RELATIONS

Meta-ultramafic rocks of the Tobacco Root Mountains occur as small bodies (2–100 m across), typically with lens-shaped map patterns, hosted by three distinct units: (1) the Spuhler Peak Metamorphic Suite—a mafic unit that consists principally of amphibolite and Ca-depleted amphibolite, with lesser quartzite and aluminous schist; (2) the Indian Creek Metamorphic Suite—quartzofeldspathic gneiss and amphibolite interlayered with many rocks that are clearly metasediments, including quartzite, marble, iron formation, and aluminous schist; and (3) the Pony–Middle Mountain Metamorphic Suite—principally quartzofeldspathic gneiss and hornblende amphibolite (see Burger, 2004, this volume, Chapter 1). Although all three suites contain meta-ultramafic pods, a significant portion of the bodies occurs within the Spuhler Peak Metamorphic Suite (Fig. 1). A number of the meta-ultramafic bodies, particularly those well exposed in the high-elevation Branham Lakes area, are flattened subparallel to the surrounding foliation. Other meta-ultramafic bodies are more equant in form and have contacts that make a high angle to the surrounding foliation. In many cases, the contacts are not exposed, as if the contact were a site of enhanced weathering. Most bodies are unfoliated and show little evidence of internal deformation. Where present, foliation tends to be parallel to that in the surrounding gneisses. Most meta-ultramafic rocks within the Indian Creek and Pony–Middle Mountain Metamorphic Suites occur at lower elevations as tree-covered hills surrounded by low-lying felsic gneisses that support only grass. Here the structural relations are obscure because outcrop is poor and ultramafic rock–gneiss contacts are not exposed. Additional field data concerning the meta-ultramafic rocks can be found in a number of theses on the Tobacco Root Mountains (Root, 1965; Burger, 1966, 1967; Hess, 1967; Gillmeister, 1972; Cordua, 1973; Hanley, 1975; Friberg, 1976; Tendall, 1978; Poulsen, 1994; Sincock, 1994; MacFarlane, 1996b; Thomas, 1996b).

Most Tobacco Root meta-ultramafic rocks have a coarse texture with prominent prismatic crystals. Randomly oriented, dark green amphibole blades 1–3 cm long highlight many samples. Where more weathered, the amphibole grains have a silvery cast and are set in a rusty matrix. A few outcrops contain large (3–6 cm), tan orthopyroxene crystals (Fig. 2) that resist weathering to form a knobby texture. An outcrop on the ridge west of Branham Lakes exhibits an unusual texture of radiating amphibole sprays up to 3 cm across and is cut by a small pegmatite. Serpentine partially replaces olivine and orthopyroxene in many bodies. The largest ultramafic body (>100 m across), which forms a tree-covered knob at the end of Granite Creek Road in the southern Tobacco Root Mountains, has a core of relatively fresh peridotite surrounded by a zone of serpentinite. Several meta-ultramafic bodies have "blackwall" reaction zones (see Sanford, 1982) along contacts with the enclosing gneisses or amphibolites. Blackwall sequences up to 0.5 m in width consist principally of anthophyllite, actinolite, chlorite, and/or biotite, commonly in near monomineralic bands. The chlorite and biotite reaction zones are typically foliated parallel to the contact. Folds were observed in a chloritic blackwall in Beuler Cirque.

Many ultramafic bodies in orogenic belts have been tectonically emplaced, and that is our interpretation for the Tobacco Root meta-ultramafic bodies (see below). This means that they are surrounded by faults. However, we have not been able to trace the faults because of the extensive deformation associated with the Big Sky orogeny that rotated and obscured many primary structures (Harms et al., 2004, this volume, Chapter 10). However, even if they were originally komatiitic lava flows or intrusions, or had another origin, their present form and distribution indicate that the meta-ultramafic units have been tectonically disrupted, as if they were stiff bodies in a more ductile matrix. Cummings and McCulloch (1992) observed that the Branham Lakes meta-ultramafic rocks are associated with shear zones

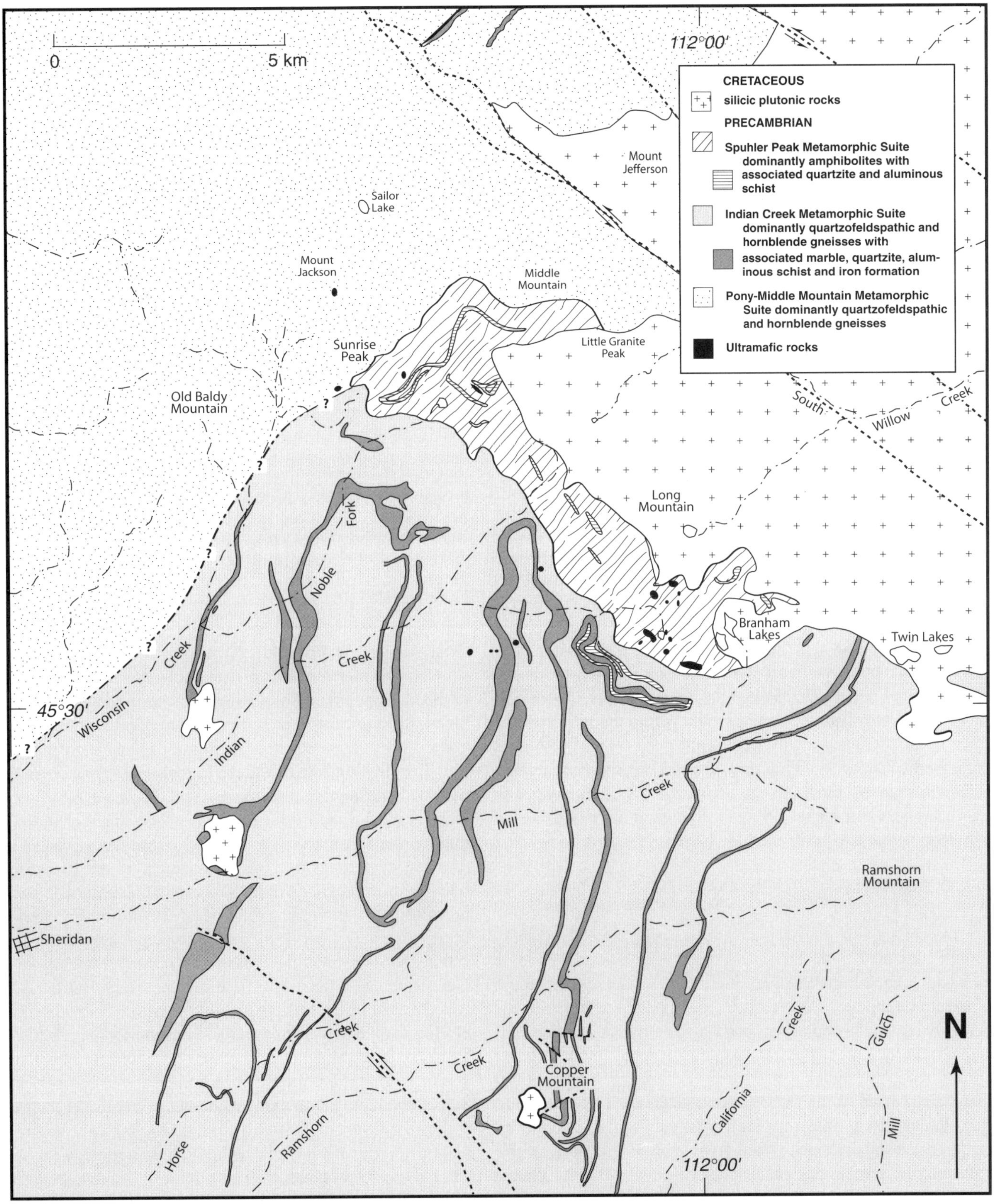

Figure 1. Geologic map of the central portion of the Tobacco Root Mountains, modified from Vitaliano et al. (1979). Meta-ultramafic bodies in and around the Spuhler Peak Metamorphic Suite are shown in black. Additional meta-ultramafic bodies in the northern and southern parts of the range are displayed on the original Vitaliano et al. (1979) map (see reprinted map and text accompanying this volume).

Figure 2. An outcrop of meta-ultramafic rock near Branham Lakes showing elongated, tan, weathered, orthopyroxene crystals in a matrix of gray green magnesiohornblende. The pocketknife in the photo is 8.5 cm long.

Figure 3. Photomicrograph in cross-polarized light of a Spuhler Peak Metamorphic Suite–hosted meta-ultramafic rock that contains large (>1 cm) grains of orthopyroxene. The left part of the field of view consists of the amphiboles magnesiohornblende (darker) and anthophyllite (lighter). They are replacing a large orthopyroxene (Opx) crystal. Many magnesiohornblende (Mg-Hbl) crystals are zoned and have tremolite rims. Chlorite (Chl) is also visible. Sample JBB-93-2 from west of Branham Lakes.

within the Spuhler Peak Metamorphic Suite. Similar structural characteristics were observed by Desmarais (1981) for the meta-ultramafic pods of the adjacent Ruby Range, which we believe experienced a metamorphic history similar to the Tobacco Root Mountains. He described boudin-shaped bodies that are concordant with respect to deformational structures of the host gneisses, and that most commonly occur within zones of intense plastic deformation. In cases where there is structural discordance between meta-ultramafic rocks and host gneisses, Desmarais (1981) suggested tectonic emplacement within the noses of isoclinal folds. The occurrence of ultramafic pods within fold hinges or as apparent boudins within shear zones is a common feature in other Precambrian terranes (e.g., Paktunc, 1984; Srikantappa et al., 1984; Jahn and Zhang, 1984; Occhipinti et al., 1998; Rosing and Rose, 1993; Kalsbeek and Manatschal, 1999).

PETROGRAPHY

In thin section the meta-ultramafic rocks typically display complex textures, with fine-grained, generally hydrous minerals such as magnesiohornblende, tremolite, anthophyllite, magnesiocummingtonite, talc, chlorite, phlogopite, and serpentine overgrowing and/or replacing coarse orthopyroxene, olivine, and magnesiohornblende. Spinel and magnetite are minor constituents of most samples. Figure 3 shows a typical texture in cross-polarized light. Much of the field of view in Figure 3 is filled with part of a cm-sized orthopyroxene crystal (dark gray) covered by randomly oriented elongate or fibrous amphiboles. Many of the light-colored amphiboles are anthophyllite, whereas the darker ones in this cross-polarized image are calcic amphiboles that are rich in magnesium (magnesiohornblende; see Table 1). Most of the minerals are colorless or very pale, reflecting low iron contents. Pale green magnesiohornblende grains typically have color and interference-color bands, reflecting their chemical zonation from magnesiohornblende (core) toward tremolite (rim). Figure 4 shows a simpler texture, in plane-polarized light, consisting of rounded, light-colored olivine crystals set among magnesiohornblende crystals of various sizes. This appears to be an equilibrium texture, not a replacement of high-grade minerals during cooling.

Textures of the meta-ultramafic rocks possess features in common with metamorphic rocks of other compositions that surround them in the Tobacco Root Mountains (see Cheney et al., 2004a, this volume, Chapter 6). Remnants of a coarse-textured, high-*P-T* (pressure-temperature) mineral assemblage are overgrown by finer-grained minerals produced during decompression and cooling. Based on grain size, the high-*P-T* assemblage would have been principally orthopyroxene, olivine, magnesiohornblende, and spinel. We found no evidence of clinopyroxene or garnet in this assemblage; nor were these minerals found in the adjacent Ruby Range by Desmarais (1981). However, a few occurrences of clinopyroxene or garnet have been reported by others for the Tobacco Root Mountains (see next section). Anthophyllite is common among the fine-grained minerals; talc is less abundant. Serpentine is not particularly abundant, except in a few ultramafic bodies. Evidence of deformation of orthopyroxene and olivine in the meta-ultramafic samples is limited, with most fabric, if present, defined by amphiboles. This contrasts with Alpine-type peridotites that retain features of intracrystalline deformation induced during emplacement. We think that the absence of deformation textures is evidence that these rocks were completely recrystallized during metamorphism of the Tobacco Root terrane.

TABLE 1. CHEMICAL COMPOSITIONS OF MINERALS IN META-ULTRAMAFIC ROCKS

Sample	KEJ-20	KEJ-20	KEJ-20	KEJ-20	KEJ-20	KEJ-20	KEJ-20	KEJ-20	TBR-68b	TBR-68b	TBR-68b
Host	SPMS	SPMS	SPMS	SPMS	SPMS	SPMS	SPMS	SPMS	SPMS	SPMS	SPMS
12T	042123E	042123E	042123E	042123E	042123E	042123E	042123E	042123E	042125E	042125E	042125E
UTM	504063N	504063N	504063N	504063N	504063N	504063N	504063N	504063N	504058N	504058N	504058N
Mineral	Ol	Ol	Opx	Opx	Ant	Ant	Tr	Chl	Opx	Ant	Mg-Hbl
Oxide wt%											
SiO_2	39.64	39.65	55.72	53.70	55.96	58.52	58.70	32.02	54.14	59.93	51.44
Al_2O_3	0.03	0.00	0.86	0.79	0.41	0.51	0.12	15.93	2.93	0.09	7.00
TiO_2	0.00	0.05	0.02	0.00	0.07	0.02	0.04	0.03	0.07	0.00	0.32
MgO	42.63	42.30	30.99	31.57	28.41	27.38	23.46	31.71	28.74	27.04	20.20
*FeO	17.77	17.82	12.15	12.68	11.79	10.79	3.88	6.05	12.34	9.88	6.29
MnO	0.70	0.79	0.76	0.84	0.96	0.96	0.43	0.12	0.68	0.91	0.43
CaO	0.00	0.04	0.20	0.20	0.30	0.43	11.42	0.00	0.22	0.53	10.59
Na_2O	0.14	0.18	0.21	0.14	0.08	0.28	0.12	0.11	0.19	0.29	1.06
K_2O	0.00	0.00	0.00	0.00	0.01	0.00	0.00	0.00	0.00	0.00	0.06
Total	100.92	100.83	100.91	99.91	98.00	98.90	98.17	85.97	99.32	98.68	97.39
Elements per Formula											
O	4	4	6	6	23	23	23	7	6	23	23
Si	1.00	1.00	1.96	1.92	7.74	7.95	8.01	1.55	1.94	8.10	7.21
Al	0.00	0.00	0.04	0.03	0.07	0.08	0.02	0.91	0.12	0.02	1.16
Ti	0.00	0.00	0.00	0.00	0.01	0.00	0.00	0.00	0.00	0.00	0.03
Mg	1.60	1.60	1.63	1.69	5.86	5.54	4.77	2.29	1.53	5.45	4.22
Fe	0.38	0.38	0.36	0.38	1.36	1.23	0.44	0.24	0.37	1.12	0.74
Mn	0.02	0.02	0.02	0.03	0.11	0.11	0.05	0.00	0.02	0.10	0.05
Ca	0.00	0.00	0.01	0.01	0.04	0.06	1.67	0.00	0.01	0.08	1.59
Na	0.01	0.01	0.01	0.01	0.02	0.07	0.03	0.01	0.01	0.08	0.29
K	0.00	0.00	0.00	0.01	0.00	0.00	0.00	0.00	0.00	0.00	0.01
Total Cations	3.00	3.00	4.02	4.07	15.22	15.05	15.00	5.00	4.01	14.93	15.30
Mg/(Mg+Fe)	0.81	0.81	0.82	0.82	0.81	0.82	0.92	0.90	0.81	0.83	0.85

(continued)

Chemical compositions of representative minerals from meta-ultramafic samples hosted by each of the three metamorphic suites are listed in Table 1. These data were collected at Amherst College using a Link/Oxford energy-dispersive X-ray spectrometer and mineral standards. Olivine compositions are typically Fo_{81-83} with associated orthopyroxene compositions slightly more magnesian (En_{82-84}). Amphiboles are also magnesium-rich with anthophyllite closely matching the Mg/(Mg + Fe) value (X_{Mg}) of adjacent olivine; magnesiohornblende has a higher X_{Mg} value, and tremolite is even more magnesian ($X_{Mg} \geq 0.90$). Where present, green chromian spinels have X_{Mg} of 0.41–0.55 and Cr/(Cr + Al) of 0.29–0.08. Most chlorites have low Al and a high X_{Mg} (≥0.90). The data in Table 1 are quite similar to mineral analyses reported by Tendall (1978) for the Tobacco Root Mountains and Desmarais (1981) for the Ruby Range.

METAMORPHISM

The observed coarse-grained mineral assemblage of orthopyroxene, olivine, and magnesiohornblende and the absence of clinopyroxene in our samples places an important constraint on the *P-T* path of these rocks. The "clinopyroxene-in" reaction

$$\mathrm{Fo + Tr = Di + En + H_2O}$$

for the Mg-rich portion of the $CaO-MgO-SiO_2-H_2O$ (CMSH) system is shown in Figure 5 (reaction 3), as calculated by Spear (1993) from the Berman (1988) thermodynamic database. Also shown in Figure 5 (dashed line at slightly higher temperature) is a CMSH version of this reaction calculated by Schmädicke (2000, Figure 12 therein) using the Holland and Powell (1998) database (1993 version). In either case, the maximum temperature during the high-pressure stage (≥1.0 GPa) is limited to ~800 °C by the absence of clinopyroxene with orthopyroxene.

Another constraint on the *P-T* path is given by the presence of post-peak anthophyllite. The bulk compositions determined for the Tobacco Root meta-ultramafic rocks (see next section) are shown in Figure 5 as black triangles on a composition diagram of the system diopside ($Ca_2Mg_2Si_4O_{12}$), quartz (Si_6O_{12}), forsterite ($Mg_6Si_3O_{12}$). Because most of the iron in the rocks is contained in the minerals on this diagram, FeO was added to MgO in calculating the plotting positions of the bulk compositions. An average value of these compositions is shown with an "X" on the other ternary phase assemblage diagrams to indicate the expected mineral assemblages. As is evident from Figure 5, anthophyllite

TABLE 1. CHEMICAL COMPOSITIONS OF MINERALS IN META-ULTRAMAFIC ROCKS (*continued*)

Sample	RBT-28	RBT-28	RBT-28	RBT-28	RBT-28	TBR-84	TBR-85	TBR-86	TBR-87	TBR-88	TBR-89
Host	PMMMS	PMMMS	PMMMS	PMMMS	PMMMS	ICMS	ICMS	ICMS	ICMS	ICMS	ICMS
12T	042803E	042803E	042803E	042803E	042803E	041663E	041663E	041663E	041663E	041663E	041663E
UTM	506327N	506327N	506327N	506327N	506327N	504011N	504011N	504011N	504011N	504011N	504011N
Mineral	Ol	Opx	Spl	Mg-Hbl	Chl	Opx	Spl	Spl	Mg-Hbl	Mg-Hbl	Chl
Oxide wt%											
SiO_2	39.51	55.97	0.11	51.70	42.12	55.20	0.21	0.14	52.99	52.70	50.40
Al_2O_3	0.06	2.28	53.87	6.12	0.76	1.90	53.75	52.25	5.90	7.03	4.00
TiO_2	0.00	0.17	0.05	0.58	0.02	0.07	0.08	0.05	0.35	0.57	0.07
MgO	43.29	31.15	14.66	19.98	30.66	29.61	13.11	12.04	20.25	19.96	31.52
*FeO	15.86	10.87	21.97	4.85	8.93	12.03	20.85	22.47	5.01	4.55	4.42
MnO	0.43	0.42	0.30	0.19	0.74	0.22	0.10	0.14	0.08	0.07	0.05
CaO	0.02	0.21	0.00	12.50	0.37	0.35	0.07	0.01	11.89	12.56	0.05
Na_2O	0.27	0.24	0.53	0.54	0.30	0.00	0.19	0.25	0.39	0.69	0.14
K_2O	0.00	0.00	0.02	0.20	0.03	0.00	0.00	0.00	0.12	0.09	0.07
Cr_2O_3	N.D.	N.D.	7.37	0.45	0.00	0.20	9.96	10.23	0.66	0.35	0.13
Total	99.44	101.32	98.88	97.12	83.93	99.58	98.32	97.58	97.65	98.57	90.85
Elements per Formula											
O	4	6	4	23	7	6	4	4	23	23	7
Si	1.00	1.95	0.00	7.26	2.10	1.96	0.01	0.00	7.37	7.27	2.22
Al	0.00	0.09	1.75	1.01	0.04	0.08	1.75	1.73	0.97	1.14	0.21
Ti	0.00	0.00	0.00	0.06	0.00	0.00	0.00	0.00	0.04	0.06	0.00
Mg	1.64	1.61	0.60	4.18	2.28	1.57	0.54	0.51	4.20	4.10	2.07
Fe	0.34	0.32	0.51	0.57	0.37	0.36	0.48	0.53	0.58	0.53	0.16
Mn	0.01	0.01	0.01	0.02	0.03	0.01	0.00	0.00	0.01	0.01	0.00
Ca	0.00	0.01	0.00	1.88	0.02	0.01	0.00	0.00	1.77	1.86	0.00
Na	0.01	0.02	0.03	0.15	0.03	0.00	0.01	0.01	0.11	0.18	0.01
K	0.00	0.00	0.00	0.04	0.00	0.00	0.00	0.00	0.02	0.02	0.00
Cr	N.D.	N.D.	0.16	0.05	0.00	0.01	0.22	0.23	0.07	0.04	0.00
Total Cations	3.00	4.01	3.05	15.23	4.89	4.00	3.01	3.02	15.14	15.20	4.68
Mg/(Mg+Fe)	0.83	0.84	0.54	0.88	0.86	0.81	0.53	0.49	0.88	0.89	0.93

Note: Data from Thomas (1996b). SPMS—Spuhler Peak Metamorphic Suite; PMMMS—Pony–Middle Mountain Metamorphic Suite; ICMS—Indian Creek Metamorphic Suite; Ol—olivine; Opx—orthopyroxene; Ant—anthophyllite; Tr—tremolite; Chl—chlorite; Mg-Hbl—magnesiohornblende.

*FeO—total iron as FeO

Figure 4. Photomicrograph in plane-polarized light of a Spuhler Peak Metamorphic Suite–hosted meta-ultramafic rock that contains rounded olivine (Ol) crystals in a matrix of coarse magnesiohornblende (Mg-Hbl). The grain boundaries and fractures are dusted with small crystals of magnetite and perhaps other oxide minerals. Sample RBT-22 collected west of Branham Lakes.

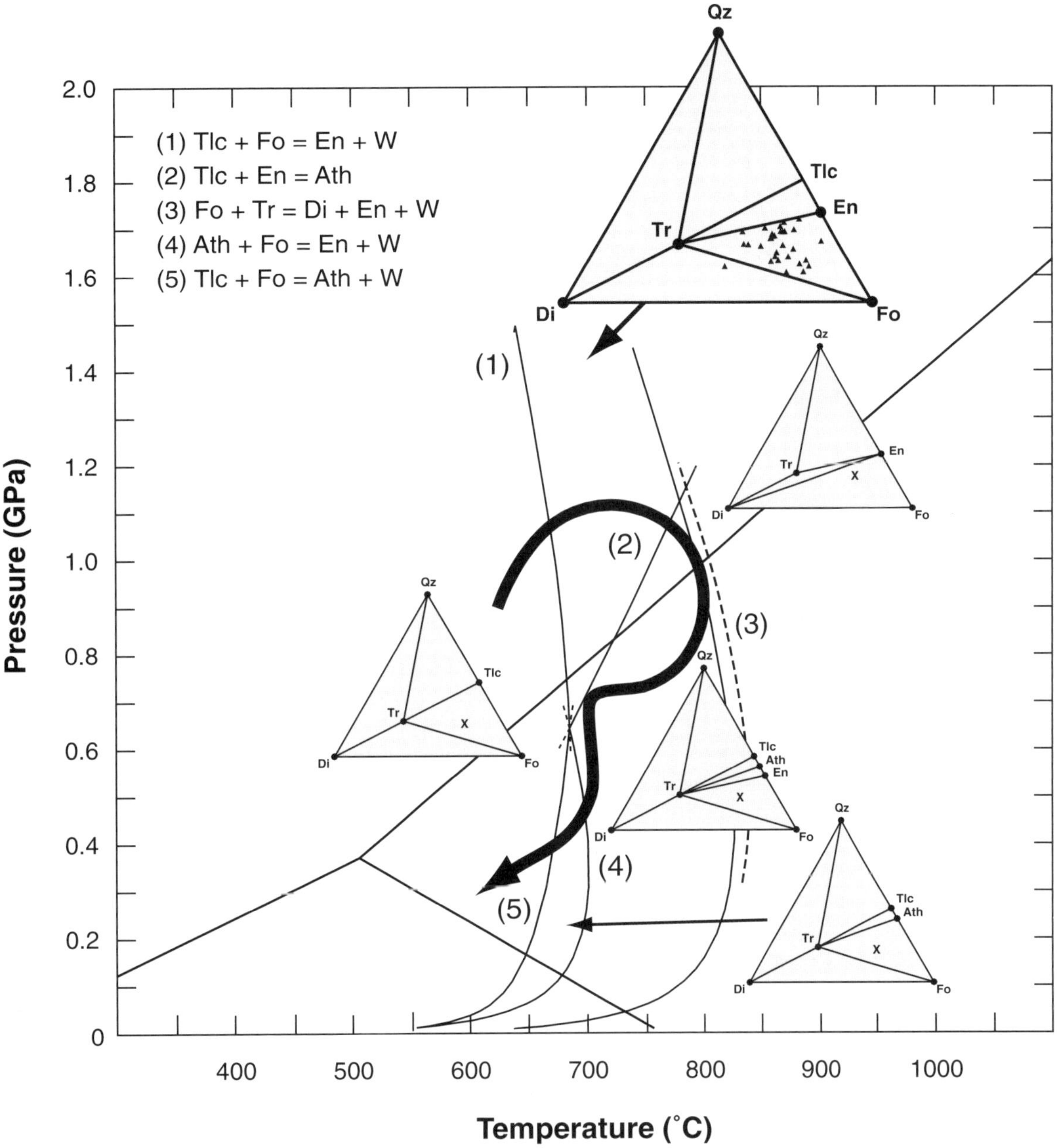

Figure 5. Pressure-temperature diagram showing selected end-member chemical reactions for ultramafic rocks during metamorphism. Minerals involved are quartz (Qz), forsterite (Fo), diopside (Di), tremolite (Tr), enstatite (En), talc (Tlc), anthophyllite (Ath), and water (W). Locations of the reactions were calculated by Spear (1993) for $P_{H_2O} = P_{total}$ using the thermodynamic database of Berman (1988). Mineral assemblages are shown on a composition diagram in oxygen units with $Ca_2Mg_2Si_4O_{12}$ (Di), Si_6O_{12} (Qz), and $Mg_6Si_3O_{12}$ (Fo) at the corners. Bulk compositions of Tobacco Root meta-ultramafic rocks (Table 2) are shown on one diagram and are represented by an X on other diagrams (see text). The *P-T* path shown by the dark arrow is proposed for the Tobacco Root Mountains based on the petrology of other rock types (Cheney et al., 2004a, this volume, Chapter 6). The absence of clinopyroxene with orthopyroxene in the meta-ultramafic rocks indicates that the *P-T* path passes to the left of reaction 3. The presence of anthophyllite is consistent with a *P-T* path that includes temperatures between 650 and 700 °C at pressures below 0.6 GPa.

+ olivine stably coexist in rocks with the "X" bulk composition only at pressures below ~0.6 GPa and at temperatures between 650 and 700 °C. We conclude, therefore, that the meta-ultramafic rocks passed through these conditions during cooling.

The Precambrian rocks of the Tobacco Root Mountains have been metamorphosed twice and contain mineral assemblages reflecting upper-amphibolite to granulite facies conditions (Reid, 1963). Cheney et al. (2004b, this volume, Chapter 8) and Mueller et al. (2004, this volume, Chapter 9) demonstrate that these metamorphic events occurred at 2.45 Ga and 1.78–1.72 Ga. Cheney et al. (2004a, this volume, Chapter 6) propose a *P-T* path for the 1.78–1.72 metamorphism (the Big Sky orogeny), which is reproduced in Figure 5 as a heavy black line. The clockwise *P-T* path is based in part on the superposition of fine-grained, 700 °C, 0.6–0.8 GPa mineral assemblages (e.g., garnet + cordierite) on coarse-grained, 750 °C, 1.0 GPa mineral assemblages (e.g., kyanite + orthopyroxene). The unusual "kink" in the *P-T* path is based on the occurrence of garnet necklaces surrounding orthopyroxene in a plagioclase-rich rock (produced by near isobaric cooling) and the nearby occurrence of orthopyroxene + plagioclase pseudomorphs after garnet in a hornblende-rich rock (produced by near isothermal decompression). See Cheney et al. (2004a, this volume, Chapter 6) for details. Interestingly, this independently determined path is consistent with the two principal constraints provided by the meta-ultramafic rocks.

Hess (1967), Cordua (1973), Tendall (1978), and Wilson and Hyndman (1990) all reported diopside as a minor phase in meta-ultramafic rocks containing orthopyroxene. This could be due to higher metamorphic temperatures, but we think it is more likely due to locally lower values of a_{H_2O}. Gillmeister (1976) reported garnet ($Py_{40}Alm_{40}Gr_{17}Sp_3$) in a more iron-rich ultramafic rock consisting of orthopyroxene (En_{76}), olivine (Fo_{74}), and magnesiohornblende from the central Tobacco Roots. Cordua (1973) and Tendall (1978) also reported garnet. We did not observe garnet in any meta-ultramafic sample, so we suspect that the presence of garnet is due to a more iron-rich bulk composition.

WHOLE-ROCK GEOCHEMISTRY

Analytical Methods

Large samples were collected in the field with the goal of obtaining representative whole-rock chemical analyses. Two to three fist-sized or larger samples were collected from each location to ensure that the sample would be at least 20 times the size of the largest crystal or pseudomorph after all weathered portions were removed. Samples were prepared and analyzed for major and trace elements by X-ray fluorescence spectroscopy following standard procedures at Franklin and Marshall College (Sincock, 1994), Carleton College (Poulsen, 1994), or Colorado College (MacFarlane, 1996b). Rare earth element (REE) and other trace element analyses of a subset of the samples were determined by neutron activation analysis at Oregon State University under the Department of Energy–sponsored Reactor Sharing Program (http://ne.oregonstate.edu/facilities/radiation_center/naa2.html) or by inductively coupled plasma–mass spectrometry at SGS Geochemical Exploration and Research Analysis Laboratories (formerly XRAL) in Toronto (http://www.sgs.ca/geochem/index.html).

Results

Chemical analyses for 32 meta-ultramafic samples are listed in Table 2. A wide range of major and trace element concentrations is evident, with some varying by a factor of two or more. The SiO_2 content (44–54 wt%) and MgO content (21–34 wt%) of these rocks (see Fig. 6) suggests a pyroxene-rich protolith, rather than a more olivine- or hornblende-rich protolith that would have ultrabasic chemistry (Wyllie, 1967). Mg numbers (molar Mg/(Mg + Fe) values times 100) are 74 to 84. Al_2O_3 varies from 4 to 10 wt% and CaO varies from 1 to 8 wt%, with CaO/Al_2O_3 (weight percent ratio) below one for most samples (Fig. 7). Based on normative calculations, an igneous rock protolith for these meta-ultramafic samples may have consisted primarily of orthopyroxene and olivine, with subordinate clinopyroxene and plagioclase (Fig. 8; Table 2). Total alkalis account for <1 wt%. Ni (500–1400 ppm) and to some extent Cr (1900–4900 ppm) concentrations are low relative to typical mantle peridotites (Hess, 1989, p. 86). For samples with REE analyses, all are enriched relative to chondritic values (Sun and McDonough, 1989), with most light rare earth elements (LREEs) 10–30 times chondritic values and heavy REEs 2–10 times chondritic values (Fig. 9). Several Spuhler Peak Metamorphic Suite–hosted ultramafic samples and one Indian Creek Metamorphic Suite–hosted sample exhibit negative Eu anomalies. Relative to a primitive mantle composition, these rocks are enriched in a variety of trace elements by 2 to 30 times (Fig. 10), whereas several samples are depleted in K, Sr, or Eu and all are enriched in Cs. Our data are consistent with ten analyses of meta-ultramafic rocks from the Branham Lakes area of the Tobacco Root Mountains reported by Cummings and McCulloch (1992) and more than span the range of their results.

Interpretation of the Geochemical Data

Because of the Precambrian age of the metamorphism of the Tobacco Root Mountains and the MgO contents (21–34 wt%) of the meta-ultramafic rocks, we looked for evidence that their protoliths may have been komatiitic magmas. The bulk chemical compositions straddle the boundary between ultramafic komatiites and basaltic komatiites on a Jensen (1976) cation plot (Fig. 11), as well as on other discriminant diagrams. Although no spinifex textures were observed, it is possible that they may have been destroyed during the granulite facies metamorphism experienced by these rocks. However, several other features of these rocks are not consistent with a komatiitic origin. Chondrite-normalized REE patterns for komatiites are typically LREE depleted (e.g., Arndt et al., 1977; Arndt and Nisbet, 1982), whereas the Tobacco Root rocks are LREE enriched. Crystal fractionation in komatiites is dominated by olivine, whereas the chemical

TABLE 2. WHOLE-ROCK CHEMICAL ANALYSES OF META-ULTRAMAFIC ROCKS

Host	SPMS	SPMS	SPMS	ICMS	ICMS	ICMS	PMMMS	PMMMS	PMMMS
12T	042126E	042148E	041704E	041722E	041729E	041839E	041426E	042392E	042803E
UTM	504059N	504148N	504553N	504014N	504015N	504203N	504841N	506541N	506327N
Sample	WAM-2b	WAM-6a	WAM-22b	WAM-18a	WAM-20b	WAM-31	WAM-8a	WAM-9b	WAM-13a
Oxide (wt%)									
SiO_2	51.02	47.94	52.83	52.93	51.56	48.09	50.45	49.60	46.80
TiO_2	0.29	0.36	0.34	0.37	0.39	0.27	0.39	0.26	0.24
Al_2O_3	6.54	6.72	5.63	5.42	7.84	6.69	5.09	5.12	5.09
*FeO	8.59	11.19	10.18	12.07	10.26	8.45	11.85	12.63	11.20
MnO	0.54	0.37	0.12	0.16	0.17	0.18	0.25	0.31	0.27
MgO	27.60	29.42	24.88	26.79	21.80	25.20	22.92	27.51	32.85
CaO	5.15	3.82	5.13	2.10	6.83	9.74	7.34	4.51	3.38
Na_2O	0.19	0.14	0.77	0.06	0.98	0.80	1.26	0.04	0.15
K_2O	0.03	0.00	0.07	0.06	0.11	0.50	0.39	0.00	0.00
P_2O_5	0.04	0.04	0.04	0.04	0.05	0.08	0.06	0.03	0.03
Total	100.00	100.00	100.00	100.00	100.00	100.00	100.00	100.00	100.00
Trace Elements (ppm)									
Sc	18.3	17.4	20.4	17.0	21.7	29.1	17.7	24.1	14.8
V	86	126	121	142	154	118	106	122	77
Cr	2020	2850	3309	3304	2140	2935	2038	2363	3561
Co	104	102	126	153	102	88	111	118	128
Ni	771	1180	1320	1270	702	840	918	670	1150
Zn	62	95	72	77	58	66	70	72	71
Rb	4	4	2	4	2	19	11	3	5
Sr	13	17	21	12	6	35	120	38	22
Y	8	11	15	10	11	12	15	8	9
Zr	36	48	39	49	48	33	51	29	33
Nb	6	4	7	4	4	3	5	3	1
Cs	N.D.[†]	0.55	0.52	0.45	N.D.	3.00	1.32	0.54	N.D.
La	3.7	8.7	4.1	7.9	7.3	10.1	7.7	2.9	5.7
Ce	9.0	19.9	10.5	15.4	15.5	23.7	19.3	6.5	12.4
Nd	4.1	8.0	5.9	7.2	8.1	12.8	10.8	3.5	5.9
Sm	0.90	1.33	1.16	1.49	1.39	2.48	2.06	0.63	0.86
Eu	0.13	0.26	0.18	0.13	0.44	0.64	0.60	0.20	0.28
Tb	0.14	0.18	0.24	0.26	0.25	0.29	0.30	0.14	0.14
Yb	0.68	0.81	0.99	1.18	1.04	0.92	0.97	0.59	0.58
Lu	0.10	0.10	0.16	0.21	0.17	0.09	0.15	0.10	0.08
Hf	0.89	1.19	0.96	1.73	1.41	1.01	1.38	0.58	0.70
Ta	0.12	0.22	0.17	0.15	0.17	0.09	0.16	0.06	0.08
Th	1.45	2.50	1.21	2.50	2.30	2.49	1.55	1.06	1.66
CIPW Weight Norms									
Or	0.2	0.0	0.4	0.4	0.7	3.0	2.3	0.0	0.0
Ab	1.6	1.2	6.5	0.5	8.3	6.8	10.7	0.3	1.3
An	16.9	17.7	11.7	10.2	16.7	13.2	7.1	13.8	13.2
C	0.0	0.0	0.0	1.5	0.0	0.0	0.0	0.0	0.0
Di	6.7	0.8	10.8	0.0	13.5	27.6	23.3	6.7	2.7
Hy	54.3	44.2	56.3	81.8	43.8	8.2	25.8	52.9	34.9
Ol	19.7	35.4	13.5	4.9	16.2	40.7	30.1	25.7	47.4
Il	1.0	1.0	1.0	1.0	1.0	1.0	1.0	0.0	0.0
Ap	0.0	0.0	0.0	0.0	0.0	0.0	0.0	0.0	0.0

(continued)

TABLE 2. WHOLE-ROCK CHEMICAL ANALYSES OF META-ULTRAMAFIC ROCKS *(continued)*

Host	SPMS	SPMS	SPMS	SPMS	SPMS	SPMS	SPMS	SPMS	SPMS	SPMS	SPMS	SPMS
12T	042133E	042073E	042080E	042075E	041546E	041547E	041493E	042125E	042123E	042089E	042120E	042141E
UTM	504171N	504063N	504057N	504064N	504658N	504660N	504591N	504058N	504063N	504049N	504139N	504138N
Sample	CP-16	CP-23	CP-29	CP-30	CP-54	CP-55	CP-58	WAM-2d	WAM-2e	WAM-3c	WAM-5a	WAM-6c
Oxide (wt%)												
SiO_2	44.90	50.93	45.77	49.59	51.22	52.57	48.21	50.52	45.68	47.97	50.68	51.18
TiO_2	0.25	0.27	0.24	0.21	0.33	0.29	0.30	0.29	0.25	0.36	0.36	0.25
Al_2O_3	5.70	5.57	4.54	4.58	5.82	4.26	5.62	6.86	4.82	6.90	6.83	6.63
FeO*	10.65	12.18	11.14	12.06	9.12	9.41	10.15	10.58	11.46	10.80	10.66	10.31
MnO	0.37	0.52	0.39	0.27	0.12	0.15	0.15	0.51	0.41	0.44	0.49	0.40
MgO	31.48	25.49	34.27	32.62	27.28	28.44	30.55	26.14	33.55	28.08	25.54	28.18
CaO	5.31	4.01	4.14	1.30	4.89	4.00	5.59	4.67	3.39	5.20	4.77	2.96
Na_2O	0.30	0.48	0.14	0.03	0.69	0.43	0.16	0.34	0.07	0.12	0.56	0.07
K_2O	0.09	0.09	0.05	0.04	0.20	0.09	0.05	0.06	0.33	0.05	0.05	0.00
P_2O_5	0.03	0.02	0.02	0.08	0.06	0.06	0.02	0.04	0.04	0.07	0.05	0.03
Total	99.08	99.57	100.69	100.79	99.74	99.72	100.80	100.00	100.00	100.00	100.00	100.00
Trace Elements (ppm)												
V	N.D.†	N.D.	N.D.	N.D.	N.D.	N.D.	N.D.	120	93	107	171	113
Cr	2720	N.D.	2790	2450	2550	N.D.	N.D.	2091	3813	2739	1944	3416
Co	N.D.	N.D.	N.D.	N.D.	N.D.	N.D.	N.D.	109	112	90	93	107
Ni	1260	N.D.	1120	1290	1140	N.D.	N.D.	N.D.	N.D.	N.D.	N.D.	N.D.
Zn	N.D.	N.D.	N.D.	N.D.	N.D.	N.D.	N.D.	90	69	71	88	145
Rb	4	N.D.	5	4	4	N.D.	N.D.	6	18	5	6	5
Sr	25	N.D.	8	7	22	N.D.	N.D.	19	11	25	13	15
Y	8	N.D.	3	2	6	N.D.	N.D.	6	6	7	13	4
Zr	27	N.D.	25	28	41	N.D.	N.D.	34	33	52	41	31
Nb	3	N.D.	3	4	4	N.D.	N.D.	4	7	4	4	3
CIPW Weight Norms												
Or	0.5	0.5	0.3	0.2	1.2	0.6	0.3	0.4	2.0	0.3	0.3	0.0
Ab	2.5	4.1	1.2	0.3	5.8	3.7	1.4	2.9	0.6	1.0	4.7	0.6
An	13.9	12.8	11.6	5.9	12.2	9.4	14.5	17.0	11.9	18.1	16.0	14.5
C	0.0	0.0	0.0	2.3	0.0	0.0	0.0	0.0	0.0	0.0	0.0	1.2
Di	9.8	5.6	7.0	0.0	9.3	8.0	10.5	4.7	3.7	5.8	5.9	0.0
Hy	17.0	56.3	23.2	61.0	46.7	60.6	33.7	53.3	26.3	40.3	51.0	67.4
Ol	54.8	19.7	56.9	30.4	23.8	16.8	39.9	21.2	55.0	33.6	21.3	15.8
Il	0.5	0.5	0.5	0.4	0.6	0.6	0.6	1.0	0.0	1.0	1.0	0.0
Ap	0.1	0.0	0.0	0.2	0.1	0.1	0.0	0.0	0.0	0.0	0.0	0.0

(continued)

variation of the Tobacco Root rocks cannot be explained by olivine fractionation (see below). The CaO/Al_2O_3 (weight percent ratio) values for komatiites are typically ~1.0 or higher, whereas the Tobacco Root meta-ultramafic rocks have CaO/Al_2O_3 values less than 1.0 (Fig. 7). REE data are also inconsistent with typical upper-mantle protoliths, such as Alpine-type peridotites and those found in typical ophiolites, which are LREE depleted. Moreover, the observed Mg numbers (74–84) are much lower than expected for mantle rocks (88–92).

To identify alternative protoliths, the covariation of elements or their oxides was explored. Weight percent ratios of TiO_2 to Zr, V, and CaO were all found to be nearly constant, in spite of large variations in the concentrations of the individual elements and the wide geographic distribution of the ultramafic rocks. The most convincing ratio is Zr/TiO_2, which is shown in Figure 12. We interpret this to mean that (1) the meta-ultramafic rocks distributed across the Tobacco Root Mountains have a common or related origin and (2) these elements were immobile or conserved in the processes that produced the observed variety of ultramafic rock compositions. We feel confident in these conclusions because our analyses, from several labs, are consistent and match those of Cummings and McCulloch (1992).

With conserved elements identified, it is possible to quantify the relative variations of other elements and to test hypotheses regarding the cause of those variations (Pearce, 1968; Stanley and Russell, 1989) by using one of the conserved elements to

TABLE 2. WHOLE-ROCK CHEMICAL ANALYSES OF META-ULTRAMAFIC ROCKS (*continued*)

Host	SPMS	SPMS	SPMS	SPMS	SPMS	SPMS	SPMS	ICMS	ICMS	PMMMS	PMMMS
12T	042140E	042146E	041704E	041546E	042116E	042060E	042056E	041671E	041663E	042419E	042776E
UTM	503997N	503994N	504553N	504658N	504020N	504041N	504041N	504011N	504011N	506470N	506277N
Sample	WAM-7f	WAM-7g	WAM-23a	WAM-26	WAM-27a	WAM-29a	WAM-29c	WAM-19a	WAM-19b	WAM-10a	WAM-14a
Oxide (wt%)											
SiO_2	45.94	49.40	53.92	52.56	49.67	46.06	52.05	48.68	46.24	47.51	46.62
TiO_2	0.26	0.38	0.23	0.29	0.31	0.30	0.23	0.36	0.37	0.29	0.20
Al_2O_3	5.34	7.79	4.37	4.54	7.17	7.20	5.47	7.31	8.24	5.99	5.66
FeO*	12.50	10.59	10.41	9.46	11.26	13.20	12.33	13.58	13.54	10.67	12.46
MnO	0.32	0.47	0.15	0.14	0.55	0.45	0.52	0.24	0.29	0.39	0.35
MgO	31.28	23.71	26.86	28.63	25.29	27.85	24.99	23.57	25.93	28.96	30.27
CaO	4.01	6.90	3.49	3.86	4.96	4.43	3.85	6.05	5.35	6.07	4.30
Na_2O	0.28	0.54	0.54	0.45	0.67	0.41	0.49	0.12	0.00	0.07	0.12
K_2O	0.02	0.14	0.01	0.04	0.07	0.05	0.04	0.05	0.00	0.01	0.00
P_2O_5	0.04	0.06	0.02	0.03	0.04	0.06	0.02	0.04	0.04	0.04	0.02
Total	100.00	100.00	100.00	100.00	100.00	100.00	100.00	100.00	100.00	100.00	100.00
Trace Elements (ppm)											
V	94	125	100	88	145	135	132	154	156	95	72
Cr	3174	1832	3631	4061	1970	3271	2828	1764	2688	3427	2596
Co	133	82	129	116	107	115	100	116	115	103	122
Ni	N.D.†	N.D.	N.D.	N.D.	N.D.	N.D.	N.D.	N.D.	N.D.	N.D.	N.D.
Zn	74	61	76	104	74	81	95	60	77	69	79
Rb	4	3	7	6	0	4	8	1	1	5	0
Sr	19	23	20	30	18	13	17	13	20	98	21
Y	3	12	6	8	14	6	0	8	10	6	2
Zr	30	35	35	44	35	36	26	33	38	41	24
Nb	3	5	5	8	2	2	2	4	5	3	2
CIPW Weight Norms											
Or	0.1	0.8	0.1	0.2	0.4	0.3	0.2	0.3	0.0	0.1	0.0
Ab	2.4	4.6	4.6	3.8	5.7	3.5	4.2	1.0	0.0	0.6	1.0
An	13.3	18.4	9.5	10.3	16.4	17.7	12.6	19.3	22.5	16.0	14.9
C	0.0	0.0	0.0	0.0	0.0	0.0	0.0	0.0	0.0	0.0	0.0
Di	5.1	12.4	6.2	6.9	6.4	3.2	5.1	8.5	3.1	11.1	5.0
Hy	27.8	37.7	69.9	60.6	43.3	29.3	63.0	46.3	37.0	33.7	34.0
Ol	50.8	25.2	9.4	17.5	27.2	45.4	14.4	23.9	36.7	38.0	44.6
Il	0.0	1.0	0.0	1.0	1.0	1.0	0.0	1.0	1.0	1.0	0.0
Ap	0.0	0.0	0.0	0.0	0.0	0.0	0.0	0.0	0.0	0.0	0.0

Note: Data from Polson (1994) and MacFarlane (1996b). MacFarlane data are normalized to 100% without loss on ignition (LOI). SPMS—Spuhler Peak Metamorphic Suite; ICMS—Indian Creek Metamorphic Suite; PMMMS—Pony–Middle Mountain Metamorphic Suite; UTM—Universal Transverse Mercator.

*FeO—total iron as FeO.

†N.D.—no data.

normalize the data. This must be done with caution because of statistical correlations introduced by using a common denominator on both axes (e.g., Rollinson and Roberts, 1986). Because these are ultramafic rocks, olivine fractionation is a logical mechanism to explore for compositional variation. If olivine fractionation is the principal cause of silica variation, then a plot of the molar ratio (MgO + FeO)/TiO_2 versus SiO_2/TiO_2 should yield a trend with a slope of two. As shown in Figure 13, the slope of the trend line is close to one, which would be expected for orthopyroxene fractionation. Plotting other combinations of oxides to look for plagioclase and for plagioclase + clinopyroxene fractionation (also on Fig. 13) indicates that there is little variation of CaO, Al_2O_3, or Na_2O relative to TiO_2.

It is clear from Figure 13 that the majority of the chemical variation of SiO_2 relative to a conserved TiO_2 is due to the removal or accumulation of one or more Mg-Fe silicates. With a slope near one, orthopyroxene is probably important. Additional tests, not involving Pearce element ratio diagrams,

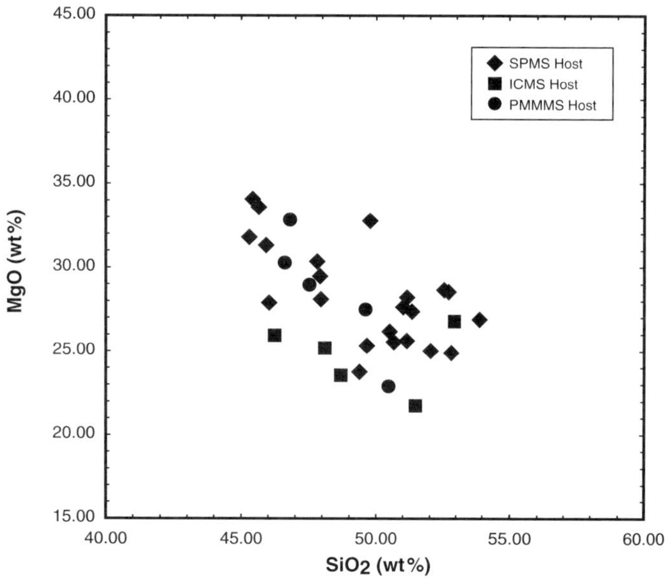

Figure 6. Variation diagram showing weight percent MgO vs. weight percent SiO_2. Different symbols are used for meta-ultramafic rocks that occur in different hosts. SPMS—Spuhler Peak Metamorphic Suite; ICMS—Indian Creek Metamorphic Suite; PMMMS—Pony–Middle Mountain Metamorphic Suite.

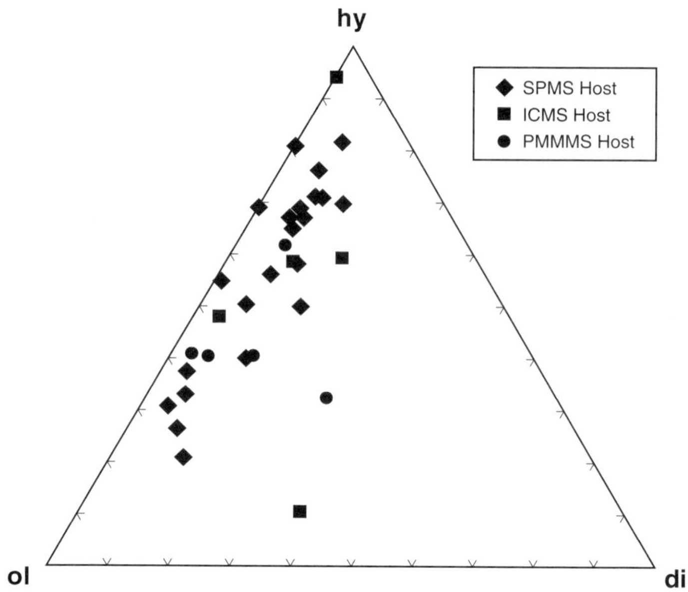

Figure 8. Meta-ultramafic rock chemical analyses are shown in terms of the relative proportions of the mafic CIPW weight norm minerals olivine (ol), hypersthene (hy), and diopside (di). The data are consistent with a protolith rich in orthopyroxene and olivine with subordinate clinopyroxene (and plagioclase—see Table 2). SPMS—Spuhler Peak Metamorphic Suite; ICMS—Indian Creek Metamorphic Suite; PMMMS—Pony–Middle Mountain Metamorphic Suite.

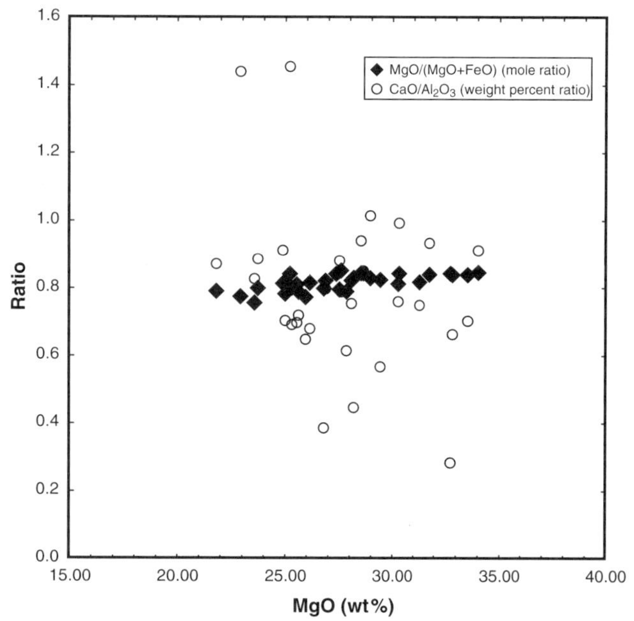

Figure 7. Variation diagram showing weight percent ratio CaO/Al_2O_3 vs. weight percent MgO and the mole ratio MgO/(MgO + FeO) vs. weight percent MgO.

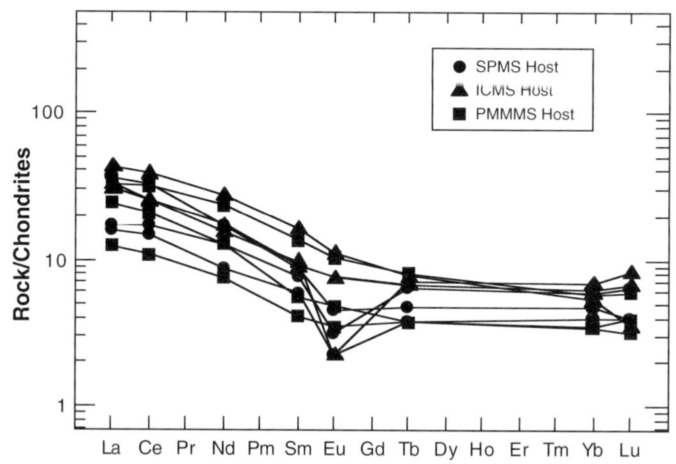

Figure 9. Rare earth data for the Tobacco Root meta-ultramafic rocks normalized to the chondrite values of Sun and McDonough (1989). All samples are enriched in the light rare earth elements and several show negative Eu anomalies. SPMS—Spuhler Peak Metamorphic Suite; ICMS—Indian Creek Metamorphic Suite; PMMMS—Pony–Middle Mountain Metamorphic Suite.

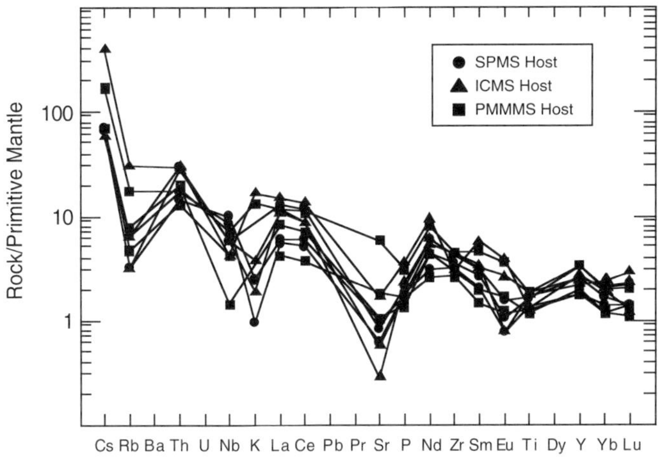

Figure 10. Trace element data for the Tobacco Root meta-ultramafic rocks normalized to the primitive mantle values of Sun and McDonough (1989). Note that several samples are depleted in K, Sr, or Eu and all are enriched in Cs. SPMS—Spuhler Peak Metamorphic Suite; ICMS—Indian Creek Metamorphic Suite; PMMMS—Pony–Middle Mountain Metamorphic Suite.

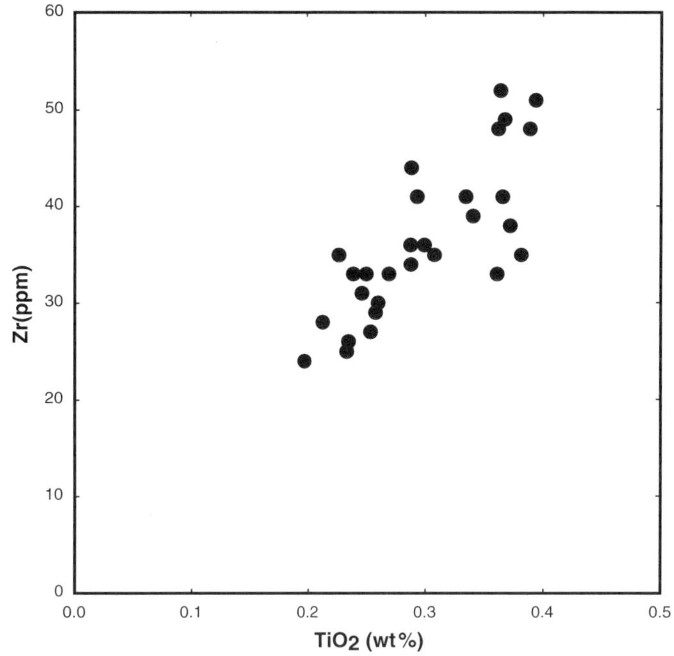

Figure 12. Variation of Zr (ppm) vs. TiO_2 (wt%) for analyzed Tobacco Root meta-ultramafic rocks. Although there is some scatter, the ratio of the concentrations of these two components is nearly constant. We interpret this to mean that both are incompatible with any fractionating phase.

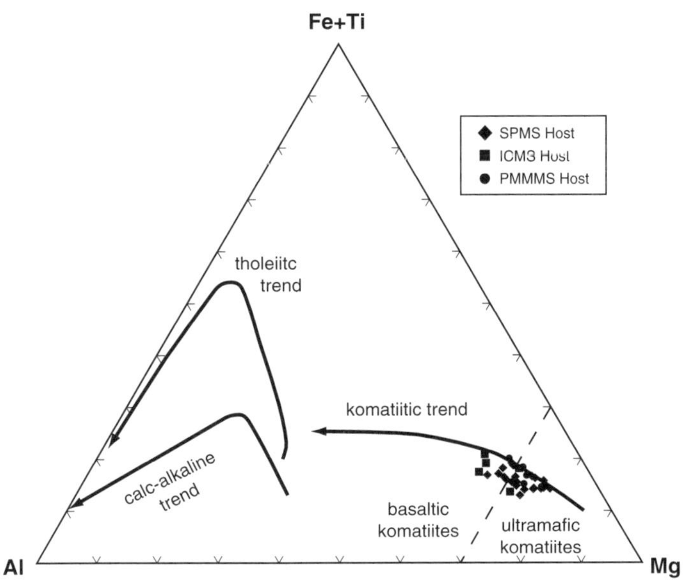

Figure 11. Meta-ultramafic rock chemical analyses are shown in terms of the relative cation proportions of Al, Fe + Ti, and Mg. Compositional fields for komatiites and fractional crystallization trends are those proposed by Jensen (1976). SPMS—Spuhler Peak Metamorphic Suite; ICMS—Indian Creek Metamorphic Suite; PMMMS—Pony–Middle Mountain Metamorphic Suite.

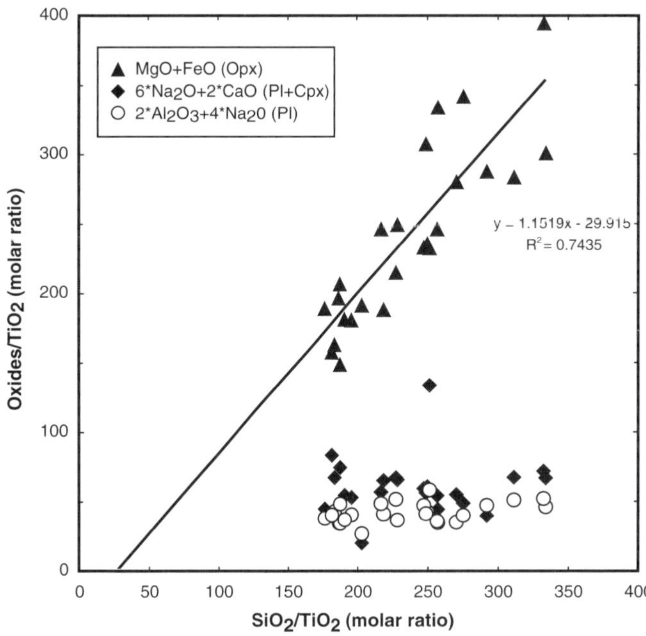

Figure 13. A variation diagram showing combinations of oxides in mole units vs. molar SiO_2, all divided by molar TiO_2. Combinations of elements chosen were selected to test for possible fractional crystallization of the minerals listed in parentheses. Both SiO_2 and FeO + MgO vary by a factor of two relative to TiO_2, whereas the quantities of CaO, Na_2O, and Al_2O_3 show little variation relative to TiO_2. Therefore, removal or addition of one or more Mg-Fe-rich, Ca-Al-poor minerals must be responsible for the chemical variation shown. If this mineral were olivine, the slope of the trend should be two. The observed slope (1.15) indicates that most of the variation is due to orthopyroxene addition or subtraction. Opx—orthopyroxene; Pl—plagioclase; Cpx—clinopyroxene.

support this conclusion. Figure 14 shows the meta-ultramafic rock compositions on a molar MgO + FeO, SiO_2, CaO + Al_2O_3 diagram. It is evident from this diagram that these rocks contain substantial amounts of normative orthopyroxene (see also Fig. 8). Indeed, we conclude that the protoliths of the Tobacco Root meta-ultramafic rocks were cumulates rich in orthopyroxene, which was also the preferred interpretation of Tendall (1978) based on modal data. Bulk compositions are appropriate for an orthopyroxene + olivine cumulate with intercumulus basaltic liquid (Fig. 14). Minor and trace element data are also consistent with orthopyroxene addition or subtraction as the major source of chemical variation. In the file of distribution coefficients for low-calcium pyroxene compiled by Roger Nielson for the GERM Kd Database (http://earthref.org/databases/KDD/index.html), only Cr, Ni, Sc, Co, and Zn (of the elements analyzed) have Kd values greater than one. For the Tobacco Root meta-ultramafic rocks, each of these elements varies in the manner expected for compatible elements: (1) Their ratios relative to TiO_2 decrease as SiO_2/TiO_2 decreases. (2) Their weight percentage decreases as the weight percentage of MgO decreases (see Figs. 15 and 16). Other trace elements, notably V, Y, and Zr, show incompatible trends.

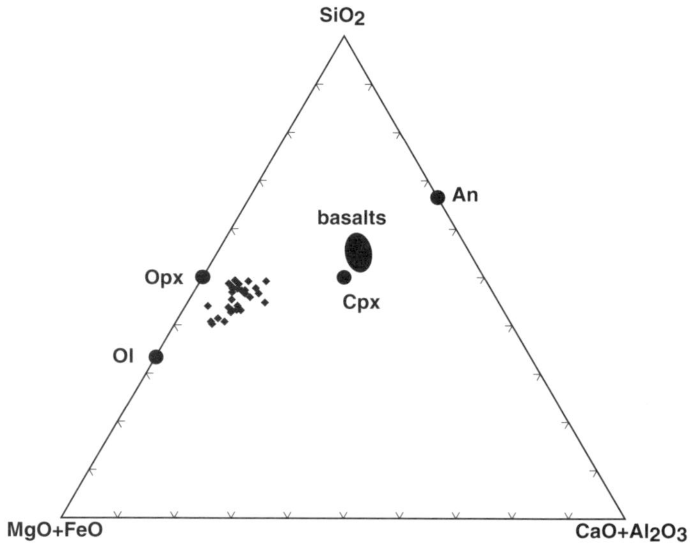

Figure 14. Meta-ultramafic rock chemical analyses are shown in terms of the relative mole proportions of MgO + FeO, SiO_2, and CaO + Al_2O_3. Compositions of olivine (Ol), orthopyroxene (Opx), clinopyroxene (Cpx), anorthite (An), and typical basalts are shown for reference.

DISCUSSION

The geochemistry of the Tobacco Root meta-ultramafic rocks has significant implications regarding their origin and tectonic setting. Low Mg numbers and LREE enrichment are inconsistent with an upper-mantle origin. Major and trace element data match the patterns expected for orthopyroxene-rich cumulates. Such rocks occur in oceanic settings, but orthopyroxene is generally subordinate to olivine and clinopyroxene, and the rocks are not LREE enriched (e.g., Meyer et al., 1989; Ross and Elthon, 1993). Orthopyroxene cumulates occur in suprasubduction zone ophiolites (Pearce et al., 1984), but the geochemical characteristics of the suprasubduction zone ophiolites, such as the typical TiO_2 values (<0.05 wt%), are not good matches with the Tobacco Root data (TiO_2 of 0.2–0.4 wt%). Abundant orthopyroxene is more typical of the crystallization of a continental tholeiite, such as the Stillwater igneous complex, than an oceanic tholeiite (Campbell, 1985). Campbell (1985) argued that additional silica assimilated by continental tholeiites directs the olivine fractionation path to cross the orthopyroxene reaction boundary, rather than to intersect the clinopyroxene cotectic (in the system olivine-clinopyroxene-silica) as in oceanic tholeiites. Elevated pressure shifts the olivine to orthopyroxene reaction curve away from silica on the same diagram, and also can lead to orthopyroxene fractionation. Some Archean, high-magnesium basalts show flat to light element–enriched REE patterns (Sun and Nesbitt, 1978; Sun, 1984) and silica enrichment, possibly reflecting an origin similar to the Tobacco Root rocks. Other Precambrian cumulate rocks from a continental setting also are rich in orthopyroxene (e.g., Srikantappa et al., 1984).

Continental tholeiites commonly occur in rift settings where fractional crystallization can lead to mafic rocks that match the geochemistry of the Tobacco Root meta-ultramafic rocks (Crawford et al., 1984). Interestingly, the 2.06 Ga metamorphosed mafic dikes and sills (MMDS) that occur in the Tobacco Root Mountains are believed to have been intruded in a rift setting and sample a magma source that has undergone considerable fractional crystallization of orthopyroxene, as well as clinopyroxene

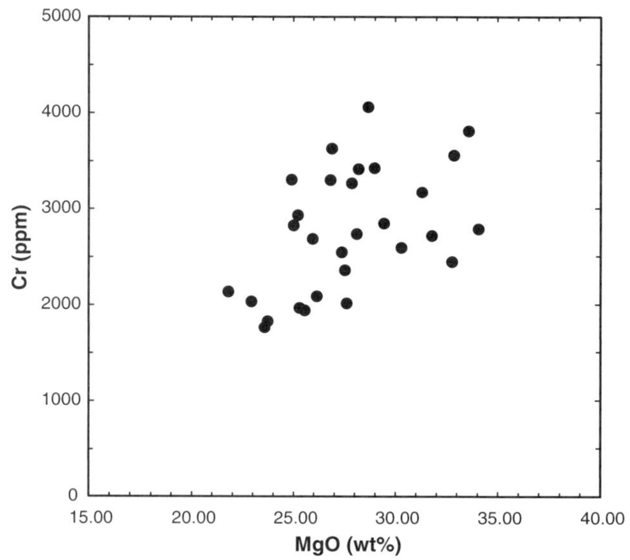

Figure 15. Variation diagram showing ppm Cr vs. weight percent MgO. Cr increases as MgO increases, which is consistent with the accumulation of orthopyroxene in which Cr is concentrated.

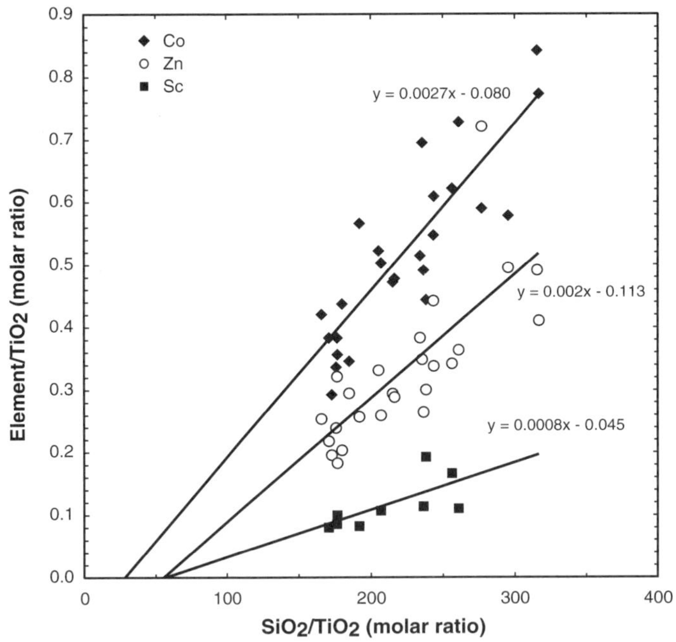

Figure 16. A variation diagram showing elements in mole units vs. molar SiO_2, all divided by molar TiO_2. Each of these elements is compatible with orthopyroxene and each ratio increases as SiO_2/TiO_2 increases, consistent with the accumulation of orthopyroxene.

and plagioclase (Brady et al., 2004, this volume, Chapter 5). Thus, it is possible that the MMDS and the meta-ultramafic rocks are related through the process of fractional crystallization. Both rock groups have similar light element–enriched REE patterns, with the MMDS values higher on average by a factor of four or more. Many of the ultramafic bodies occur in the Spuhler Peak Metamorphic Suite, which contains a significant volume of mafic rocks. However, the REE values for the Spuhler Peak Metamorphic Suite mafic rocks are similar in magnitude to the meta-ultramafic REE values (~10× chondrite), but with flat or LREE-depleted patterns (Burger et al., 2004, this volume, Chapter 3). So the only local candidate likely related by fractionation to the meta-ultramafic rock magmatic suite appears to be the MMDS. Cordua (1973) suggested that the MMDS and meta-ultramafic rocks might be related because they commonly occur near one another in the Tobacco Root Mountains. However, we believe that any current juxtaposition is tectonic and not directly related to the possibility of a common origin.

Meta-ultramafic rocks occur as isolated, podiform bodies in all three major Precambrian rock suites of the Tobacco Root Mountains. Based on the geochemical evidence for a common or related origin and on other descriptive data presented in this paper, we conclude that these bodies were tectonically emplaced rather than being small ultramafic intrusions. Elsewhere in this volume (Brady et al., 2004, this volume, Chapter 5), it is demonstrated that the Spuhler Peak Metamorphic Suite was not present for the 2.45 Ga metamorphism of the Pony–Middle Mountain Metamorphic Suite and the Indian Creek Metamorphic Suite, or for the 2.06 Ga intrusion of the MMDS. Therefore, it appears that the current distribution of ultramafic rocks was accomplished during the Big Sky orogeny at 1.78–1.72 Ga, which was probably also responsible for the emplacement of the Spuhler Peak Metamorphic Suite. If both the meta-ultramafic rocks and the Spuhler Peak Metamorphic Suite were tectonically emplaced ca. 1.78 Ga, it makes sense that many of the meta-ultramafic bodies studied in the field were found in or near to the Spuhler Peak Metamorphic Suite. Perhaps future detailed structural work will be able to use the distribution of meta-ultramafic rocks to unravel more of the complex deformation that must have accompanied the Big Sky orogeny.

ACKNOWLEDGMENTS

We would like to thank the Keck Geology Consortium, its member colleges, and especially the W.M. Keck Foundation for providing support for this project. Additional funding to K.E. Johnson was provided by a Summer Faculty Fellowship from the University of New Orleans. Instrumental neutron activation analysis data were provided by Oregon State University under the Department of Energy's Reactor Sharing Program. Other members of the Tobacco Root Project for 1995–1996 who deserve special thanks for assistance in locating and sampling previously "uncharted" ultramafic rocks in all three metamorphic suites are Kathleen DeGraff, Heidi Mohlman, David Owen, Caroline Tuit, Jennifer Martin, Jillian Hirst, Wilfredo Chaparro, Steven Kranenburg, Jason Cox, and Karl Wegeman. We thank especially Gary Ernst and Edward Stoddard, who improved the manuscript with their timely and incisive reviews.

REFERENCES CITED

Arndt, N.T., and Nisbet, E.G., 1982, What is a komatiite?, *in* Arndt, N.T., and Nisbet, E.G., eds., Komatiites: London, George Allen and Unwin, p. 19–27.

Arndt, N.T., Naldrett, A.J., and Pyke, D.R., 1977, Komatiitic and iron-rich tholeiitic lavas of Munro Township, Northeast Ontario: Journal of Petrology, v. 18, p. 319–369.

Berman, R.G., 1988, Internally consistent thermodynamic data for minerals in the system Na_2O-K_2O-CaO-MgO-FeO-Fe_2O_3-Al_2O_3-SiO_2-TiO_2-H_2O-CO_2: Journal of Petrology, v. 29, no. 2, p. 445–522.

Brady, J.B., Mohlman, H.K., Harris, C., Carmichael, S.K., Jacob, L.K., and Chaparro, W.R., 2004, General geology and geochemistry of metamorphosed Proterozoic mafic dikes and sills, Tobacco Root Mountains, Montana, *in* Brady, J.B., et al., eds., Precambrian geology of the Tobacco Root Mountains, Montana: Boulder, Colorado, Geological Society of America Special Paper 377, p. 89–104 (this volume).

Burger, H.R., 1966, Structure, petrology, and economic geology of the Sheridan district, Madison County, Montana [Ph.D. thesis]: Bloomington, Indiana University, 156 p.

Burger, H.R., 1967, Bedrock geology of the Sheridan district, Madison County, Montana: Montana Bureau of Mines and Geology Memoir 41, 22 p.

Burger, H.R., 2004, General geology and tectonic setting of the Tobacco Root Mountains, Montana, *in* Brady, J.B., et al., eds., Precambrian geology of the Tobacco Root Mountains, Montana: Boulder, Colorado, Geological Society of America Special Paper 377, p. 1–14 (this volume).

Burger, H.R., Peck, W.H., Johnson, K.E., Tierney, K.A., Poulsen, C.J., Cady, P., Lowell, J., MacFarlane, W.A., Sincock, M.J., Archuleta, L.L., Pufall, A., and Cox, M.J., 2004, Geology and geochemistry of the Spuhler Peak

Metamorphic Suite, *in* Brady, J.B., et al., eds., Precambrian geology of the Tobacco Root Mountains, Montana: Boulder, Colorado, Geological Society of America Special Paper 377, p. 47–70 (this volume).

Campbell, J.H., 1985, The difference between oceanic and continental tholeiites; a fluid dynamic explanation: Contributions to Mineralogy and Petrology, v. 91, p. 37–43.

Cheney, J.T., Brady, J.B., Tierney, K.A., DeGraff, K.A., Mohlman, H.K., Frisch, J.D., Hatch, C.E., Steiner, M.L., Carmichael, S.K., Fisher, R.G.M., Tuit, C.B., Steffen, K.J., Cady, P., Lowell, J., Archuleta, L.L., Hirst, J., Wegmann, K.W., and Monteleone, B., 2004a, Proterozoic metamorphism of the Tobacco Root Mountains, Montana, *in* Brady, J.B., et al., eds., Precambrian geology of the Tobacco Root Mountains, Montana: Boulder, Colorado, Geological Society of America Special Paper 377, p. 105–129 (this volume).

Cheney, J.T., Webb, A.A.G., Coath, C.D., and McKeegan, K.D., 2004b, In situ ion microprobe $^{207}Pb/^{206}Pb$ dating of monazite from Precambrian metamorphic suites, Tobacco Root Mountains, Montana, *in* Brady, J.B., et al., eds., Precambrian geology of the Tobacco Root Mountains, Montana: Boulder, Colorado, Geological Society of America Special Paper 377, p. 151–179 (this volume).

Cordua, W.S., 1973, Precambrian geology of the southern Tobacco Root Mountains, Madison County, Montana [Ph.D. thesis]: Bloomington, Indiana University, 247 p.

Crawford, A.J., Cameron, W.E., and Keays, R.R., 1984, The association of boninite low-Ti andesite-tholeiite in the Heatcote Greenstone Belt, Victoria; ensimatic setting for the early Lachlan fold belt: Australian Journal of Earth Sciences, v. 31, p. 161–175.

Cummings, M.L., and McCulloch, W.R., 1992, Geochemistry and origin of amphibolite and ultramafic rocks, Branham Lakes area, Tobacco Root Mountains, southwestern Montana, *in* Bartholomew, M.J., et al., eds., Basement tectonics 8: Characterization and comparison of ancient and Mesozoic continental margins—Proceedings of the 8th International Conference on Basement Tectonics (Butte, Montana, 1988): Dordrecht, Netherlands, Kluwer Academic Publishers, p. 323–340.

Desmarais, N.R., 1981, Metamorphosed Precambrian ultramafic rocks in the Ruby Range, Montana: Precambrian Research, v. 16, p. 67–101.

Friberg, N., 1976, Petrology of a metamorphic sequence of upper amphibolite facies in the central Tobacco Root Mountains, southwestern Montana [Ph.D. thesis]: Bloomington, Indiana University, 146 p.

Garihan, J.M., 1979, Geology and structure of the central Ruby Range, Madison County, Montana: Geological Society of America Bulletin, v. 90, p. 323–326 (Part I), p. 695–788 (Part II).

Gillmeister, N.M., 1972, Petrology of Precambrian rocks in the central Tobacco Root Mountain, Madison County, Montana [Ph.D. thesis]: Cambridge, Massachusetts, Harvard University, 201 p.

Gillmeister, N.M., 1976, Garnet-bearing metamorphosed ultramafic rocks from southwestern Montana: Eos (Transactions, American Geophysical Union), v. 57, no. 4, p. 338.

Hanley, T.B., 1975, Structure and petrology of the northwestern Tobacco Root Mountains, Madison County, Montana [Ph.D. thesis]: Bloomington, Indiana University, 289 p.

Harms, T.A., Burger, H.R., Blednick, D.G., Cooper, J.M., King, J.T., Owen, D.R., Lowell, J., Sincock, M.J., Kranenburg, S.R., Pufall, A., and Picornell, C.M., 2004, Character and origin of Precambrian fabrics and structures in the Tobacco Root Mountains, Montana, *in* Brady, J.B., et al., eds., Precambrian geology of the Tobacco Root Mountains, Montana: Boulder, Colorado, Geological Society of America Special Paper 377, p. 203–226 (this volume).

Heinrich, E.W., 1963, Paragenesis of clinohumite and associated minerals from Wolf Creek, Montana: American Mineralogist, v. 48, p. 597–613.

Hess, D.F., 1967, Geology of pre-Beltian rocks in the central and southern Tobacco Root Mountains (Montana), with reference to superposed effects of the Laramide age Tobacco Root Batholith [Ph.D. thesis]: Bloomington, Indiana University, 333 p.

Hess, P.C., 1989, Origins of igneous rocks: Cambridge, Massachusetts, Harvard University Press, 336 p.

Holland, T.J.B., and Powell, R., 1998, An internally consistent thermodynamic data set for phases of petrological interest: Journal of Metamorphic Geology, v. 16, no. 3, p. 309–343.

Jahn, B.M., and Zhang, Z.Q., 1984, Radiometric ages (Rb-Sr, Sm-Nd, U-Pb) and REE geochemistry of Archean granulite gneisses from Eastern Hebei Province, China, *in* Kroner, A., et al., eds., Archean geochemistry: Berlin, Springer Verlag, p. 205–230.

Jensen, L.S., 1976, A new cation plot for classifying subalkalic volcanic rocks: Ontario Geological Survey Miscellaneous Paper 66, 22 p.

Johnson, K.E., Burger, H.R., Brady, J.B., Cheney, J.T., Harms, T.A., Thomas, R.B., and MacFarlane, W.A., 1996, Archean meta-ultramafics in the Tobacco Root Mountains, SW Montana: An evaluation of the metamorphic history and protolith: Geological Society of America Abstracts with Programs, v. 28, no. 7, p. 493.

Kalsbeek, F., and Manatschal, G., 1999, Geochemistry and tectonic significance of peridotitic and metakomatiitic rocks from the Ussuit area, Nagssugtoqidian orogen, West Greenland: Precambrian Research, v. 94, p. 101–120.

MacFarlane, W.A., 1996a, A Geochemical and petrographic analysis of the meta-ultramafic rocks in the Tobacco Root Mountains, southwest Montana, *in* Mendelson, C.V., and Mankiewicz, C., eds., Ninth Keck research symposium in geology: Williamstown, Massachusetts, Williams College, p. 106–109.

MacFarlane, W.A., 1996b, A geochemical and petrographic analysis of the meta-ultramafic rocks in the Tobacco Root Mountains, southwest Montana [B.A. thesis]: Colorado Springs, Colorado College, 132 p.

Merrill, G.P., 1895, Notes on some eruptive rocks from Gallatin, Jefferson, and Madison counties, Montana: Proceedings of the United States National Museum, p. 637–673.

Meyer, P.S., Dick, H.J.B., and Thompson, G., 1989, Cumulate gabbros from the southwest Indian Ridge, 54° S, 7°16′ E: Implications for magmatic processes at a slow spreading ridge: Contributions to Mineralogy and Petrology, v. 103, p. 44–63.

Mueller, P.A., Burger, H.R., Wooden, J.L., Heatherington, A.L., Mogk, D.W., and D'Arcy, K., 2004, Age and evolution of the Precambrian crust of the Tobacco Root Mountains, Montana, *in* Brady, J.B., et al., eds., Precambrian geology of the Tobacco Root Mountains, Montana: Boulder, Colorado, Geological Society of America Special Paper 377, p. 181–202 (this volume).

Occhipinti, S.A., Swager, C.P., and Pirajno, F., 1998, Structural-metamorphic evolution of the Paleoproterozoic Bryah and Padbury Groups during the Capricorn orogeny, Western Australia: Precambrian Research, v. 90, p. 141–158.

Paktunc, A.D., 1984, Metamorphism of the ultramafic rocks of the Thompson Mine, Thompson Nickel Belt, northern Manitoba: Canadian Mineralogist, v. 22, p. 77–91.

Pearce, J.A., Lippard, S.J., and Roberts, S., 1984, Characteristics and tectonic significance of supra-subduction zone ophiolites, *in* Kokelaar, B.P., and Howells, M.F., eds., Marginal basin geology; volcanic and associated sedimentary and tectonic processes in modern and ancient marginal basins: Geological Society [London] Special Publication 16, p. 74–94.

Pearce, T.H., 1968, A contribution to the theory of variation diagrams: Contributions to Mineralogy and Petrology, v. 19, p. 142–157.

Poulsen, C.J., 1994, Origin, metamorphic history, and tectonic setting of Archean rocks from the Spuhler Peak Assemblage, Tobacco Root Mountains, Montana [B.A. thesis]: Northfield, Minnesota, Carleton College, 54 p.

Reid, R.R., 1963, Metamorphic rocks of the northern Tobacco Root Mountains, Madison County, Montana: Geological Society of America Bulletin, v. 74, p. 293–305.

Rollinson, H.R., and Roberts, C.R., 1986, Ratio correlations and major element mobility in altered basalts and komatiites: Contributions to Mineralogy and Petrology, v. 93, p. 89–97.

Root, F.K., 1965, Structure, petrology, and mineralogy of pre-Beltian metamorphic rocks of the Pony-Sappington area, Madison County, Montana [Ph.D. thesis]: Bloomington, Indiana University, 184 p.

Rosing, M.T., and Rose, N.M., 1993, The role of ultramafic rocks in regulating the concentrations of volatile and nonvolatile components during deep crustal metamorphism: Chemical Geology, v. 108, p. 187–200.

Ross, K., and Elthon, D., 1993, Cumulates from strongly depleted mid-ocean-ridge basalt: Nature, v. 365, p. 826–829.

Sanford, R.F., 1982, Growth of ultramafic reaction zones in greenschist to amphibolite facies metamorphism: American Journal of Science, v. 282, p. 543–616.

Schmädicke, E., 2000, Phase relations in peridotitic and pyroxenitic rocks in the model systems CMASH and NCMASH: Journal of Petrology, v. 41, p. 69–86.

Sincock, M.J., 1994, Structural and metamorphic history of an intensely deformed zone, lower Branham Lake, southern Tobacco Root Mountains, Montana [B.A. thesis]: Lancaster, Pennsylvania, Franklin and Marshall College, 31 p.

Spear, F.S., 1993, Metamorphic phase equilibria and pressure-temperature-time paths, Mineralogical Society of America Monographs: Washington, D.C., Mineralogical Society of America, 799 p.

Srikantappa, C., Hormann, P.K., and Raith, M., 1984, Petrology and geochemistry of layered ultramafic to mafic complexes from the Archean craton of Karnataka, southern India, *in* Kroner, A., et al., eds., Archean geochemistry: Berlin, Springer Verlag, p. 138–160.

Stanley, C.R., and Russell, J.K., 1989, Petrologic hypothesis testing with Pearce element ratio diagrams; derivation of diagram axes: Contributions to Mineralogy and Petrology, v. 103, p. 78–89.

Sun, S.S., 1984, Geochemical characteristics of Archaean ultramafic and mafic volcanic rocks; implications for mantle composition and evolution, *in* Kroener, A., et al., eds., Archaean geochemistry: The origin and evolution of the Archaean continental crust: Berlin, Springer Verlag, p. 25–46.

Sun, S.S., and McDonough, W.F., 1989, Chemical and isotopic systematics of oceanic basalts; implications for mantle composition and processes, *in* Saunders, A.D., and Norry, M.J., eds., Magmatism in the ocean basins: Geological Society [London] Special Publication 42, p. 313–345.

Sun, S.S., and Nesbitt, R.W., 1978, Petrogenesis of Archaean ultrabasic and basic volcanics; evidence from rare earth elements: Contributions to Mineralogy and Petrology, v. 65, p. 301–325.

Tendall, B.A., 1978, Mineralogy and petrology of Precambrian ultramafic bodies from the Tobacco Root Mountains, Madison County, Montana [M.S. thesis]: Bloomington, Indiana University, 127 p.

Thomas, R.B., 1996a, Metamorphic history of Archean ultramafic pods in the Tobacco Root Mountains, southwestern Montana, *in* Mendelson, C.V., and Mankiewicz, C., eds., Ninth Keck research symposium in geology: Williamstown, Massachusetts, Williams College, p. 82–85.

Thomas, R.B., 1996b, Metamorphic history of Archean ultramafic pods of the Tobacco Root Mountains, southwestern Montana [B.A. thesis]: Williamstown, Massachusetts, Williams College, 125 p.

Vitaliano, C.J., Burger, H.R., Cordua, W.S., Hanley, T.B., Hess, D.P., and Root, F.K., 1979, Explanatory text to accompany geologic map of southern Tobacco Root Mountains, Madison County, Montana: Geological Society of America Map and Chart Series MC31, scale 1:62,500, 1 sheet, 8 p. text.

Wilson, M.L., and Hyndman, D.W., 1990, Tectonic interpretation of an Archean lithologic package enclosing iron-formation in the southern Tobacco Root and northern Ruby Ranges of southwestern Montana, USA, *in* Chauvel, J.J., ed., Ancient banded iron formations; regional presentations: Athens, Greece, Theophrastus Publishing, p. 27–61.

Wyllie, P.J., 1967, Ultramafic and ultrabasic rocks, *in* Wyllie, P.J., ed., Ultramafic and related rocks: New York and London, John Wiley & Sons, p. 1–7.

Manuscript Accepted by the Society September 12, 2003

Printed in the USA

Geological Society of America
Special Paper 377
2004

General geology and geochemistry of metamorphosed Proterozoic mafic dikes and sills, Tobacco Root Mountains, Montana

John B. Brady
Department of Geology, Smith College, Northampton, Massachusetts 01063, USA

Heidi K. Mohlman
Department of Geology, Amherst College, Amherst, Massachusetts 01002, USA

Caroline Harris
Pomona College, Claremont, California 91711-6339, USA

Sarah K. Carmichael
Lisa J. Jacob
Department of Geology, Smith College, Northampton, Massachusetts 01063, USA

Wilfredo R. Chaparro
University of Puerto Rico, Mayaguez, Puerto Rico, 00681

ABSTRACT

Just over two billion years ago, basaltic magma intruded rocks of the Pony–Middle Mountain Metamorphic Suite and the Indian Creek Metamorphic Suite that now crop out in the Tobacco Root Mountains of Montana near the northwestern margin of the Wyoming province. Numerous examples can be found of mafic dikes that crosscut layering and gneissic banding, demonstrating that the host rocks were metamorphosed to form the gneissic texture prior to intrusion of the dikes. Although many of the intrusions appear to be sills that followed compositional layering, close inspection reveals low-angle discordance in nearly every case, consistent with rotation of dikes by shearing into nearly layer-parallel orientations. The mafic dikes do not intrude the adjacent Spuhler Peak Metamorphic Suite, which we interpret to mean that the Spuhler Peak Metamorphic Suite was not present at the time of intrusion. These dikes and sills were metamorphosed along with their host rocks during the Big Sky orogeny, a major orogenic event at 1.77 Ga that is documented in this volume. The fine-grained, garnet-bearing, rusty-weathering metamorphosed mafic dikes and sills (MMDS) generally have a distinctive appearance in the field. Some MMDS were clearly folded or boudinaged during metamorphism, but many show only weak foliation and still have sharp contacts with their host rocks. A decrease in grain size commonly occurs on the margins of the MMDS and is believed to be a metamorphic texture that developed from the chilled margins of intrusions into cold rocks.

Chemically, the MMDS are subalkaline tholeiites that have been modified by significant fractional crystallization, as evidenced by molar Mg/(Mg + Fe) values that vary from 0.6 to 0.3 and TiO_2 values that vary from 0.6 to 3.0 wt%. The ratio of TiO_2 to P_2O_5 is virtually constant, indicating a common origin for the MMDS and that these two components were similarly incompatible with the fractionating minerals. The weight

Brady, J.B., Mohlman, H.K., Harris, C., Carmichael, S.K., Jacob, L.K., and Chaparro, W.R., 2004, General geology and geochemistry of metamorphosed Proterozoic mafic dikes and sills, Tobacco Root Mountains, Montana, *in* Brady, J.B., Burger, H.R., Cheney, J.T., and Harms, T.A., eds., Precambrian geology of the Tobacco Root Mountains, Montana: Boulder, Colorado, Geological Society of America Special Paper 377, p. 89–104. For permission to copy, contact editing@geosociety.org.

percentages of both change by a factor of five, which would require crystallization and removal of 80% of the liquid if there were a single magma source. Based on the major element chemistry, plagioclase, orthopyroxene (or pigeonite), and clinopyroxene were the major fractionating phases, with little involvement of olivine. Trace element variations are consistent with this interpretation. For example, there is no clear correlation between Ni and MgO, which would be expected if there were significant olivine fractionation. Measured rare earth element (REE) data for the MMDS are all at least ten times chondrite REE values. In most cases, light rare earth elements (LREEs) are enriched relative to heavy rare earth elements (HREEs). These data are consistent with a model of intrusion of the MMDS into cool, metamorphic rocks at 2.06 Ga during continental rifting of the Wyoming province. The resulting ocean basin closed at 1.77 Ga during the Big Sky orogeny, emplacing rocks of the Spuhler Peak Metamorphic Suite and slivers of an ultramafic, orthopyroxene cumulate, deforming all units together, and metamorphosing all rocks to at least upper amphibolite facies conditions.

Keywords: fractional crystallization, subalkaline tholeiites, Wyoming province.

INTRODUCTION

One of the most striking and important features of the metamorphic rocks of the Tobacco Root Mountains is the widespread occurrence of metamorphosed mafic dikes and sills (MMDS). These fine-grained, dark rocks are easily noticed where they sharply truncate the metamorphic banding of light-colored, quartzofeldspathic gneisses (Fig. 1). From a distance, they appear to be unmetamorphosed basalt. However, upon closer examination, the presence of fine-grained garnet crystals reveals that these ancient intrusions have also been metamorphosed. Because the MMDS clearly postdate the metamorphism that produced the gneissic banding they crosscut, but predate the metamorphism that produced the garnet crystals they contain, the timing and nature of MMDS intrusion are critical features of the Precambrian history of the Tobacco Root Mountains.

The MMDS intrude the two major suites of metamorphic rocks that constitute the bulk of the Precambrian rocks of the Tobacco Root Mountains in the northwestern Wyoming province. The host rocks are (1) the Indian Creek Metamorphic Suite and (2) the Pony–Middle Mountain Metamorphic Suite. These units are described in detail by Burger (2004, this volume, Chapter 1) and throughout this volume. Monazite dating (Cheney et al., 2004b, this volume, Chapter 8) and zircon dating (Mueller et al., 2004, this volume, Chapter 9) demonstrate that these rocks were metamorphosed at 2.45 Ga and at 1.77 Ga. The Indian Creek Metamorphic Suite consists of rocks that are clearly metasediments, including marble and iron formation, but also includes rocks that are meta-igneous, including amphibolite. The Indian Creek Metamorphic Suite rocks were called Cherry Creek metamorphic rocks by Reid (1963) and others. The Pony–Middle Mountain Metamorphic Suite consists principally of quartzofeldspathic gneisses and hornblende amphibolites. These are the rocks that Reid (1963) and others have called Pony metamorphic rocks. The MMDS do not intrude a third important rock package, the Spuhler Peak Metamorphic Suite, which is a mafic unit that consists principally of amphibolite and Ca-depleted amphibolite, with lesser quartzite and aluminous schist. These are the rocks that Gillmeister (1972) called the Spuhler Peak Formation. The map distribution of the host rocks in the central Tobacco Root Mountains is shown in Figure 2. The MMDS are too small to appear on a map of this scale, but they are widely distributed throughout the Indian Creek and Pony–Middle Mountain Metamorphic Suites. There is no persistent map-scale orientation of the MMDS because they have been deformed and reoriented along with their host rocks.

Many previous workers have recognized the unique character of the MMDS and the constraints they provide for the geologic history of the Tobacco Root Mountains. Tansley et al. (1933) observed "dark dikes and sills" that crosscut the other metamorphic rocks. Reid (1963) emphasizes the "metabasalt dikes and sills" as an important time marker in his sequence of metamorphic and deformation events. Cordua (1973) and Hanley (1975) each devote a chapter of their Ph.D. theses to the MMDS, which they call "metabasites," and both obtained several whole-rock chemical analyses of them. Vitaliano et al. (1979; see reprinted map and text accompanying this volume) show a number of the larger MMDS occurrences as "orthoamphibolites" on their geologic map of the southern Tobacco Root Mountains. Hanley and Vitaliano (1983) provide petrographic details for the "Archean mafic dikes" or "metabasites." Prismatic zircon grains within the MMDS yield a weighted mean $^{207}Pb/^{206}Pb$ age of 2.060 ± 0.006 Ga, which is interpreted to be the age of intrusion (Mueller et al., 2004, this volume, Chapter 9). A second population of smaller, rounded zircon grains have an error-weighted mean $^{207}Pb/^{206}Pb$ age of 1.763 ± 0.008 Ga, which is interpreted to be the age of the metamorphism of these dikes (Mueller et al., 2004, this volume, Chapter 9). This metamorphism is part of a regional-scale event that affected the northwestern Wyoming province, called the Big Sky orogeny by Harms et al. (2004b, this volume, Chapter 11).

We present here whole-rock chemical analyses of MMDS samples, including data for rare earth elements (REEs) and a variety of other trace elements, collected as part of Keck Geology Consortium undergraduate research projects (Brady et al., 1998). Jacob (1994) analyzed nine MMDS samples as part of a larger

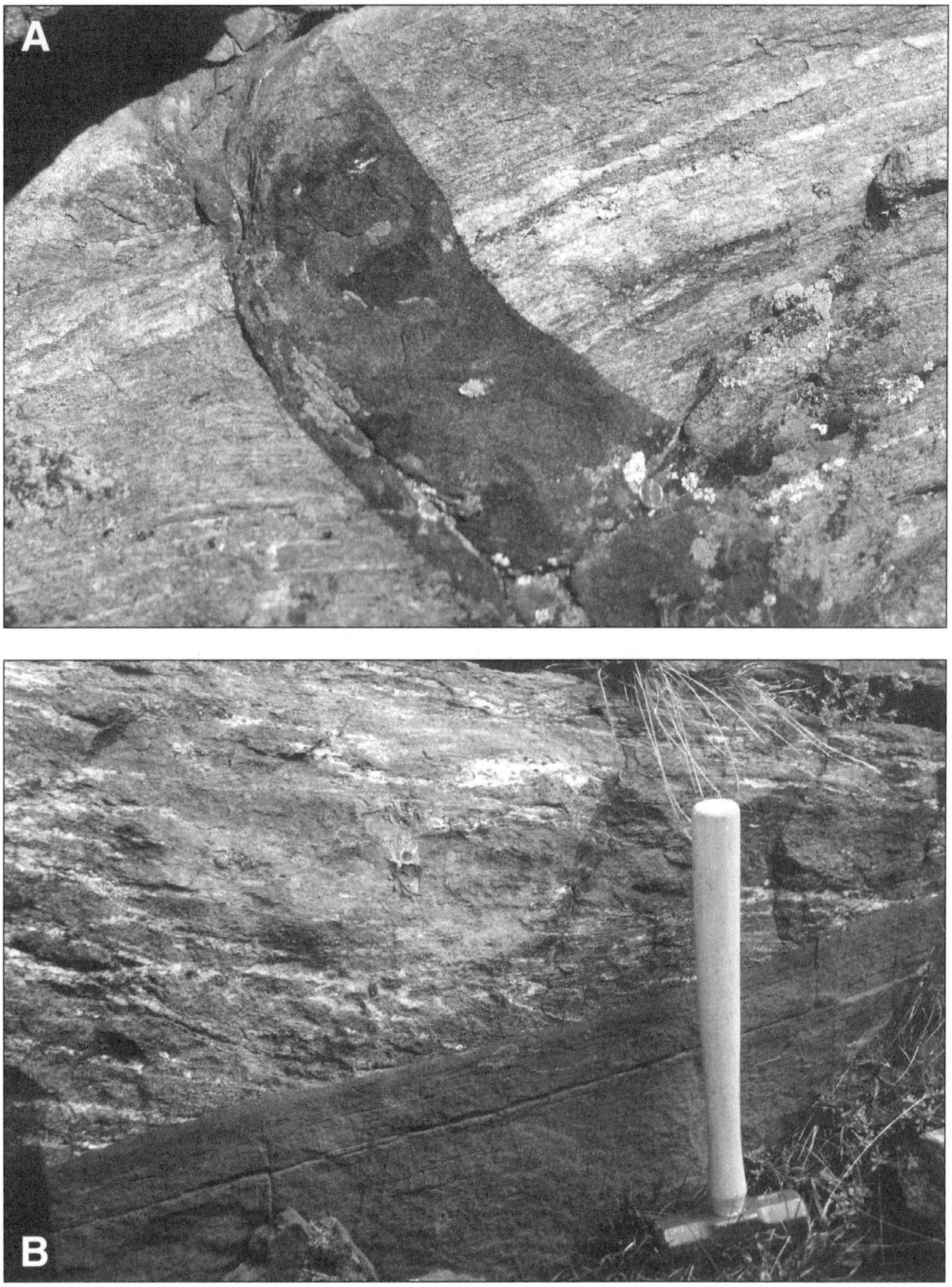

Figure 1. A: A narrow, fine-grained metamorphosed mafic dike (MMDS) cuts Indian Creek Metamorphic Suite gneissic banding in an outcrop near Ramshorn Creek in the southwestern Tobacco Root Mountains. A penny is in the lower right for scale. B: An MMDS dike of unknown thickness crosscuts Indian Creek Metamorphic Suite layering on Thompson Ridge at an angle <30°. Note the weak foliation in the MMDS parallel to its contact with the Indian Creek Metamorphic Suite. Hammer is 40 cm high.

geochemical study of the rocks in the vicinity of Noble Lake. Harris (1998) presented whole-rock chemical data for 14 MMDS samples collected at localities across the Tobacco Root Mountains. These data, along with the wet chemical analyses of Cordua (1973) and Hanley (1975), are consistent with a subalkaline, tholeiitic basalt protolith for the MMDS. The MMDS show a considerable range of enrichment in iron and incompatible elements. We argue here that the chemical variations can be attributed to extensive fractional crystallization of plagioclase, orthopyroxene, and clinopyroxene in the magmatic source of the MMDS.

FIELD RELATIONS

The MMDS of the Tobacco Root Mountains have a characteristic field appearance that permits all but the thinnest of them to be distinguished from other dark metamorphic rocks. Most MMDS features have been described by previous workers and we will use some of their words to aid this description. Perhaps the most distinctive MMDS are those dikes that "… cut across layering in surrounding gneisses at high angles" (Hanley and Vitaliano, 1983) (see Fig. 1). Indeed, because the textures and

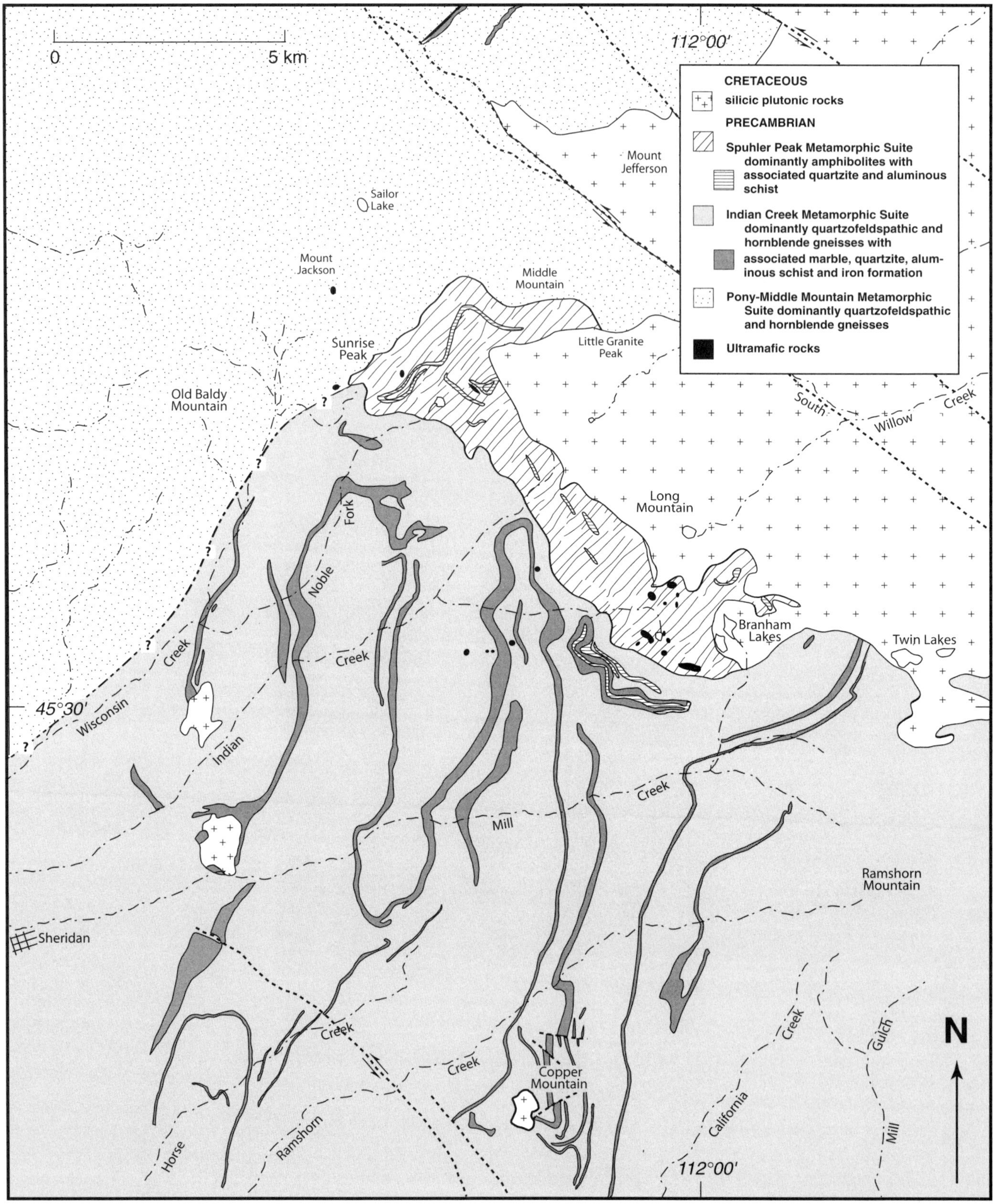

Figure 2. Geologic map of the central portion of the Tobacco Root Mountains. Modified from Vitaliano et al. (1979).

mineralogy of MMDS that have been hydrated can be similar to those of mafic layers in the host gneisses, crosscutting contacts are the only guarantee that a mafic unit is a MMDS. This is true for sills as well as dikes. "Close examination of what appear to be sills generally reveals low-angle discordance to gneissic layering" (Hanley and Vitaliano, 1983) (see Fig. 3). "They are parallel or nearly parallel to the schistosity of the enclosing rocks, locally cutting it at angles of 1–10°" (Reid, 1963). Apparently the fracturing that permitted the intrusion of these sills was not limited to layer-parallel breaks, which is consistent with emplacement into cold rocks that had previously been metamorphosed to high grade. Rotation by shearing during metamorphism may have reduced significantly the contact angle of the MMDS, causing more of them to appear sill-like (Passchier et al., 1990, p. 79).

The MMDS "… are dark rocks, weathering a characteristic dark mottled brown. They are generally resistant to erosion and stand out in sharp relief relative to the amphibolitic and quartzo-feldspathic country rock. Individual intrusives are not large; dikes rarely extend for more than 750 m and generally do not exceed 7–8 m in thickness" (Cordua, 1973, p. 94). MMDS "… are fine- to medium-grained and dark brown to black; they have granoblastic, clustered granoblastic, and microflaser textures" (Vitaliano et al., 1979). Many have distinctive, small (≤1 mm across) garnet crystals that are visible in hand sample and are unlike the larger and less numerous garnet crystals of host rock mafic layers. Other minerals commonly visible include white plagioclase and black clinopyroxene and hornblende. Amphibole and biotite are more abundant near the margins of the MMDS, which indicates that during metamorphism more water was able to reach the margins than the middle of the MMDS. Unlike many unmetamorphosed basalts, the MMDS do not attract a magnet; ilmenite is the major oxide phase. Modal variations are summarized in Hanley and Vitaliano (1983).

Figure 3. A: Metamorphosed mafic dikes and sills (MMDS) outcrop as dark sills in these Pony–Middle Mountain Metamorphic Suite gneisses in a cliff wall southwest of Middle Mountain. The mafic sills truncate the gneissic banding at low angles. One of the sills thins dramatically where the other sills have a nearly constant thickness, so this may be due to original intrusive thickness rather than deformation. B: Close-up of a thin MMDS sill that outcrops near Noble Lake showing a low-angle discordance with the adjacent layering.

Although the MMDS have been metamorphosed to at least the upper amphibolite facies, remnants of primary igneous features can still be seen in outcrop. For example, contacts between the MMDS and their host rocks are invariably sharp. The margins of the MMDS are likely to be fine-grained relative to their centers, with metamorphic minerals apparently mimicking the grain size distribution in an igneous protolith with chilled margins and a coarser, more slowly cooled center (Fig. 4). Hanley and Vitaliano (1983) suggest that there are "… relict phenocrysts of plagioclase, and less commonly of hypersthene and augite…." In several places, a second, younger MMDS has intruded an older MMDS. In all cases we examined, the younger dike was narrow and fine-grained relative to the wider, coarser-grained dike being intruded (Fig. 5). These features suggest that the host MMDS was relatively cool by the time a second dike appeared. "Locally, the sills enclose xenoliths of gneiss or amphibolitic granulite" (Reid, 1963). In some of the widest MMDS, metamorphic textures mimicking primary igneous layering can be observed (Fig. 6). The preservation

Figure 4. Narrow metamorphosed mafic dike (MMDS) from Sunrise Basin with white clusters of plagioclase crystals that are smaller near the edge of the sill and are nearly invisible in what was once a chilled margin. Note that the plagioclase clusters exhibit a lineation that is parallel to the sill boundary. The footbag in the photo is 4 cm wide.

Figure 5. The contact between a coarse-grained metamorphosed mafic sill (MMDS) in Sunrise Basin and a later, fine-grained MMDS dike. Both MMDS rocks consist of the same minerals. Clusters of white plagioclase crystals are visible in both rocks, but garnet and pyroxene crystals in the fine-grained dike are too small to be seen in the photo. The contact is located just to the left of the pocketknife, which is included for scale.

Figure 6. A layered texture in a metamorphosed mafic sill (MMDS) outcrop of a thick sill with coarse grain size in Spuhler Gulch. The layers are defined by a concentration of mafic minerals at the bottom. This texture is believed to be the result of compositional variations that resulted from crystal settling during slow crystallization of the original sill. Hammer head is 12 cm long.

of igneous textures in these high-grade metamorphic rocks may be due to a lack of water during metamorphism, both in the basalt dikes and in the host rocks that were previously metamorphosed.

The igneous features described above have survived significant deformation of the MMDS during metamorphism. "Foliation and lineation exist to varying degrees in most…" MMDS (Hanley and Vitaliano, 1983) (see Fig. 4) and may be parallel to their contacts with the rocks they intruded, instead of parallel to the foliation in those rocks. These fabrics are visible in outcrop principally due to the color contrast of clusters of small, white plagioclase crystals and their matrix. "Many of the dikes have been sheared, folded, or intruded by pegmatite dikes" (Vitaliano et al., 1979) (see Figs. 7 and 8). "Many [MMDS] bodies are deformed into boudins and these are commonly stretched into laminae. Foliation within deformed [MMDS] may be folded near the ends of the boudins" (Hanley and Vitaliano, 1983). Fold axes for MMDS folds are parallel to the axes of folds in their host rocks, indicating that both the MMDS and their host rocks were deformed together during the 1.77 Ga metamorphism (see Harms et al., 2004a, this volume, Chapter 10).

Finally, MMDS intrude both the Pony–Middle Mountain and Indian Creek Metamorphic Suites, but neither we nor any previous worker has found a MMDS in the Spuhler Peak Metamorphic Suite, even though MMDS crop out within a few meters of the suite's contact (Gillmeister, 1972). The Spuhler Peak Metamorphic Suite is composed predominantly of mafic rocks and many of these rocks have chemical compositions similar to those of the MMDS (see Burger et al., 2004, this volume, Chapter 3). Nevertheless, we did not observe any crosscutting features for rocks in the Spuhler Peak Metamorphic Suite that would make them candidates to be

Figure 7. Folded metamorphosed mafic dikes and sills. A: A large, tight fold on a 200-m-high cliff face in Sunrise Basin. B: A small fold of a Sunrise Basin sill that crosscuts gneissic banding both in the fold and more prominently out of view along the sill. Note the dark margin that is rich in hornblende and biotite. A 10-cm-long pocketknife is in the photo for scale.

Figure 8. A coarse metamorphosed mafic dike (MMDS) from Spuhler Gulch with a fabric developed by shearing. The intensity of shearing increases to the left in the photo.

MMDS. Gillmeister (1972) suggested that the MMDS might be feeder dikes for some of the Spuhler Peak Metamorphic Suite. Although this is permitted by the chemistry of the MMDS, it is not consistent with other evidence that has been interpreted to suggest a marine setting for the origin of the Spuhler Peak Metamorphic Suite (Burger et al., 2004, this volume, Chapter 3) in contrast with the continental setting that we propose for the MMDS.

WHOLE-ROCK CHEMISTRY

The chemical compositions of the MMDS may be used to demonstrate important features of their igneous origins with implications for the history of the Tobacco Root Mountains. The available data include 23 whole-rock chemical analyses of MMDS obtained as part of our study (Jacob, 1994; Harris, 1998; Brady et al., 1998) and 16 whole-rock chemical analyses of MMDS from unpublished Ph.D. theses (Cordua, 1973; Hanley, 1975) (see Table 1). Chemical analyses were obtained using a variety of analytical techniques at several laboratories. Cordua (1973) and Hanley (1975) obtained only major element data using standard wet chemical procedures described in those theses. Major elements and some trace elements were determined using X-ray fluorescence equipment at the University of Massachusetts (Jacob, 1994) and at Franklin and Marshall College (Harris, 1998). Analytical procedures are described in Rhodes (1996) and Mertzman (2000), respectively. Many of the trace element data were determined using nuclear activation analysis at the Oregon State University Radiation Center (Harris, 1998), following procedures described at http://ne.oregonstate.edu/facilities/radiation_center/naa2.html, or at Boston College (Jacob, 1994) following procedures described in Cardoza et al. (1990). In all cases, chemical analyses are reported on an anhydrous basis with total iron given as FeO. We believe that the robustness of the data is demonstrated by the consistency of results obtained from the different laboratories.

Based on their major element chemistry, the protoliths of the MMDS were subalkaline, tholeiitic basalts (Fig. 9). Nearly all samples have SiO_2 contents between 48% and 52% by weight, with the alkali sum Na_2O + K_2O less than 4%. However, the molar ratio MgO/(MgO + FeO) is more variable, ranging from 0.65 to 0.33. TiO_2, P_2O_5 and a number of trace elements (e.g., Zr, Nb, Ce, Nd) vary by a factor of four or more. These and other data are consistent with significant fractional crystallization in the evolution of the MMDS magmas.

Although the weight percents of both TiO_2 and P_2O_5 cover a wide range of values (Table 1), the ratio TiO_2/P_2O_5 is nearly constant for the MMDS data set (Fig. 10). We interpret this to mean that (1) the MMDS do indeed have a common magmatic origin, (2) both TiO_2 and P_2O_5 are largely incompatible with the fractionating minerals, and (3) either TiO_2 or P_2O_5 can be used as a normalizing component to explore quantitatively the chemical evolution of the MMDS (Pearce, 1968). For example, the Mg# (molar Mg/(Mg + Fe)) is plotted in Figure 11 as a function of MgO/TiO_2. If TiO_2 is a highly incompatible component, then both the Mg# and the value of MgO/TiO_2 should decline during fractional crystallization, which they do. Indeed, the largest value of both TiO_2 and P_2O_5 is five times the smallest value, indicating that as much as 80% of the original liquid was removed by fractional crystallization if there was a single magmatic source of the MMDS. Like the Stillwater intrusion, where Fe-Ti oxides appear only in the uppermost exposed layers, it appears that Fe-Ti oxides were not stable during the fractional crystallization sampled by these dikes. We have used Mg#, MgO/TiO_2, SiO_2/TiO_2, or a largely incompatible element such as Zr or P_2O_5 as a measure of the degree of fractional crystallization on various composition diagrams, with essentially the same result.

Several lines of evidence can be used to identify the minerals that were removed from the MMDS source by fractional crystallization. Olivine is commonly an early crystallizing phase for basalts, but the MMDS chemistry appears to span a fractionation history that postdates the importance of olivine. For example, there is no clear correlation between Ni and MgO, which would be expected if there were significant olivine fractionation (Fig. 12). Also, the SiO_2 content stays approximately constant as the TiO_2 content increases, instead of the increase of SiO_2 that normally accompanies removal of significant olivine (Fig. 13). Al_2O_3, CaO, and MgO all decrease as the Mg# decreases (Fig. 14). Plagioclase fractionation is the likely cause of the decrease of Al_2O_3. Because the value of CaO/Al_2O_3 is nearly constant (Fig. 15), it is unlikely that clinopyroxene is particularly important. Therefore, we are left with a low-Ca pyroxene as the best fractionation candidate to explain the variation in MgO (and Mg#).

The plausibility of fractional crystallization of various combinations of olivine, orthopyroxene (or pigeonite), plagioclase, and clinopyroxene was also tested using the procedures of Stanley and Russell (1989) with TiO_2 as the normalizing component and SiO_2/TiO_2 as the index of fractionation. Results of several models are shown in Figure 16. Each test is designed so that if all of the variation of SiO_2/TiO_2 in the liquid could be explained

TABLE 1. WHOLE-ROCK MAJOR AND TRACE ELEMENT ANALYSES FOR MMDS SAMPLES

Source	Hanley	Harris	Cordua	Harris	Harris	Harris	Harris	Harris	Cordua	Hanley	Jacob	Cordua	Hanley
Sample	16-7-68-10	CH4-2b	E-71-624	CH2-3	CH12-1	CH10-4a	CH9-4b	KJS34a	E-71-597	9-8-67-5	RB-C	E-71-479	CC10
Host	PMMMS	PMMMS	ICMS	ICMS	ICMS	PMMMS	PMMMS	PMMMS	ICMS	PMMMS	ICMS	ICMS	PMMMS
12T	◊[§]	042095	◊	041762	041795	041291	041499	041645	◊	◊		◊	◊
UTM	◊	504827	◊	503940	504186	505006	504784	505465	◊	◊		◊	◊
Oxide (wt%)													
SiO_2	47.99	48.34	48.62	49.28	49.39	49.59	49.61	49.65	49.72	49.75	49.84	49.86	49.89
TiO_2	3.00	2.84	0.82	1.64	2.02	1.22	2.05	1.34	1.15	0.60	1.89	2.39	1.27
Al_2O_3	13.12	12.46	14.04	12.66	12.26	13.39	12.31	13.97	15.19	17.93	12.29	13.51	14.37
*FeO	16.69	17.13	10.15	16.08	17.46	13.34	16.35	13.97	12.42	9.28	16.71	14.70	12.29
MnO	0.29	0.26	0.17	0.28	0.28	0.24	0.27	0.23	0.22	0.17	0.28	0.23	0.23
MgO	5.45	5.13	10.68	5.48	4.92	7.19	5.14	6.55	7.44	7.35	5.59	5.76	7.37
CaO	9.62	9.84	11.90	10.08	9.77	11.16	9.93	11.01	10.58	13.80	9.50	9.82	11.74
Na_2O	2.42	1.59	1.33	1.85	2.23	1.85	2.21	1.96	1.68	1.52	2.31	2.53	2.30
K_2O	0.85	0.35	1.34	0.65	0.49	0.40	0.50	0.40	0.52	0.26	1.44	0.76	0.37
P_2O_5	0.34	0.33	0.04	0.13	0.20	0.11	0.20	0.13	0.09	0.04	0.18	0.26	0.14
Total	99.77	98.27	99.09	98.12	99.02	98.50	98.58	99.22	99.01	100.70	100.04	99.82	99.97
Trace elements (ppm)													
Sc	•[†]	37.9	•	44.0	41.2	44.3	43.7	47.4	•	•	•	•	•
V	•	373	•	347	315	280	315	258	•	•	424	•	•
Cr	•	143	•	155	77	219	185	60	•	•	68	•	•
Co	•	57.0	•	63.9	55.3	57.1	52.6	56.5	•	•	•	•	•
Ni	•	227	•	268	167	219	121	101	•	•	60	•	•
Cu	•	205	•	•	270	197	244	192	•	•	•	•	•
Zn	•	151	•	145	149	125	121	123	•	•	174	•	•
Ga	•	23.2	•	•	21.5	17.6	21.8	19.7	•	•	•	•	•
As	•	3.3	•	2.0	2.4	2.0	1.6	1.0	•	•	•	•	•
Se	•	1.8	•	<1.9	0.7	<1.4	<1.7	<1.7	•	•	•	•	•
Rb	•	22	•	37	19	29	25	<9	•	•	•	•	•
Sr	•	246	•	235	158	188	106	102	•	•	92	•	•
Y	•	38	•	•	36	25	37	25	•	•	•	•	•
Zr	•	222	•	69	135	111	73	<90	•	•	121	•	•
Nb	•	19.6	•	•	11.9	7.1	11.9	7.9	•	•	•	•	•
Sb	•	0.67	•	0.74	0.00	0.00	0.22	0.00	•	•	4.0	•	•
Cs	•	1.2	•	1.4	0.8	1.1	1.4	0.2	•	•	•	•	•
Ba	•	113	•	113	193	108	218	72	•	•	137	•	•
La	•	24.5	•	9.2	12.8	7.5	7.8	4.5	•	•	5.0	•	•
Ce	•	56	•	26	33	18	18	12	•	•	19	•	•
Nd	•	29	•	16	22	10	12	•	•	•	•	•	•
Sm	•	7.30	•	4.01	5.22	2.92	2.59	2.56	•	•	•	•	•
Eu	•	2.23	•	1.28	1.69	1.03	0.96	0.91	•	•	•	•	•
Tb	•	1.18	•	0.80	1.08	0.64	0.61	0.61	•	•	•	•	•
Yb	•	3.5	•	2.4	3.5	2.3	2.5	2.6	•	•	•	•	•
Lu	•	0.50	•	0.32	0.44	0.30	0.32	0.27	•	•	•	•	•
Hf	•	5.60	•	2.98	3.99	2.33	2.03	1.91	•	•	•	•	•
Ta	•	1.32	•	0.47	0.66	0.42	0.34	0.23	•	•	•	•	•
W	•	1.9	•	1.0	1.8	0.9	0.5	4.2	•	•	•	•	•
Os (ppb)	•	<1.1	•	1.3	<1.3	<1.3	<1.1	<1.2	•	•	•	•	•
Au (ppb)	•	0.01	•	0.01	0.01	0.02	0.01	0.00	•	•	•	•	•
Hg	•	<0.03	•	0.06	<0.03	0.06	<0.03	<0.03	•	•	•	•	•
Pb	•	9.1	•	•	8.9	8.4	9.1	8.5	•	•	•	•	•
Th	•	2.8	•	2.0	1.8	1.1	1.6	0.7	•	•	•	•	•
U	•	1.2	•	0.5	0.6	0.7	0.4	<0.8	•	•	•	•	•

(continued)

TABLE 1. WHOLE-ROCK MAJOR AND TRACE ELEMENT ANALYSES FOR MMDS SAMPLES (*continued*)

Source	Harris	Jacob	Hanley	Hanley	Jacob	Cordua	Harris	Jacob	Hanley	Harris	Cordua	Jacob	Hanley
Sample	CH9-4a	LJ34	SL90B	26-7-67-5	LJ16	E-71-288	CH11-2	LJ29	57903	CH10-4b	E-71-29b	LJ25	16-8-67-1A
Host	PMMMS	ICMS	PMMMS	PMMMS	ICMS	ICMS	ICMS	ICMS	PMMMS	PMMMS	ICMS	ICMS	PMMMS
12T	041499	041631	◊	◊	041814	◊	041623	041644	◊	041291	◊	041631	◊
UTM	504784	504509	◊	◊	503330	◊	503347	504517	◊	505006	◊	504509	◊
Oxide (wt%)													
SiO_2	49.91	50.13	50.16	50.18	50.28	50.29	50.31	50.32	50.44	50.48	50.74	50.78	50.81
TiO_2	1.12	1.90	0.72	1.57	1.37	1.30	0.90	1.14	1.10	1.11	2.60	2.10	0.88
Al_2O_3	13.71	12.56	14.49	13.06	14.11	14.14	13.52	14.45	15.32	14.17	14.84	14.23	14.65
*FeO	13.44	16.62	11.84	16.66	12.52	13.23	14.33	13.01	10.30	14.24	13.96	14.13	12.78
MnO	0.23	0.28	0.20	0.28	0.22	0.23	0.25	0.25	0.17	0.24	0.19	0.27	0.22
MgO	6.34	5.38	8.28	5.59	6.98	7.72	6.50	6.42	8.97	6.08	6.00	4.81	7.08
CaO	10.61	8.67	12.16	9.56	11.50	10.60	10.77	10.50	11.47	10.08	8.62	8.73	11.62
Na_2O	1.97	2.85	1.86	2.21	2.44	1.83	2.01	2.52	2.00	2.06	2.75	3.36	1.89
K_2O	0.53	0.79	0.23	0.35	0.39	0.66	0.28	0.51	0.48	0.43	0.13	1.21	0.23
P_2O_5	0.10	0.18	0.41	0.12	0.13	0.09	0.07	0.10	0.11	0.11	0.43	0.25	0.07
Total	97.97	99.35	100.35	99.58	99.94	100.09	98.95	99.22	100.36	99.02	100.26	99.88	100.23
Trace elements (ppm)													
Sc	43.8	45.3	•	•	41.6	•	48.7	41.7	•	40.9	•	•	•
V	281	422	•	•	287	•	284	295	•	237	•	277	•
Cr	177	66	•	•	163	•	154	87	•	181	•	36	•
Co	56.7	50.7	•	•	68.4	•	52.1	55.9	•	53.6	•	•	•
Ni	169	55	•	•	92	•	201	75	•	129	•	52	•
Cu		•	•	•	•	•	122	•	•	147	•	•	•
Zn	125	159	•	•	110	•	132	187	•	121	•	149	•
Ga	•	•	•	•	•	•	17.5	•	•	17.4	•	•	•
As	2.5	•	•	•	•	•	1.7	•	•	2.7	•	•	•
Se	2.0	•	•	•	•	•	<1.7	•	•	<1.4	•	•	•
Rb	17	37	•	•	20	•	11	22	•	13	•	•	•
Sr	141	160	•	•	330	•	99	178	•	200	•	269	•
Y	•	•	•	•	•	•	22	•	•	27	•	•	•
Zr	70	126	•	•	88	•	75	71	•	83	•	185	•
Nb	•	•	•	•	•	•	3.7	•	•	8.2	•	•	•
Sb	1.00	5.0	•	•	8.0	•	0.00	4.0	•	0.00	•	17.0	•
Cs	1.2	0.2	•	•	0.7	•	0.5	0.4	•	0.3	•	•	•
Ba	202	214	•	•	310	•	64	101	•	195	•	384	•
La	8.5	7.7	•	•	11.4	•	3.3	5.6	•	9.6	•	19.0	•
Ce	20	20	•	•	26	•	9	13	•	24	•	53	•
Nd	14	13	•	•	15	•	6	9	•	12	•	•	•
Sm	3.03	4.38	•	•	3.54	•	2.16	2.60	•	3.02	•	•	•
Eu	1.03	1.47	•	•	1.20	•	0.82	0.97	•	1.07	•	•	•
Tb	0.68	0.95	•	•	0.57	•	0.57	0.59	•	0.71	•	•	•
Yb	2.5	4.2	•	•	1.9	•	2.3	1.9	•	2.8	•	•	•
Lu	0.29	0.66	•	•	0.25	•	0.28	0.32	•	0.29	•	•	•
Hf	2.37	3.69	•	•	2.47	•	1.66	1.84	•	2.56	•	•	•
Ta	0.38	0.22	•	•	0.82	•	0.17	0.30	•	0.48	•	•	•
W	1.6	•	•	•	•	•	<1	•	•	1.5	•	•	•
Os (ppb)	0.6	•	•	•	•	•	0.4	•	•	0.6	•	•	•
Au (ppb)	0.01	•	•	•	•	•	0.01	•	•	0.01	•	•	•
Hg	<0.03	•	•	•	•	•	<0.03	•	•	<0.03	•	•	•
Pb	•	•	•	•	•	•	5.7	•	•	8.0	•	•	•
Th	2.0	1.3	•	•	1.2	•	0.5	0.4	•	1.7	•	•	•
U	0.4	•	•	•	6.6	•	<1.4	•	•	0.8	•	•	•

(*continued*)

TABLE 1. WHOLE-ROCK MAJOR AND TRACE ELEMENT ANALYSES FOR MMDS SAMPLES (*continued*)

Source	Jacob	Harris	Hanley	Harris	Hanley	Harris	Harris	Hanley	Jacob	Hanley	Harris	Jacob	Jacob
Sample	JB93-75	CH1-3a	20-8-67-7	CH1-3b	SL97	CH4-2a	CH3-1	MH31	LJ77	10-8-67-2A	CH5-1	LJ9	LJ7
Host	ICMS	PMMMS	PMMMS	PMMMS	PMMMS	PMMMS	ICMS	PMMMS	ICMS	PMMMS	PMMMS	ICMS	ICMS
12T	041407	041498	◊	041498	◊	042095	041573	◊	041571	◊	041810	041238	041562
UTM	504614	504771	◊	504771	◊	504827	504556	◊	504562	◊	504900	502847	504570
Oxide (wt%)													
SiO_2	50.82	50.96	50.96	51.02	51.05	51.06	51.26	51.43	51.47	51.59	52.03	52.58	52.70
TiO_2	1.15	1.06	0.71	1.00	1.07	0.78	0.86	1.23	1.03	0.81	1.51	1.84	1.04
Al_2O_3	15.67	14.76	14.73	12.80	14.50	13.85	14.35	14.00	13.66	14.93	13.38	13.79	14.01
*FeO	12.33	12.14	11.82	15.16	13.37	11.58	12.11	13.41	12.68	10.77	13.80	15.04	12.03
MnO	0.23	0.21	0.22	0.28	0.23	0.20	0.21	0.25	0.24	0.19	0.22	0.25	0.22
MgO	5.62	6.45	7.63	6.37	7.30	7.44	6.86	6.39	6.96	7.34	5.39	4.35	6.56
CaO	10.31	10.65	12.01	10.27	9.32	11.57	11.16	10.14	9.96	11.58	9.12	8.54	10.64
Na_2O	3.00	2.03	1.83	1.97	2.48	1.91	1.92	2.27	2.70	2.52	2.14	2.29	2.18
K_2O	0.39	0.64	0.23	0.23	0.39	0.40	0.51	0.49	0.89	0.45	0.97	1.29	0.54
P_2O_5	0.10	0.11	0.05	0.08	0.92	0.07	0.08	0.12	0.08	0.07	0.16	0.21	0.10
Total	99.62	99.01	100.19	99.17	100.63	98.86	99.32	99.73	99.67	100.25	98.73	100.16	100.03
Trace elements (ppm)													
Sc	•	39.0	•	48.5	•	45.6	43.2	•	•	•	8.3	•	46.3
V	312	226	•	313	•	246	226	•	299	•	286	375	283
Cr	59	245	•	60	•	30	180	•	103	•	13	29	101
Co	•	48.4	•	57.9	•	45.3	46.4	•	•	•	42.9	•	55.0
Ni	56	159	•	98	•	90	118	•	59	•	14	43	150
Cu	•	100	•	170	•		127	•	•	•	87	•	•
Zn	109	112	•	129	•	135	110	•	112	•	54	155	113
Ga	•	17.9	•	18.1	•	•	17.8	•	•	•	18.6	•	•
As	•	5.4	•	1.1	•	1.3	1.0	•	•	•	1.8	•	•
Se	•	<1.4	•	<1.9	•	<1.5	<1.4	•	•	•	1.9	•	•
Rb	•	40	•	<9	•	67	22	•	•	•	19	•	21
Sr	132	174	•	98	•	317	183	•	130	•	177	151	390
Y	•	26	•	26	•	•	20	•	•	•	29	•	•
Zr	72	74	•	<93	•	86	132	•	60	•	273	172	100
Nb	•	8.1	•	4.6	•	•	5.9	•	•	•	10.6	•	•
Sb	3.0	0.00	•	0.00	•	0.45	0.00	•	4.0	•	0.00	8.0	5.0
Cs	•	1.3	•	0.2	•	7.9	1.1	•	•	•	0.9	•	0.8
Ba	63	268	•	177	•	768	163	•	175	•	357	367	310
La	3.0	10.3	•	4.4	•	21.1	7.3	•	3.0	•	16.8	23.0	9.7
Ce	19	21	•	13	•	41	18	•	12	•	37	48	23
Nd	•	12	•	<6.9	•	25	10	•	•	•	23	•	13
Sm	•	3.04	•	2.48	•	6.49	2.38	•	•	•	4.98	•	3.07
Eu	•	1.00	•	0.92	•	2.11	0.88	•	•	•	1.41	•	1.09
Tb	•	0.67	•	0.64	•	1.05	0.53	•	•	•	0.80	•	0.58
Yb	•	2.7	•	2.9	•	3.5	2.3	•	•	•	2.8	•	2.4
Lu	•	0.34	•	0.37	•	0.49	0.24	•	•	•	0.36	•	0.35
Hf	•	2.55	•	1.96	•	3.66	2.01	•	•	•	10.90	•	2.76
Ta	•	0.42	•	0.23	•	0.47	0.29	•	•	•	1.17	•	0.81
W	•	<2	•	1.0	•	0.9	1.4	•	•	•	0.8	•	•
Os (ppb)	•	0.8	•	<1.2	•	0.7	<1.2	•	•	•	0.7	•	•
Au (ppb)	•	0.01	•	0.01	•	0.02	0.01	•	•	•	0.01	•	•
Hg	•	<0.03	•	0.03	•	<0.03	<0.03	•	•	•	<0.02	•	•
Pb	•	7.7	•	6.3	•	•	6.9	•	•	•	10.4	•	•
Th	•	2.1	•	0.7	•	2.5	1.5	•	•	•	11.4	•	1.9
U	•	0.5	•	<1.7	•	0.9	0.6	•	•	•	0.8	•	6.9

Note: Data from Cordua (1973), Hanley (1975), Harris (1998), and Jacob (1994). MMDS—metamorphosed dikes and sills; PMMMS—Pony–Middle Mountain Metamorphic Suite; ICMS—Indian Creek Metamorphic Suite.

*FeO: total iron as FeO.

[§]◊—see Hanley (1975) for location.

[†]•—no data.

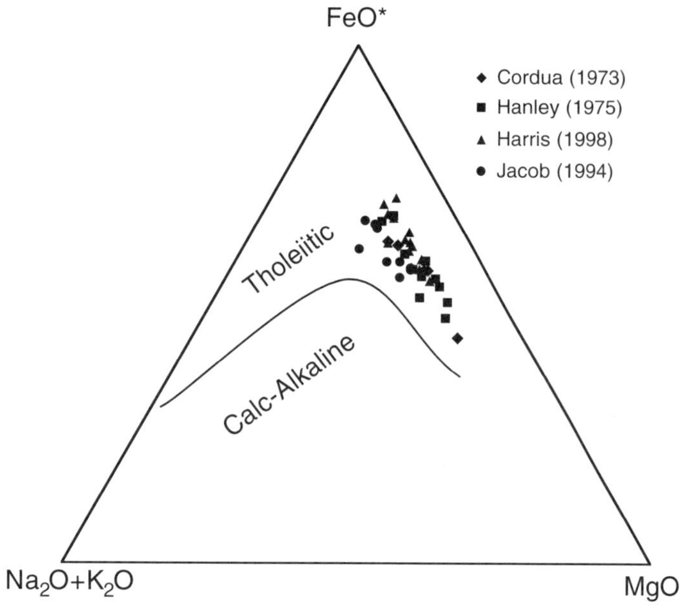

Figure 9. Metamorphosed mafic dikes and sills analyses in weight percent oxides are shown on the AFM diagram of Irvine and Baragar (1971). FeO* means that all iron is plotted as FeO. The data show a tholeiitic fractionation trend of iron enrichment at constant alkalis.

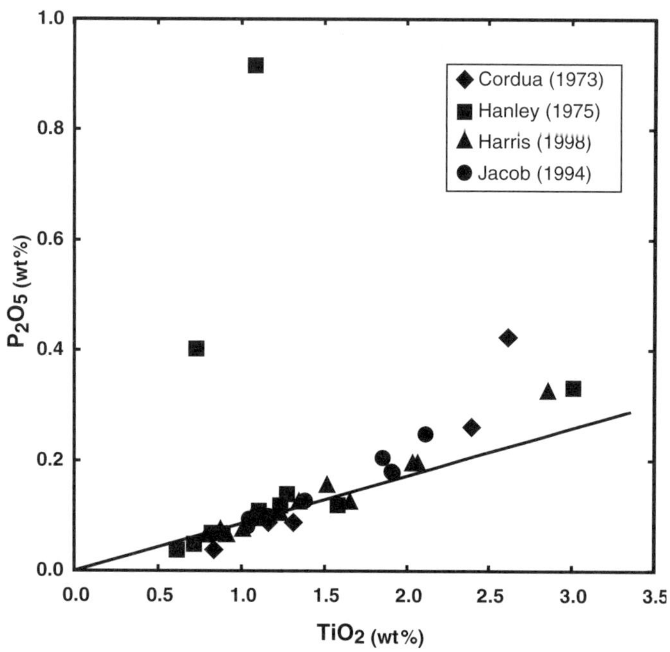

Figure 10. Weight percent P_2O_5 vs. weight percent TiO_2 for the metamorphosed mafic dikes and sills samples listed in Table 1. With the exception of three samples, all fall very close to a straight line that passes through the origin. These data are consistent with both TiO_2 and P_2O_5 being largely incompatible with any fractionating minerals and with P_2O_5 being slightly more incompatible than TiO_2.

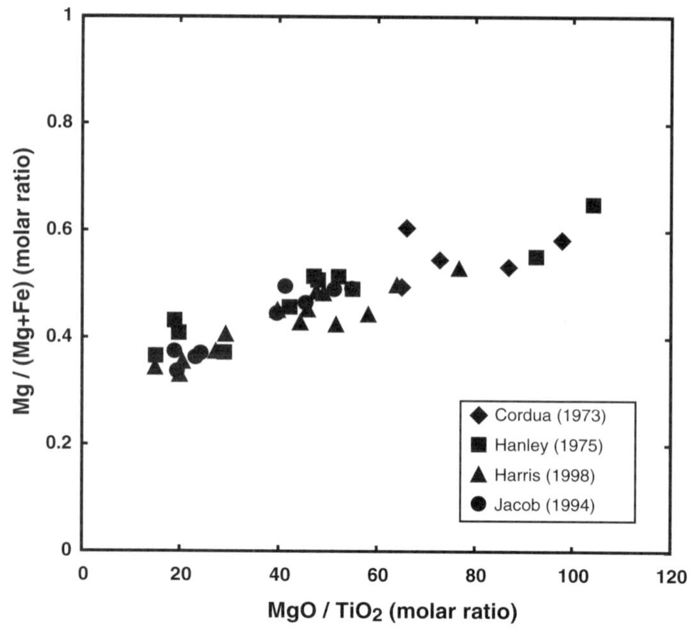

Figure 11. The Mg#, which is the molar ratio of Mg to (Mg + Fe), of the MMDS samples is shown as a function of MgO/TiO_2. A decrease in MgO/TiO_2 is matched by a decrease in Mg#, consistent with fractional crystallization of one or more Mg-Fe silicates with Ti as an incompatible element.

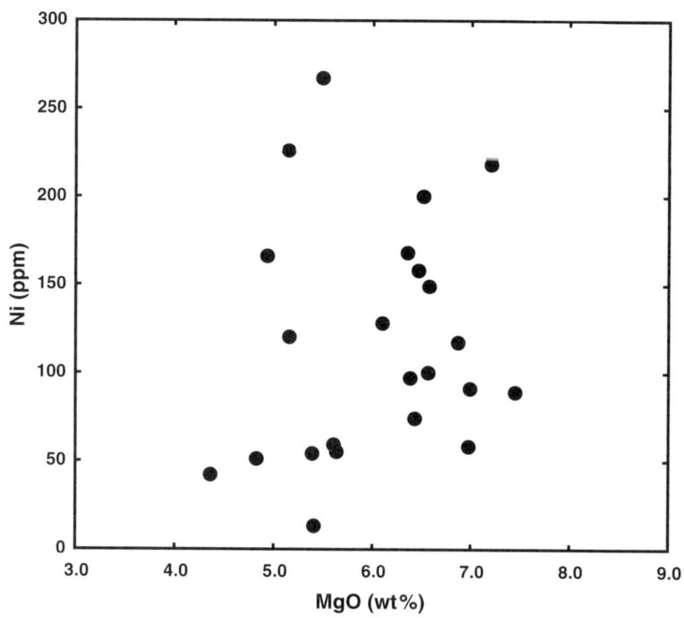

Figure 12. Ni content in parts per million is shown as a function of MgO content in weight percent. The absence of correlation between these numbers is interpreted to mean that olivine is not an important fractionating phase in the evolution of metamorphosed mafic dikes and sills magmas represented by these samples.

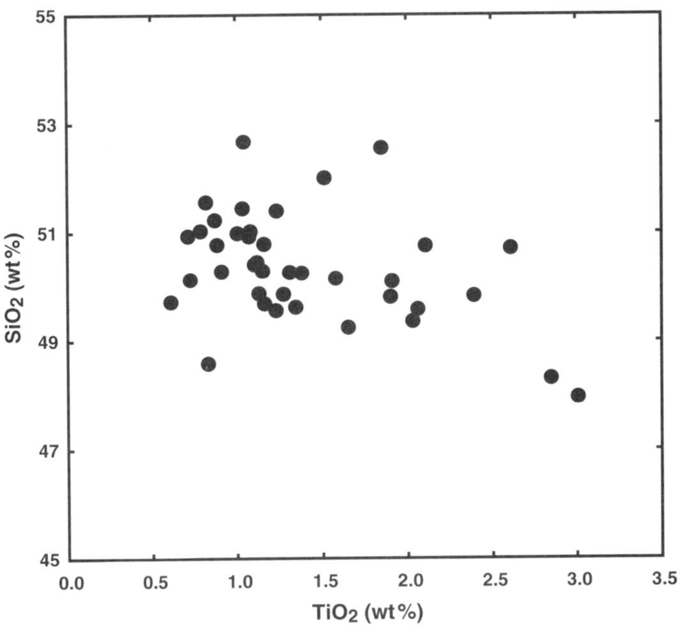

Figure 13. SiO_2 content in weight percent is shown as a function of TiO_2 content in weight percent. Although there is considerable scatter, the weight percent of SiO_2 decreases somewhat as the weight percent of TiO_2 increases during fractional crystallization. The weight percent of SiO_2 should increase if olivine is a major fractionating phase.

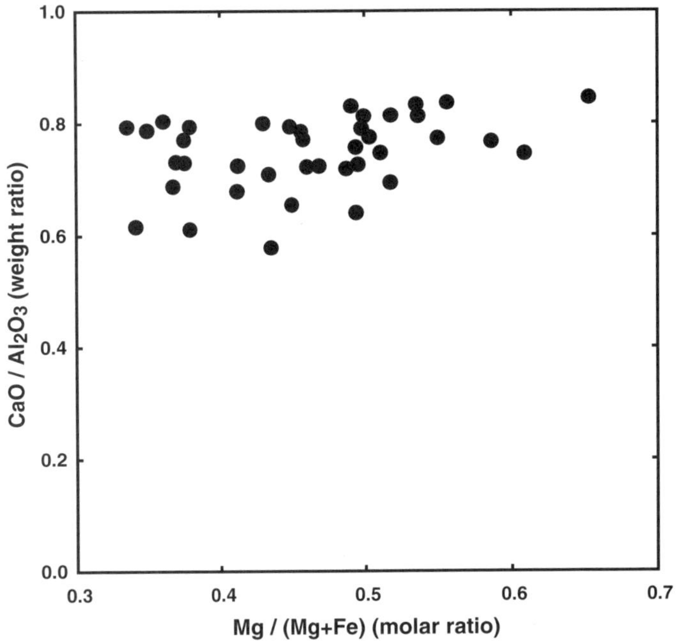

Figure 15. The CaO/Al_2O_3 weight ratio is shown as a function of the Mg# (molar Mg/(Mg + Fe)). The value of CaO/Al_2O_3 is nearly constant, decreasing only slightly as the Mg# decreases during fractional crystallization. This pattern is interpreted to mean that there is only a modest amount of clinopyroxene removal from the metamorphosed mafic dikes and sills liquids.

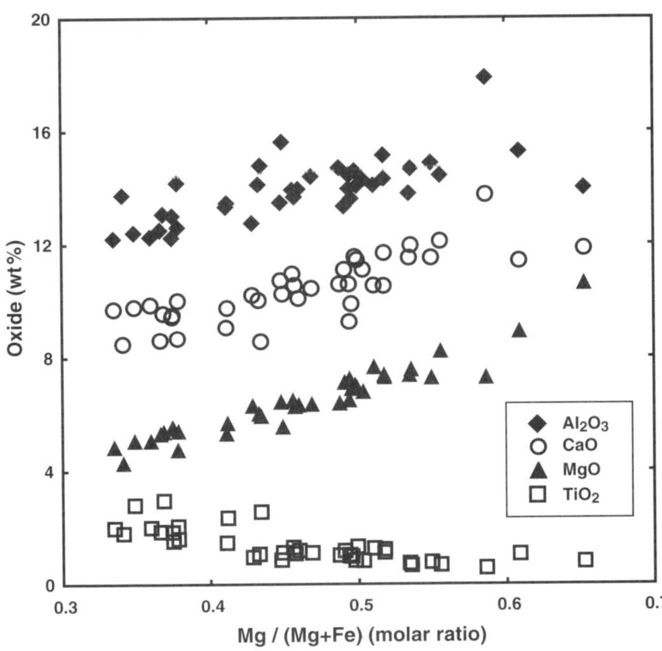

Figure 14. Weight percentages of Al_2O_3, CaO, MgO, and TiO_2 are shown as a functions of Mg# (molar Mg/(Mg + Fe)) for the metamorphosed mafic dikes and sills samples. Al_2O_3, CaO, and MgO decrease as Mg# decreases which is consistent with their removal during fractional crystallization as major components of the fractionation phases. TiO_2 increases as Mg# decreases, which shows that it is highly incompatible with the fractionating phases.

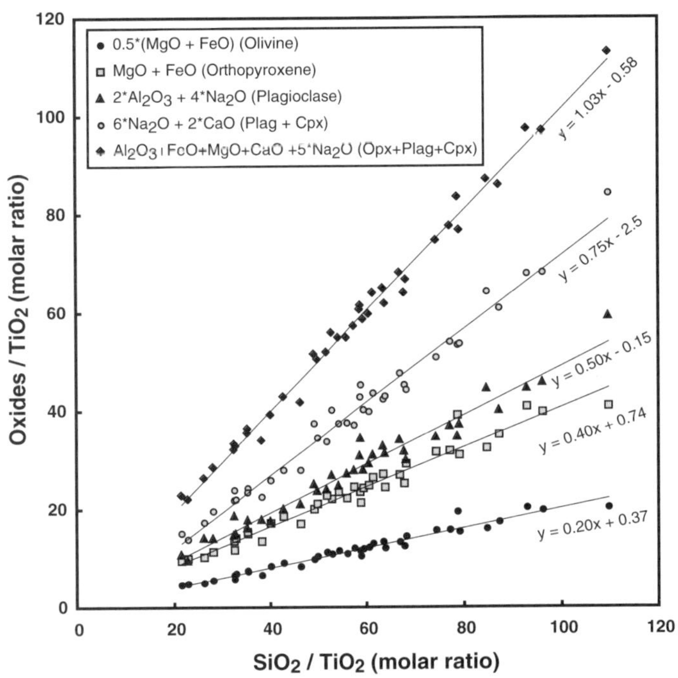

Figure 16. Pearce element ratio diagram showing several tests for fractional crystallization using TiO_2 as a normalizing component for oxide molar ratios. The oxide combinations in the key are divided by TiO_2 on the ordinate. The test for orthopyroxene, plagioclase, and clinopyroxene has a slope of one, meaning that all of the extensive chemical variation of SiO_2/TiO_2 can be explained by fractional crystallization of these phases.

by the crystallization of the named minerals, the chemical data would plot on a line with a slope of one (Stanley and Russell, 1989). Fractional crystallization of olivine could explain only a small portion of the observed change in SiO_2/TiO_2 (slope = 0.2), whereas fractional crystallization of orthopyroxene could account for 40% of the change in SiO_2/TiO_2 (slope = 0.4). Plagioclase crystallization can account for half of the change in SiO_2/TiO_2 (slope = 0.5). So removal of primarily orthopyroxene and plagioclase is consistent with this alternative test. A test for the combination orthopyroxene (or pigeonite) + plagioclase + clinopyroxene yields a slope of one. This result does not rule out the possibility of olivine as a fractionating phase, but it is likely that the crystallization of olivine is minor relative to the other three minerals.

Trace element data support this conclusion. The proportion of nearly all trace elements increases as Mg# decreases, probably because their bulk distribution coefficients are all less than one. However, those trace elements that have higher distribution coefficients for orthopyroxene or plagioclase or clinopyroxene show less increase than elements that are less compatible with those minerals. These include Sc, Co, and Zn (orthopyroxene—GERM Kd Database http://earthref.org/databases/KDD/index.html), Ga, Eu, and Pb (plagioclase—Dunn and Sen, 1994; Kravuchuk et al., 1981; Paster et al., 1974), and Sc, V, and Y (clinopyroxene—Paster et al., 1974; Jenner et al., 1994; Skulski et al., 1994). Sr variation is scattered, although it should be incorporated into plagioclase during fractionation (Matsui et al., 1977; McKenzie and O'Nions, 1991). Cr and Ni, elements that have large distribution coefficients for olivine (Nabelek, 1980; Klöck and Palme, 1988; Villemant et al., 1981), show a much larger and less systematic percentage change in value. We believe that this observation is consistent with little fractional crystallization of olivine in the range of MMDS magmas.

Analyzed REEs for the MMDS from Harris (1998) are shown on a chondrite-normalized spider diagram (Nakamura, 1974) in Figure 17. All samples have REE values that are at least ten times chondritic abundances, which is expected for mantle-derived basalts. For most MMDS samples, the LREEs are enriched relative to the HREEs. These REE trends are similar to those of continental tholeiites. Transitional to subalkalic tholeiites of the Rio Grande Rift and the East African Rift in Ethiopia have REE patterns like those of Figure 17 (Wilson, 1989, p. 360). Although fractional crystallization of orthopyroxene and clinopyroxene could be responsible for some LREE enrichment, we believe that the REE data are consistent with the continental rift setting indicated by the field relations of these dikes.

DISCUSSION

The MMDS provide key evidence for the interpretation of the geologic history of the Precambrian rocks of the Tobacco Root Mountains. They intrude the Indian Creek and Pony–Middle Mountain Metamorphic Suites, but they do not intrude the adjacent Spuhler Peak Metamorphic Suite. We interpret the absence of the MMDS in the Spuhler Peak Metamorphic Suite to mean that the suite was not present at the time of intrusion. The MMDS cut across the 2.45 Ga foliation of the Indian Creek and Pony–Middle Mountain Metamorphic Suites (Cheney et al., 2004b, this volume, Chapter 8) and have metamorphic textures that appear to identify initial chilled margins. We interpret this evidence to mean that the MMDS were intruded into cool continental crust consisting of Indian Creek and Pony–Middle Mountain Metamorphic Suites rocks, well after the 2.45 Ga metamorphism that produced the crosscut foliation. A plausible scenario would be intrusion during rifting that produced a passive continental margin to the Wyoming province west of the current location of the Tobacco Root Mountains. Euhedral zircons in the MMDS date their intrusion, and therefore this proposed rifting, at 2.06 Ga (Burger et al., 1999; Mueller et al., 2004, this volume, Chapter 9). Closing of the ocean basin created during the rifting led to emplacement of the Spuhler Peak Metamorphic Suite, tectonic mixing of slivers of orthopyroxene cumulate rocks that may have been produced during fractional crystallization of MMDS magmas, deformation of all units together, and the high-grade metamorphism of all units. The continental scale of this collision is documented by the upper amphibolite to granulite facies metamorphic conditions required to produce the MMDS mineral assemblage garnet + clinopyroxene + plagioclase + ilmenite + quartz ± orthopyroxene ± hornblende (Cheney et al., 2004a, this volume, Chapter 6). Recognizing the scale of this event, Harms et al. (2004b, this volume, Chapter 11) have called it the Big Sky orogeny. Mueller et al. (2004, this volume, Chapter 9) and Cheney et al. (2004b, this volume, Chapter 8) provide zircon and monazite ages, respectively, of 1.77 Ga for the metamorphic peak of the Big Sky orogeny.

The geochemistry of the MMDS is consistent with this interpretation. They are subalkaline, tholeiitic basalts similar to basalts

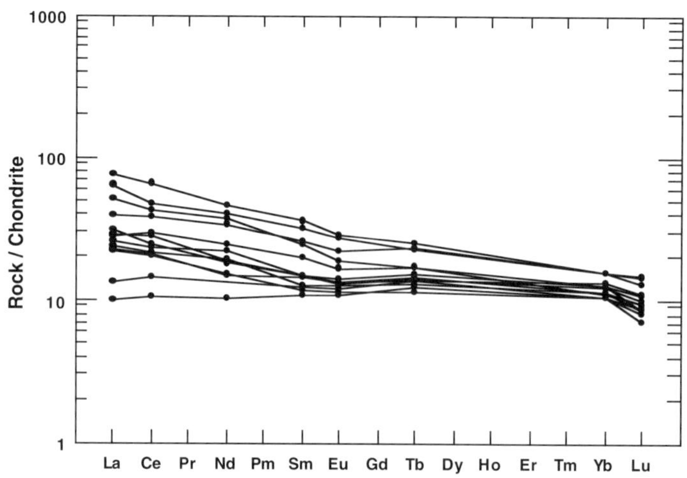

Figure 17. Chondrite-normalized rare earth element (REE) diagram for MMDS. All data are at least ten times chondrite values and most samples are enriched in the light REEs relative to the heavy REEs.

that occur in modern continental rift settings. The continental rift appears to have persisted long enough for significant fractional crystallization of the basalts to occur and for earlier MMDS to cool prior to intrusion by later MMDS. Plagioclase and orthopyroxene appear to be the major fractionating phases responsible for the observed chemical variations. Orthopyroxene is more likely to be in the crystallization sequence of a continental tholeiite than in an oceanic tholeiite (Campbell, 1985). The higher silica observed in continental tholeiites may cause the fractionation path to move from olivine to the orthopyroxene reaction boundary, rather than to the clinopyroxene cotectic (in the system olivine-clinopyroxene-silica) as in oceanic tholeiites. Elevated pressure also can lead to orthopyroxene fractionation by shifting the orthopyroxene reaction curve away from silica on the same diagram. Neither more primitive nor more evolved alkaline magmas that may have been associated with the rifting occur in the Tobacco Root Mountains, although they might occur farther west closer to the center of the rift. Johnson et al. (2004, this volume, Chapter 4) demonstrate that meta-ultramafic rocks distributed throughout the Tobacco Root Mountains are slices of orthopyroxene-rich cumulate rocks. It is geochemically possible that these cumulates resulted from the fractional crystallization of the MMDS.

Interestingly, rifting of a similar age (2.01 Ga) is postulated for the southeastern margin of the Wyoming province, where a more extensive suite of metamorphosed mafic dikes occurs (the Kennedy dikes), including rocks with a much more primitive composition (Condie et al., 1969; Snyder et al., 1991; Cox et al., 2000). The Kennedy dikes were also metamorphosed at ca. 1.77 Ga during the Medicine Bow orogeny. The similarities of mafic dikes associated with continental rifting and their metamorphism during a subsequent collision is not surprising, but the near coincidence in time of these several events on opposite sides of the Wyoming province is curious indeed.

ACKNOWLEDGMENTS

The authors thank the W.M. Keck Foundation and the colleges of the Keck Geology Consortium for their financial support of this work. Stan Mertzman of Franklin and Marshall College graciously performed whole-rock, XRF analyses on short notice for Caroline Harris's senior thesis. We thank Tom Hanley for providing advice, samples, data, and helpful comments on the text. The manuscript was also improved due to an insightful review by Kevin Chamberlain and Ron Frost.

REFERENCES CITED

Brady, J.B., Carmichael, S., Burger, H.R., Harris, C., Cheney, J.T., and Harms, T.A., 1998, Thermobarometry and geochemistry of metamorphosed mafic dikes and sills (MMDS) of the Tobacco Root Mountains, Montana: Geological Society of America Abstracts with Programs, v. 30, no. 7, p. 97.

Brady, J.B., Burger, H.R., Cheney, J.T., and Harms, T.A., editors, 2004, Precambrian geology of the Tobacco Root Mountains, Montana: Geological Society of America Special Paper 377, 256 p.

Burger, H.R., 2004, General geology and tectonic setting of the Tobacco Root Mountains, *in* Brady, J.B., et al., eds, Precambrian geology of the Tobacco Root Mountains, Montana: Boulder, Colorado, Geological Society of America Special Paper 377, p. 1–14 (this volume).

Burger, H.R., Brady, J.B., Cheney, J.T., Harms, T.A., Mueller, P., Heatherington, A., and Wooden, J., 1999, Evidence for a major, Early Proterozoic orogenic event in the Tobacco Root Mountains of southwestern Montana: Geological Society of America Abstracts with Programs, v. 31, no. 7, p. 177–178.

Burger, H.R., Peck, W.H., Johnson, K.E., Tierney, K.A., Poulsen, C.J., Cady, P., Lowell, J., and MacFarlane, W.A., Sincock, M.J., Archuleta, L.L., Pufall, A., and Cox, M.J., 2004, Geology and geochemistry of the Spuhler Peak Metamorphic Suite, *in* Brady, J.B., et al., eds, Precambrian geology of the Tobacco Root Mountains, Montana: Boulder, Colorado, Geological Society of America Special Paper 377, p. 47–70 (this volume).

Campbell, I.H., 1985, The difference between oceanic and continental tholeiites: A fluid dynamic explanation: Contributions to Mineralogy and Petrology, v. 91, p. 37–43.

Cardoza, K.D., Hepburn, J.C., and Hon, R., 1990, Geochemical constraints on the paleotectonic settings of two Late Proterozoic mafic volcanic suites, Boston-Avalon Zone, eastern Massachusetts, *in* Socci, A.D., et al., eds., Geology of the composite Avalon Terrane of southern New England: Geological Society of America Special Paper 245, p. 113–131.

Cheney, J.T., Brady, J.B., Tierney, K.A., DeGraff, K.A., Mohlman, H.K., Frisch, J.D., Hatch, C.E., Steiner, M.L., Carmichael, S.K., Fisher, R.G.M., Tuit, C.B., Steffen, K.J., Cady, P., Lowell, J., Archuleta, L.L., Hirst, J., Wegmann, K.W., and Monteleone, B., 2004a, Proterozoic metamorphism of the Tobacco Root Mountains, Montana, *in* Brady, J.B., et al., eds., Precambrian geology of the Tobacco Root Mountains, Montana: Boulder, Colorado, Geological Society of America Special Paper 377, p. 105–129 (this volume).

Cheney, J.T., Webb, A.A.G., Coath, C.D., and McKeegan, K.D., 2004b, In situ ion microprobe $^{207}Pb/^{206}Pb$ dating of monazite from Precambrian metamorphic suites, Tobacco Root Mountains, Montana, *in* Brady, J.B., et al., eds., Precambrian geology of the Tobacco Root Mountains, Montana: Boulder, Colorado, Geological Society of America Special Paper 377, p. 151–179 (this volume).

Condie, K.C., Barsky, C.K., and Mueller, P.A., 1969, Geochemistry of Precambrian diabase dikes from Wyoming: Geochimica et Cosmochimica Acta, v. 33, no. 11, p. 1371–1388.

Cordua, W.S., 1973, Precambrian geology of the southern Tobacco Root Mountains, Madison County, Montana [Ph.D. thesis]: Bloomington, Indiana University, 248 p.

Cox, D.M., Frost, C.F., and Chamberlain, K.R., 2000, 2.01 Ga Kennedy dike swarm, southeastern Wyoming: Record of a rifted margin along the southern Wyoming province: Rocky Mountain Geology, v. 35, p. 7–30.

Dunn, T., and Sen, C., 1994, Mineral/matrix partition coefficients for orthopyroxene, plagioclase, and olivine in basaltic to andesitic systems: A combined analytical and experimental study: Geochimica et Cosmochimica Acta, v. 58, p. 717–733.

Gillmeister, N.M., 1972, Petrology of Precambrian rocks in the central Tobacco Root Mountains, Madison Country, Montana [Ph.D. thesis]: Cambridge, Massachusetts, Harvard University, 210 p.

Hanley, T.B., 1975, Structure and petrology of the northwestern Tobacco Root Mountains, Madison County, Montana [Ph.D. thesis]: Bloomington, Indiana University, 289 p.

Hanley, T.B., and Vitaliano, C.J., 1983, Petrography of Archean mafic dikes of the Tobacco Root Mountains, Madison County, Montana: Northwest Geology, v. 12, p. 43–55.

Harms, T.A., Burger, H.R., Blednick, D.G., Cooper, J.M., King, J.T., Owen, D.R., Lowell, J., Sincock, M.J., Kranenburg, S.R., Pufall, A., and Picornell, C.M., 2004a, Character and origin of Precambrian fabrics and structures in the Tobacco Root Mountains, Montana, *in* Brady, J.B., et al., eds., Precambrian geology of the Tobacco Root Mountains, Montana: Boulder, Colorado, Geological Society of America Special Paper 377, p. 203–226 (this volume).

Harms, T.A., Brady, J.B., Burger, H.R., and Cheney, J.T., 2004b, Advances in the geology of the Tobacco Root Mountains, Montana, and their implications for the history of the northern Wyoming province, *in* Brady, J.B., et al., eds., Precambrian geology of the Tobacco Root Mountains, Montana: Boulder, Colorado, Geological Society of America Special Paper 377, p. 227–243 (this volume).

Harris, C., 1998, Geochemistry of metamorphosed mafic dikes, Tobacco Root Mountains, Montana [B.A. thesis]: Claremont, California, Pomona College, 58 p.

Irvine, T.N., and Baragar, W.R.A., 1971, A guide to the chemical classification of the common volcanic rocks: Canadian Journal of Earth Sciences, v. 8, p. 523–548.

Jacob, L.J., 1994, A geochemical and geochronological characterization of the Indian Creek Metamorphic Suite in the Noble Lake area, south-central Tobacco Root Mountains, southwestern Montana [B.A. thesis]: Northampton, Massachusetts, Smith College, 73 p.

Jenner, G.A., Foley, S.F., Jackson, S.E., Green, T.H., Fryer, B.J., and Longerich, H.P., 1994, Determination of partition coefficients for trace elements in high pressure-temperature experimental run products by laser ablation microprobe-inductively coupled plasma-mass spectrometry (LAM-ICP-MS): Geochimica et Cosmochimica Acta, v. 58, p. 5099–5103.

Johnson, K.E., Brady, J.B., MacFarlane, W.A., Thomas, R.B., Poulsen, C., and Sincock, M.J., 2004, Precambrian meta-ultramafic rocks from the Tobacco Root Mountains, Montana, *in* Brady, J.B., et al., eds., Precambrian geology of the Tobacco Root Mountains, Montana: Boulder, Colorado, Geological Society of America Special Paper 377, p. 71–87 (this volume).

Klöck, W., and Palme, H., 1988, Partitioning of siderophile and chalcophile elements between sulfide, olivine, and glass in a naturally reduced basalt from Disko Island, Greenland: Lunar and Planetary Science, v. 18, p. 471–483.

Kravuchuk, I.K., Chernysheva, I., and Urosov, S., 1981, Element distribution between plagioclase and groundmass as an indicator for crystallization conditions of the basalts in the southern vent of Tolbachik: Geochemical International, v. 17, p. 18–24.

Matsui, Y., Onuma, N., Nagasawa, H., Higuchi, H., and Banno, S., 1977, Crystal structure control in trace element partition between crystal and magma: Bulletin de la Société française de Minéralogie et Cristallographie, v. 100, p. 324–328.

McKenzie, D., and O'Nions, R.K., 1991, Partial melt distributions from inversion of rare earth element concentrations: Journal of Petrology, v. 32, p. 1021–1091.

Mertzman, S.A., 2000, K-Ar results from the southern Oregon–northern California Cascade Range: Oregon Geology, v. 62, no. 4, p. 99–122.

Mueller, P.A., Burger, H.R., Wooden, J.L., Heatherington, A.L., Mogk, D.W., and D'Arcy, K., 2004, Age and evolution of the Precambrian crust of the Tobacco Root Mountains, Montana, *in* Brady, J.B., et al., eds., Precambrian geology of the Tobacco Root Mountains, Montana: Boulder, Colorado, Geological Society of America Special Paper 377, p. 181–202 (this volume).

Nabelek, P.I., 1980, Nickel partitioning between olivine and liquid in natural basalts: Henry's Law behavior: Earth and Planetary Science Letters, v. 48, p. 293–302.

Nakamura, N., 1974, Determination of REE, Ba, Fe, Mg, Na and K in carbonaceous and ordinary chondrites: Geochimica et Cosmochimica Acta, v. 38, p. 757–775.

Passchier, C.W., Myers, J.S., and Kroener, A., 1990, Field geology of high-grade gneiss terrains: Berlin, Springer Verlag, 150 p.

Paster, T.P., Schauwecher, D.S., and Haskin, L.A., 1974, The behavior of some trace elements during solidification of the Skjergaard layered series: Geochimica et Cosmochimica Acta, v. 38, p. 1549–1577.

Pearce, T.H., 1968, A contribution to the theory of variation diagrams: Contributions to Mineralogy and Petrology, v. 19, p. 142–157.

Reid, R.R., 1963, Metamorphic rocks of the northern Tobacco Root Mountains, Madison County, Montana: Geological Society of America Bulletin, v. 74, p. 293–306.

Rhodes, J.M., 1996, Geochemical stratigraphy of lava flows sampled by the Hawaiian Scientific Drilling Project: Journal of Geophysical Research, v. 101, p. 11,729–11,746.

Skulski, T., Minarik, W., and Watson, E.B., 1994, High-pressure experimental trace-element partitioning between clinopyroxene and basaltic melts: Chemical Geology, v. 117, p. 127–147.

Snyder, G.L., Hall, R.P., Hughes, D.J., and Ludwig, K.R., 1991, Early Precambrian basic rocks of the USA, *in* Hall, R.P., and Hughes, D.J., eds., Early Precambrian basic magmatism: Glasgow, Blackie, p. 191–220.

Stanley, C.R., and Russell, J.K., 1989, Petrologic hypothesis testing with Pearce element ratio diagrams: Derivation of diagram axes: Contributions to Mineralogy and Petrology, v. 103, p. 78–89.

Tansley, W., Schafer, P.A., and Hart, L.H., 1933, A geological reconnaissance of the Tobacco Root Mountains, Madison County, Montana: Montana Bureau of Mines and Geology Memoir 9, 57 p.

Villemant, B., Jaffrezic, H., Joran, J.-L., and Treuil, M., 1981, Distribution coefficients of major and trace elements; fractional crystallization in the alkali basalt series of Chaine des Puys (Massif Central, France): Geochimica et Cosmochimica Acta, v. 45, p. 1997–2016.

Vitaliano, C.J., Burger, H.R., Cordua, W.S., Hanley, T.B., Hess, D.F., and Root, F.K., 1979, Explanatory text to accompany the geologic map of the southern Tobacco Root Mountains, Madison County, Montana: Geological Society of America Map and Chart Series MC31, scale 1:62,500, 1 sheet, 8 p. text.

Wilson, M., 1989, Igneous petrogenesis: London, Unwin Hyman, 466 p.

Manuscript Accepted by the Society September 12, 2003

Printed in the USA

Geological Society of America
Special Paper 377
2004

Proterozoic metamorphism of the Tobacco Root Mountains, Montana

John T. Cheney
Department of Geology, Amherst College, Amherst, Massachusetts 01002, USA

John B. Brady
Department of Geology, Smith College, Northampton, Massachusetts 01063, USA

Kara A. Tierney
Kathleen A. DeGraff
Heidi K. Mohlman
Jessica D. Frisch
Christine E. Hatch
Michael L. Steiner
Department of Geology, Amherst College, Amherst, Massachusetts 01002, USA

Sarah K. Carmichael
Robin G.M. Fisher
Department of Geology, Smith College, Northampton, Massachusetts 01063, USA

Carrie B. Tuit
Beloit College, Beloit, Wisconsin 53511, USA

Kurt J. Steffen
Carleton College, Northfield, Minnesota 55057, USA

Pamela Cady
Josh Lowell
Colorado College, Colorado Springs, Colorado 80903, USA

LeAndra L. Archuleta
Jillian Hirst
Pomona College, Claremont, California 91711-6339, USA

Karl W. Wegmann
Whitman College, Walla Walla, Washington 99362, USA

Brian Monteleone
College of Wooster, Wooster, Ohio 44691-2363, USA

ABSTRACT

Textures and mineral assemblages of metamorphic rocks of the Tobacco Root Mountains are consistent with metamorphism of all rocks during the Big Sky orogeny (1.77 Ga) at relatively high pressure (*P* >1.0 GPa) followed by differential

Cheney, J.T., Brady, J.B., Tierney, K.A., DeGraff, K.A., Mohlman, H.K., Frisch, J.D., Hatch, C.E., Steiner, M.L., Carmichael, S.K., Fisher, R.G.M., Tuit, C.B., Steffen, K.J., Cady, P., Lowell, J., Archuleta, L.L., Hirst, J., Wegmann, K.W., and Monteleone, B., 2004, Proterozoic metamorphism of the Tobacco Root Mountains, Montana, *in* Brady, J.B., Burger, H.R., Cheney, J.T., and Harms, T.A., eds., Precambrian geology of the Tobacco Root Mountains, Montana: Boulder, Colorado, Geological Society of America Special Paper 377, p. 105–129. For permission to copy, contact editing@geosociety.org.

reequilibration on a clockwise *P-T* path at lower pressures (0.6–0.8 GPa). The highest pressures are documented by coarse-grained kyanite and orthopyroxene in aluminous orthoamphibolites, which require $P \geq 1.0$ GPa. Other higher-pressure mineral assemblages of note include kyanite + orthoamphibole and kyanite + K-feldspar. Abundant textural evidence for partial melting in pelitic and basaltic rocks includes leucosomes, very large (several cm across) porphyroblasts of garnet, and an absence of primary (foliation-defining) muscovite. Partial to complete overprinting of the coarse-textured, high-pressure assemblages by lower-pressure assemblages and textures occurred across the Tobacco Root Mountains, especially where assisted by deformation and the availability of water. In aluminous rocks, sillimanite bundles typically replace kyanite, and garnet may be rimmed by cordierite + orthopyroxene symplectite or, in quartz-absent rocks, sapphirine + spinel + cordierite symplectite. Orthoamphibolites with partial pseudomorphs of garnet by cordierite are common. Garnet necklaces surround orthopyroxene in orthopyroxene-plagioclase gneisses, whereas orthopyroxene + plagioclase pseudomorphs of garnet occur in nearby hornblende amphibolites. These features appear to require nearly isobaric cooling at pressures near 0.8 GPa, followed by nearly isothermal decompression at temperatures near 700 °C. The resulting *P-T* path is believed to be the result of tectonic denudation late in the orogenic cycle. Quartz-plagioclase-garnet-hornblende amphibolites occur throughout the Tobacco Root Mountains. Near-rim mineral compositions from these rocks have been used to calculate *T*s of 650–750 °C at *P*s of 0.7–0.9 GPa across the terrane. There is no systematic variation in calculated *P* and *T* between units nor geographically within units; differences appear to reflect variations in thermometer closure possibly due to the availability of water during cooling. Field relations involving metamorphosed mafic dikes, as well as geochronological data from monazite and zircon, demonstrate that some rocks were first metamorphosed at high temperatures and pressures at 2.45 Ga. However, we have not identified mineral assemblages that can be assigned unequivocally to this earlier event.

Keywords: Tobacco Root Mountains, Precambrian, metamorphism, *P-T*-t path.

INTRODUCTION

The metamorphic history of the Tobacco Root Mountains has been studied by a number of previous workers (e.g., Tansley et al., 1933; Reid, 1957, 1963; Root, 1965; Burger, 1966, 1969; Hess, 1967; Gillmeister, 1972b, 1972a; Cordua, 1973; Hanley, 1975, 1976; Friberg, 1976; Immega and Klein, 1976; Vitaliano et al., 1979, see reprinted map and text accompanying this volume; Hanley and Vitaliano, 1983). Many of the important features were first elucidated by Reid (1963) in a paper summarizing his careful field and petrographic work on the northern portion of the range. Reid (1963) postulated the occurrence of "three major metamorphic episodes" (*a*, *b*, *c*) and distinguished five deformations (1–5). In unraveling this history, Reid (1963, p. 293) emphasized the significance of the "metabasalt dikes and sills which cut foliation, tight isoclinal folds, migmatite patterns, and other structures of an older metamorphism *a* of unknown but probably high grade." Previous workers (Tansley et al., 1933) called for two metamorphisms based on their interpretation that the Cherry Creek metasediments (our Indian Creek Metamorphic Suite; see below) contained boulders of Pony gneiss (our Pony–Middle Mountain Metamorphic Suite) and apparently ignoring the constraints of the metamorphosed basalt dikes and sills (our MMDS).

Reid (1963, p. 297) argues that tight, isoclinal folding of compositional layering (deformation no. 1) and an axial plane schistosity occurred during metamorphism *a*, but that the "mineral assemblages of metamorphism *a* have been completely destroyed by later metamorphic recrystallization." He calls for the intrusion of the MMDS after metamorphism *a*, because they crosscut its foliation and isoclinal folds (see Figure 1 *in* Brady et al., 2004b, this volume, Chapter 5). Reid's (1963) metamorphism *b* is assigned to the granulite facies based on the presence of hypersthene in mafic rocks. This metamorphism affects all rocks, including the MMDS, with a shearing deformation (deformation no. 2) that first destroys and then reconstitutes the axial plane schistosity of deformation no. 1. His metamorphism *c* is assigned to the amphibolite facies and is based on mineral assemblages that "appear texturally to be superimposed upon the granulite facies assemblages" (Reid, 1963, p. 296). He calls for additional shear deformation (no. 3) during metamorphism *c* because "new hornblende and biotite … grew parallel to the axial planes of the isoclinal folds of deformation no. 1…." In addition to the three main metamorphic events, Reid (1963, p. 300) observed the incomplete development of replacement assemblages belonging to the epidote-amphibolite facies (his "retrogressive metamorphism *d*") and the greenschist facies (his "retrogressive metamorphism *e*"). Reid's Table 1 lists the important mineral assemblages that support his interpretation.

Subsequent workers have elaborated upon Reid's (1963) analysis, examining rocks in the central and southern portions of the range, giving additional structural and petrographic details, describing additional mineral assemblages, distinguishing a new map unit (our Spuhler Peak Metamorphic Suite) that did not experience metamorphism *a*, reporting mineral compositions, and using mineral compositions to constrain temperatures and pressures of metamorphism. Much of this research was completed under the supervision of Charles Vitaliano by graduate students at Indiana University. Their results are summarized nicely in Vitaliano et al. (1979) and will not be listed again here. We acknowledge our debt to all these scientists, whose detailed maps, theses, and publications made our own studies possible.

Our work on the metamorphism of Tobacco Root rocks began in 1993, has included portions of 17 undergraduate theses (Archuleta, 1994b; Cady, 1994b; Fisher, 1994b; Lowell, 1994a; Tierney, 1994a; DeGraff, 1996a; Hirst, 1996; Mohlman, 1996a; Tuit, 1996c; Wegmann, 1996a; Carmichael, 1998b; Frisch, 1998b; Hatch, 1998b; Monteleone, 1998a; Steffen, 1998b; Steiner, 1999; Rodriguez, 2002), and has been partially documented through a series of abstracts (Archuleta, 1994a; Brady et al., 1998a; Carmichael et al., 1998; Carmichael, 1998a; Cady, 1994a; Cheney et al., 1994, 1996, 1998; DeGraff et al., 1996; DeGraff, 1996b; Fisher, 1994a; Frisch and Cheney, 1998; Frisch, 1998a; Hatch, 1998a; Hatch and Cheney, 1998; Lowell, 1994b; Martin, 1996; Mohlman, 1996b; Mohlman and Cheney, 1996; Monteleone, 1998b; Steffen, 1998a, 1998c; Tierney, 1994a, 1994b; Tuit, 1996a, 1996b; Wegmann, 1996b). Field observations and suites of samples were collected during the summers of 1993, 1995, and 1997, as part of individual projects focusing on one rock unit or rock type. Our analysis builds on previous work by adding new petrographic and chemical data and by using those data to refine the pressure-temperature (*P-T*) path followed during metamorphism.

In this paper and throughout this volume (see Burger, 2004, this volume, Chapter 1), the metamorphic rocks of the Tobacco Root Mountains have been divided into four suites. (1) The Indian Creek Metamorphic Suite consists of many rocks that are clearly metasediments, including quartzite, marble, iron formation, and aluminous schist, but also includes rocks that are meta-igneous, including amphibolite and quartzofeldspathic gneiss. The Indian Creek Metamorphic Suite rocks were called Cherry Creek metamorphic rocks by Reid (1963) and others. (2) The Pony–Middle Mountain Metamorphic Suite consists principally of quartzofeldspathic gneisses and hornblende amphibolites (see Mogk et al., 2004, this volume, Chapter 2). These are the rocks that Reid (1963) and others have called Pony metamorphic rocks. (3) The Spuhler Peak Metamorphic Suite is a mafic unit that consists principally of amphibolite and Ca-depleted amphibolite, with lesser quartzite and aluminous schist (see Burger et al., 2004, this volume, Chapter 3). These are the rocks that Gillmeister (1972b) called the Spuhler Peak Formation. (4) The metamorphosed mafic dikes and sills (MMDS) are fine-grained, metabasalt rocks that intrude the Pony–Middle Mountain Metamorphic Suite and the Indian Creek Metamorphic Suite, but not the Spuhler Peak Metamorphic Suite (see Brady et al., 2004b, this volume, Chapter 7). They are the metabasalt dikes and sills of Reid (1963) and the metabasites of (Hanley and Vitaliano, 1983). Ultramafic rocks are also present, but these are considered separately in Johnson et al. (2004, this volume, Chapter 4).

All four rock suites have mineral assemblages that record upper amphibolite facies metamorphism. Examples of important observed assemblages are shown in Figure 1 on metamorphic facies diagrams. In the following pages, we use rim compositions of minerals in these assemblages along with several geothermobarometers to calculate temperatures and pressures for the metamorphism that produced them. The upper amphibolite assemblages correspond with metamorphism *c* of Reid (1963), which partly to completely overprints evidence for an earlier metamorphism at higher pressure and at least locally at higher temperature (Reid's metamorphism *b* or metamorphism *a*). Features that appear to have been produced during an earlier event include (1) coarse porphyroblasts of garnet, orthopyroxene, gedrite, and other minerals inconsistent with the grain size of the amphibolite facies minerals; (2) the presence of kyanite (partially pseudomorphed by sillimanite in some cases) with K-feldspar inclusions, requiring a minimum *T* and *P* of 700 °C and 0.8 GPa (Spear et al., 1999); (3) the presence of coarse kyanite and orthopyroxene in the same thin section, requiring a minimum *P* of 1.0 GPa (see below); (4) the presence of orthopyroxene in some rocks of basaltic composition, requiring a minimum *T* of ~800 °C (Pattison, 2003); (5) a variety of corona and necklace textures that are consistent with chemical reactions during decompression at temperatures above 700 °C; and (6) abundant textural evidence for partial melting in pelitic and basaltic rocks, including leucosomes, very large (several cm across) porphyroblasts of garnet, and an absence of primary (foliation-defining) muscovite.

Elsewhere in this volume (Brady et al., 2004b, this volume, Chapter 7; Mueller et al., 2004, this volume, Chapter 9) it has been demonstrated that the MMDS were intruded at 2.07 Ga into the Indian Creek and Pony–Middle Mountain Metamorphic Suites (but not the Spuhler Peak Metamorphic Suite). Because both the MMDS and the Spuhler Peak Metamorphic Suite show evidence of a high-pressure (>0.8 GPa) metamorphism followed by a lower-pressure (~0.6 GPa) metamorphism during the 1.77 Ga Big Sky orogeny, we believe that the Indian Creek and Pony–Middle Mountain Metamorphic Suites also experienced the same two periods of metamorphism. Throughout this paper we will refer to the Big Sky high-pressure event as M1 (Reid's [1963] episode *b*) and the Big Sky lower-pressure event as M2 (Reid's [1963] episode *c*). Metamorphic features that predate M1 textures will be called pre-M1 or M0 (Reid's [1963] episode *a*). In some cases (e.g., monazite growth) we can demonstrate that the pre-M1 features occurred during M0 at 2.45 Ga (Cheney et al., 2004, this volume, Chapter 8); in other cases, we can say with certainty only that the features are pre-M1. Similarly, there are some post-M2 retrograde features of uncertain age (Reid's

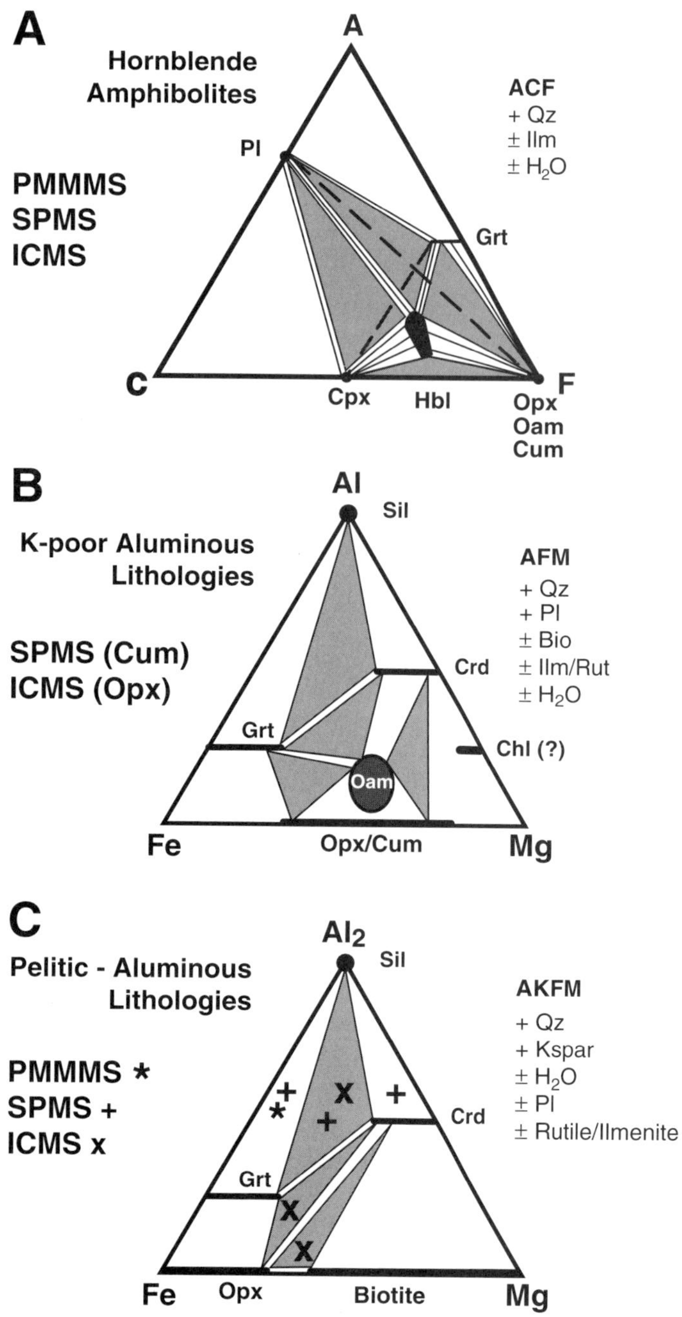

Figure 1. Schematic representation of observed rim assemblages for many Tobacco Root metamorphic rocks are shown as gray triangles on typical metamorphic assemblage diagrams. Additional minerals are listed as projection phases (+) or as possibly present (±). ACF: A—Al, C—Ca, F—Fe + Mg. PMMMS—Pony–Middle Mountain Metamorphic Suite; SPMS—Spuhler Peak Metamorphic Suite; ICMS—Indian Creek Metamorphic Suite; Qz—quartz; Ilm—ilmenite; Opx—orthopyroxene; Oam—orthoamphibole; Cum—cummingtonite; Cpx—clinopyroxene; Hbl—hornblende; Pl—plagioclase; Bio—biotite; Rut—rutile; Sil—sillimanite; Crd—cordierite; Chl—chlorite; Kspar—K-feldspar.

[1963] episodes *d* and *e*). Possibilities include a late Big Sky heating event, heating during ca. 1.4 Ga rifting (Brady et al., 1998b), or contact metamorphic heating during intrusion of the Cretaceous Tobacco Root batholith.

TEMPERATURES AND PRESSURES FROM GEOTHERMOBAROMETERS

Garnet-Hornblende-Plagioclase-Quartz Amphibolites

Amphibolite and amphibolite gneiss are common metamorphic rocks in the Tobacco Root Mountains, occurring in all the major units (Indian Creek Metamorphic Suite, Pony–Middle Mountain Metamorphic Suite, Spuhler Peak Metamorphic Suite, and MMDS). Although textures and modes vary considerably, many samples contain the mineral assemblage hornblende + plagioclase + quartz ± garnet ± augite ± cummingtonite ± biotite along with a variety of other minerals in minor proportions. We have used chemical data from these rocks along with the garnet-hornblende geothermometer of Graham and Powell (1984) and the garnet-hornblende-plagioclase-quartz (Mg) geobarometer of Kohn and Spear (1990) to evaluate metamorphic conditions across the Tobacco Root Mountains using a consistent mineral assemblage.

Chemical data were collected for the critical minerals in rocks with a great variety of textures and a range of bulk composition. See Cady (1994b), DeGraff (1996a), Mohlman (1996a), Wegmann (1996a), Carmichael (1998b), Frisch (1998b), and Steffen (1998b) for petrographic details. In most rocks, hornblende and clinopyroxene (if present) are chemically homogeneous. Plagioclase may be slightly zoned or homogenous. No consistent zoning pattern was observed, except in the MMDS where the anorthite content of plagioclase increases slightly toward the rims of grains, by up to 5 mol%. Garnet grains show only moderate chemical zoning. In the Pony–Middle Mountain and Indian Creek Metamorphic Suites, zoning of Ca in garnet is patchy in some samples. Ca zoning is balanced principally by Fe zoning, with Mg and Mn comparatively unzoned. We interpret this chemical signature as growth of new garnet at 1.77 Ga on old garnet cores in rocks that also experienced metamorphism at 2.45 Ga. In the Spuhler Peak Metamorphic Suite, zoned garnets typically show increasing Mg and decreasing Ca from core to rim, with an Fe increase at the rim. We interpret this zoning as prograde growth zoning during the 1.77 Ga metamorphism, with a rim modified during cooling. Garnets in the MMDS are homogeneous and small (0.1–0.2 mm), possibly recording growth at comparatively high temperature in these dry rocks.

Typical chemical analyses for garnet, plagioclase, and hornblende from one sample each of the four major units are given in Table 1. Chemical analyses of minerals were obtained principally by energy-dispersive X-ray fluorescence (EDS-XRF) at Amherst College and Smith College using Link/Oxford EDS hardware and software and ZAF corrections based on mineral standards. Garnet compositions are 50%–60% almandine and 20%–30%

TABLE 1. GARNET-HORNBLENDE-PLAGIOCLASE-CLINOPYROXENE ANALYSES

Sample	KJS-9a	KJS-9a	KJS-9a	KJS-9a	KJS-9a	JDF-38a2	JDF-38a2	JDF-38a2	JDF-38a2	JDF-38a2	JDF-38a2
Unit	SPMS	SPMS	SPMS	SPMS	SPMS	ICMS	ICMS	ICMS	ICMS	ICMS	ICMS
UTM	504045N	504045N	504045N	504045N	504045N	504402N	504402N	504402N	504402N	504402N	504402N
12T	042018E	042018E	042018E	042018E	042018E	041338E	041338E	041338E	041338E	041338E	041338E
Mineral	Grt	Grt	Hbl	Pl	Pl	Grt	Grt	Hbl	Pl	Pl	Cpx
	core	rim	unzoned	core	rim	core	rim	unzone	core	rim	unzoned
Oxide (wt%)											
SiO_2	38.37	38.48	45.94	49.18	50.43	38.52	37.96	42.28	53.00	56.23	52.79
Al_2O_3	21.02	21.19	12.18	31.94	31.30	20.79	20.53	12.43	29.15	27.47	2.03
TiO_2	N.D.†	N.D.	1.27	N.D.	N.D.	0.08	0.23	1.67	0.03	0.00	0.21
MgO	5.03	5.47	12.34	N.D.	N.D.	3.41	3.44	8.30	0.07	0.00	10.47
FeO*	23.40	25.14	12.43	0.00	0.26	24.25	24.29	18.18	0.25	0.57	13.73
MnO	2.40	1.96	0.16	N.D.	N.D.	1.15	1.35	0.09	0.00	0.00	0.15
CaO	8.89	7.14	11.10	15.23	13.65	11.82	11.46	11.59	11.40	8.10	20.90
Na_2O	N.D.	N.D.	0.92	2.58	3.15	N.D.	N.D.	1.65	4.65	6.18	0.66
K_2O	N.D.	N.D.	0.43	0.34	0.34	N.D.	N.D.	0.78	0.35	0.91	0.01
Total	99.11	99.38	96.77	99.27	99.13	100.02	99.26	96.97	98.90	99.46	100.95
Elements per Formula											
O	12	12	23	8	8	12	12	23	8	8	6
Si	3.02	3.02	6.72	2.26	2.31	3.02	3.01	6.43	2.43	2.55	1.95
Al	1.95	1.96	2.10	1.73	1.69	1.92	1.92	2.23	1.57	1.47	0.10
Ti	N.D.	N.D.	0.14	N.D.	N.D.	0.00	0.01	0.19	0.00	0.00	0.01
Mg	0.59	0.64	2.69	N.D.	N.D.	0.40	0.41	1.88	0.01	0.00	0.63
Fe	1.54	1.65	1.52	0.00	0.01	1.59	1.61	2.31	0.01	0.02	0.35
Mn	0.16	0.13	0.02	N.D.	N.D.	0.08	0.09	0.01	0.00	0.00	0.01
Ca	0.75	0.60	1.74	0.75	0.67	0.99	0.97	1.89	0.56	0.39	0.91
Na	N.D.	N.D.	0.26	0.23	0.28	N.D.	N.D.	0.49	0.41	0.54	0.07
K	N.D.	N.D.	0.08	0.02	0.02	N.D.	N.D.	0.15	0.02	0.05	0.00
Total Cations	8.01	8.00	15.27	4.99	4.98	8.00	8.02	15.58	5.01	5.02	4.03
Mg/Mg + Fe)	0.28	0.28	0.64			0.20	0.20	0.45			0.64

(continued)

grossular. Plagioclase is much more variable, ranging from 25% to 75% anorthite. Hornblende has a significant edenite component and, except for Spuhler Peak Metamorphic Suite samples, has more Fe than Mg. These mineral compositions meet the criteria listed by Kohn and Spear (1990) for use with their geobarometer. Calculated temperatures and pressures using rim compositions for the samples in Table 1 and other rocks are shown in Figure 2. Because there is little chemical zoning, core compositions yield similar results. Most temperatures are in the interval 600–750 °C. Pressures range from 0.5 to 0.9 GPa. Uncertainties in the individual data points are based on observed variations of the mineral compositions in single thin sections. Friberg (1994) reported similar pressures and temperatures for Spuhler Peak Metamorphic Suite amphibolites using this assemblage. These conditions are in the kyanite zone, although nearby pelitic rocks contain sillimanite. This may mean that the thermobarometer is inaccurate, but we suspect that reequilibration of the relevant minerals was not achieved during the decompression that produced the sillimanite (see below).

Although there is a considerable range of calculated temperatures and pressures, we can find no system to the scatter. We were not able to document geographic regions with consistently higher temperatures and/or pressures nor to find any regular variation with sample elevation (Frisch, 1998b). Similarly, there seems to be no systematic variation with strain. Rodriguez (2002) found that amphibolite samples of all fabric types gave kyanite zone temperatures and pressures consistent with those given in Figure 2. We believe that the range in temperatures and pressures shown on Figure 2 results from an array of garnet and hornblende compositions that record the complex reaction history of each sample. This history is differentially preserved in the minerals and includes both retrograde exchange reactions and net transfer reactions. The temperature variation within a given sample is typically on the order of ±25 °C and reflects the chemical homogeneity of the garnet. All iron was treated as ferrous for this analysis. Accordingly some of the inter-sample temperature variation may be due to variation in ferric iron among amphiboles from different samples. Variation in ferric iron does not significantly affect the geobarometer of Kohn and Spear (1990) because the calibration is not strongly dependent on the Mg/Fe ratio of the amphibole. We believe that these temperatures and pressures were recorded in the mineral compositions during a

TABLE 1. GARNET-HORNBLENDE-PLAGIOCLASE-CLINOPYROXENE ANALYSES (*continued*)

Sample	JDF-20a2	JDF-20a2	JDF-20a2	JDF-20a2	JDF-20a2	SKC-1-3a	SKC-1-3a	SKC-1-3a	SKC-1-3a	SKC-1-3a	SKC-1-3a
Unit	PMMMS	PMMMS	PMMMS	PMMMS	PMMMS	MMDS	MMDS	MMDS	MMDS	MMDS	MMDS
UTM	505076N	505076N	505076N	505076N	505076N	504771N	504771N	504771N	504771N	504771N	504771N
12T	041476E	041476E	041476E	041476E	041476E	041498E	041498E	041498E	041498E	041498E	041498E
Mineral	Grt	Grt	Hbl	Pl	Pl	Grt	Grt	Hbl	Pl	Pl	Cpx
	core	rim	unzoned	core	rim	core	rim	unzoned	core	rim	unzoned
Oxide (wt%)											
SiO_2	37.86	36.92	41.46	59.97	58.73	37.61	38.43	42.03	63.33	62.08	52.79
Al_2O_3	20.44	20.10	12.15	24.84	25.59	20.32	20.82	11.61	24.24	25.03	2.03
TiO_2	0.25	0.00	2.14	0.11	0.05	0.05	0.04	2.17	0.04	0.03	0.21
MgO	3.15	2.68	8.18	0.02	0.07	3.37	3.32	7.99	0.01	0.06	10.47
*FeO	27.63	26.90	19.14	0.31	0.43	29.41	29.62	20.01	0.11	0.36	13.73
MnO	3.12	4.21	0.22	0.04	0.01	0.95	0.95	0.05	0.00	0.00	0.15
CaO	7.83	7.79	11.15	6.94	7.67	7.19	7.04	11.18	5.35	6.25	20.90
Na_2O	N.D.	N.D.	1.35	7.51	7.38	N.D.	N.D.	1.77	8.43	8.06	0.66
K_2O	N.D.	N.D.	1.79	0.24	0.12	N.D.	N.D.	0.98	0.21	0.08	0.01
Total	98.60	97.58	99.98	100.05	98.91	100.21	97.79	101.71	101.94	100.95	100.95
Elements per Formula											
O	12	12	23	8	8	12	12	23	8	8	6
Si	3.01	3.00	6.34	2.68	2.63	3.03	3.04	6.42	2.76	2.71	1.98
Al	1.92	1.92	2.19	1.31	1.35	1.93	1.94	2.09	1.24	1.29	0.09
Ti	0.02	0.00	0.25	0.00	0.00	0.00	0.00	0.25	0.00	0.00	0.01
Mg	0.37	0.32	1.86	0.00	0.01	0.40	0.39	1.82	0.00	0.00	0.59
Fe	1.84	1.83	2.45	0.01	0.02	1.98	1.96	2.56	0.00	0.01	0.43
Mn	0.21	0.29	0.03	0.00	0.00	0.06	0.06	0.01	0.00	0.00	0.00
Ca	0.67	0.68	1.83	0.33	0.37	0.62	0.60	1.83	0.25	0.29	0.84
Na	N.D.	N.D.	0.40	0.65	0.64	N.D.	N.D.	0.52	0.71	0.68	0.05
K	N.D.	N.D.	0.35	0.01	0.01	N.D.	N.D.	0.19	0.01	0.00	0.00
Total Cations	8.04	8.04	15.70	4.99	5.03	8.02	7.99	15.69	4.98	5.00	3.99
Mg/(Mg + Fe)	0.17	0.15	0.43			0.17	0.17	0.42			0.58

Note: Data from Carmichael (1998b), Frisch (1998b), Steffen (1998b). SPMS—Spuhler Peak Metamorphic Suite; ICMS—Indian Creek Metamorphic Suite; PMMMS—Pony–Middle Mountain Metamorphic Suite; MMDS—metamorphosed mafic dikes and sills; Grt—garnet; Hbl—hornblende; Pl—Plagioclase; Cpx—Clinopyroxene.

*FeO—total iron as FeO.

†N.D.—no data.

decompression part (M2) of the *P-T* path (see below) following the highest-pressure phase (M1) of the Big Sky orogeny.

Garnet-Clinopyroxene-Hornblende-Plagioclase-Quartz Amphibolites

Some of the amphibolite samples from the Indian Creek Metamorphic Suite, Pony–Middle Mountain Metamorphic Suite, and MMDS have clinopyroxene. See Table 1 for examples of clinopyroxene compositions. Pattison (2003) reviewed experimental and natural data for this assemblage and finds considerable uncertainty regarding its stability. Most examples he reports occur at pressures from 0.7 to 1.2 GPa and from 700 to 800 °C. The assemblage garnet + clinopyroxene + plagioclase + quartz is the basis of the geobarometer of Moecher et al. (1988) and can be used with the garnet-clinopyroxene geothermometer of Powell (1985) to calculate temperatures and pressures. Results of this calculation for representative samples are shown in Figure 3. On average these pressures are higher than those obtained using the hornblende-bearing assemblages for the same rocks. However, the pressures are lowered if ferric iron corrections are made to the analyses, so there may be no significant difference from the hornblende results. Ferric iron is an issue here because calibrations used by Moecher et al. (1988) and Powell (1985) were based on ferric iron–free experimental data, whereas the Kohn and Spear (1990) and Graham and Powell (1984) calibrations were based on natural assemblages that would have contained some ferric iron. Hess and Vitaliano (1990) report similar temperatures and somewhat lower pressures for this assemblage in MMDS samples using different thermobarometer calibrations.

Other thermobarometers were applied to Tobacco Root rocks where the assemblages were appropriate. Aluminosilicate-bearing pelitic rocks are not widely distributed across the Tobacco Root Mountains, and we did not systematically evaluate the thermobarometry in these rocks. Different calibrations of garnet-biotite thermometers were applied to a few samples of

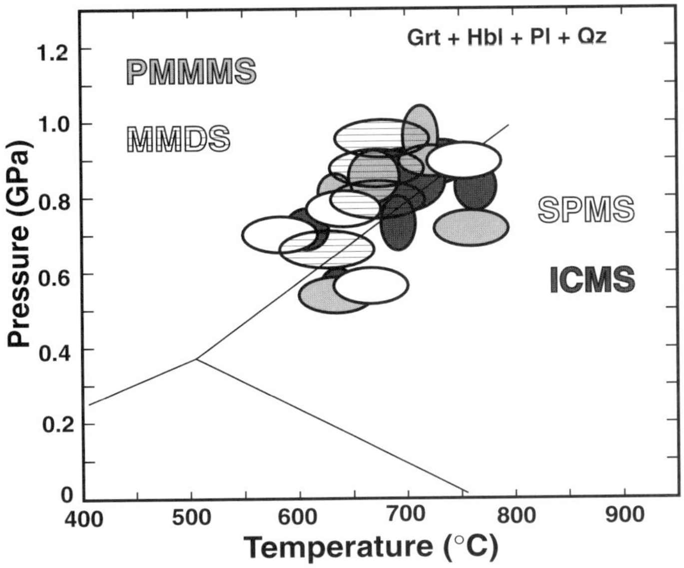

Figure 2. Pressures and temperatures for garnet (Grt) + hornblende (Hbl) + plagioclase (Pl) + quartz (Qz) amphibolites calculated with the Graham and Powell (1984) geothermometer and the Kohn and Spear (1990) geobarometer. Uncertainty ellipses are derived from observed compositional variations of minerals in the samples studied. The aluminosilicate triple point of Holdaway (1971) is shown for reference. PMMMS—Pony–Middle Mountain Metamorphic Suite, MMDS—metamorphosed mafic dikes and sills, SPMS—Spuhler Peak Metamorphic Suite, ICMS—Indian Creek Metamorphic Suite.

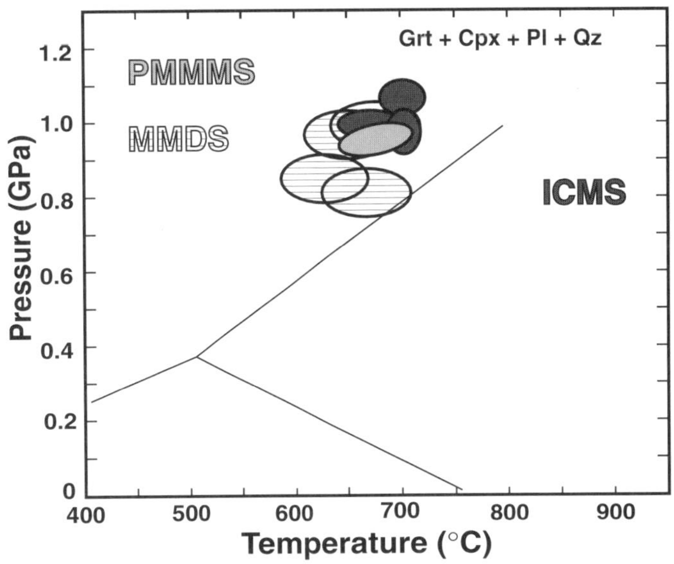

Figure 3. Pressures and temperatures for garnet (Grt) + clinopyroxene (Cpx) + plagioclase (Pl) + quartz (Qz) amphibolites calculated with the Powell (1985) geothermometer and the Moecher et al. (1988) geobarometer. Uncertainty ellipses are derived from observed compositional variations of minerals in the samples studied. No correction was made for possible small quantities of ferric iron. PMMMS—Pony–Middle Mountain Metamorphic Suite, MMDS—metamorphosed mafic dikes and sills, ICMS—Indian Creek Metamorphic Suite.

several rock types where appropriate. The resulting temperatures are typically 25–50 °C lower for the calibration of Patino Douce et al. (1993) than for the calibration of Hodges and Spear (1982). The lower temperatures given by the Patino Douce et al. (1993) geothermometer reflect the correction for the relatively high titanium contents of these biotites. The near-rim (typically within 30 µm of the edge) temperatures from the aluminous rocks are consistent with and overlap those discussed above for the garnet amphibolites (Fisher, 1994b; Monteleone, 1998b). However, garnet-biotite temperatures calculated from hornblende-bearing rocks are generally higher, some by as much as 100 °C (with the Patino Douce et al., 1993, calibration), than hornblende-garnet temperatures from the same rock. Spear and Markussen (1997) reported similar discrepancies for high-grade hornblende-bearing rocks. Reasons for this discrepancy may include different closure temperatures for biotite and hornblende, inappropriate garnet and/or biotite compositions for the thermometer calibrations, or incorrect determination of biotite and amphibole composition as discussed by Spear and Markussen (1997). A number of samples contained orthopyroxene. Although the presence of orthopyroxene in mafic rocks is used to recognize granulite facies conditions, the garnet + orthopyroxene + plagioclase + quartz assemblage yielded temperatures and pressures similar to those of Figures 2 and 3. In sum, geothermobarometers record upper amphibolite facies conditions of 0.5–0.9 GPa and 600–750 °C for rim compositions of mafic rocks throughout the Tobacco Root Mountains. Whatever their detailed metamorphic *P-T* path, it appears that all rocks passed through these conditions. More information about the *P-T* path of these rocks can be obtained from reaction textures in rocks of less common bulk compositions, to which we now turn.

CRITICAL ASSEMBLAGES AND REACTION TEXTURES

Orthoamphibolite

Some of the best evidence for the complex metamorphic history of the Tobacco Root Mountains occurs in aluminous orthoamphibolites, which have preserved a variety of diagnostic mineral assemblages and reaction textures. Coarse-textured (>1 cm) orthoamphibolites are a common and distinctive component of the Spuhler Peak Metamorphic Suite, but orthoamphibolites have also been mapped in the Indian Creek Metamorphic Suite. For example, the highest pressures for Tobacco Root metamorphism are documented by coarse-grained kyanite and orthopyroxene in aluminous orthoamphibolites, which require pressures above ~1.0 GPa as shown in Figure 4. Although we have not observed these minerals in physical contact, they occur in the same thin section and both have the coarse texture that we believe represents growth

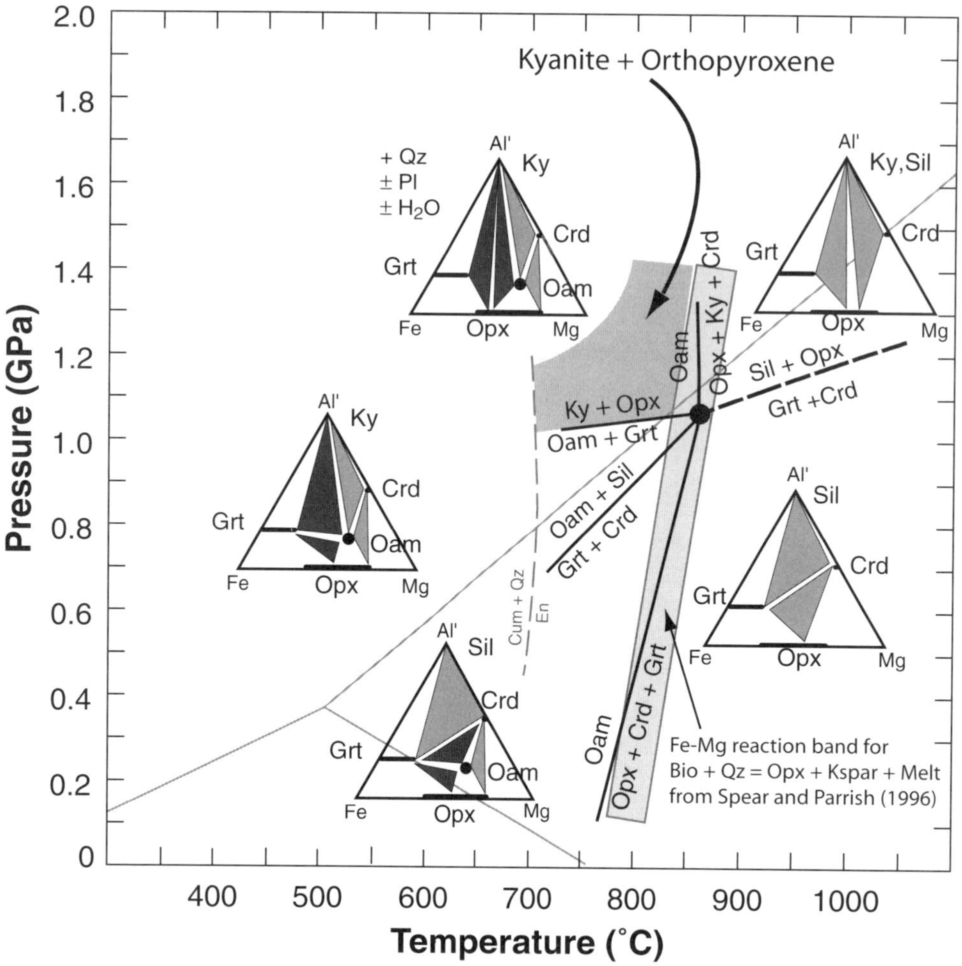

Figure 4. *P-T* diagram showing kyanite + orthopyroxene stability and an invariant point from a schematic petrogenetic grid for cordierite (Crd) + orthoamphibole (Oam) rocks and for garnet (Grt) + orthopyroxene (Opx) + orthoamphibole (Oam) rocks with kyanite (Ky) or sillimanite (Sil). The grid is slightly modified from that of Harley (1985). In particular, the location for the reaction Grt + Crd = Sil + Opx was calculated using the data of Holland and Powell (1998). The other reactions radiating from the invariant point were adjusted to fit the calculated curve. $P_{H_2O} = P_{total}$ is assumed; therefore pressures are maximum values for hydrous cordierite. A band is shown for the reaction biotite (Bio) + Qz = Opx + K-feldspar (Ksp) + Melt for varying Fe/Mg values from Spear and Parrish (1996). The Holdaway (1971) aluminosilicate triple point and the curve (dashed) marking the lower temperature limit of enstatite (En), calculated from the database of Holland and Powell (1998), are shown for reference.

during M1. The range of *P-T* conditions permitted by the coexistence of kyanite + orthopyroxene (Fig. 4) is limited at low pressure by the intersection of the sillimanite + orthopyroxene curve with the kyanite stability field. The low temperature boundary is provided by the low temperature limit of enstatite stability. The absence of orthopyroxene + K-feldspar or orthopyroxene-bearing migmatites in either the aluminous gneisses or the orthoamphibolites and the abundance of biotite in these rocks requires temperatures less than the dehydration melting reaction of Spear and Parrish (1996) for biotite + quartz, as shown in Figure 4.

The important minerals in these rocks can be shown on a modified AFM diagram (Spear, 1993, Chapter 13 therein) that is a projection from quartz, water, and if present, plagioclase. Figure 4 also shows a schematic invariant point and associated chemical reactions for the orthoamphibolites using this modified AFM diagram to illustrate the chemography. The location of the invariant point is modified slightly from Harley (1985) to be consistent with the sillimanite + orthopyroxene curve, located on Figure 4 assuming $P_{H_2O} = P_{total}$ and using the thermodynamic data set of Holland and Powell (1998). The upper stability limit of orthoamphibole for $P_{H_2O} = P_{total}$ as shown on Figure 4 is consistent with observed mineral occurrences in the Tobacco Root Mountains. Because there is abundant orthoamphibole in Tobacco Root rocks, with no textural evidence of its instability at high temperature, we believe the orthoamphibole-out reaction is an upper temperature bound to the *P-T* path.

Garnet-Cordierite Textures

A particularly impressive Spuhler Peak Metamorphic Suite sequence that includes orthoamphibolite with golf ball–sized (1–5 cm across) garnets is well exposed along the ridge running west from Thompson Peak (Fig. 5). The coarse grain size of these rocks, their absence of fabric, and the high concentration of refractory minerals leads us to believe that these rocks underwent partial melting during metamorphism. Many garnet crystals are roughly dodecahedral in form and are red brown in color, except along fractures and in rims where they have the blue gray color of cordierite. Coarse (1 mm) brown biotite flakes may wrap around garnet porphyroblasts, whereas fine, brown green biotite shows no orientation and may be intergrown with cordierite. The orthoamphibole gedrite occurs as euhedral, tabular grains up to 15 cm in length. Many samples have a characteristic play of

Figure 5. Orthoamphibolite with large garnets from the Spuhler Peak Metamorphic Suite along the ridge running west from Thompson Peak. The garnet crystals are replaced partially by cordierite on their margins and along cracks. The hammer in the photo is 33 cm long.

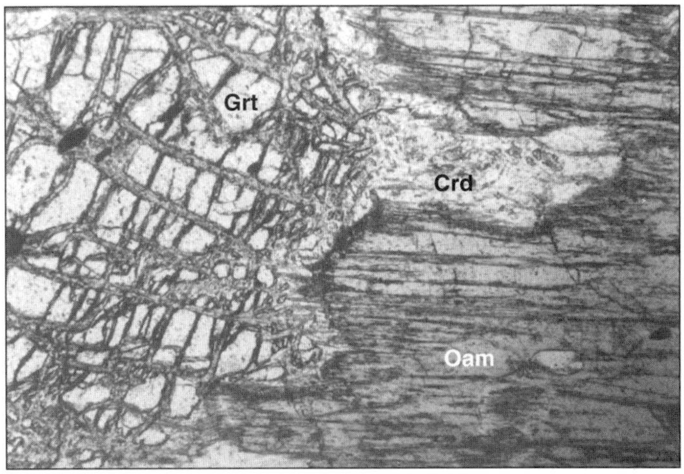

Figure 6. Photomicrograph in plane-polarized light of cordierite (Crd) replacing garnet (Grt) and as a moat between garnet and orthoamphibole (Oam). Sample KAT-44. The field of view is 5 mm wide.

colors due to anthophyllite exsolution lamellae. In thin section, the presence of cordierite can be confirmed by the presence of yellow pleochroic haloes around zircon and monazite inclusions. Cordierite occurs principally as a textural replacement of garnet along fractures and where it is in contact with gedrite (see Fig. 6). Chemical analyses of the minerals in a sample with cordierite are given in Table 2.

The presence of cordierite in equilibrium with garnet and gedrite limits the maximum pressure and temperature to 1.0 GPa and 800 °C (Fig. 4) at $P_{H_2O} = P_{total}$, and to lower temperatures and pressures if $P_{H_2O} < P_{total}$. We believe the introduction of cordierite into the orthoamphibolites begins with the decompression reaction

$$\text{Oam} + \text{Ky} = \text{Grt} + \text{Crd} + H_2O \quad (1),$$

where Oam is orthoamphibole, Ky is kyanite, Grt is garnet, and Crd is cordierite. With the addition of Mn and/or Ca in the garnet, discontinuous reaction 1 becomes continuous. As shown in Table 2, the amount of grossular component is less than 2% and the spessartine component is less than 1% for garnet from these rocks. The modal amount of garnet is also much greater than that of cordierite. Accordingly, effects on reaction 1 produced by the addition of these minor amounts of Mn and Ca are expected to be relatively small. Moreover, the extent of this reaction is limited in most orthoamphibolites by the small quantity of kyanite permitted by the bulk composition. Rare gedrite-bearing rocks contain kyanite or partial to complete sillimanite pseudomorphs after kyanite. The aluminosilicate (Als) is in all cases rimmed by cordierite, attesting to its previous equilibrium with orthoamphibole and to lower-temperature and higher-pressure conditions than reaction 1. These coronas are virtually identical to those that occur in the aluminous gneisses discussed below. In addition, we see little evidence remaining for the growth of garnet with cordierite; garnet appears to be replaced by cordierite. Therefore, consumption of garnet and growth of cordierite requires one or more additional reactions. Possible cordierite-producing reactions for these rocks include

$$\text{Grt} + \text{Qz} + H_2O = \text{Crd} + \text{Oam} \quad (2)$$

and

$$\text{Grt} + \text{Kspar} + \text{Liq} = \text{Crd} + \text{Bio} \quad (3).$$

Reaction 2 (Qz is quartz) would occur as a continuous reaction as the stable garnet, in the assemblage Grt + Crd + Oam, becomes more iron-rich with falling temperature. However, reaction 2 is limited by the supply of water. Therefore, a reaction involving a melt phase (Liq) such as reaction 3 (Bio is biotite; Kspar is K-feldspar) is more likely because there is other evidence (e.g., Fig. 7) of partial melting having occurred in the Spuhler Peak Metamorphic Suite. Reactions such as (3) could occur with falling temperature (Spear et al., 1999); and fine-grained, low-Ti, green biotite that is commonly present with cordierite may be the result of this reaction.

Orthopyroxene + Cordierite Symplectite

Orthoamphibolite samples containing fine-scale symplectite textures around garnet were collected from a Granite Creek locality (UTM 502873N, 12T 043149E) in the southern Tobacco Roots and from the Leggat Mountain ridge above Quartz Creek (UTM 504011N, 12T 041680E) in the central Tobacco Roots. In the field, these rocks are distinguished by large (2–5 cm),

TABLE 2. CORDIERITE-BEARING ORTHOAMPHIBOLITE

Mineral	Grt core	Grt core	Grt rim	Crd unzoned	Crd unzoned	Ged unzoned	Ged unzoned	Pl core	Pl rim
Oxide (wt%)									
SiO_2	38.81	38.56	38.01	49.14	48.66	48.88	47.63	61.20	59.78
Al_2O_3	21.59	21.43	21.32	32.82	32.49	10.69	10.44	23.90	25.26
TiO_2	0.00	0.00	0.00	0.00	0.00	0.47	0.42	0.00	0.00
MgO	9.53	9.51	5.20	10.83	10.45	18.59	16.89	0.00	0.00
FeO*	26.64	26.80	33.11	5.35	5.27	17.36	19.72	0.00	0.00
MnO	0.72	0.61	1.10	0.00	0.00	0.16	0.28	0.00	0.00
CaO	1.85	2.35	2.02	0.00	0.00	0.41	0.44	5.70	6.83
Na_2O	0.00	0.00	0.00	N.D.	0.58	1.31	1.43	8.13	7.92
K_2O	N.D.†	N.D.	N.D.	N.D.	N.D.	0.00	0.00	0.09	0.08
Total	99.14	99.27	100.75	98.14	97.45	97.87	97.25	99.02	99.86
Elements per Formula									
O	12	12	12	18	18	23	23	8	8
Si	3.01	2.99	3.00	5.00	5.00	6.96	6.93	2.74	2.67
Al	1.97	1.96	1.98	3.94	3.93	1.80	1.79	1.26	1.33
Ti	0.00	0.00	0.00	0.00	0.00	0.05	0.05	0.00	0.00
Mg	1.10	1.10	0.61	1.64	1.60	3.95	3.66	0.00	0.00
Fe	1.73	1.74	2.18	0.46	0.45	2.07	2.40	0.00	0.00
Mn	0.05	0.04	0.07	0.04	0.01	0.02	0.04	0.00	0.00
Ca	0.15	0.20	0.17	N.D.	N.D.	0.06	0.07	0.27	0.33
Na	N.D.	N.D.	N.D.	N.D.	N.D.	0.36	0.40	0.71	0.69
K	N.D.	N.D.	N.D.	N.D.	N.D.	0.00	0.00	0.01	0.01
Total Cations	8.01	8.03	8.02	11.08	10.98	15.27	15.33	4.98	5.01
Mg/(Mg + Fe)	0.39	0.39	0.22	0.78	0.78	0.66	0.60	N.D.	N.D.

Note: Data from Tierney (1994b). Sample: KAT-30, Spuhler Peak Metamorphic Suite (SPMS), UTM 504200N, 12T 041951E. Grt—garnet; Crd—cordierite; Ged—gedrite; Pl—plagioclase.
*FeO—Total iron as FeO.
†N.D.—no data.

Figure 7. Glacially polished outcrop of Spuhler Peak Metamorphic Suite (SPMS) amphibolite showing "wispy" leucosomes consisting principally of plagioclase and quartz. This texture is a common one in the SPMS. Chemically, these outcrops tend to be more iron-rich than amphibolite outcrops with a uniform distribution of mafic and felsic minerals. The hammer in the photo is 40 cm long.

irregularly shaped, pale purple garnet (Grt) porphyroblasts in a coarse (1 cm) matrix of tan gedrite (Ged), quartz (Qz), and biotite (Bio). In thin section, orthopyroxene (Opx), kyanite (Ky), and rutile (Rut) are also clearly part of the coarse equilibrium assemblage. Garnet cores commonly have inclusions of staurolite (Sta) and kyanite (Hatch, 1998b, p. 31), whereas garnet rims have inclusions of the matrix minerals. In several samples, garnet grains are surrounded by a fine-grained, symplectite intergrowth of orthopyroxene and cordierite (Crd). We interpret the coarse-grained assemblage as M1, the symplectite assemblage as M2, and the inclusions in the garnet cores as early M1 or pre-M1.

Symplectite bands 0.2 to 5 mm wide contain wormy, rod-shaped crystals of orthopyroxene that radiate outward from garnet grains in a matrix of cordierite (see Fig. 8). Where present, quartz inclusions in the garnet are also surrounded by the Opx + Crd symplectite. In some cases, a thin band of orthopyroxene directly surrounds quartz grains near garnet. Typical compositions of the minerals in the Opx + Crd symplectite samples are given in Table 3. All of the minerals are chemically homogeneous except for garnet, which is slightly depleted in Mg near the rim. Similar Opx + Crd symplectite textures have been described in rocks from South Africa (Van Reenen, 1986), Antarctica (Harley,

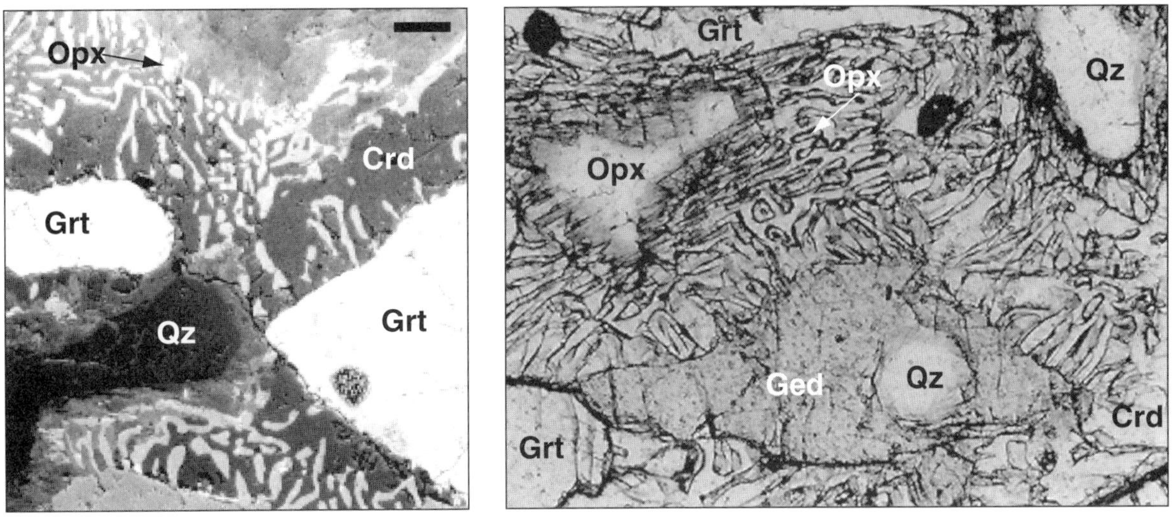

Figure 8. Orthopyroxene (Opx) + cordierite (Crd) symplectite in sample TBR-252H from Granite Creek. On the left is a backscattered electron image (scale bar = 100 μm). On the right is a plane-polarized light photomicrograph (width = 1.5 mm). The symplectite occurs between garnet (Grt) and quartz (Qz), which are never in contact in this sample.

1989), India (Raith et al., 1997), Sri Lanka (Kriegsman and Schumacher, 1999), and elsewhere. These workers and others interpret the symplectite textures to be a result of the reaction

$$\text{Grt} + \text{Qz} = \text{Opx} + \text{Crd} \quad (4),$$

which is believed (Harley, 1989; Spear, 1993, p. 390) to occur during nearly isothermal decompression at pressures ~0.8 GPa for garnets with the Mg/(Mg + Fe) value (0.5) of those in Table 3 at relatively high temperatures (~700 °C). This texture can also form with cooling or heating as pressure decreases, but the absence of chemical zoning in the symplectite minerals precludes quantitative assessment of the *P-T* path. Hensen and Harley (1990) evaluated experimental data for garnet-orthopyroxene equilibria and calculated isopleths for garnet and orthopyroxene as a function of temperature and pressure. Using the compositions of garnet and orthopyroxene from Table 3, both of Hensen and Harley's (1990) models (their Figures 2.13 and 2.14) yield temperatures of 700 to 770 °C and pressures of 0.75 to 0.9 GPa.

Sapphirine + Spinel + Cordierite Symplectite

Two orthoamphibolite samples from the Granite Creek locality were observed to have symplectite bands that surround kyanite or sillimanite and separate it from gedrite and biotite. These bands consist of fine-grained intergrowths of elongated sapphirine (Spr) grains and spinel (Spl) in a matrix of cordierite (see Fig. 9). Many of the sapphirine and spinel crystals are aligned roughly perpendicular to the kyanite contact and extend only partway across the band surrounding the aluminosilicate mineral. Sillimanite replaces kyanite in some of these samples. The chemical compositions (Table 4) of the minerals in the Spr + Spl + Crd symplectite sample are very similar to those of the Opx

TABLE 3. OPX + CRD SYMPLECTITE

Mineral	Grt core	Grt rim	Crd unzoned	Opx unzoned	Ged unzoned
Oxide (wt%)					
SiO_2	39.90	39.79	50.35	53.23	44.37
Al_2O_3	22.34	22.33	33.77	4.27	18.12
TiO_2	0.00	0.03	0.02	0.07	0.56
MgO	13.45	12.94	12.35	26.21	19.45
FeO*	22.66	22.89	2.01	14.92	11.91
MnO	0.20	0.23	0.03	0.05	0.04
CaO	0.69	0.97	N.D.	0.05	0.27
Na_2O	N.D.†	N.D.	N.D.	N.D.	1.83
K_2O	N.D.	N.D.	N.D.	N.D.	0.02
Total	99.24	99.18	98.53	98.80	96.57
Elements per Formula					
O	12	12	18	6	23
Si	3.01	3.01	5.02	1.93	6.27
Al	1.98	1.99	3.97	0.18	3.02
Ti	0.00	0.00	0.00	0.00	0.06
Mg	1.51	1.46	1.84	1.41	4.10
Fe	1.43	1.45	0.17	0.45	1.41
Mn	0.01	0.01	0.00	0.00	0.01
Ca	0.06	0.08	0.00	0.00	0.04
Na	N.D.	N.D.	0.03	0.00	0.50
K	N.D.	N.D.	N.D.	N.D.	0.00
Total Cations	8.00	8.00	11.02	3.98	15.41
Mg/(Mg+Fe)	0.51	0.50	0.92	0.76	0.74

Note: Data from Hatch (1998a). Sample: TBR-252H, Indian Creek Metamorphic Suite (ICMS), UTM 504011N, 12T 041680E. Grt—garnet; Crd—cordierite; Opx—orthopyroxene; Ged—gedrite.
*FeO—total iron as FeO.
†N.D.—no data.

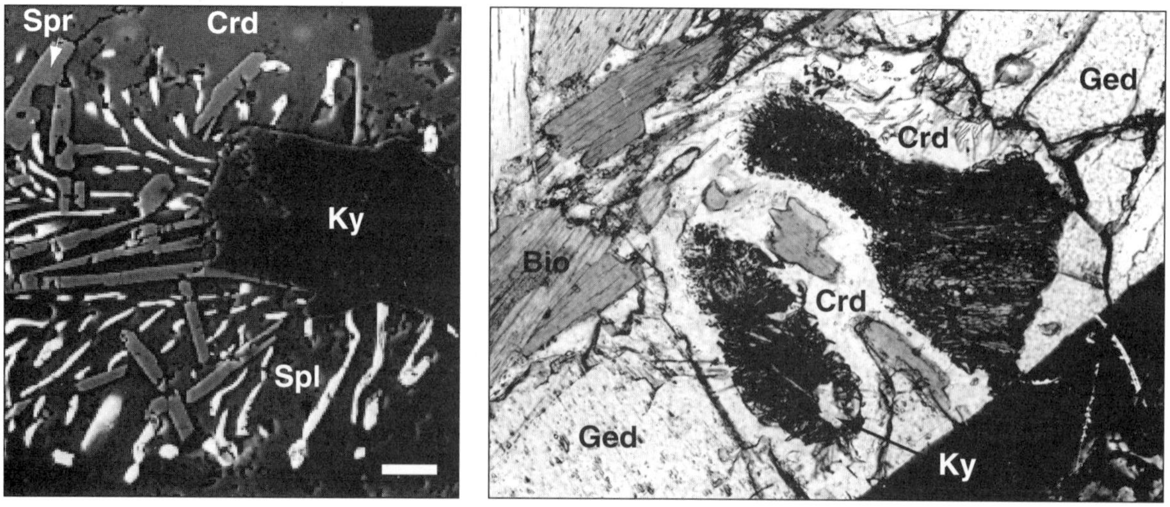

Figure 9. Sapphirine (Spr) + Spinel (Spl) + cordierite (Crd) symplectite in sample CDH-97-51E, also from Granite Creek. On the left is a backscattered electron image (scale bar = 20 μm). On the right is a plane-polarized light photomicrograph (width = 2 mm). The symplectite occurs between kyanite (Ky) and gedrite (Ged) or biotite (Bio).

TABLE 4. SPR + SPL + CRD SYMPLECTITE MINERAL COMPOSITIONS

Mineral	Grt core	Grt rim	Crd unzoned	Opx unzoned	Ged unzoned	Bio unzoned	Spr unzoned	Ky unzoned	Spl normalized
Oxide (wt%)									
SiO_2	40.20	39.97	49.95	53.26	43.93	39.44	12.00	36.84	N.D.
Al_2O_3	22.55	22.31	33.81	3.94	18.15	16.99	64.86	62.63	54.55
TiO_2	0.02	0.04	0.03	0.07	0.63	3.93	0.09	0.03	0.32
MgO	13.84	12.97	12.32	25.71	19.08	18.49	16.46	0.10	9.83
FeO*	22.74	23.44	2.24	16.37	12.43	8.48	5.58	0.21	30.18
MnO	0.23	0.24	0.02	0.07	0.08	0.06	0.01	0.10	2.11
CaO	0.66	0.81	0.01	0.04	0.28	0.07	N.D.	N.D.	N.D.
Na_2O	N.D.[†]	N.D.	N.D.	N.D.	1.92	0.53	N.D.	N.D.	N.D.
K_2O	N.D.	N.D.	N.D.	N.D.	0.01	7.81	N.D.	N.D.	N.D.
ZnO	N.D.	N.D.	N.D.	N.D.	N.D.	N.D.	N.D.	N.D.	3.01
Total	100.24	99.79	98.37	99.45	96.51	95.79	98.99	99.91	99.99
Elements per Formula									
O	12	12	18	6	23	11	10	5	4
Si	3.00	3.01	5.00	1.93	6.24	2.80	0.71	1.00	N.D.
Al	1.98	1.98	3.99	0.17	3.04	1.42	4.55	2.00	1.83
Ti	0.00	0.00	0.00	0.00	0.07	0.21	0.00	0.00	0.01
Mg	1.54	1.45	1.84	1.39	4.04	1.96	1.46	0.00	0.42
Fe	1.42	1.48	0.19	0.50	1.48	0.50	0.28	0.00	0.72
Mn	0.01	0.02	0.00	0.00	0.01	0.00	0.00	0.00	0.05
Ca	0.05	0.07	0.02	0.00	0.04	0.01	N.D.	N.D.	N.D.
Na	N.D.	N.D.	N.D.	N.D.	0.53	0.07	N.D.	N.D.	N.D.
K	N.D.	N.D.	N.D.	N.D.	N.D.	0.71	N.D.	N.D.	N.D.
Zn	N.D.	N.D.	N.D.	N.D.	N.D.	N.D.	N.D.	N.D.	0.06
Total Cations	8.00	8.00	11.03	3.98	15.44	7.68	7.01	3.01	3.08
Mg/(Mg + Fe)	0.52	0.50	0.91	0.74	0.73	0.80	0.84	N.D.	0.37

Note: Data from Hatch (1998a). Sample: CEH-97-51E, Indian Creek Metamorphic Suite (ICMS), UTM 504011N, 12T 041680E. Spr—sapphirine; Spl—spinel; Crd—cordierite; Grt—garnet; Opx—orthopyroxene; Ged—gedrite; Bio—biotite; Ky—kyanite.

*FeO—total iron as FeO.

[†]N.D.—no data.

+ Crd symplectite sample (see Table 3). The principal difference seems to be the absence of quartz.

Sapphirine symplectites have been described from many localities (e.g., Lal et al., 1984; Windley et al., 1984; Harley and Hensen, 1990; Kihle and Bucher-Nurminen, 1992; Mohan et al., 1996; Ouzegane and Boumaza, 1996; Godard and Mabit, 1998; Kriegsman and Schumacher, 1999; Elvevold and Gilotti, 2000), but we are not aware of any other occurrences of sapphirine in the Wyoming province. In addition, the details of the Tobacco Root assemblages are not identical to those previously described in the literature. In particular, most sapphirine symplectites in mafic rocks surround sillimanite, rather than kyanite, and occur in amphibole-free rocks that require temperatures higher than the upper stability limit of gedrite (~800 °C). Many involve minerals that are not part of the Tobacco Root assemblages, such as corundum or garnet, or do not have spinel in the symplectite. Descriptions of sapphirine symplectites that do surround kyanite typically are for more felsic rocks with plagioclase as part of the symplectite.

Kihle and Bucher-Nurminen (1992) describe a Spr + Crd symplectite (no spinel) that separates sillimanite and gedrite in feldspar-absent metamorphic rocks from southern Norway, and Windley et al. (1984) describe the same texture from the Limpopo Belt of South Africa. Both papers call for the reaction

$$\text{Ged} + \text{Sil} = \text{Crd} + \text{Spr} \quad (5).$$

Lal et al. (1984) do have Spr + Spl + Crd symplectites that separate sillimanite and gedrite in rocks they studied from Kiranur, southern India. They suggest the reaction

$$\text{Ged} + \text{Sil} + \text{V} = \text{Crd} + \text{Spr} + \text{Spl} + \text{Ab} \quad (6)$$

that produces spinel but also feldspar, which is lacking in the Tobacco Root rocks.

The similarities of the Montana Spr + Spl + Crd symplectites to other sapphirine symplectite occurrences, such as the fine grain size, coronal textures, and volume increases, support the interpretation that these features result from decompression reactions. However, it is difficult to balance a reaction with only the minerals in the observed symplectite assemblage, so perhaps one or more important phases are missing. This could be the case if the Spr + Spl + Crd symplectite grew during a decompression melting reaction involving aluminosilicate (Ky or Sil), gedrite, and biotite as the pressure fell from the kyanite to the sillimanite stability field at temperatures between 700 and 800 °C.

Aluminous Gneisses

Garnet-Cordierite-Sillimanite Textures

Aluminous gneisses occur in the Spuhler Peak Metamorphic Suite that typically contain the assemblage quartz + biotite + sillimanite + garnet + plagioclase. The absence of primary (foliation-defining) muscovite and K-feldspar from these rocks is interpreted to be the result of partial melting. In many samples, cordierite + biotite appear to have replaced garnet as a rim and along fractures (Fig. 10). In the same samples, cordierite also appears to be replacing sillimanite. Chemical analyses of the minerals in an aluminous gneiss with cordierite are given in Table 5. These textures might be explained by the continuous reaction

$$\text{Grt} + \text{Sil} + \text{Qz} + \text{H}_2\text{O} = \text{Crd} \quad (7).$$

As with previously discussed reactions, reaction 7 requires the addition of water if cordierite is hydrous. Alternatively, if a partial melt is present, the reaction

$$\text{Grt} + \text{Kspar} + \text{Liq} = \text{Crd} + \text{Bio} \quad (3)$$

could produce cordierite and biotite, which commonly occur together in these rocks (Fig. 10). Figure 11 shows a manganese X-ray map of a garnet being replaced by cordierite + biotite. Note the increase in Mn content near the edges of the garnet fragments. We have used the program GIBBS (version 4.7, 2002) of Spear and Menard (1989) to model the chemical variations of garnet being replaced by cordierite by the continuous reaction 7. Figure 12 shows the resulting Mn isopleths (spessartine mole fractions), which are nearly isobaric. Based on this model, measured garnet zoning in the aluminous gneisses

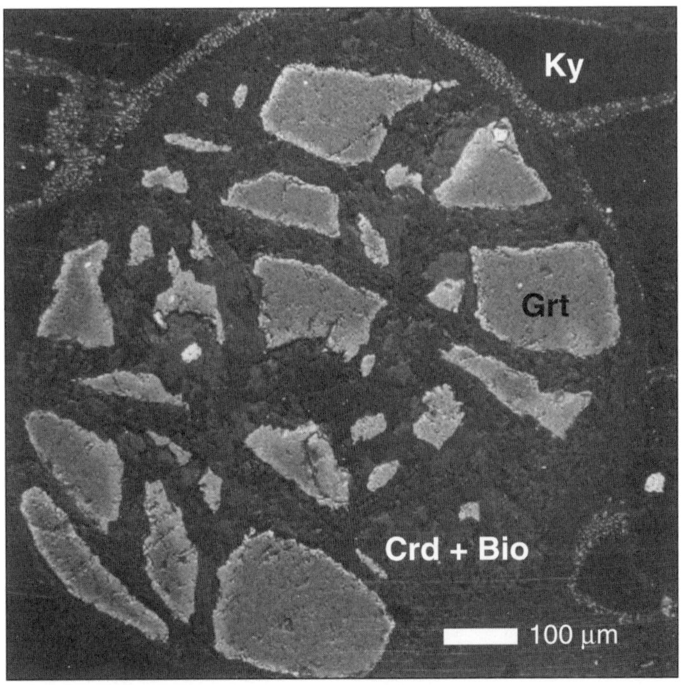

Figure 10. Backscattered electron image of a garnet (Grt) grain being replaced by cordierite (Crd) and biotite (Bio). The thin bright rims on each garnet fragment are manganese-enriched and pyrope-depleted. Sample KAT-41.

TABLE 5. CORDIERITE-BEARING ALUMINOUS GNEISS

Mineral	Grt core	Grt core	Grt rim	Grt rim	Crd unzoned	Crd unzoned	Pl core	Pl rim	Bio core	Bio rim
Oxide (wt%)										
SiO_2	38.20	38.05	36.78	37.89	49.49	49.18	57.49	59.09	37.46	36.74
Al_2O_3	21.14	21.46	20.63	21.03	33.52	33.44	28.21	26.10	17.99	17.97
TiO_2	0.00	0.00	0.00	0.00	0.00	0.00	0.00	0.00	2.07	2.23
MgO	7.26	7.56	3.11	5.95	9.65	8.98	0.00	0.00	14.26	13.86
FeO*	26.68	26.32	26.77	27.03	6.48	6.88	0.00	0.00	14.05	14.49
MnO	4.56	4.56	10.09	6.13	0.84	0.89	0.00	0.00	0.34	0.42
CaO	1.74	1.65	1.51	1.74	N.D.†	N.D.	9.94	7.42	0.00	0.00
Na_2O	N.D.	N.D.	N.D.	N.D.	N.D.	N.D.	6.24	7.66	1.12	0.00
K_2O	N.D.	N.D.	N.D.	N.D.	N.D.	N.D.	0.00	0.09	8.49	8.58
Total	99.58	99.59	98.89	99.77	99.98	99.36	101.88	100.36	94.29	94.29
Elements per Formula										
O	12	12	12	12	18	18	8	8	11	11
Si	3.00	2.99	3.00	3.00	4.98	4.99	2.53	2.63	5.49	5.51
Al	1.96	1.99	1.98	1.96	3.98	4.00	1.47	1.37	3.17	3.12
Ti	0.00	0.00	0.00	0.00	0.00	0.00	0.00	0.00	0.25	0.23
Mg	0.85	0.88	0.38	0.70	1.45	1.36	0.00	0.00	3.09	3.13
Fe	1.75	1.73	1.83	1.79	0.55	0.58	0.00	0.00	1.81	1.73
Mn	0.30	0.30	0.70	0.41	0.07	0.08	0.00	0.00	0.05	0.04
Ca	0.15	0.14	0.13	0.15	N.D.	N.D.	0.47	0.35	0.00	0.00
Na	N.D.	N.D.	N.D.	N.D.	N.D.	N.D.	0.53	0.66	0.00	0.32
K	N.D.	N.D.	N.D.	N.D.	N.D.	N.D.	0.00	0.01	1.64	1.59
Total Cations	8.02	8.02	8.01	8.02	11.03	11.01	5.00	5.02	13.86	14.07
Mg/(Mg + Fe)	0.33	0.34	0.17	0.28	0.73	0.70	N.D.	N.D.	N.D.	0.64

Note: Data from Tierney (1994b). Sample: KAT-42, Spuhler Creek Metamorphic Suite (SPMS), UTM 504188N, 12T 041977E. Grt—garnet; Crd—cordierite; Pl—plagioclase; Bio—biotite.

*FeO—total iron as FeO.

†N.D.—no data.

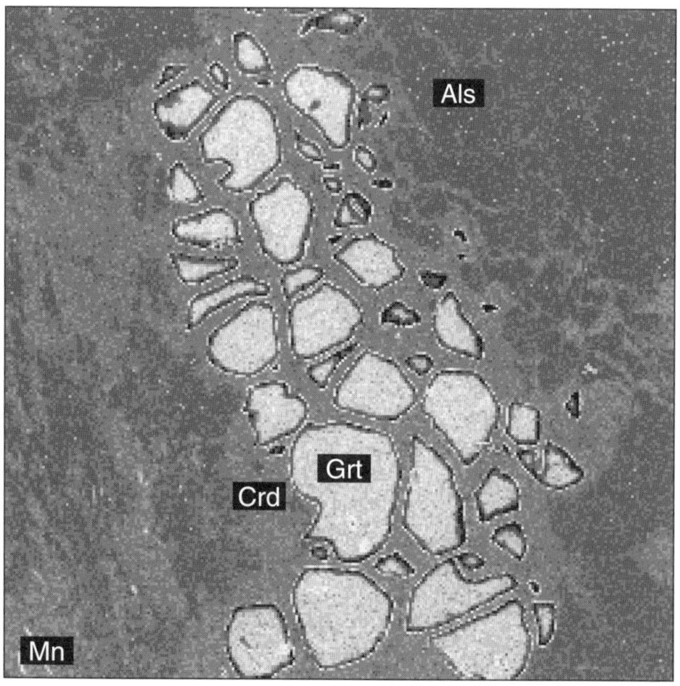

Figure 11. Manganese (Mn) X-ray map for sample KAT-42 showing Mn-enrichment at the rims of garnet (Grt) islands remaining from a single garnet crystal being replaced by cordierite (Crd). The thin "light" rim on each garnet fragment is enriched in Mn, as also shown by the analyses in Table 5. Als—aluminosilicate. Field of view is 5 mm.

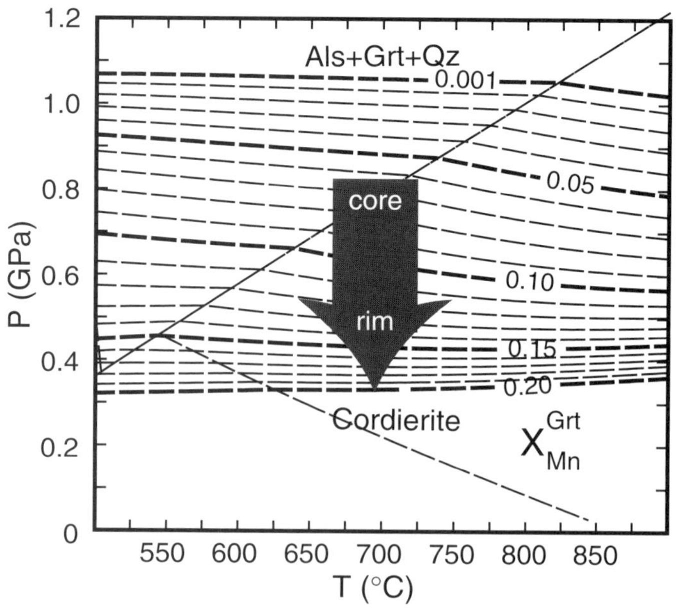

Figure 12. Manganese (Mn) isopleths (mole fraction spessartine) for garnet during the continuous reaction aluminosilicate (Als) + garnet (Grt) + quartz (Qz) = cordierite (Crd) calculated using GIBBS (version 4.7, 2002) of Spear and Menard (1989). The Mn isopleths are virtually parallel to the garnet mode isopleths (not shown) in the MnKFMASH system at constant bulk composition. The large arrow shows the compositional zoning for garnets in cordierite-bearing aluminous gneisses such as KAT-42 in Table 5 at temperatures suggested by geothermometry and other mineral assemblages and textures.

appears to have formed as the pressure fell from 0.8 to 0.4 GPa. The development of the thin Mn-rich rims on the garnet fragments (Fig. 10) is consistent with initial garnet consumption to produce the fragments by crystallizing leucosome via reaction 3 during cooling, followed by a pressure decrease to form the rims on the fragments via reaction 2.

In several aluminous gneiss samples containing kyanite, we observed narrow (20–40 µm) symplectite coronas around both garnet and kyanite (see Figs. 13 and 14). Garnet is rimmed by cordierite and spinel, whereas kyanite is rimmed by anorthite + spinel. The occurrence of these rims on sillimanite pseudomorphs after kyanite as well as on surviving kyanite suggest that these features in both the aluminous gneisses and the orthoamphibolites (discussed above) formed after the sillimanite grew. Because these coronas are surrounded by cordierite in both cases, the coronas may be a fairly late feature caused by the continued breakdown of garnet and kyanite when the supply of other reactants was limited by the surrounding cordierite.

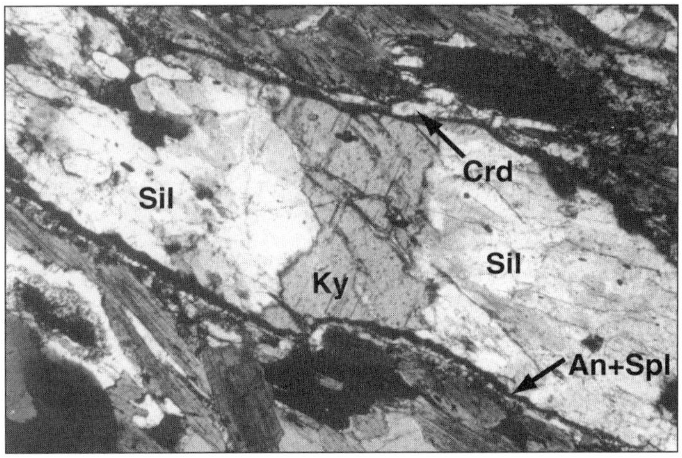

Figure 13. Photomicrograph in cross-polarized light of an anorthite (An) + spinel (Spl) symplectite rim on a sillimanite (Sil) pseudomorph of kyanite (Ky) in sample KAT-41, an aluminous gneiss from the Spuhler Peak Metamorphic Suite. The dark symplectite rim is in turn surrounded by cordierite (Crd). The field of view is 5 mm wide.

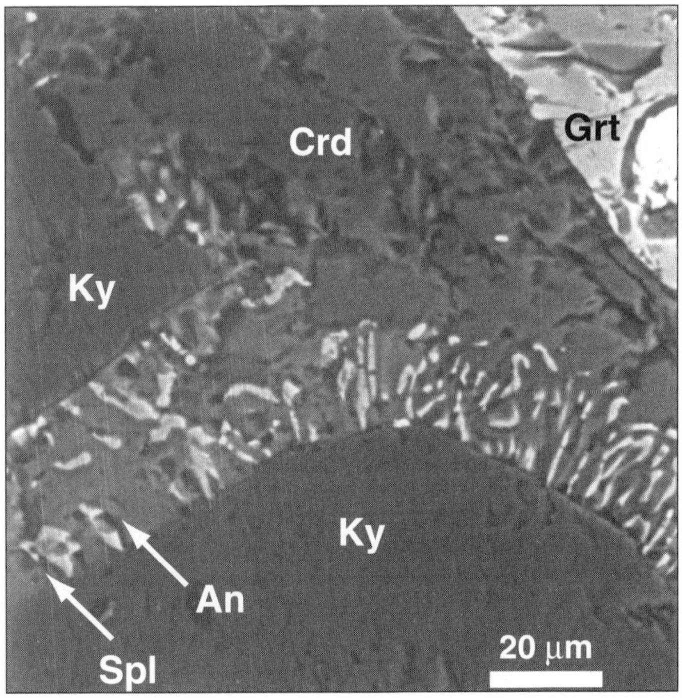

Figure 14. Backscattered electron image of an anorthite (An) + spinel (Spl) symplectite rim on kyanite (Ky) in sample KAT-41, an aluminous gneiss from the Spuhler Peak Metamorphic Suite. The symplectite rim is surrounded by cordierite (Crd).

Orthopyroxene-Bearing Gneiss

Garnet Necklace Textures

Orthopyroxene-bearing, plagioclase + quartz rocks are locally abundant in the Indian Creek Metamorphic Suite. One distinctive occurrence is on the ridge (UTM 504028N, 12T 041774E) north of Quartz Creek near Leggat Mountain. In this location, mafic aggregates (up to 1 cm across) of orthopyroxene, garnet, and clinopyroxene occur in a plagioclase-quartz matrix. These aggregates are distinguished by orthopyroxene megacrysts (up to 8 mm in diameter) that are separated from the plagioclase-quartz matrix by "necklaces" or coronas of small (0.2 mm), euhedral garnet crystals with associated clinopyroxene, quartz, and hornblende (see Figs. 15 and 16). The orthopyroxene megacrysts have a fractured appearance and contain inclusions of garnet and plagioclase, but not hornblende. Garnet occurs in the plagioclase matrix, as well as in the necklaces and as inclusions in orthopyroxene. Subhedral clinopyroxene and quartz appear to be part of the reaction texture and generally occur on the orthopyroxene side of the necklaces, although quartz also is part of the plagioclase-quartz matrix. Hornblende grains appear to overgrow orthopyroxene near garnet necklaces and contain inclusions of orthopyroxene, clinopyroxene, quartz, and garnet. Ilmenite occurs as an accessory mineral throughout this rock, whereas rutile is found only in the mafic aggregates.

Chemical compositions of the minerals in sample TBR-181c are given in Table 6. The orthopyroxene megacrysts are unzoned. Plagioclase grains in the matrix are nearly homogeneous (An_{49}), with a small anorthite enrichment (An_{52}) at their rims. Plagioclase inclusions in the orthopyroxene are much more albitic (An_{38}) and contain more Fe and Mg. Individual garnets in the matrix have flat compositional profiles with a slight enrichment in grossular and depletion in pyrope at their rims. Necklace garnets have a slight asymmetrical zoning with more grossular on the plagioclase side and more almandine on the orthopyroxene side.

We interpret the garnet corona texture as a product of a reaction between orthopyroxene and plagioclase (Pl) that occurred as the metamorphic conditions changed from M1 to M2. The corona texture clearly indicates that garnet is a product of the reaction, which led one of us (Tuit, 1996a) initially to argue that

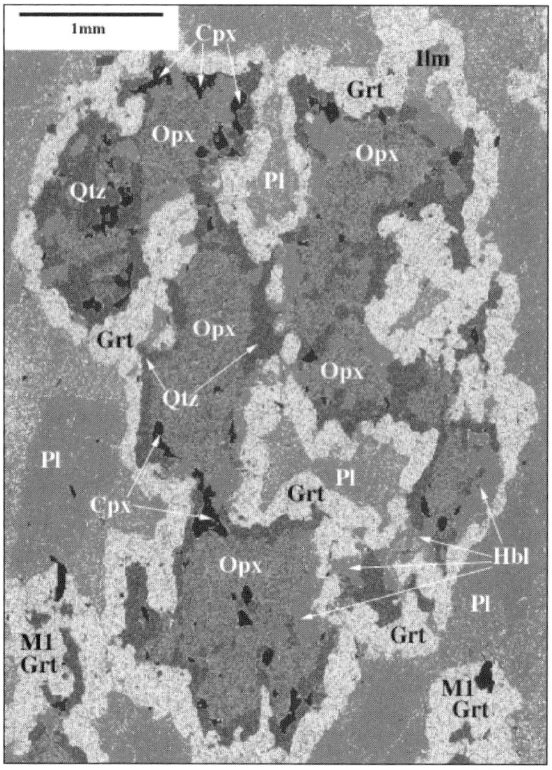

Figure 15. A calcium X-ray map of an orthopyroxene-rich mafic aggregate in a plagioclase-quartz matrix from sample TBR-181c. Lighter shades represent more calcium. Note the calcium-rich garnet bands that separate orthopyroxene from plagioclase.

the reaction was the result of an increase in pressure. However, a decrease in pressure is consistent with other petrologic data, so we believe the garnet-producing reaction

$$\text{Opx} + \text{Pl} = \text{Grt} + \text{Cpx} + \text{Qz} \quad (8)$$

was crossed during falling temperature, rather than rising pressure. Spear and Markussen (1997) studied this reaction in rocks of similar composition and texture from the Adirondack Mountains of New York. They argued for nearly isobaric cooling (−6 bar/°C) based on chemical zoning of minerals and Gibbs method modeling. Minerals in our rocks have similar zoning: Orthopyroxene typically has decreasing Al^{tot} toward the rim; clinopyroxene decreases in Fe/(Fe + Mg) and Al^{tot} and increases in Ca toward the rim; garnet Fe/(Fe + Mg) increases from core to rim. Our samples lack the zoning of Fe/(Fe + Mg) in orthopyroxene and anorthite in plagioclase that Spear and Markussen (1997) observed. The temperature and pressure at which the garnet necklaces were produced depends on composition. For a −6 bar/°C *P-T* cooling path, Spear and Markussen (1997) calculate the following stoichiometry for reaction 8:

$$4\ \text{Opx} + 4\ \text{Pl} = 5\ \text{Grt} + 1\ \text{Cpx} + 1\ \text{Qz} \quad (9).$$

The small modal amounts of clinopyroxene and quartz (relative to garnet) found in the Montana necklaces (Fig. 15) are consistent with the stoichiometry of reaction 9. The garnet necklace texture and its interpretation form an important *P-T* path constraint, especially when contrasted with the garnet pseudomorphs observed in nearby garnet-hornblende amphibolites, as described in the next section.

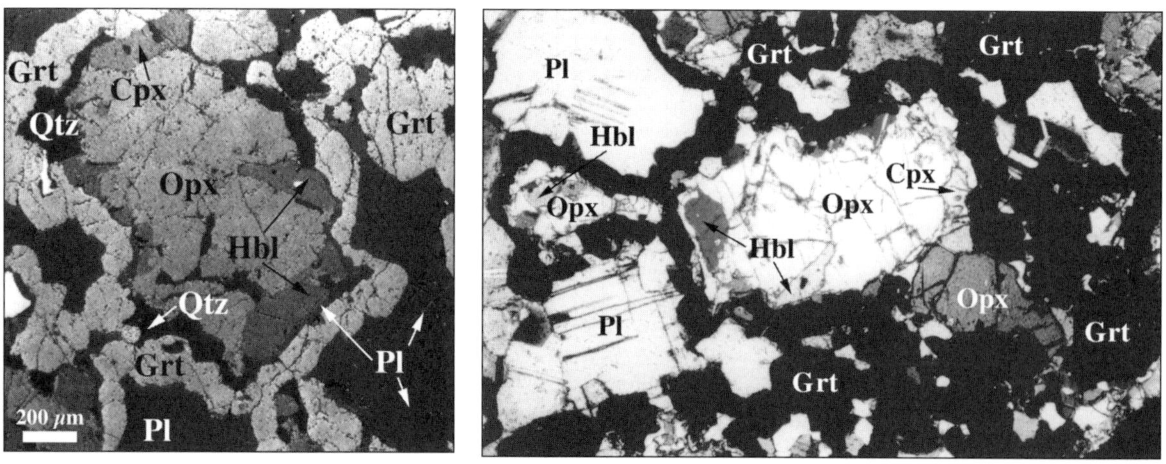

Figure 16. Details of the garnet necklace reaction texture in sample TBR-181c from the Quartz Creek area. On the left is a backscattered electron image (scale bar = 200 μm). On the right is a cross-polarized light photomicrograph (field of view is 4 mm). Bands of polycrystalline garnet with quartz and clinopyroxene separate orthopyroxene and plagioclase. Hornblende growth is believed to postdate the garnet-producing reaction.

TABLE 6. GARNET NECKLACE COMPOSITIONS

Mineral	Grt mafic side	Grt felsic side	Grt mafic side	Grt felsic side	Opx core	Opx rim	Opx core	Opx rim	Cpx core	Cpx rim
Oxide (wt%)										
SiO_2	38.78	38.90	39.27	39.33	54.01	53.51	53.59	52.99	54.66	54.12
Al_2O_3	21.30	21.43	21.83	21.90	1.18	1.10	1.37	1.10	1.60	1.46
TiO_2	N.D.†	N.D.	N.D.	N.D.	0.06	0.09	0.23	0.04	0.31	0.24
MgO	7.07	7.20	7.25	7.24	21.90	21.70	21.62	20.87	14.22	14.26
FeO*	25.17	24.38	25.77	24.30	23.00	23.41	23.94	23.57	7.75	7.58
MnO	0.79	0.56	0.64	0.66	0.20	0.24	0.19	0.19	0.11	0.02
CaO	6.62	7.32	6.77	7.39	0.54	0.45	0.48	0.50	22.36	23.13
Na_2O	N.D.	N.D.	N.D.	N.D.	0.58	0.25	0.30	0.00	0.63	0.41
Total	99.73	99.78	101.54	100.83	101.46	100.74	101.72	99.26	101.64	101.22
Elements per Formula										
O	12	12	12	12	6	6	6	6	6	6
Si	3.01	3.00	2.98	2.99	1.98	1.98	1.97	1.99	1.99	1.98
Al	1.95	1.95	1.95	1.97	0.05	0.05	0.06	0.05	0.07	0.06
Ti	N.D.	N.D.	N.D.	N.D.	0.00	0.00	0.01	0.00	0.01	0.01
Mg	0.82	0.83	0.82	0.82	1.20	1.20	1.18	1.17	0.77	0.78
Fe	1.63	1.58	1.64	1.55	0.71	0.73	0.74	0.74	0.24	0.23
Mn	0.05	0.04	0.04	0.04	0.01	0.01	0.01	0.01	0.00	0.00
Ca	0.55	0.61	0.55	0.60	0.02	0.02	0.02	0.02	0.87	0.91
Na	N.D.	N.D.	N.D.	N.D.	0.04	0.02	0.02	0.00	0.04	0.03
Total Cations	8.00	8.00	7.98	7.97	3.96	4.00	4.00	3.98	3.99	3.99
Mg/(Mg + Fe)	0.33	0.34	0.33	0.35	0.63	0.62	0.62	0.61	0.77	0.77

Mineral	Pl core	Pl rim	Pl core	Pl rim	Pl opx inclu.	Hbl core	Hbl rim	Ilm unzoned	Rut unzoned
Oxide (wt%)									
SiO_2	55.49	54.52	54.96	54.67	55.34	46.74	47.58	N.D.	0.20
Al_2O_3	27.15	27.81	28.82	29.03	24.54	10.18	7.82	N.D.	0.01
TiO_2	0.11	0.08	0.13	0.12	0.11	1.68	1.36	52.48	99.55
MgO	0.11	0.07	0.28	0.40	1.57	13.51	14.82	0.06	N.D.
*FeO	0.08	0.37	0.09	0.32	1.91	10.67	10.11	46.47	0.08
MnO	N.D.	N.D.	N.D.	N.D.	N.D.	0.05	0.07	0.56	0.13
CaO	10.22	10.54	11.14	11.33	7.58	11.72	11.86	N.D.	N.D.
Na_2O	5.94	5.63	5.84	5.17	6.85	1.47	1.01	N.D.	N.D.
K_2O	0.26	0.26	0.25	0.17	0.18	0.69	0.51	N.D.	N.D.
Cr_2O_3	N.D.	N.D.	N.D.	N.D.	N.D.	N.D.	N.D.	N.D.	0.23
Total	99.35	99.29	101.50	101.20	98.07	96.70	95.13	99.56	100.20
Elements per Formula									
O	8	8	8	8	8	23	23	3	2
Si	2.52	2.48	2.45	2.44	2.56	6.82	7.03	N.D.	0.00
Al	1.45	1.49	1.52	1.53	1.34	1.75	1.36	N.D.	0.00
Ti	0.00	0.00	0.00	0.00	0.00	0.18	0.15	1.00	0.99
Mg	0.01	0.00	0.02	0.03	0.11	2.94	3.26	0.00	0.00
Fe	0.00	0.01	0.00	0.01	0.07	1.30	1.25	0.99	0.00
Mn	N.D.	N.D.	N.D.	N.D.	N.D.	0.01	0.01	0.01	0.00
Ca	0.50	0.51	0.53	0.54	0.38	1.83	1.88	N.D.	0.00
Na	0.52	0.50	0.50	0.45	0.61	0.42	0.29	N.D.	0.00
K	0.02	0.02	0.01	0.01	0.01	0.13	0.10	N.D.	0.00
Cr	N.D.	N.D.	N.D.	N.D.	N.D.	N.D.	N.D.	N.D.	0.00
Total Cations	5.02	5.02	5.05	5.02	5.08	15.38	15.32	2.00	1.00
Mg/(Mg + Fe)	N.D.	N.D.	N.D.	N.D.	N.D.	0.69	0.72	N.D.	N.D.

Note: Data from Steiner (1999). Sample: TBR-181c, Indian Creek Metamorphic Suite (ICMS), UTM 504028N, 12T 041774E. Grt—garnet; Opx—orthopyroxene; Cpx—clinopyroxene; Pl—plagioclase; Hbl—hornblende; Ilm—ilmenite; Rut—rutile.

*FeO—total iron as FeO.

†N.D.—no data.

Garnet-Hornblende Amphibolite

Orthopyroxene + Plagioclase Pseudomorphs of Garnet

A few samples of garnet-hornblende amphibolite in the Spuhler Peak Metamorphic Suite contain dodecahedral pseudomorphs after garnet consisting principally of orthopyroxene, plagioclase, and ilmenite (see Fig.17). These pseudomorphs are surrounded by thin coronas of plagioclase + quartz ± orthopyroxene and then hornblende. Very fine needles of cummingtonite may occur in the pseudomorph core and in both the orthopyroxene and the hornblende coronas, although their very fine grain size suggests that their growth may postdate much of the pseudomorphing reaction. Small biotite grains also occur throughout. Quartz is absent from the cores. Spinel and magnetite may be present in the cores. Hornblende is never in contact with garnet. In many cases, some garnet remains in the core. Where remnant garnet is present, plagioclase + orthopyroxene + ilmenite occur along cracks in the garnet as well as surrounding it.

Representative chemical analyses of the minerals in the pseudomorph texture are listed in Table 7. Plagioclase in the core of the pseudomorph is anorthite-rich (An_{70-85}), whereas plagioclase in the corona is similar to the matrix plagioclase in composition (An_{30-45}). None of the silicate product minerals has Mg/(Mg + Fe) as low as the garnet being replaced, freeing iron to make ilmenite and magnetite.

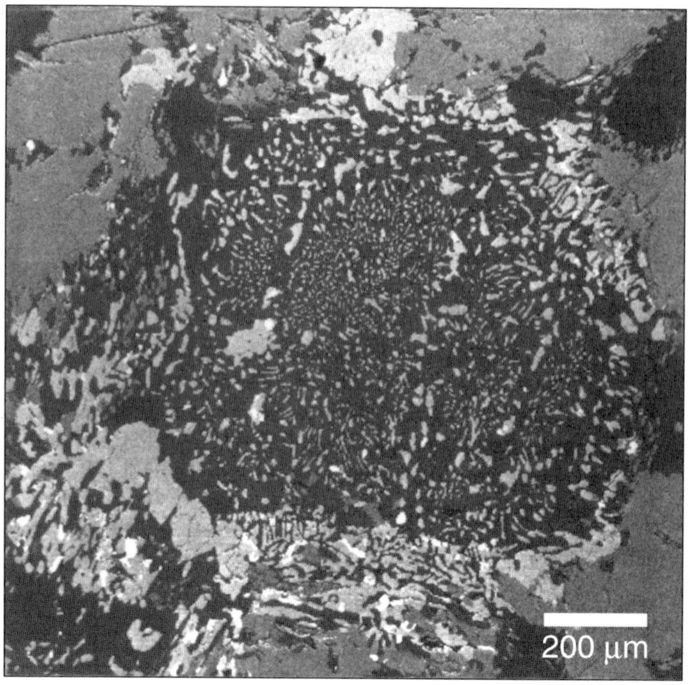

Figure 17. Backscattered electron image of an orthopyroxene + plagioclase pseudomorph of garnet with no relict garnet (sample KAD-10b). The light-colored mineral in the center is orthopyroxene; the dark mineral is plagioclase. The pseudomorph is surrounded by gray hornblende.

The consumption of garnet generally requires decreasing pressure or increasing temperature. Because garnet is always separated from hornblende in these rocks, a possible pseudomorphing reaction is

$$\text{Grt} + \text{Hbl} = \text{Opx} + \text{Pl} + \text{H}_2\text{O} \quad (10).$$

Reaction 10 is shown on Figure 18, as part of a series of reactions around a pseudoinvariant point modified from Mukhopadhyay and Bose (1994, Figure 8A therein) and from Pattison (2003, Figure 4B therein) for a rock with Mg/(Mg + Fe) = 0.65 and a_{H_2O} = 0.3. Because of the fine-grained nature of this texture, we think that it is more likely to have developed during decompression than during rising temperature. Other reactions that may be involved are

$$\text{Grt} + \text{Qz} = \text{Opx} + \text{Pl} \quad (11)$$

and

$$\text{Grt} = \text{Opx} + \text{Pl} + \text{Spl} \quad (12),$$

as suggested by Perkins and Chipera (1985) and Harley (1989). The absence of quartz from the cores implicates the quartz-consuming reaction 11. The presence of spinel suggests the simple garnet breakdown reaction 12. An experimental calibration (Bohlen et al., 1983) of the garnet + quartz reaction 11 is also shown on Figure 18 for the compositions of the minerals given in Table 7. With pressure falling from the ≥1.0 GPa of M1 to the 0.5–0.7 GPa of M2, garnet may have reacted with hornblende (Hbl) and/or quartz and then continued to break down according to reaction 12 as it was separated from the other reactants. Cummingtonite may be produced from orthopyroxene and quartz during falling temperature if water is available. We observed cummingtonite between orthopyroxene and quartz in a texture that is consistent with this reaction (Fig. 19).

The occurrence of garnet necklaces on orthopyroxene in one rock (TBR-181c) within 2 km of another rock (KAD-10b2) with orthopyroxene + plagioclase pseudomorphs of garnet places interesting constraints on the *P-T* path. Because the necklaces occur in the Indian Creek Metamorphic Suite and the pseudomorphs occur in the Spuhler Peak Metamorphic Suite, one might be tempted to conclude that they had a different metamorphic history. However, both textures appear to be late in the metamorphic cycle when other metamorphic and structural evidence supports a common history. The Indian Creek Metamorphic Suite rocks are more magnesian and less calcic, so the garnet necklaces may have developed at higher *T* and *P*. A possible *P-T* path that could have produced these textures is shown in Figure 18. At higher pressure, the reaction

$$\text{Opx} + \text{Pl} = \text{Grt} + \text{Cpx} + \text{Qz} \quad (8)$$

is crossed as temperature decreases at nearly constant pressure (e.g., −6 bar/°C). When the temperature falls to ~700 °C, the pressure is lowered so that the reaction

TABLE 7. GARNET PSEUDOMORPH ANALYSES

Unit	SPMS	SPMS	SPMS	SPMS	SPMS	SPMS	SPMS	SPMS
Mineral	Grt	Grt	Opx	Opx	Pl	Pl	Cum	Hbl
	core	rim	core	corona	core	rim	corona	matrix
Oxide (wt%)								
SiO_2	37.53	36.67	51.53	52.26	47.84	57.39	55.05	46.04
Al_2O_3	20.44	20.38	1.32	0.55	33.37	27.01	1.09	9.65
TiO_2	0.09	0.03	0.02	0.15	0.02	0.01	0.06	1.23
MgO	4.23	3.27	17.14	18.02	0.09	0.00	18.03	12.33
*FeO	23.62	24.59	27.58	27.72	0.51	0.36	22.16	16.50
MnO	4.65	9.59	1.95	1.07	0.03	0.01	0.97	0.35
CaO	8.30	3.69	0.45	0.48	16.64	9.19	0.64	10.55
Na_2O	†N.D.	N.D.	0.62	0.73	2.37	6.74	0.60	1.78
K_2O	N.D.	N.D.	0.00	0.00	0.01	0.04	0.00	0.35
Total	98.85	98.22	100.60	100.98	100.88	100.74	98.60	98.78
Elements per Formula								
O	12	12	6	6	8	12	23	23
Si	2.99	2.99	1.98	1.99	2.18	2.56	7.93	6.76
Al	1.92	1.96	0.06	0.03	1.79	1.42	0.19	1.67
Ti	0.00	0.00	0.00	0.01	0.00	0.00	0.01	0.14
Mg	0.50	0.40	0.98	1.02	0.01	0.00	3.87	2.70
Fe	1.57	1.68	0.88	0.88	0.02	0.01	2.67	2.03
Mn	0.31	0.66	0.06	0.03	0.00	0.00	0.12	0.04
Ca	0.71	0.32	0.02	0.02	0.81	0.44	0.10	1.66
Na	N.D.	N.D.	0.05	0.05	0.21	0.58	0.17	0.51
K	N.D.	N.D.	0.00	0.00	0.00	0.00	0.00	0.07
Total Cations	8.01	8.01	4.02	4.03	5.03	5.02	15.05	15.56
Mg/(Mg + Fe)	0.24	0.19	0.53	0.54	N.D.	N.D.	0.59	0.57

Note: Data from DeGraff (1996). Sample: KAD-10b2, Spuhler Peak Metamorphic Suite (SPMS), UTM 504200N, 12T 042086E). Grt—garnet; Opx—orthopyroxene; Pl—plagioclase; Cum—cummingtonite; Hbl—hornblende.

*FeO—total iron as FeO.

†N.D.—no data.

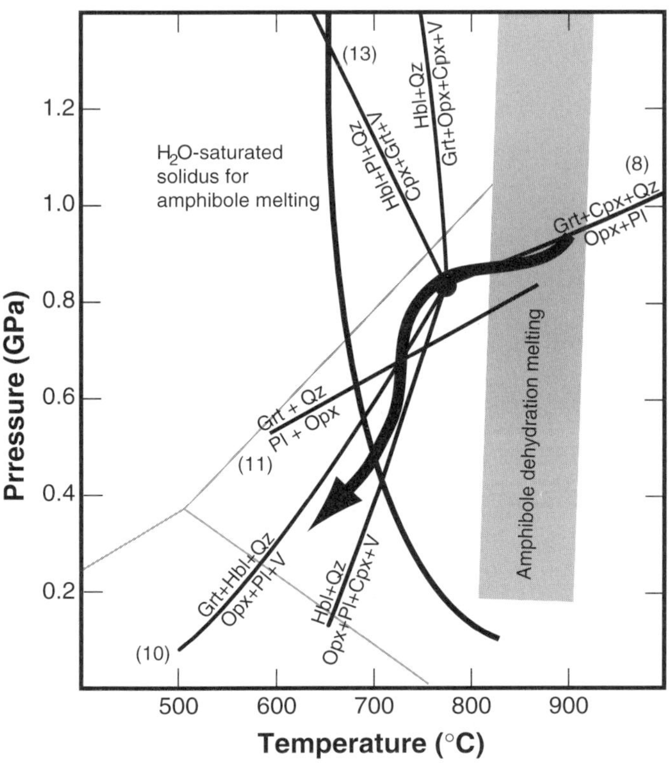

Figure 18. Pressure and temperature estimates for various reactions and a possible *P-T* path (heavy solid line) as discussed in the text. The pseudoinvariant point is based on a bulk composition Mg/(Mg + Fe) value of 0.65, a_{H_2O} of 0.3, and is modified from Figure 8 of Mukhopadhyay and Bose (1994) and Figures 2 and 4B of Pattison (2003). The exact locations of these reaction curves and their intersection are uncertain (see Pattison, 2003), although their relative positions are reasonable and place constraints on the *P-T* path of the Tobacco Root rocks. The curved line is the start of H_2O-saturated melting of amphibole determined by Wolf and Wyllie (1995). The bulk composition–dependent beginning of amphibole dehydration melting is shown as a gray band based Pattison (2003). The single reaction line (11) for the garnet-pseudomorphing reaction garnet (Grt) + quartz (Qz) = plagioclase (Pl) + orthopyroxene (Opx) observed in sample KAD-10b2 is taken from the geobarometer of Bohlen et al. (1983) using the mineral compositions from Table 7 along with a ferric iron correction.

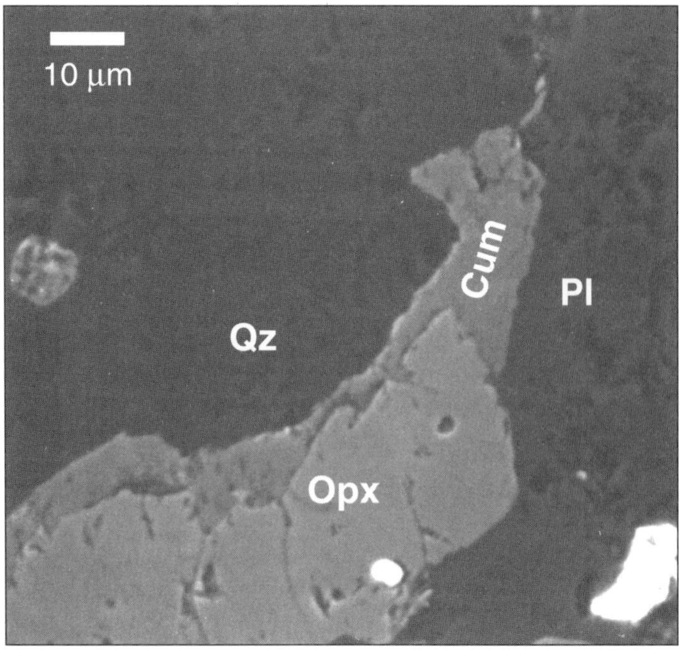

Figure 19. Backscattered electron image of cummingtonite (Cum) separating orthopyroxene (Opx) from quartz (Qz). Sample JB93-50.

$$\text{Grt} + \text{Hbl} = \text{Opx} + \text{Pl} + \text{H}_2\text{O} \quad (10)$$

is crossed at nearly constant temperature. Reactions 11 and 12 may then occur as the pressure is lowered further. If this scenario is correct, nearly isothermal decompression of 0.2 to 0.3 GPa must have occurred late in the Big Sky orogeny.

MMDS Garnet Necklace Texture

A variety of intriguing textures occur in the metamorphosed basalt dikes and sills (MMDS). The MMDS are typically quite fine-grained and granular, with most minerals (plagioclase, clinopyroxene, garnet, hornblende, quartz) 0.1–0.2 mm across. However, they can have a clustered granoblastic texture (Hanley and Vitaliano, 1983) with garnet necklaces separating plagioclase from clinopyroxene and hornblende (Fig. 20). These textures are best preserved in samples that have little foliation. The geometry of the necklaces could be produced by a chemical reaction between plagioclase and clinopyroxene, but the garnet composition (20% grossular, Table 1) does not fall between that of plagioclase and clinopyroxene. A possible explanation is that the rock evolved from a plagioclase-hornblende amphibolite via the reaction

$$\text{Hbl} + \text{Pl} + \text{Qz} = \text{Cpx} + \text{Grt} + \text{H}_2\text{O} \quad (13)$$

with the garnet growing next to the plagioclase (the aluminum source) and the clinopyroxene replacing the hornblende. Hornblende was certainly present during garnet growth because it occurs as inclusions in garnet grains. Reaction 13 is included in the amphibolite invariant point shown in Figure 18. To produce the observed texture, the reaction would be crossed as the temperature and pressure rose during the first half of the clockwise *P-T* path. The actual pressure and temperature of this reaction for the MMDS would have been a function of both water activity and the Mg/(Mg + Fe) value for the rock. The presence of hornblende in the rock today would then mean either that the reaction was incomplete or that some of the clinopyroxene has reacted back to hornblende. The fine grain size of the MMDS probably results from the low activity of water and perhaps an absence of melt during metamorphism.

Metapelite and Marble

Metapelites are not particularly common and typically occur as thin units sandwiched between marbles in the Indian Creek Metamorphic Suite. Where pelitic schists do occur, they are commonly graphitic and contain abundant sillimanite (up to 20 mod%) along with garnet, biotite, and quartz ± plagioclase ± orthoclase. The sillimanite occurs both dispersed in the matrix and as blue green bundles that appear to be pseudomorphs after kyanite. In a few samples of Spuhler Peak Metamorphic Suite (Hirst, 1996, p. 43) and Pony–Middle Mountain Metamorphic Suite (Fisher, 1994b, p. 41), K-feldspar occurs in textural equilibrium with kyanite or as inclusions in kyanite. Muscovite is conspicuously absent as a foliation-defining phase, suggesting that muscovite (Mus) dehydration melting has occurred via the reaction

$$\text{Mus} + \text{Pl} + \text{Qz} = \text{Als} + \text{Kspar} + \text{Melt} \quad (14)$$

Garnets may contain inclusions of staurolite and kyanite. We interpret both the staurolite and kyanite as M1. In some Indian

Figure 20. Photomicrograph in cross-polarized light of a metamorphosed basalt dikes and sills sample (SKC-11-2) showing a necklace of garnet grains (extinct) surrounding a single plagioclase grain and separating it from adjacent augite and hornblende grains. Note that the quartz inclusions in the garnets are radial to the plagioclase. Scale bar = 0.2 mm.

Creek Metamorphic Suite rocks only, sillimanite inclusions are found inside garnet indicating that at least some garnet growth occurred in the sillimanite stability field (Fisher, 1994b, Figure 7 therein; Monteleone, 1998a). This sillimanite and the enclosing garnet are possibly remnants from the pre-M1 metamorphism at 2450 Ma (Cheney et al., 2004, this volume, Chapter 8). The occurrence of sillimanite inclusions in garnet requires late garnet growth that normally occurs with increasing temperature in the sillimanite stability field. This raises the possibility of a very different *P-T* path for this older metamorphism, as originally suggested for M1-M2 by Tuit (1996a).

Marble, which best demonstrates the supracrustal origin of the Indian Creek Metamorphic Suite, also contains evidence of at least two episodes of metamorphism. In some samples, the high-grade assemblage calcite + diopside + forsterite can be observed, although the forsterite is commonly replaced partially by serpentine. Nearby, quartz pods and quartzite layers in the marble are typically rimmed by tremolite, which also occurs as coarse (2–4 cm), randomly oriented blades in the calcite + dolomite marble.

PRESSURE-TEMPERATURE-TIME EVOLUTION OF THE TOBACCO ROOT MOUNTAINS

The mineralogy, texture, and chemistry of the metamorphic rocks of the Tobacco Root Mountains can be used to document a fairly detailed pressure-temperature-time (*P-T*-t) history for the northwestern edge of the Wyoming province. This metamorphic history began (M0) near the start of the Proterozoic era at 2.45 Ga (see Cheney et al., 2004, this volume, Chapter 8) with a high-grade metamorphic event (Reid [1963] metamorphism *a*, Vitaliano et al. [1979] M1). This first metamorphism produced coarse (cm-scale) gneissic banding in many rocks and may have led to partial melting. Some of the minerals in the rocks today may have grown during M0, but in most cases it is not possible to prove that they did not grow or recrystallize during a subsequent metamorphism (M1 or M2). The rocks were cooled enough following M0 to permit the intrusion of basaltic dikes (MMDS) at high angles to foliation and to produce chilled margins in the dikes (Brady et al., 2004b, this volume, Chapter 7). Although these dikes, dated at 2.07 Ga (Mueller et al., 2004, this volume, Chapter 9), occur throughout the Tobacco Root Mountains and in the adjacent Highland Range, they do not intrude the Spuhler Peak Metamorphic Suite. In addition, there is no evidence of monazite or zircon growth in the Spuhler Peak Metamorphic Suite at 2.45 Ga (Cheney et al., 2004, this volume, Chapter 8; Mueller et al., 2004, this volume, Chapter 9), so we believe that all the metamorphic features of the Spuhler Peak Metamorphic Suite and the MMDS were produced during the Big Sky orogeny at 1.77 Ga.

All of the metamorphic rock suites in the Tobacco Roots have at least some samples that have both higher-pressure and lower-pressure mineral assemblages. Large (cm-scale) porphyroblasts of minerals such as kyanite, orthopyroxene, clinopyroxene, orthoamphibole, and garnet define a texture that is overprinted by later, fine-grained minerals such as sillimanite, cordierite, sapphirine, orthopyroxene, hornblende, biotite, and cummingtonite. We believe that these two textures (assigned to M1 and M2, respectively) developed during a single metamorphism at 1.77 Ga along a clockwise *P-T* path (Fig. 21). Although the actual *P-T* path for different areas in the Tobacco Root Mountains may deviate from the proposed summary path, many of the features used to define the path can be found across the range.

Large garnet crystals in the Indian Creek Metamorphic Suite contain staurolite and kyanite inclusions, although staurolite does not occur as a matrix mineral. Thus, the metamorphic record begins in the staurolite-kyanite zone. The upper stability limit of staurolite (Spear, 1993) is shown on Figure 21. Rocks of the Indian Creek Metamorphic Suite contain coarse orthopyroxene and kyanite in the same thin section, which we believe (Fig. 4) is evidence for pressures above 1.0 GPa and temperatures over 700 °C. We also observed Kspar inclusions in large kyanite crystals of the Spuhler Peak Metamorphic Suite, which are consistent with a minimum pressure of 0.8 GPa and a minimum temperature of 700 °C (Spear et al., 1999). The coarse grain size of the high-pressure assemblages (Fig. 5), migmatitic field relations (Fig. 7), and the virtual absence of muscovite from aluminous rocks are all consistent with partial melting during metamorphism. Recognizing that the amount of melting depends on many factors but particularly on the bulk composition of even specific protoliths, an amphibolite minimum melting curve (Wolf and Wyllie, 1995) is placed on Figure 21 for reference. In high-grade metamorphic terranes, melts will form once the temperature exceeds that of the water-saturated melting curve. The amount of this water-saturated melt will be small as it is controlled by the vanishingly small amount of water in such rocks (Spear et al., 1999). At temperatures greater than the water-saturated melting temperature for each rock composition, all dehydration reactions are melting reactions. The occurrence of abundant biotite in many bulk compositions, and the lack of orthopyroxene + K-feldspar or orthopyroxene-bearing migmatites limits the maximum temperature to 850 °C at ~0.7 GPa, as shown on Figure 4 and also on Figure 21. The partial to complete sillimanite pseudomorphs after kyanite that occur in garnet + orthoamphibole rocks from the Spuhler Peak Metamorphic Suite require that the *P-T* path pass to the low-temperature side of the invariant point on Figure 4. Similarly, the abundance of orthoamphibole in the Spuhler Peak and Indian Creek Metamorphic Suites limits the *P-T* path of most samples to temperatures below the upper stability of orthoamphibole (Fig. 4).

Garnet necklaces (~300 μm across, Fig.15) surround orthopyroxene crystals in Opx + Pl gneisses. Their coronal texture and size is believed to result from nearly isobaric cooling at ~0.8 GPa. Nearby, Opx + Pl pseudomorphs of garnet in hornblende amphibolites (Fig. 17; Table 7) appear to record nearly isobaric decompression. The juxtaposition of these two textures can be explained by an abrupt change in *P-T* trajectory (Fig. 21) that is consistent with other minerals and textures. For example, fine-grained pseudomorphs and symplectite reaction textures developed in a number of rocks during decompression at high temperatures. Orthopyroxene + cordierite symplectites rimming

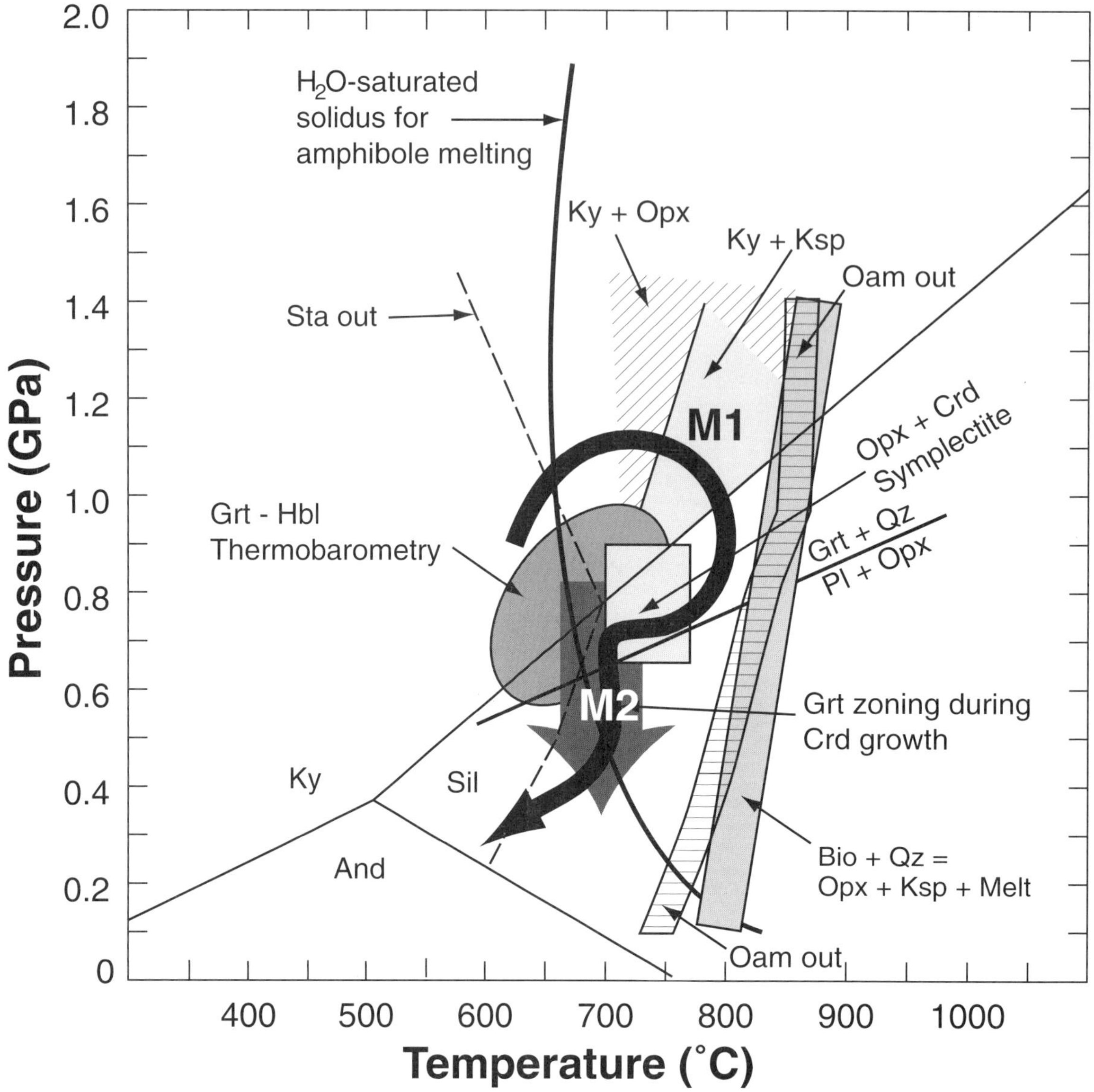

Figure 21. Summary diagram showing constraints on the pressure-temperature evolution of the metamorphic rocks of the Tobacco Root Mountains as outlined in this paper. Specific fields outlined on the diagram are (1) the temperatures and pressures (dashed line) above which staurolite is no longer stable (Spear, 1993), (2) a dehydration melting curve for amphibolite (Fig. 18), (3) the lower part (triangular region) of the stability field of kyanite + orthopyroxene (Fig. 4), (4) the temperatures and pressures (line with hatch marks) above which orthoamphibole is no longer stable (Fig. 4), (5) the lower part (triangular region) of the stability field of kyanite + K-feldspar resulting from dehydration melting (Spear et al., 1999), (6) the range of temperatures and pressures (dark oval) calculated from the garnet-hornblende-plagioclase-quartz thermobarometer (Fig. 2), (7) garnet-orthopyroxene equilibria field (rectangular box) for the orthopyroxene + cordierite symplectite of Table 3 (Hensen and Harley, 1990), (8) the calculated location (solid line) of the garnet pseudomorphing reaction garnet + quartz = plagioclase + orthopyroxene (Fig. 18), (9) the dehydration melting of biotite + quartz provides a maximum temperature as there are no orthopyroxene + K-feldspar rocks in the Tobacco Root Mountains, and (10) a calculated decompression path (wide arrow) resulting from cordierite growth based on observed Mn zoning of garnet (Fig. 12). A possible *P-T* path based on these constraints for the 1.77 Ga Big Sky orogeny is shown by the heavy black line. To produce both the garnet necklaces on orthopyroxene and the orthopyroxene + plagioclase pseudomorphs of garnet in nearby rocks, a nearly isobaric cooling at high pressure must be followed by a nearly isothermal decompression. We believe these and other features are related to tectonic unroofing during extension late in the metamorphic cycle. Sta—staurolite; And—andalusite.

garnet in Indian Creek Metamorphic Suite samples (Fig. 8; Table 3) developed at 700–770 °C and 0.75–0.9 GPa, according to the calculations of Hensen and Harley (1990). Chemical zoning of Mn in garnet grains (Figs. 10 and 11; Table 2) being pseudomorphed by cordierite is consistent with pressure falling from 0.8 to 0.5 GPa at high temperature (Fig. 12). A large proportion of the mafic samples studied are decorated with extremely fine-grained, needlelike sprays of cummingtonite that may have grown upon fairly rapid decompression, followed by fairly rapid cooling.

The high pressures and temperatures recorded in Tobacco Root metamorphic rocks require deep burial in a continental-scale collision. Clearly, the Big Sky orogeny matches other Himalayan-scale events. And like some other large orogenies, we see abundant evidence for significant decompression at high temperatures (Spear, 1993, Chapter 21 therein). Reaction constraints involving garnet and orthopyroxene (Fig. 18) make tectonic denudation the likely cause for this decompression, both for the observed metamorphic textures and for the subsequent rapid cooling leading to similar radiometric ages for zircon, monazite, hornblende and biotite (Brady et al., 2004a, this volume, Chapter 5; Cheney et al., 2004, this volume, Chapter 8; Mueller et al., 2004, this volume, Chapter 9).

ACKNOWLEDGMENTS

The authors thank the W.M. Keck Foundation and the colleges that form the Keck Geology Consortium for the support of this work over many years. We also thank the many faculty members at those colleges who advised the undergraduate theses that form the basis for our results. We appreciate very much the thoughtful reviews of the manuscript by John C. Schumacher and Frank S. Spear.

REFERENCES CITED

Archuleta, L.L., 1994a, The mineralogy and petrology of garnet-amphibole rocks from the Spuhler Peak Formation along Thompson Peak ridge, Tobacco Root Mountains, SW Montana, *in* Wilson, M.A., ed., Seventh Keck research symposium in geology: San Antonio, Texas, p. 60–63.

Archuleta, L.L., 1994b, Mineralogy, petrography, and geochemistry including geothermobarometry of an aluminous enclave rock and host rocks within the Spuhler Peak Formation, Tobacco Root Mountains, southwestern Montana [B.A. thesis]: Claremont, California, Pomona College, 75 p.

Bohlen, S.R., Wall, V.J., and Boettcher, A.L., 1983, Experimental investigations and geological applications of equilibria in the system FeO-TiO_2-Al_2O_3-SiO_2-H_2O: American Mineralogist, v. 68, p. 1049–1058.

Brady, J.B., Carmichael, S., Burger, H.R., Harris, C., Cheney, J.T., and Harms, T.A., 1998a, Thermobarometry and geochemistry of metamorphosed mafic dikes and sills (MMDS) of the Tobacco Root Mountains, Montana: Geological Society of America Abstracts with Programs, v. 30, no. 7, p. 97.

Brady, J.B., Cheney, J.T., Rhodes, A.L., Vasquez, A., Green, C., Duvall, M., Kogut, A., Kaufman, L., and Kovaric, D., 1998b, Isotope geochemistry of Proterozoic talc occurrences in Archean marbles of the Ruby Mountains, southwest Montana, USA: Geological Materials Research, v. 2, p. 41.

Brady, J.B., Kovaric, D.N., Cheney, J.T., Jacob, L.J., and King, J.T., 2004a, $^{40}Ar/^{39}Ar$ ages of metamorphic rocks from the Tobacco Root Mountains region, Montana, *in* Brady, J.B., et al., eds., Precambrian geology of the Tobacco Root Mountains, Montana: Boulder, Colorado, Geological Society of America Special Paper 377, p. 131–149 (this volume).

Brady, J.B., Mohlman, H.K., Harris, C., Carmichael, S.K., Jacob, L.K., and Chaparro, W.R., 2004b, General geology and geochemistry of metamorphosed Proterozoic mafic dikes and sills, Tobacco Root Mountains, Montana, *in* Brady, J.B., et al., eds., Precambrian geology of the Tobacco Root Mountains, Montana: Boulder, Colorado, Geological Society of America Special Paper 377, p. 89–104 (this volume).

Burger, H.R., 1966, Structure, petrology, and economic geology of the Sheridan District, Madison County, Montana [Ph.D. thesis]: Bloomington, Indiana University, 156 p.

Burger, H.R., 1969, Structural evolution of the southwestern Tobacco Root Mountains, Montana: Geological Society of America Bulletin, v. 80, p. 1329–1341.

Burger, H.R., 2004, General geology and tectonic setting of the Tobacco Root Mountains, Montana, *in* Brady, J.B., et al., eds., Precambrian geology of the Tobacco Root Mountains, Montana: Boulder, Colorado, Geological Society of America Special Paper 377, p. 1–14 (this volume).

Burger, H.R., Peck, W.H., Johnson, K.E., Tierney, K.A., Poulsen, C.J., Cady, P., Lowell, J., MacFarlane, W.A., Sincock, M.J., Archuleta, L.L., Pufall, A., and Cox, M.J., 2004, Geology and geochemistry of the Spuhler Peak Metamorphic Suite, *in* Brady, J.B., et al., eds., Precambrian geology of the Tobacco Root Mountains, Montana: Boulder, Colorado, Geological Society of America Special Paper 377, p. 47–70 (this volume).

Cady, P., 1994a, Petrographic analysis of amphibolites from the Spuhler Peak Formation, Indian Creek Metamorphic Suite, and Pony–Middle Mountain Metamorphic Complex, Tobacco Root Mountains, Montana, *in* Wilson, M.A., ed., Seventh Keck research symposium in geology: San Antonio, Texas, p. 64–66.

Cady, P., 1994b, Petrographic and geochemical analysis of amphibolites from the Tobacco Root Mountains, Montana [B.A. thesis]: Colorado Springs, Colorado College, 86 p.

Carmichael, S.K., 1998a, Geothermobarometry of metamorphosed mafic dikes and sills, Tobacco Root Mountains, southwestern Montana, *in* Mendelson, C.V., and Mankiewicz, C., eds., Eleventh Keck research symposium in Geology: Amherst, Massachusetts, Amherst College, p. 168–171.

Carmichael, S.K., 1998b, Geothermobarometry of metamorphosed mafic dikes and sills, Tobacco Root Mountains, southwestern Montana [B.A. thesis]: Northampton, Massachusetts, Smith College, 51 p.

Carmichael, S.K., Brady, J.B., Cheney, J.T., and Mohlman, H.K., 1998, Metamorphic history of mafic dikes and sills in the Tobacco Root Mountains, southwestern Montana: Geological Society of America Abstracts with Programs, v. 30, no. 1, p. 10.

Carrington, D.P., and Harley, S.L., 1995, Partial melting and phase relations in high-grade metapelites; an experimental petrogenetic grid in the KFMASH system: Contributions to Mineralogy and Petrology, v. 120, no. 3–4, p. 270–291.

Cheney, J.T., Brady, J.B., Burger, H.R., Fisher, R., Jacob, L., King, J.T., Tierney, K.A., Peck, W.H., Poulsen, C.J., Cady, P., Lowell, J., Sincock, M.J., and Archuleta, L.L., 1994, Metamorphic evolution of Archean rocks in the Tobacco Root Mountains of SW Montana: Geological Society of America Abstracts with Programs, v. 26, no. 7, p. 226.

Cheney, J.T., Brady, J.B., Burger, H.R., Harms, T.A., Johnson, K.E., DeGraff, K.A., King, J.F., Mohlman, H.K., Tuit, C.B., Hirst, J., and Wegmann, K.W., 1996, Tectonometamorphic evolution of Archean amphibolites from the Tobacco Root Mountains of Montana: Geological Society of America Abstracts with Programs, v. 28, no. 7, p. 45.

Cheney, J.T., Hatch, C.E., Harms, T.A., Brady, J.B., Burger, H.R., and Monteleone, B.D., 1998, Sapphirine symplectite bearing kyanite-orthopyroxene orthoamphibolite from the Archean Indian Creek Metamorphic Suite (ICMS), Tobacco Root Mountains, Montana: Geological Society of America Abstracts with Programs, v. 30, no. 7, p. 97.

Cheney, J.T., Webb, A.A.G., Coath, C.D., and McKeegan, K.D., 2004, In situ ion microprobe $^{207}Pb/^{206}Pb$ dating of monazite from Precambrian metamorphic suites, Tobacco Root Mountains, Montana, *in* Brady, J.B., et al., eds., Precambrian geology of the Tobacco Root Mountains, Montana: Boulder, Colorado, Geological Society of America Special Paper 377, p. 151–179 (this volume).

Cordua, W.S., 1973, Precambrian geology of the southern Tobacco Root Mountains, Madison County, Montana [Ph.D. thesis]: Bloomington, Indiana University, 247 p.

DeGraff, K., 1996a, Petrology of amphibolites from the Archean Spuhler Peak Metamorphic Suite of the Tobacco Root Mountains, SW Montana [B.A. thesis]: Amherst, Massachusetts, Amherst College, 74 p.

DeGraff, K., 1996b, Petrology of amphibolites from the Archean Spuhler Peak metamorphic suite of the Tobacco Root Mountains, southwestern

Montana, *in* Mendelson, C.V., and Mankiewicz, C., eds., Ninth Keck research symposium in geology: Williamstown, Massachusetts, Williams College, p. 94–97.

DeGraff, K.A., Tierney, K.A., Cheney, J.T., and Brady, J.B., 1996, Petrology of amphibolites from the Archean Spuhler Peak metamorphic suite (SPMS) of the Tobacco Root Mountains, Montana: Geological Society of America Abstracts with Programs, v. 28, no. 3, p. 48.

Elvevold, S., and Gilotti, J.A., 2000, Pressure-temperature evolution of retrogressed kyanite eclogites, Weinschenk Island, Northeast Greenland Caledonides: Lithos, v. 53, no. 2, p. 127–147.

Fisher, R., 1994a, Deducing metamorphic histories first step: Garnet zoning in metapelites, *in* Wilson, M.A., ed., Seventh Keck research symposium in geology: San Antonio, Texas, p. 67–70.

Fisher, R.G.M., 1994b, Comparative metamorphic histories of three high-grade gneiss terrains in the Tobacco Root Mountains of southwest Montana [B.A. thesis]: Northampton, Massachusetts, Smith College, 69 p.

Friberg, N., 1976, Petrology of a metamorphic sequence of upper-amphibolite facies in the central Tobacco Root Mountains, southwestern Montana [Ph.D. thesis]: Bloomington, Indiana University, 146 p.

Friberg, L.M., 1994, Geothermobarometry of Precambrian metamorphic rocks from the Tobacco Root Mountains of southwestern Montana: Geological Society of America Abstracts with Programs, v. 26, no. 5, p. 16.

Frisch, J.D., 1998a, Geothermobarometry of garnet amphibolites from the Indian Creek and Pony–Middle Mountain metamorphic suites, Tobacco Root Mountains, Montana, *in* Mendelson, C.V., and Mankiewicz, C., eds., Eleventh Keck research symposium in Geology: Amherst, Massachusetts, Amherst College, p. 172–175.

Frisch, J.D., 1998b, Geothermobarometry of Garnet-Amphibolites form the Indian Creek and Pony Middle Mountain Metamorph8ic Suites, Tobacco Root Mountains, Montana [B.A. thesis]: Amherst, Massachusetts, Amherst College, 85 p.

Frisch, J.D., and Cheney, J.T., 1998, Geothermobarometry of garnet-amphibolites from the Indian Creek and Pony–Middle Mountain metamorphic suites, Tobacco Root Mountains, Montana: Geological Society of America Abstracts with Programs, v. 30, no. 1, p. 20.

Gillmeister, N.M., 1972a, Cherry Creek Group–Pony Group relationship in the central Tobacco Root Mountains, Madison County, Montana: Northwest Geology, v. 1, p. 21–24.

Gillmeister, N.M., 1972b, Petrology of Precambrian rocks in the central Tobacco Root Mountain, Madison County, Montana: Cambridge, Massachusetts, Harvard University, Cambridge, 201 p.

Godard, G., and Mabit, J.-L., 1998, Peraluminous sapphirine formed during retrogression of a kyanite-bearing eclogite from Pays de Leon, Armorican Massif, France: Lithos, v. 43, no. 1, p. 15–29.

Graham, C.M., and Powell, R., 1984, A garnet-hornblende geothermometer; calibration, testing, and application to the Pelona Schist, southern California: Journal of Metamorphic Geology, v. 2, no. 1, p. 13–31.

Hanley, T.B., 1975, Structure and petrology of the northwestern Tobacco Root Mountains, Madison County, Montana [Ph.D. thesis]: Bloomington, Indiana University, 289 p.

Hanley, T.B., 1976, Stratigraphy and structure of the Central Fault Block, northwestern Tobacco Root Mountains, Madison County, Montana, *in* Anonymous, ed., Guidebook, The Tobacco Root Geological Society 1976 Field Conference: Butte, Montana, Montana Bureau of Mines and Geology Special Publication 73, p. 7–14.

Hanley, T.B., and Vitaliano, C.J., 1983, Petrography of Archean mafic dikes of the Tobacco Root Mountains, Madison County, Montana: Northwest Geology, v. 12, p. 43–55.

Harley, S.L., 1985, Paragenetic and mineral-chemical relationships in orthoamphibole-bearing gneisses from Enderby Land, East Antarctica; a record of Proterozoic uplift: Journal of Metamorphic Geology, v. 3, no. 2, p. 179–200.

Harley, S.L., 1989, The origins of granulites: A metamorphic perspective: Geological Magazine, v. 126, p. 215–247.

Harley, S.L., and Hensen, B.J., 1990, Archaean and Proterozoic high-grade terranes of East Antarctica (40–80 degrees E); a case study of diversity in granulite facies metamorphism, *in* Ashworth, J.R., and Brown, M., eds., High-temperature metamorphism and crustal anatexis: The Mineralogical Society Series 2, p. 320–370.

Hatch, C.E., 1998a, Sapphirine symplectites in orthoamphibolites of the Archean Indian Creek metamorphic suite (ICMS), Tobacco Root Mountains, Montana, *in* Mendelson, C.V., and Mankiewicz, C., eds., Eleventh Keck research symposium in Geology: Amherst, Massachusetts, Amherst College, p. 180–183.

Hatch, C.E., 1998b, Sapphirine symplectities in orthoamphibole gneiss of the Archean Indian Creek Metamorphic Suite, Tobacco Root Mountains, Montana [B.A. thesis]: Amherst, Massachusetts, Amherst College, 110 p.

Hatch, C.E., and Cheney, J.T., 1998, Geothermobarometry in the Archean Indian Creek Metamorphic Suite (ICMS) rocks in the Tobacco Root Mountains, Montana: Geological Society of America Abstracts with Programs, v. 30, no. 1, p. 24.

Hensen, B.J., and Harley, S.L., 1990, Graphical analysis of P-T-X relations in granulite facies metapelites, *in* Ashworth, J.R., and Brown, M., eds., High-temperature metamorphism and crustal anatexis: The Mineralogical Society Series 2, p. 19–56.

Hess, D.F., 1967, Geology of pre-Beltian rocks in the central and southern Tobacco Root Mountains (Montana), with reference to superposed effects of the Laramide age Tobacco Root Batholith [Ph.D. thesis]: Bloomington, Indiana University, 333 p.

Hess, D.F., and Vitaliano, C.J., 1990, Metabasites; an indicator of Late Archean geologic history in the Tobacco Root Mountains, Madison County, Montana: Geological Society of America Abstracts with Programs, v. 22, no. 5, p. 13.

Hirst, J., 1996, Mineralogy, petrology, and geothermobarometry of aluminous Archean rocks in the Spuhler Peak metamorphic suite of the Tobacco Root Mountains, southwestern Montana [B.A. thesis]: Claremont, California, Pomona College, 64 p.

Hodges, K.V., and Spear, F.S., 1982, Geothermometry, geobarometry and the Al_2SiO_5 triple point at Mt. Moosilauke, New Hampshire: American Mineralogist, v. 67, no. 11–12, p. 1118–1134.

Holdaway, M.J., 1971, Stability of andalusite and the aluminum silicate phase diagram: American Journal of Science, v. 271, p. 97–131.

Holland, T.J.B., and Powell, R., 1998, An internally consistent thermodynamic data set for phases of petrological interest: Journal of Metamorphic Geology, v. 16, no. 3, p. 309–343.

Immega, I.P., and Klein, C., Jr., 1976, Mineralogy and petrology of some metamorphic Precambrian iron-formations in southwestern Montana: American Mineralogist, v. 61, no. 11–12, p. 1117–1144.

Johnson, K.E., Brady, J.B., MacFarlane, W.A., Thomas, R.B., Poulsen, C., and Sincock, M.J., 2004, Precambrian meta-ultramafic rocks from the Tobacco Root Mountains, Montana, *in* Brady, J.B., et al., eds., Precambrian geology of the Tobacco Root Mountains, Montana: Boulder, Colorado, Geological Society of America Special Paper 377, p. 71–87 (this volume).

Kihle, J., and Bucher-Nurminen, K., 1992, Orthopyroxene-sillimanite-sapphirine granulites from the Bamble granulite terrane, southern Norway: Journal of Metamorphic Geology, v. 10, no. 5, p. 671–683.

Kohn, M.J., and Spear, F.S., 1990, Two new geobarometers for garnet amphibolites, with applications to southeastern Vermont: American Mineralogist, v. 75, no. 1–2, p. 89–96.

Kriegsman, L.M., and Schumacher, J.C., 1999, Petrology of sapphirine-bearing and associated granulites from central Sri Lanka: Journal of Petrology, v. 40, no. 8, p. 1211–1239.

Lal, R.K., Ackermand, D., Raith, M., Raase, P., and Seifert, F., 1984, Sapphirine-bearing assemblages from Kiranur, southern India; a study of chemographic relationships in the Na_2O-FeO-MgO-Al_2O_3-SiO_2-H_2O system: Neues Jahrbuch für Mineralogie, Abhandlungen, v. 150, no. 2, p. 121–152.

Lowell, J., 1994a, A metamorphic and deformational history of the Spuhler Peak Formation, Tobacco Root Mountains, Montana [B.A. thesis]: Colorado Springs, Colorado College, 100 p.

Lowell, J., 1994b, Petrological and structural constraints on the history of the Spuhler Peak Formation near Noble Lake, Tobacco Root Mountains, Montana, *in* Wilson, M.A., ed., Seventh Keck research symposium in geology: San Antonio, Texas, p. 78–81.

Martin, J., 1996, Petrology of mafic and felsic gneisses of the Indian Creek metamorphic suite, Tobacco Root Mountains, southwestern Montana, *in* Mendelson, C.V., and Mankiewicz, C., eds., Ninth Keck research symposium in geology: Williamstown, Massachusetts, Williams College, p. 110–113.

Moecher, D.P., Essene, E.J., and Anovitz, L.M., 1988, Calculation and application of clinopyroxene-garnet-plagioclase-quartz geobarometers: Contributions to Mineralogy and Petrology, v. 100, no. 1, p. 92–106.

Mogk, D.W., Burger, H.R., Mueller, P.A., D'Arcy, K., Heatherington, A.L., Wooden, J.L., Abeyta, R.L., Martin, J., and Jacob, L.J., 2004, Geochemistry of quartzofeldspathic gneisses and metamorphic mafic rocks of the Indian Creek and Pony–Middle Mountain Metamorphic Suites, Tobacco

Root Mountains, Montana, *in* Brady, J.B., et al., eds., Precambrian geology of the Tobacco Root Mountains, Montana: Boulder, Colorado, Geological Society of America Special Paper 377, p. 15–46 (this volume).

Mohan, A., Prakash, D., and Motoyoshi, Y., 1996, Decompressional P-T history in sapphirine-bearing granulites from Kodaikanal, southern India: Journal of Southeast Asian Earth Sciences, v. 14, no. 3–4, p. 231–243.

Mohlman, H.K., 1996a, Evolution of Archean metamorphosed mafic dikes in the Tobacco Root Mountains, SW Montana [B.A. thesis]: Amherst, Massachusetts, Amherst College, 82 p.

Mohlman, H.K., 1996b, Tectonometamorphic evolution of metamorphosed mafic dikes in the Archean rocks of the Tobacco Root Mountains, Montana, *in* Mendelson, C.V., and Mankiewicz, C., eds., Ninth Keck research symposium in geology: Williamstown, Massachusetts, Williams College, p. 114–117.

Mohlman, H.K., and Cheney, J.T., 1996, Tectonometamorphic evolution of metamorphosed mafic dikes (MMD) in the Archean rocks of the Tobacco Root Mountains, Montana: Geological Society of America Abstracts with Programs, v. 28, no. 3, p. 83.

Monteleone, B., 1998a, A petrologic and mathematical analysis of metamorphism in the Indian Creek metamorphic suite and the Spuhler Peak metamorphic suite, Tobacco Root Mountains, southwestern Montana [B.A. thesis]: Wooster, Ohio, College of Wooster, 56 p.

Monteleone, B., 1998b, The petrology and geothermobarometry of the Indian Creek metamorphic suite and the Spuhler Peak metamorphic suite, Tobacco Root Mountains, southwestern Montana, *in* Mendelson, C.V., and Mankiewicz, C., eds., Eleventh Keck research symposium in Geology: Amherst, Massachusetts, Amherst College, p. 184–187.

Mueller, P.A., Burger, H.R., Wooden, J.L., Heatherington, A.L., Mogk, D.W., and D'Arcy, K., 2004, Age and evolution of the Precambrian crust of the Tobacco Root Mountains, Montana, *in* Brady, J.B., et al., eds., Precambrian geology of the Tobacco Root Mountains, Montana: Boulder, Colorado, Geological Society of America Special Paper 377, p. 181–202 (this volume).

Mukhopadhyay, B., and Bose, M.K., 1994, Transitional granulite-eclogite facies metamorphism of basic supracrustal rocks in a shear zone complex in the Precambrian shield of South India: Mineralogical Magazine, v. 58, no. 1(390), p. 97–118.

Ouzegane, K., and Boumaza, S., 1996, An example of ultrahigh-temperature metamorphism; orthopyroxene-sillimanite-garnet, sapphirine-quartz and spinel-quartz parageneses in Al-Mg granulites from In Hihaou: Journal of Metamorphic Geology, v. 14, no. 6, p. 693–708.

Patino Douce, A.E., Johnston, A.D., and Rice, J.M., 1993, Octahedral excess mixing properties in biotite; a working model with applications to geobarometry and geothermometry: American Mineralogist, v. 78, no. 1–2, p. 113–131.

Pattison, D.R.M., 2003, Petrogenetic significance of orthopyroxene-free garnet + clinopyroxene + plagioclase ± quartz-bearing metabasites with respect to the amphibolite and granulite facies: Journal of Metamorphic Geology, v. 21, no. 1, p. 21–34.

Perkins, D., and Chipera, S.J., 1985, Garnet-orthopyroxene-plagioclase-quartz barometry: Refinement and application to the English River subprovince and the Minnesota River Valley: Contributions to Mineralogy and Petrology, v. 89, p. 69–80.

Powell, R., 1985, Regression diagnostics and robust regression in geothermometer/geobarometer calibration; the garnet-clinopyroxene geothermometer revisited: Journal of Metamorphic Geology, v. 3, no. 3, p. 231–243.

Raith, M., Karmakar, S., and Brown, M., 1997, Ultra-high-temperature metamorphism and multistage decompressional evolution of sapphirine granulites from the Palni Hills Ranges, southern India: Journal of Metamorphic Geology, v. 15, no. 3, p. 379–399.

Reid, R.R., 1957, Bedrock geology of the north end of the Tobacco Root Mountains, Madison County, Montana: Montana Bureau of Mines and Geology Memoir 36, p. 27.

Reid, R.R., 1963, Metamorphic rocks of the northern Tobacco Root Mountains, Madison County, Montana: Geological Society of America Bulletin, v. 74, p. 293–305.

Rodriguez, M.L., 2002, Analysis of the relationship between strain and metamorphic equilibrium as preserved in Archean amphibolites from the Tobacco Root Mountains, Montana [B.A. thesis]: Amherst, Massachusetts, Amherst College, 72 p.

Root, F.K., 1965, Structure, petrology, and mineralogy of pre-Beltian metamorphic rocks of the Pony-Sappington area, Madison County, Montana [Ph.D. thesis]: Bloomington, Indiana University, 184 p.

Spear, F.S., 1993, Metamorphic phase equilibria and pressure-temperature-time paths, Mineralogical Society of America Monographs: Washington, D.C., Mineralogical Society of America, 799 p.

Spear, F.S., and Markussen, J.C., 1997, Mineral zoning, P-T-X-M phase relations, and metamorphic evolution of some Adirondack granulites, New York: Journal of Petrology, v. 38, no. 6, p. 757–783.

Spear, F.S., and Menard, T., 1989, Program GIBBS: A generalized Gibbs method algorithm: American Mineralogist, v. 74, p. 942–943.

Spear, F.S., and Parrish, R.R., 1996, Petrology and cooling rates of the Valhalla Complex, British Columbia, Canada: Journal of Petrology, v. 37, no. 4, p. 733–765.

Spear, F.S., Kohn, M.J., and Cheney, J.T., 1999, P-T paths from anatectic pelites: Contributions to Mineralogy and Petrology, v. 134, no. 1, p. 17–32.

Steffen, K.J., 1998a, Garnet-amphibolite geothermobarometry of the Archean metamorphic suites in the Tobacco Root Mountains, Montana; implications for crustal evolution of the northern Wyoming province: Geological Society of America Abstracts with Programs, v. 30, no. 6, p. 37.

Steffen, K.J., 1998b, Geothermbarometry of garnet-amphibolites of the Archean Spuhler Peak and Pony–Middle Mountain Metamorphic Suites of the Tobacco Root Mountains, southwest Montana [B.A. thesis]: Northfield, Minnesota, Carleton College, 101 p.

Steffen, K.J., 1998c, Geothermobarometry of Archean garnet-amphibolites; Pony–Middle Mountain and Spuhler Peak metamorphic suites, Tobacco Root Mountains, southwest Montana, *in* Mendelson, C.V., and Mankiewicz, C., eds., Eleventh Keck research symposium in Geology: Amherst, Massachusetts, Amherst College, p. 192–195.

Steiner, M.L., 1999, Metamorphic evolution of Indian Creek Metamorphic Suite garnet+orthopyroxene assemblages, Tobacco Root Mountains, Montana [B.A. thesis]: Amherst, Massachusetts, Amherst College, 93 p.

Tansley, W., Schafer, P.A., and Hart, L.H., 1933, A geological reconnaissance of the Tobacco Root Mountains, Madison County, Montana: Montana Bureau of Mines and Geology Memoir 9, p. 57.

Tierney, K.A., 1994a, The origin and evolution of the Spuhler Peak Formation along the western ridge of Thompson Peak, Tobacco Root Mountains, Montana [B.A. thesis]: Amherst, Massachusetts, Amherst College, 87 p.

Tierney, K.A., 1994b, The origin and evolution of the Spuhler Peak Formation along the western ridge of Thompson Peak, Tobacco Root Mountains, Montana, *in* Wilson, M.A., ed., Seventh Keck research symposium in geology: San Antonio, Texas, p. 94–97.

Tuit, C.B., 1996a, Geothermobarometry of the Indian Creek Metamorphic Suite, Tobacco Root Mountains, Montana: Geological Society of America Abstracts with Programs, v. 28, no. 7, p. 357–358.

Tuit, C.B., 1996b, Geothermobarometry of the Indian Creek metamorphic suite, Tobacco Root Mountains, Montana, *in* Mendelson, C.V., and Mankiewicz, C., eds., Ninth Keck research symposium in geology: Williamstown, Massachusetts, Williams College, p. 86–89.

Tuit, C.B., 1996c, Geothermometry of the Indian Creek Metamorphic Suite, Tobacco Root Mountains, Montana [B.S. thesis]: Beloit, Wisconsin, Beloit College, 43 p.

Van Reenen, D.D., 1986, Hydration of cordierite and hypersthene and a description of the retrograde orthoamphibole isograd in the Limpopo belt, South Africa: American Mineralogist, v. 71, p. 900–915.

Vitaliano, C.J., Burger, H.R., Cordua, W.S., Hanley, T.B., Hess, D.P., and Root, F.K., 1979, Explanatory text to accompany geologic map of southern Tobacco Root Mountains, Madison County, Montana: Geological Society of America Map and Chart Series MC31, scale 1:62,500, 1 sheet, 8 p. text.

Wegmann, K.W., 1996a, Metamorphic evolution of the Archean Pony Middle Mountain Metamorphic Suite, Tobacco Root Mountains, southwestern Montana [B.A. thesis]: Walla Walla, Washington, Whitman College, 74 p.

Wegmann, K.W., 1996b, Metamorphic evolution of the Archean Pony Middle Mountain metamorphic suite, Tobacco Root Mountains, southwestern Montana, *in* Mendelson, C.V., and Mankiewicz, C., eds., Ninth Keck research symposium in geology: Williamstown, Massachusetts, Williams College, p. 122–125.

Windley, B.F., Ackermand, D., and Herd, R.K., 1984, Sapphirine/kornerupine-bearing rocks and crustal uplift history of the Limpopo Belt, Southern Africa: Contributions to Mineralogy and Petrology, v. 86, p. 342–358.

Wolf, M.B., and Wyllie, P.J., 1995, Liquid segregation parameters from amphibolite dehydration melting experiments: Journal of Geophysical Research, B, Solid Earth and Planets, v. 100, no. 8, p. 15,611–15,621.

Manuscript Accepted by the Society September 12, 2003

Printed in the USA

Geological Society of America
Special Paper 377
2004

$^{40}Ar/^{39}Ar$ ages of metamorphic rocks from the Tobacco Root Mountains region, Montana

John B. Brady
Dana N. Kovaric
Department of Geology, Smith College, Northampton, Massachusetts 01063, USA

John T. Cheney
Department of Geology, Amherst College, Amherst, Massachusetts 01002, USA

Lisa J. Jacob
Department of Geology, Smith College, Northampton, Massachusetts 01063, USA

J. Toby King
Department of Geology, Amherst College, Amherst, Massachusetts 01002, USA

ABSTRACT

Measurements of 60 single-grain, UV laser microprobe $^{40}Ar/^{39}Ar$ total gas ages for hornblende from metamorphic rocks of the Tobacco Root Mountains in southwest Montana yield a mean age of 1.71 ± 0.02 Ga. Measurements of $^{40}Ar/^{39}Ar$ step-heating plateau ages of three bulk hornblende samples from the Tobacco Root Mountains metamorphic rocks average 1.70 ± 0.02 Ga. We believe that these and the K/Ar or $^{40}Ar/^{39}Ar$ ages reported by previous workers are cooling ages from a 1.78 to 1.72 Ga, upper-amphibolite to granulite facies, regional metamorphism (Big Sky orogeny) that affected the northwestern portion of the Wyoming province, including the Tobacco Root Mountains and adjacent ranges. Based on the $^{40}Ar/^{39}Ar$ data, this 1.78–1.72 Ga metamorphism must have achieved temperatures greater than ~500 °C to reset the hornblende $^{40}Ar/^{39}Ar$ ages of samples from the Indian Creek Metamorphic Suite, which was previously metamorphosed at 2.45 Ga, and of the crosscutting metamorphosed mafic dikes and sills (MMDS), which were intruded at 2.06 Ga. Biotite and hornblende from the Tobacco Root Mountains appear to give the same $^{40}Ar/^{39}Ar$ or K/Ar age (within uncertainty), indicating that the rocks cooled rapidly through the interval from 500 to 300 °C. This is consistent with a model of the Big Sky orogeny that includes late-stage tectonic denudation that leads to decompression and rapid cooling. A similar cooling history is suggested by our data for the Ruby Range. Three biotite samples from the Ruby Range yield $^{40}Ar/^{39}Ar$ step-heating plateau ages with a mean of 1.73 ± 0.02 Ga, identical to the best-estimate (near-plateau) age for a hornblende from the same rocks. Two samples of the orthoamphibole, gedrite, from the Tobacco Root Mountains were studied, but did not have enough K to yield a reliable $^{40}Ar/^{39}Ar$ age. Several biotite and three hornblende samples from the region yield $^{40}Ar/^{39}Ar$ dates significantly younger than 1.7 Ga. We believe these samples were partially reset during contact metamorphism by Cretaceous (75 Ma) intrusive rocks. Hydrothermal alteration associated with ca. 1.4 Ga rifting led to growth of muscovite with that age in the Ruby Range, but this alteration was apparently not hot enough to reset biotite and hornblende ages there.

Keywords: Wyoming province, Big Sky orogeny, Proterozoic era.

Brady, J.B., Kovaric, D.N., Cheney, J.T., Jacob, L.J., and King, J.T., 2004, $^{40}Ar/^{39}Ar$ ages of metamorphic rocks from the Tobacco Root Mountains region, Montana, *in* Brady, J.B., Burger, H.R., Cheney, J.T., and Harms, T.A., eds., Precambrian geology of the Tobacco Root Mountains, Montana: Boulder, Colorado, Geological Society of America Special Paper 377, p. 131–149. For permission to copy, contact editing@geosociety.org.

INTRODUCTION

The Tobacco Root Mountains of southwestern Montana (Fig. 1) are cored by metamorphic rocks of the Wyoming province that are believed to have a geologic history spanning at least 3 Ga. Sensitive high-resolution ion microprobe (SHRIMP) dating of individual Tobacco Root Mountain zircons by Mueller et al. (1998) yielded detrital zircon core ages of 3.2 to 3.9 Ga, with none younger than 2.9 Ga. Although Rb/Sr studies by Mueller and Cordua (1976) and by James and Hedge (1980) yielded a whole-rock age of 2.7 Ga for the metamorphic rocks of this region, consistent with similar ages reported for the central Wyoming province, recent work on U and Pb isotopes of monazite and zircon (Krogh et al., 1997; Burger et al., 1999; Cheney et al., 1999; Dahl et al., 1999; Roberts et al., 2002; Dahl et al., 2002; Cheney et al., 2004b, this volume, Chapter 8; Mueller et al., 2004, this volume, Chapter 9) has identified metamorphic events in the Tobacco Root Mountains at 2.45 Ga and between 1.78 and 1.72 Ga with no clear evidence for a 2.7 Ga event.

Early K/Ar studies of biotite in southwest Montana by Hayden and Wehrenberg (1959, 1960) demonstrated that the Tobacco Root Mountains rocks experienced a heating event at 1.7 Ga that was not observed in the Archean rocks of the Beartooth Plateau. Giletti and Gast (1961) using Rb/Sr data for micas and Giletti (1966, 1968) using K/Ar and Rb/Sr data for micas and whole rocks showed that the entire northwestern portion of the Wyoming province was heated at 1.6–1.7 Ga, including the Tobacco Root Mountains and adjacent Ruby Range and Highland Mountains. Giletti (1966) presented a map of the Wyoming province showing a line that divided the 1.6–1.7 Ga terrane from the rest of the Wyoming province. Marvin and Dobson (1979) also reported 1.7 Ga K/Ar ages for muscovite in a pegmatite and for hornblende in gneiss from the northern Tobacco Root Mountains. However, because the studies of Mueller and Cordua (1976)

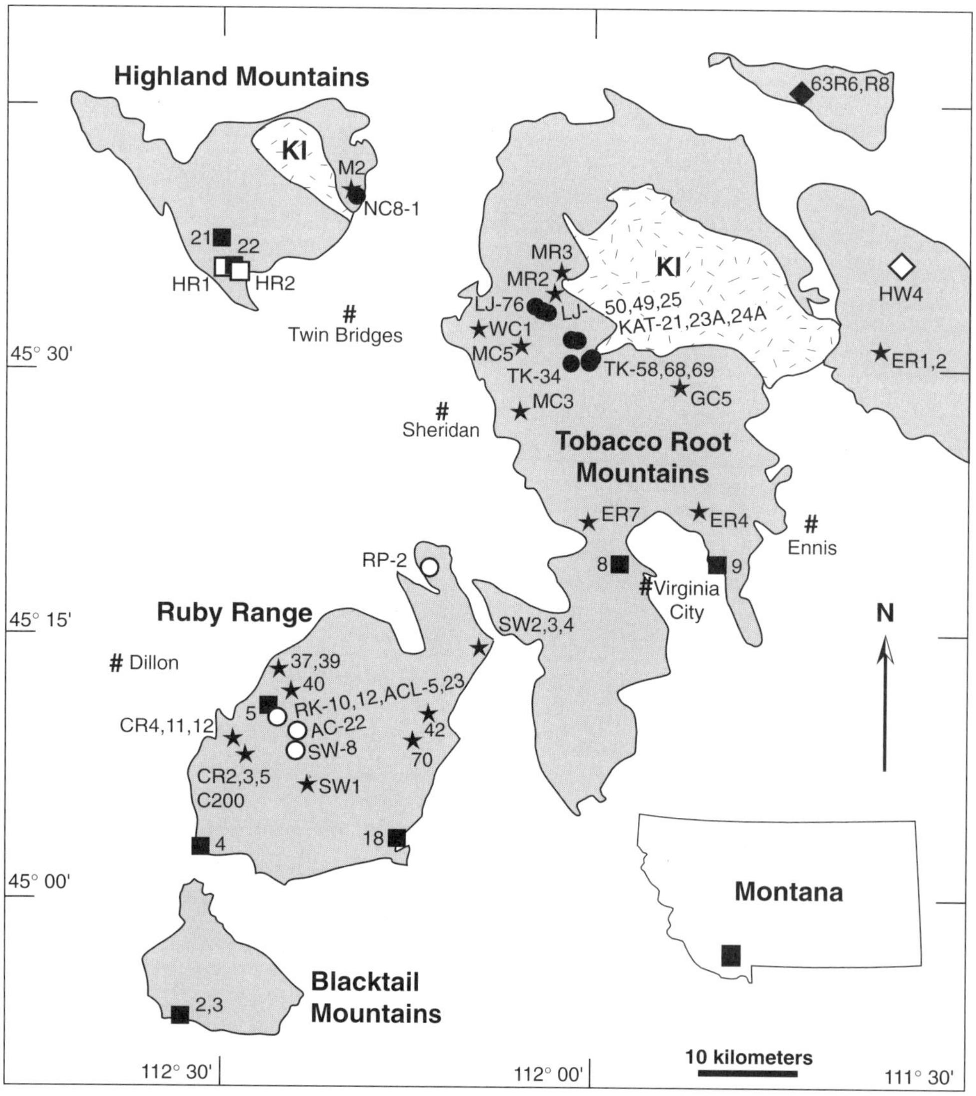

Figure 1. An outline map of the outcrop distribution of Precambrian metamorphic rocks (in gray) and adjacent Cretaceous intrusive rocks (KI) in southwest Montana. Also shown are the locations of samples used for K/Ar and $^{40}Ar/^{39}Ar$ dating. Solid circles—samples newly described in this paper. Open circles—samples of Brady et al. (1998). Solid squares—samples of Giletti (1966). Open squares—samples of Harlan et al. (1996). Solid diamond—samples of Marvin and Dobson (1979). Open diamond—sample of Hayden and Wehrenberg (1960). Stars—samples of Roberts et al. (2002).

and James and Hedge (1980) showed 2.7 Ga Rb/Sr data that were apparently not reset at 1.6–1.7 Ga, it was generally believed that the 1.6–1.7 Ga metamorphism was a low-grade, patchy, greenschist facies event (e.g., Mueller and Cordua, 1976; Berg, 1979; Dahl, 1979) with little accompanying deformation or fabric development. Indeed, most of the data were from micas, which can be crystallized or have their K/Ar clocks reset at relatively low temperatures (300–400 °C, McDougall and Harrison, 1988).

More recently, O'Neill et al. (1988a, 1988b) described 1.8–1.9 Ga euhedral rims on zircon crystals in a Highland Mountains gneiss dome, which they believe demonstrate a significant metamorphism accompanied by penetrative deformation—certainly more significant than a static regional greenschist facies heating. Erslev and Sutter (1990) and Harlan et al. (1996) also argued that the Proterozoic heating recorded by mica and amphibole K/Ar data was a major, regional event involving significant deformation.

It was in this context that we obtained a number of $^{40}Ar/^{39}Ar$ isometric ages that further constrain the tectonometamorphic history of the region as part of an ongoing study of the metamorphic rocks of southwest Montana (Brady et al., 1994; Kovaric et al., 1996; Brady et al., 1998). In the following pages, we present our $^{40}Ar/^{39}Ar$ data and compare them to previously published $^{40}Ar/^{39}Ar$ and K/Ar data, including the extensive $^{40}Ar/^{39}Ar$ data presented by Roberts et al. (2002). We join the other authors cited previously to argue that the 1.7 Ga ages determined for the potassium-bearing minerals of the Tobacco Root Mountains and adjacent ranges date their cooling at the end of a major orogenic event that included regional-scale, granulite facies metamorphism. We argue further that the similar K/Ar and $^{40}Ar/^{39}Ar$ ages obtained for hornblende, biotite, and muscovite are evidence for rapid cooling, consistent with a model of decompression due to tectonic unroofing as suggested by Cheney et al. (2004a, this volume, Chapter 6).

SAMPLES

The Tobacco Root Mountains samples were collected by Jacob (1994), King (1994), and Tierney (1994) from amphibolites and gneisses while studying the metamorphism, geochemistry, and structure of the Spuhler Peak Metamorphic Suite and adjacent Indian Creek Metamorphic Suite (see Fig. 1). The Spuhler Peak Metamorphic Suite is a mafic unit—largely of basaltic composition but with significant portions of Ca-poor, Mg-Fe–rich orthoamphibole gneisses and layers of Al-rich quartzite—that outcrops principally along the southwestern margin of the Tobacco Root batholith (see also Burger, 2004, this volume, Chapter 1; Vitaliano et al., 1979a, see reprinted map and text accompanying this volume). The Indian Creek Metamorphic Suite is a quartzofeldspathic gneiss package that also includes marbles, metamorphosed iron formation, and other metasediments as well as felsic meta-igneous rocks, and occupies the southern portion of the Tobacco Root Mountains (see Vitaliano et al., 1979b; Mogk et al., 2004, this volume, Chapter 2). Amphiboles were also dated from weakly foliated and locally folded metamorphosed mafic dikes and sills (MMDS) that crosscut the gneissic layering of the Indian Creek Metamorphic Suite but do not occur in the Spuhler Peak Metamorphic Suite. The goal of our sample collection was to identify differences (if any) in the $^{40}Ar/^{39}Ar$ data for similar minerals in the three rock groups (Spuhler Peak Metamorphic Suite, Indian Creek Metamorphic Suite, MMDS), so our Tobacco Root samples are all from a comparatively small area in the central Tobacco Root Mountains where all three units occur near one another. As a consequence, the Tobacco Root samples studied are all within five kilometers of the Cretaceous Tobacco Root batholith and may have been heated by it. No new samples were dated from the Pony–Middle Mountain Metamorphic Suite, which is believed to have the same Proterozoic geologic history as the Indian Creek Metamorphic Suite because both have gneissic banding cut by the MMDS.

Ruby Range rocks were collected in 1990 by Green (1991), Larson (1991), and Brady et al. (1991) in pursuit of an origin for the talc deposits in the marbles there (see Brady et al., 1998). All of the Ruby Range samples were taken from gneisses, marbles, or amphibolites of the Christensen Ranch Metamorphic Suite within or adjacent to talc deposits (see James, 1990). The Highland Mountains sample was collected by Brady in 1978 from gneisses within a chlorite deposit as part of the Dillon 1° × 2° Sheet study of the U.S. Geological Survey. More detailed sample information can be found in the theses listed above as well as in Table 1.

METHODS

$^{40}Ar/^{39}Ar$ isotopic ages for these samples were determined in three different laboratories (University of California at Los Angeles [UCLA], Massachusetts Institute of Technology [MIT], University of Maine) employing contrasting techniques. Some of the Tobacco Root Mountains samples were analyzed at MIT using UV laser microprobe, single-grain $^{40}Ar/^{39}Ar$ total gas methods on irradiated hornblende grains as described in Hames and Cheney (1997). Other Tobacco Root Mountains samples were dated at UCLA by Kovaric (1996) following standard step-heating procedures of irradiated bulk mineral separates (Quidelleur et al., 1997). The Ruby Range samples (Brady et al., 1998) were analyzed at the University of Maine, also following standard step-heating procedures of irradiated bulk mineral separates (Lux et al., 1989).

RESULTS

Sample information, total gas ages, and plateau ages where available are listed for all samples in Table 1. Errors for the ages in Table 1 are reported as one standard deviation (1σ). Also compiled for comparison in Table 1 are total gas and plateau ages for previously published K/Ar and $^{40}Ar/^{39}Ar$ studies of K-bearing minerals in the Tobacco Root Mountains, the Highland Mountains, the Ruby Range, and the Blacktail Mountains. $^{40}Ar/^{39}Ar$ step-heating release spectra for individual samples are shown in Appendix Figures A1 and A2 along with the original data (Tables A1 and A2). Also in the Appendix are histograms for each sample of UV laser microprobe, single-grain, and total

TABLE 1. K/Ar DATA FOR PRECAMBRIAN ROCKS OF THE TOBACCO ROOT MOUNTAINS AND VICINITY

Data Source	Sample no.	*Latitude (°N)	Longitude (°W)	†Unit	Rock	Mineral	Total gas age (Ma)	±1σ	Plateau age (Ma)	±1σ	Lab	§Method
Tobacco Root Mountains												
Hayden and Wehrenberg (1960)	HW4	45°35'55"	111°35'17"	PMMMS	Biotite Gneiss	Biotite	1620	32	#N.D.	N.D.	Argonne	K/Ar (a)
Giletti (1966)	8	45°19'03"	112°02'10"	ICMS	Pegmatite in Gneiss	Muscovite	1640	33	N.D.	N.D.	Brown	K/Ar (b)
Giletti (1966)	9	45°19'03"	111°50'17"	PMMMS	Granitic Gneiss	Biotite	1700	34	N.D.	N.D.	Brown	K/Ar (b)
Marvin and Dobson (1979)	63R6	45°45'39"	111°43'22"	PMMMS	Pegmatite	Muscovite	1720	50	N.D.	N.D.	USGS	K/Ar
Marvin and Dobson (1979)	63R8	45°45'39"	111°43'22"	PMMMS	Bio-Hbl Gneiss	Hornblende	1740	90	N.D.	N.D.	USGS	K/Ar
Jacob (1994)	LJ-25	45°33'23"	112°04'17"	MMDS	Amphibolite	Hornblende	1694	53	N.D.	N.D.	MIT	LM-10
Jacob (1994)	LJ-49-2	45°33'12"	112°03'53"	SPMS	Amphibolite	Hornblende	1572	57	N.D.	N.D.	MIT	LM-10
Jacob (1994)	LJ-50	45°33'16"	112°03'48"	SPMS	Amphibolite	Hornblende	1730	55	N.D.	N.D.	MIT	LM-10
Jacob (1994)	LJ-76	45°33'38"	112°04'49"	ICMS	Amphibolite	Hornblende	1733	34	N.D.	N.D.	MIT	LM-10
Jacob (1994)	LJ-77	45°33'38"	112°04'49"	MMDS	Amphibolite	Hornblende	1747	80	N.D.	N.D.	MIT	LM-10
King (1994)	TK-34	45°30'24"	112°00'43"	ICMS	Amphibolite	Hornblende	1799	41	N.D.	N.D.	MIT	LM-10
King (1994)	TK-58	45°30'42"	111°59'52"	SPMS	Amphibolite	Hornblende	839	177	N.D.	N.D.	MIT	LM-9
King (1994)	TK-68	45°30'34"	112°00'22"	SPMS	Amphibolite	Hornblende	1228	376	N.D.	N.D.	MIT	LM-9
King (1994)	TK-69	45°30'27"	112°00'22"	ICMS	Amphibolite	Hornblende	1298	257	N.D.	N.D.	MIT	LM-10
Kovaric (1996)	KAT-21	45°31'44"	112°01'53"	SPMS	Bio-Hbl Amphibolite	Biotite	**129	1	N.D.	N.D.	UCLA	$^{40}Ar/^{39}Ar$
Kovaric (1996)	KAT-21	45°31'44"	112°01'53"	SPMS	Bio-Hbl Amphibolite	Hornblende	1639	26	N.D.	N.D.	UCLA	$^{40}Ar/^{39}Ar$
Kovaric (1996)	KAT-23A	45°31'45"	112°01'51"	SPMS	Amphibolite	Hornblende	1697	20	1735	13	UCLA	$^{40}Ar/^{39}Ar$
Kovaric (1996)	KAT-24A	45°31'45"	112°01'50"	SPMS	Ged Amphibolite	Mg-hornblende	1539	74	1665	37	UCLA	$^{40}Ar/^{39}Ar$
Kovaric (1996)	KAT-41	45°31'42"	112°01'29"	SPMS	Bio-Grt-Sil Gneiss	Biotite	**106	0	N.D.	N.D.	UCLA	$^{40}Ar/^{39}Ar$
Kovaric (1996)	KAT-44	45°31'41"	112°01'26"	SPMS	Ged Amphibolite	Gedrite	N.D.	N.D.	N.D.	N.D.	UCLA	$^{40}Ar/^{39}Ar$
Kovaric (1996)	KAT-51B	45°31'41"	112°01'20"	SPMS	Amphibolite	Hornblende	1680	26	1698	27	UCLA	$^{40}Ar/^{39}Ar$
Roberts et al. (2002)	TRER1	45°31'	111°37'	PMMMS	N.D.	Amphibole	1272	16	N.D.	N.D.	Open U.	LM-5
Roberts et al. (2002)	TRER2	45°31'	111°37'	PMMMS	N.D.	Amphibole	**83	4	N.D.	N.D.	Open U.	LM-4
Roberts et al. (2002)	TRER2	45°31'	111°37'	PMMMS	N.D.	Biotite	**72	8	N.D.	N.D.	Open U.	LM-17
Roberts et al. (2002)	TRER4	45°22'	111°51'	ICMS	N.D.	Biotite	1728	12	N.D.	N.D.	Open U.	LM-10
Roberts et al. (2002)	TRER7	45°21'	112°00'	ICMS	N.D.	Biotite	1713	8	N.D.	N.D.	Open U.	LM-15
Roberts et al. (2002)	TRGC5	45°29'	111°53'	ICMS	N.D.	Biotite	1692	8	N.D.	N.D.	Open U.	LM-13
Roberts et al. (2002)	TRMC3	45°28'	112°06'	ICMS	N.D.	Biotite	1752	8	N.D.	N.D.	Open U.	LM-22
Roberts et al. (2002)	TRMC5	45°31'	112°06'	ICMS	N.D.	Amphibole	1803	34	N.D.	N.D.	Open U.	LM-4
Roberts et al. (2002)	TRMR2	45°34'	112°03'	ICMS	N.D.	Biotite	1585	4	N.D.	N.D.	Open U.	LM-26
Roberts et al. (2002)	TRMR3	45°36'	112°03'	PMMMS	N.D.	Amphibole	**248	44	N.D.	N.D.	Open U.	LM-4
Roberts et al. (2002)	TRMR3	45°36'	112°03'	PMMMS	N.D.	Biotite	**80	39	N.D.	N.D.	Open U.	LM-1
Roberts et al. (2002)	TRWC1	45°32'	112°09'	ICMS	N.D.	Amphibole	1634	34	N.D.	N.D.	Open U.	LM-5
Roberts et al. (2002)	TRWC1	45°32'	112°09'	ICMS	N.D.	Biotite	1689	10	N.D.	N.D.	Open U.	LM-10

(continued)

TABLE 1. K/Ar DATA FOR PRECAMBRIAN ROCKS OF THE TOBACCO ROOT MOUNTAINS AND VICINITY (*continued*)

Data Source	Sample no.	*Latitude (°N)	Longitude (°W)	†Unit	Rock	Mineral	Total gas age (Ma)	±1σ	Plateau age (Ma)	±1σ	Lab	§Method
Ruby Range												
Giletti (1966)	4	45°02'55"	112°31'37"	DGG	Bio Gneiss	Biotite	1640	33	N.D.	N.D.	Brown	K/Ar (b)
Giletti (1966)	5	45°10'57"	112°26'15"	CCG	Mus-Bio Schist	Biotite	1580	32	N.D.	N.D.	Brown	K/Ar (b)
Giletti (1966)	5	45°10'57"	112°26'15"	CCG	Mus-Bio Schist	Biotite	1600	32	N.D.	N.D.	Brown	K/Ar (b)
Giletti (1966)	5	45°10'57"	112°26'15"	CCG	Mus-Bio Schist	Muscovite	1670	33	N.D.	N.D.	Brown	K/Ar (b)
Giletti (1966)	6	45°10'57"	112°26'15"	CCG	Bio-Hbl Gneiss	Biotite	1480	30	N.D.	N.D.	Brown	K/Ar (b)
Giletti (1966)	18	45°03'28"	112°15'36"	DGG	Bio Gneiss	Biotite	1540	31	N.D.	N.D.	Brown	K/Ar (b)
Brady et al. (1998)	RK-10	45°10'16"	112°25'32"	CRMS	Amphibolite	Hornblende	1640	9	N.D.	N.D.	Maine	$^{40}Ar/^{39}Ar$
Brady et al. (1998)	ACL-005	45°10'16"	112°25'32"	CRMS	Phlog Marble	Phlogopite	1700	16	1723	15	Maine	$^{40}Ar/^{39}Ar$
Brady et al. (1998)	SW-8	45°08'22"	112°24'03"	CRMS	Talc-Dol Marble	Phlogopite	1670	9	1716	15	Maine	$^{40}Ar/^{39}Ar$
Brady et al. (1998)	RP-2	45°18'48"	112°13'37"	CRMS	Marble	Biotite	1757	8	1764	11	Maine	$^{40}Ar/^{39}Ar$
Brady et al. (1998)	RK-12	45°10'16"	112°25'32"	CRMS	Marble	Biotite	1644	9	N.D.	N.D.	Maine	$^{40}Ar/^{39}Ar$
Brady et al. (1998)	ACL-023	45°10'16"	112°25'32"	CRMS	Marble	Biotite	1659	8	N.D.	N.D.	Maine	$^{40}Ar/^{39}Ar$
Brady et al. (1998)	JBB-AC-22	45°09'30"	112°23'55"	CRMS	Chl-Mus Gneiss	Muscovite	††1330	9	1363	3	UCLA	$^{40}Ar/^{39}Ar$
Roberts et al. (2002)	70-30-2	45°09'	112°15'	DGG	N.D.	Biotite	1744	6	N.D.	N.D.	Open U.	LM-9
Roberts et al. (2002)	RM37	45°13'	112°25'	CRMS	N.D.	Biotite	1731	4	N.D.	N.D.	Open U.	LM-23
Roberts et al. (2002)	RM39	45°13'	112°25'	CRMS	N.D.	Biotite	1784	6	N.D.	N.D.	Open U.	LM-10
Roberts et al. (2002)	RM40	45°12'	112°24'	CRMS	N.D.	Biotite	1772	4	N.D.	N.D.	Open U.	LM-24
Roberts et al. (2002)	RM42	45°10'	112°13'	DGG	N.D.	Biotite	1734	14	N.D.	N.D.	Open U.	LM-6
Roberts et al. (2002)	RMC200	45°08'	112°28'	CRMS	N.D.	Biotite	1742	4	N.D.	N.D.	Open U.	LM-16
Roberts et al. (2002)	RRCR2	45°08'	112°28'	CRMS	N.D.	Biotite	1761	4	N.D.	N.D.	Open U.	LM-35
Roberts et al. (2002)	RRCR3	45°08'	112°28'	CRMS	N.D.	Biotite	1765	6	N.D.	N.D.	Open U.	LM-58
Roberts et al. (2002)	RRCR4	45°09'	112°29'	CRMS	N.D.	Amphibole	1561	60	N.D.	N.D.	Open U.	LM-5
Roberts et al. (2002)	RRCR5	45°08'	112°28'	CRMS	N.D.	Biotite	1741	10	N.D.	N.D.	Open U.	LM-20
Roberts et al. (2002)	RRCR11	45°09'	112°29'	CRMS	N.D.	Amphibole	1557	60	N.D.	N.D.	Open U.	LM-5
Roberts et al. (2002)	RRCR12	45°09'	112°29'	CRMS	N.D.	Amphibole	1566	38	N.D.	N.D.	Open U.	LM-5
Roberts et al. (2002)	RRSW1	45°06'	112°23'	CRMS	N.D.	Amphibole	1708	14	N.D.	N.D.	Open U.	LM-10
Roberts et al. (2002)	RRSW1	45°06'	112°23'	CRMS	N.D.	Biotite	1771	32	N.D.	N.D.	Open U.	LM-6
Roberts et al. (2002)	RRSW2	45°14'	112°09'	DGG	N.D.	Biotite	1738	22	N.D.	N.D.	Open U.	LM-9
Roberts et al. (2002)	RRSW3	45°14'	112°09'	DGG	N.D.	Biotite	1774	12	N.D.	N.D.	Open U.	LM-30
Roberts et al. (2002)	RRSW4	45°14'	112°09'	DGG	N.D.	Biotite	1744	18	N.D.	N.D.	Open U.	LM-9

(*continued*)

TABLE 1. K/Ar DATA FOR PRECAMBRIAN ROCKS OF THE TOBACCO ROOT MOUNTAINS AND VICINITY (*continued*)

Data Source	Sample no.	*Latitude (°N)	Longitude (°W)	†Unit	Rock	Mineral	Total gas age (Ma)	±1σ	Plateau age (Ma)	±1σ	Lab	§Method
Highland Mountains												
Giletti (1966)	21	45°37'24"	112°30'12"	N.D.	Granite Gneiss	Muscovite	**174	4	N.D.	N.D.	Brown	K/Ar (b)
Giletti (1966)	22	45°35'40"	112°29'10"	N.D.	Biotite Schist	Biotite	**75	2	N.D.	N.D.	Brown	K/Ar (b)
Harlan et al. (1996)	HR1-C	45°34'01"	112°25'19"	N.D.	Hbl-Bio Gneiss	Hornblende	1839	5	1815	10	USGS	$^{40}Ar/^{39}Ar$
Harlan et al. (1996)	HR1-A	45°34'01"	112°25'19"	N.D.	Biotite Gneiss	Biotite	1776	8	1816	10	USGS	$^{40}Ar/^{39}Ar$
Harlan et al. (1996)	HR2-Ja	45°35'46"	112°26'49"	MMDS	Amphibolite	Hornblende	1780	5	1796	12	USGS	$^{40}Ar/^{39}Ar$
Kovaric (1996)	NC8-1	45°39'49"	112°19'21"	N.D.	Chloritized Gneiss	Muscovite	**121	2	N.D.	N.D.	UCLA	$^{40}Ar/^{39}Ar$
Roberts et al. (2002)	HLM2	45°40'	112°20'	N.D.	N.D.	Biotite	1749	6	N.D.	N.D.	Open U.	LM-9
Blacktail Mountains												
Giletti (1966)	2	44°53'20"	112°33'13"	N.D.	Bio Gneiss	Biotite	1590	32	N.D.	N.D.	Brown	K/Ar (b)
Giletti (1966)	3	44°53'20"	112°33'13"	N.D.	Bio Gneiss	Biotite	1320	150	N.D.	N.D.	Brown	K/Ar (b)

Note: Bio—biotite; Hbl—hornblende; Grt—garnet; Sil—sillimanite; Ged—gedrite; Mus—muscovite; Phlog—phlogopite; Dol—dolomite; Chl—chlorite.

* Values for older data from Berg (1979). Roberts et al. (2002) locations approximated from location map.

† Unit—Map unit where known. PMMMS—Pony–Middle Mountain Metamorphic Suite; ICMS—Indian Creek Metamorphic Suite; SPMS—Spuhler Peak Metamorphic Suite; MMDS—metamorphosed mafic dikes and sills; DGG—Dillon Granite Gneiss; CCG—Cherry Creek Gneiss; CRMS—Christensen Ranch Metamorphic Suite.

§ Method of analysis. K/Ar—traditional K/Ar analysis of bulk mineral separate; $^{40}Ar/^{39}Ar$—step heating of bulk mineral separate; LM—UV laser microprobe $^{40}Ar/^{39}Ar$—number of grains. $\lambda_e = 0.581 \times 10^{-10}\ a^{-1}$; $\lambda_\beta = 4.962 \times 10^{-10}\ a^{-1}$. Older data corrected from (a) $\lambda_e = 0.585 \times 10^{-10}\ a^{-1}$; $\lambda_\beta = 4.72 \times 10^{-10}\ a^{-1}$; (b) $\lambda_e = 0.584 \times 10^{-10}\ a^{-1}$; $\lambda_\beta = 4.72 \times 10^{-10}\ a^{-1}$.

\# N.D.—no data.

** Believed to be partially to completely reset by a 75 Ma batholith.

†† Growth during hydrothermal alteration that produced economic talc deposits.

gas ages (Fig. A3), along with the original data (Table A3). Two samples of the orthoamphibole, gedrite, from the Spuhler Peak Metamorphic Suite were studied, but did not have enough K to yield a meaningful $^{40}Ar/^{39}Ar$ age.

As many as ten separate hornblende crystals from each of nine Tobacco Root Mountains samples were dated using a UV laser microprobe $^{40}Ar/^{39}Ar$ total gas method. Data for these 89 crystals are collected in a histogram in Figure 2 (see also Fig. A3). The ages shown in gray are for the 29 crystals of samples TK-58, TK-68, TK-69, which we believe to have been partially reset by Cretaceous intrusions, based on their location and on their scattered ages. The 60 ages shown in black have a mean of 1.71 ± 0.02 Ga. Some of these individual ages are older than the metamorphic maximum that occurred between 1.78 and 1.72 Ga (Cheney et al., 2004b, this volume, Chapter 8; Mueller et al., 2004, this volume, Chapter 9), so they may have accumulated excess ^{40}Ar. Others are as young as 1.50 Ga and appear to have lost some Ar due to reheating.

Measurements of $^{40}Ar/^{39}Ar$ step-heating plateau ages of three bulk hornblende samples from the Spuhler Peak Metamorphic Suite of the Tobacco Root Mountains average 1.70 ± 0.02 Ga. Measurement of three bulk biotite samples from the Ruby Range yielded $^{40}Ar/^{39}Ar$ step-heating plateau ages with a mean of 1.73 ± 0.02 Ga, identical to the best-estimate (near-plateau) age for a hornblende sample from the same rocks. The similarity of the step-heating plateau ages to the mean of the 60 UV laser microprobe total gas ages leads us to believe that 1.71 ± 0.02 Ga is the most recent time the metamorphic rocks in the Tobacco Root Mountains and in the Ruby Range were at temperatures above the Ar closure temperature of hornblende and biotite.

Monazite and zircon studies (Krogh et al., 1997; Burger et al., 1999; Cheney et al., 1999; Dahl et al., 1999; Dahl et al., 2002; Cheney et al., 2004b, this volume, Chapter 8; Mueller et al., 2004, this volume, Chapter 9) show that there was a major, orogenic event (the Big Sky orogeny) that affected the Tobacco Root Mountains region during the Middle Proterozoic. Detailed monazite data on Pb isotopes are interpreted by Cheney et al. (2004b, this volume, Chapter 8) to mean that high-pressure metamorphism with monazite growth beginning at 1780 Ma was followed by low-pressure metamorphism with monazite growth ending at 1720 Ma (see also Cheney et al., 2004a, this volume, Chapter 6). The large cluster of K/Ar and $^{40}Ar/^{39}Ar$ ages observed between 1.73 and 1.71 Ga is consistent with cooling of these rocks through the Ar closure temperatures as part of this metamorphic pressure-temperature path.

Overall, our data are similar to the results of other investigators. A histogram of all 71 total gas $^{40}Ar/^{39}Ar$ and K/Ar ages in Table 1 is presented in Figure 3 with our data (23 samples) shown in the gray pattern. Most (55) of the ages in Figure 3 fall in a group between 1.54 and 1.84 Ga with a mean of 1.70 ± 0.07 Ga. Nine ages form a group between 72 and 248 Ma. Another seven ages fall between these two groups. Based on the data in Figure 3, there is no significant or systematic difference in age between the various Precambrian rock suites (Indian Creek

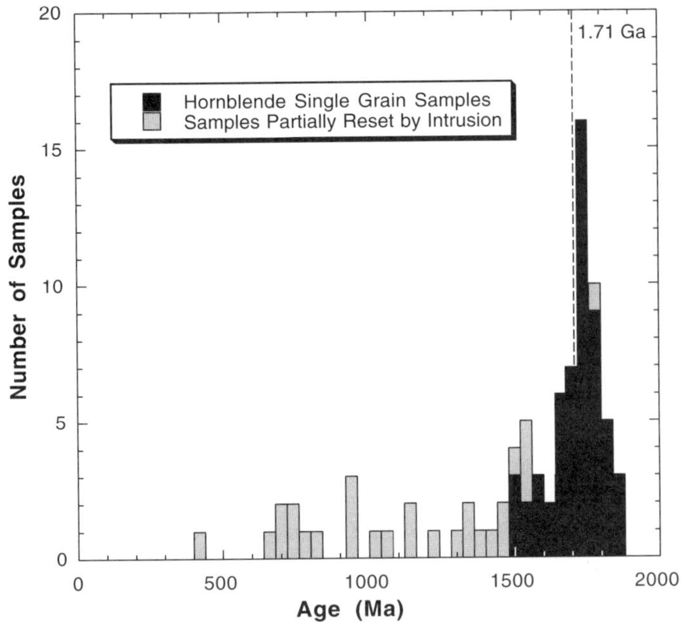

Figure 2. A histogram of 89 UV laser microprobe $^{40}Ar/^{39}Ar$ ages for single hornblende crystals from nine Precambrian metamorphic rocks of the Tobacco Root Mountains. Ten individual hornblende crystals were dated for each sample, save one. Ages in gray (29) are in rocks that are believed to have lost ^{40}Ar due to reheating by Cretaceous intrusions. Ages in black (60) cluster about a mean of 1.71 ± 0.02 Ga. Bin size is 40 m.y.

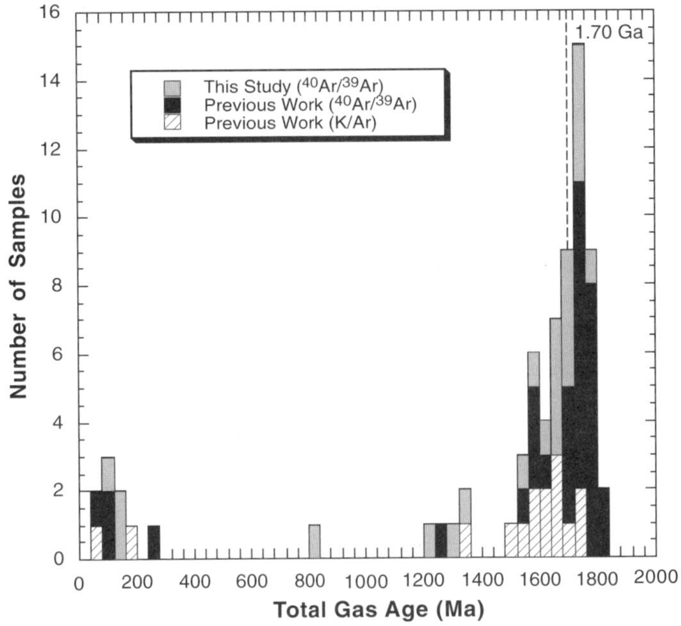

Figure 3. A histogram of the K/Ar and $^{40}Ar/^{39}Ar$ total gas ages for Precambrian metamorphic rocks of the Tobacco Root Mountains and vicinity listed in Table 1. Samples in gray are $^{40}Ar/^{39}Ar$ total gas ages determined as part of this project. Samples in black are $^{40}Ar/^{39}Ar$ total gas ages from the literature, mostly Roberts et al. (2002). Samples in stripes are K/Ar total gas ages from the literature. Bin size is 40 m.y. The large cluster of ages with a mean of 1.70 ± 0.07 Ga is believed to record cooling from a 1.78 to 1.72 Ga regional metamorphism of upper-amphibolite or granulite facies. Younger ages are believed to reflect various degrees of ^{40}Ar loss due to later heating by Late Cretaceous intrusive rocks.

Metamorphic Suite, Pony–Middle Mountain Metamorphic Suite, Spuhler Peak Metamorphic Suite, and MMDS) in the Tobacco Root Mountains. However, the scatter among the total gas ages is too large (300 m.y.) to reasonably attribute to differences in cooling age from a single orogenic event. Clearly, some samples must have been disturbed by subsequent heating to lose Ar, while other samples must have acquired excess Ar.

We interpret the nine 72–248 Ma ages to be the result of partial to complete Ar loss due to heating by nearby 75 Ma intrusive igneous rocks (Tobacco Root batholith, Hells Canyon pluton, and smaller igneous bodies). Two of these nine ages, for Highland Range micas, were reported by Giletti (1966), who first recognized this Cretaceous resetting. Eight of these nine samples are mica separates (biotite or muscovite), which can lose their Ar if heated to between 300 and 400 °C (McDougall and Harrison, 1988). In our study, both biotite and hornblende were separated and dated from one Tobacco Root Mountains Spuhler Peak Metamorphic Suite sample (KAT-21). Both minerals have disturbed $^{40}Ar/^{39}Ar$ step-heating spectra (Fig. 4) and no good age plateau (adjacent steps including >50% of the ^{39}Ar, all within 2σ of the plateau age). The hornblende yields a total gas age of 1639 ± 26 Ma, whereas the biotite gives a total gas age of 129 ± 6 Ma. Because the biotite age is close to the Tobacco Root batholith intrusion age (75 Ma, Vitaliano et al., 1980), we believe this sample was heated above 300 °C by that intrusion. Because the hornblende is only slightly

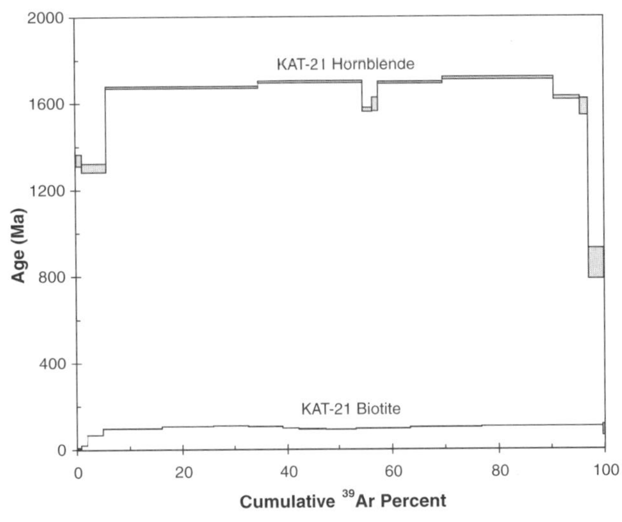

Figure 4. $^{40}Ar/^{39}Ar$ step-heating age spectra for hornblende and biotite separated from the same Tobacco Root Mountains Spuhler Peak Metamorphic Suite sample (KAT-21). The young age of the biotite is due to ^{40}Ar loss resulting from heating by Late Cretaceous (75 Ma) intrusive rocks. The hornblende gives much older age steps, with a best-estimate age of 1700 ± 9 Ma that would be a plateau without steps five and six.

disturbed, the host rock was probably not heated above 500 °C, but the hornblende total gas age may be a minimum age. Indeed, there is a "near plateau" at 1700 ± 9 Ma that is our best estimate of the metamorphic cooling age of this sample, so there may have been some Ar loss due to heating by the intrusion. Sample KAT-21 was collected approximately one kilometer from the surface exposure of the Tobacco Root batholith. Using simple thermal models for batholith emplacement (Jaeger, 1964), Kovaric (1996) demonstrated that temperatures between 300 and 500 °C are reasonable contact metamorphic temperatures for the size of the batholith and the position of the samples.

Several samples give ages between the main group with the 1.70 ± 0.07 Ga mean and the young group clearly reset by the Cretaceous intrusions. One of these (JBB-AC-22) is a muscovite sample from a zone of hydrothermal alteration that formed a talc deposit in the Ruby Range. Brady et al. (1998) argue that this age (total gas = 1.33 ± 0.01 Ga, plateau = 1.36 ± 0.01 Ga) represents the growth of muscovite during formation of the talc as part of ca. 1.4 Ga rifting. Indeed, this sample was collected in an attempt to learn the age of talc formation. Because samples of biotite (RP-2, RK-12, ACL-023), phlogopite (ACL-005, SW8), and hornblende (RK-10) near the talc deposits were not reset, the hydrothermal alteration and growth of mica must have occurred at temperatures below 300 °C. The three amphibole samples (TK-58, TK-68, TK-69) that give intermediate mean ages (839, 1228, 1298 Ma) were collected near one another about one kilometer from the Tobacco Root batholith contact and even closer to several satellite intrusions. We believe that these three samples were heated enough to lose some, but not all, of their Ar. The heating of these samples was probably due to the batholith, but these and other intermediate-age samples may have been affected by the intrusion of (unmetamorphosed) mafic dikes during the Proterozoic era at 1450 Ma (Wooden et al., 1978) or at 780 Ma (Harlan et al., 2003).

DISCUSSION

K/Ar and $^{40}Ar/^{39}Ar$ data for hornblende, biotite, and muscovite of the northwestern portion of the Wyoming province present a consistent temporal pattern, whether measured by traditional bulk K/Ar methods, by $^{40}Ar/^{39}Ar$ step heating of bulk, irradiated mineral separates, or by total gas $^{40}Ar/^{39}Ar$ UV laser spot heating of irradiated single crystals, and in a number of different labs (Fig. 3). As first observed by Hayden and Wehrenberg (1959) and more fully documented by Giletti (1966), the K/Ar systems of the metamorphic rocks of the Tobacco Root Mountains and adjacent areas record Early Proterozoic ages, unless reset by thermal effects such as those due to Late Cretaceous plutons. $^{40}Ar/^{39}Ar$ data reported here further document the nature of the Proterozoic event. The Indian Creek Metamorphic Suite, which was previously metamorphosed at 2.45 Ga (Cheney et al., 1999; 2004b, this volume, Chapter 8), and the MMDS, which were intruded at 2.06 Ga (Burger et al., 1999; Mueller et al., 2004, this volume, Chapter 9), must have achieved temperatures greater than ~500 °C (McDougall and Harrison, 1988) during the 1.78–1.72 Ga metamorphism to reset their hornblende $^{40}Ar/^{39}Ar$ ages and to record cooling by 1.71 ± 0.02 Ga. The fact that similar $^{40}Ar/^{39}Ar$ ages are obtained for hornblende from the Ruby Range and the Highland Mountains is consistent with a high-temperature, tectonothermal event of regional extent.

Interestingly, the K/Ar and $^{40}Ar/^{39}Ar$ ages of amphiboles and micas are very similar in these rocks (Fig. 5), even though the amphibole ages are believed to be set as the rock cools through ~500 °C and the mica ages are believed to be set as the rock cools through ~300 °C (McDougall and Harrison, 1988). Cooling rate, grain size, and mineral composition can affect the value of the closure temperature, but the available K/Ar and $^{40}Ar/^{39}Ar$ data appear to reflect fairly rapid cooling through the 500–300 °C temperature interval. Cheney et al. (2004a, this volume, Chapter 6) argue on the basis of mineral assemblages and reaction textures that the pressure-temperature path followed by Tobacco Root Mountains rocks during the Big Sky orogeny includes a nearly isothermal decompression segment, possibly caused by tectonic unroofing during late extension. If their model is correct, then one would expect rapid cooling to follow the removal of overburden. The observed similarity in amphibole and mica ages from the Tobacco Root Mountains lends support to this model.

Plateau ages for the three Highland Mountains samples reported by Harlan et al. (1996) are slightly older (averaging 1.81 ± 0.01 Ga) than the ages reported here for the Tobacco Root

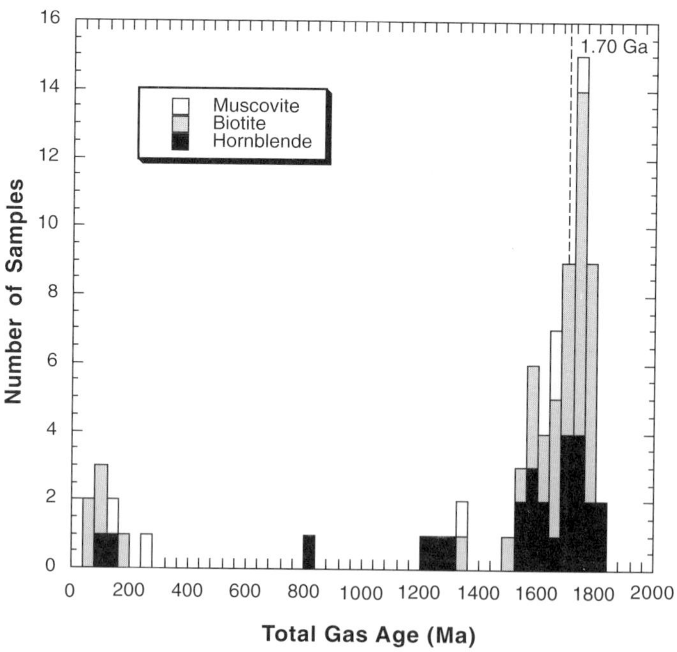

Figure 5. A histogram of K/Ar and $^{40}Ar/^{39}Ar$ total gas ages for Precambrian metamorphic rocks of the Tobacco Root Mountains and vicinity. This figure is similar to Figure 3, except that the ages are sorted by mineral rather than by investigator. Bin size is 40 m.y.

Mountains and the Ruby Range. Our three plateau ages for the Tobacco Root Mountains average 1.70 ± 0.02 Ga, whereas our three plateau ages for the Ruby Range average 1.73 ± 0.02 Ga. It is possible that some of the Harlan et al. (1996) samples held excess ^{40}Ar, giving them apparently older ages. However, it is also likely that, in an event as large as the 1.78–1.72 Ga Big Sky orogeny, rocks with different cooling histories can be found, due to their different locations in the orogen. Indeed, different levels of the crust might be juxtaposed by the extensive Cenozoic block faulting. Clearly, there is much to learn from additional isotopic studies in this region.

Recently, Roberts et al. (2002) presented $^{40}Ar/^{39}Ar$ UV laser microprobe ages for 18 samples from the Ruby Range, 13 samples from the Tobacco Root Mountains, and two samples from the Highland Mountains, examining at least ten single crystals of biotite (23 samples) or hornblende (10 samples) for each sample. This large data set, which nearly doubled the number of samples with measured $^{40}Ar/^{39}Ar$ total gas ages from this region, is included in the histograms of Figures 3 and 5. Alone, the Roberts et al. (2002) biotite data define a histogram similar to Figure 2, but with a mean age of 1.76 ± 0.02 Ga, older than the 1.71 ± 0.02 Ga $^{40}Ar/^{39}Ar$ cooling age adopted in this paper. Roberts et al. (2002) argue that all measured ages younger than 1700 Ma are for samples that have lost Ar due to the Cretaceous intrusions, and only use ages older than 1700 Ma to date cooling from the Proterozoic metamorphism, possibly biasing their results to older values. They also suggest that some of their amphibole ages that are older than 1780–1740 Ma are due to the presence of excess ^{40}Ar and, therefore, focus on their biotite data. As a consequence, they argue from their selected biotite data that these rocks cooled through the temperature interval of 350–300 °C during the interval 1780–1740 Ma.

The Roberts et al. (2002) interpretation of their data is in conflict with the abundant evidence of monazite growth in the Tobacco Root Mountains as late as 1720 Ma reported by Cheney et al. (1999; 2004b, this volume, Chapter 8), who correlate late monazite growth with a lower-pressure portion of the metamorphic pressure-temperature path at temperatures of 650–700 °C (see also Cheney et al., 2004a, this volume, Chapter 6). The Roberts et al. (2002) 1780–1740 Ma age is also somewhat in conflict with the $^{40}Ar/^{39}Ar$ plateau ages obtained in our study. Roberts et al. (2002) presented ^{207}Pb-^{206}Pb garnet step-leaching ages, probably dating monazite inclusions in the garnet, of 1820–1780 Ma for samples from this region (also in conflict with the results of Cheney et al., 2004b, this volume, Chapter 8). They argue that their 1780–1740 Ma $^{40}Ar/^{39}Ar$ total gas ages were set during cooling from 1820 to 1780 Ma regional metamorphism. One possible explanation for the conflict between the data and interpretations of this paper along with the data of Cheney et al. (2004b, this volume, Chapter 8) and that of Roberts et al. (2002) is that some of the Roberts et al. (2002) biotite samples may contain excess ^{40}Ar. Another possibility is that there may be systematic differences in the dates obtained from two or more $^{40}Ar/^{39}Ar$ labs.

ACKNOWLEDGMENTS

The authors thank the W.M. Keck Foundation and the colleges of the Keck Geology Consortium for their financial support of this work. Kovaric thanks Smith College for a Schultz Foundation summer fellowship. Dan Lux, Bill Hames, Mark Harrison, and Marty Grove all graciously shared their facilities and expertise to help us complete this project, but bear no responsibility for the errors in interpretation that remain. We thank Hazel Roberts and Pete Dahl for providing drafts of their work in progress. The manuscript was significantly improved due to excellent and helpful reviews by Steve Harlan and Pete Dahl.

REFERENCES CITED

Berg, R.B., 1979, Talc and chlorite deposits in Montana: Butte, Montana, Montana Bureau of Mines and Geology Memoir 45, 66 p.

Brady, J.B., Cheney, J.T., Duvall, M.L., Green, C., Kaufman, L., Kogut, A.I., Larson, A.C., and Vasquez, A., 1991, Metasomatic talc deposits in southwestern Montana: Geochemical evidence for deep circulation of water-rich fluids: Geological Society of America Abstracts with Programs, v. 23, no. 5, p. 263.

Brady, J.B., Burger, H., Cheney, J.T., King, J.T., Tierney, K.A., Peck, W.H., Poulsen, C.J., Cady, P., Lowell, J., Sincock, M.J., Archuleta, L.L., Fisher, R., and Jacob, L., 1994, Geochemical and $^{40}Ar/^{39}Ar$ evidence for terrane assembly in the Archean of southwestern Montana: Geological Society of America Abstracts with Programs, v. 26, no. 7, p. 232–233.

Brady, J.B., Cheney, J.T., Rhodes, A.L., Vasquez, A., Green, C., Duvall, M., Kogut, A., Kaufman, L., and Kovaric, D., 1998, Isotope geochemistry of Proterozoic talc occurrences in Archean marbles of the Ruby Mountains, southwest Montana, USA: Geological Materials Research, v. 2, p. 41.

Burger, H.R., 2004, General geology and tectonic setting of the Tobacco Root Mountains, Montana, *in* Brady, J.B., et al., eds., Precambrian geology of the Tobacco Root Mountains, Montana: Boulder, Colorado, Geological Society of America Special Paper 377, p. 89–104 (this volume).

Burger, H.R., Brady, J.B., Cheney, J.T., Harms, T.A., Mueller, P., Heatherington, A., and Wooden, J., 1999, Evidence for a major, Early Proterozoic orogenic event in the Tobacco Root Mountains of southwestern Montana: Geological Society of America Abstracts with Programs, v. 31, no. 7, p. 177–178.

Cheney, J.T., Harms, T.A., Brady, J.B., and Burger, H.R., 1999, Early Proterozoic $^{208}Pb/^{232}Th$ in situ ion probe dating of monazite from Archean metamorphic suites, Tobacco Root Mountains, Montana: Geological Society of America Abstracts with Programs, v. 31, no. 7, p. 178.

Cheney, J.T., Brady, J.B., Tierney, K.A., DeGraff, K.A., Mohlman, H.K., Frisch, J.D., Hatch, C.E., Steiner, M.L., Carmichael, S.K., Fisher, R.G.M., Tuit, C.B., Steffen, K.J., Cady, P., Lowell, J., Archuleta, L., Hirst, J., Wegmann, K.W., and Monteleone, B., 2004a, Proterozoic metamorphism of the Tobacco Root Mountains, Montana, *in* Brady, J.B., et al., eds., Precambrian geology of the Tobacco Root Mountains, Montana: Boulder, Colorado, Geological Society of America Special Paper 377, p. 105–129 (this volume).

Cheney, J.T., Webb, A.A.G., Coath, C.D., and McKeegan, K.D., 2004b, In situ ion microprobe $^{207}Pb/^{206}Pb$ dating of monazite from Precambrian metamorphic suites, Tobacco Root Mountains, Montana, *in* Brady, J.B., et al., eds., Precambrian geology of the Tobacco Root Mountains, Montana: Boulder, Colorado, Geological Society of America Special Paper 377, p. 151–179 (this volume).

Dahl, P.S., 1979, Comparative geothermometry based on major-element and oxygen isotope distributions in Precambrian metamorphic rocks from southwestern Montana: American Mineralogist, v. 64, no. 11-12, p. 1280–1293.

Dahl, P.S., Dorais, M.J., Roberts, H.J., Kelley, S.P., and Frei, R., 1999, Electron microprobe geochronometry of age-zoned monazite crystals in Archean metapelites from the Wyoming province: The nature of Pb rejuvenation and implications for regional tectonism: Geological Society of America Abstracts with Programs, v. 31, no. 7, p. 39.

Dahl, P.S., Hamilton, M.A., Terry, M.P., Roberts, H.J., Kelly, S.P., Frei, R., Jercinovic, M.J., and Williams, M.L., 2002, Comparative ion and electron microprobe dating of Wyoming province monazite, with tectonic and analytical implications: Geological Society of America Abstracts with Programs, v. 34, no. 1, p. A9.

Erslev, E.E., and Sutter, J.F., 1990, Evidence for Proterozoic mylonitization in the northwestern Wyoming province: Geological Society of America Bulletin, v. 102, p. 1681–1694.

Giletti, B.J., 1966, Isotopic ages from southwestern Montana: Journal of Geophysical Research, v. 71, no. 16, p. 4029–4036.

Giletti, B.J., 1968, Isotopic geochronology of Montana and Wyoming, *in* Hamilton, E.I., and Farquhar, R.M., eds., Radiometric dating for geologists: London, Interscience Publishers, p. 111–146.

Giletti, B.J., and Gast, P.W., 1961, Absolute age of Pre-Cambrian rocks in Wyoming and Montana: Annals of the New York Academy of Sciences, v. 91, p. 454–458.

Green, C.T., 1991, The influence of major faults on fluid movement and talc genesis in the Ruby Range, Montana [B.A. thesis]: Amherst, Massachusetts, Amherst College, 83 p.

Hames, W.E., and Cheney, J.T., 1997, On the loss of ^{40}Ar from muscovite during polymetamorphism: Innovative applications of ^{40}Ar/^{39}Ar microanalytical research: Geochimica et Cosmochimica Acta, v. 61, no. 18, p. 3863–3872.

Harlan, S.S., Heaman, L., LeCheminant, A.N., Premo, W.R., 2003, Gunbarrel mafic magmatic event: A key 780 Ma time marker for Rodinia plate reconstructions: Geology, v. 31, p. 1053–1056.

Harlan, S.S., Geissman, J.W., Snee, L.W., and Reynolds, R.L., 1996, Late Cretaceous remagnetization of Proterozoic mafic dikes, southern Highland Mountains, southwestern Montana: A paleomagnetic and ^{40}Ar/^{39}Ar study: Geological Society of America Bulletin, v. 108, p. 653–668.

Hayden, R.J., and Wehrenberg, J.P., 1959, Potassium-argon dating in western Montana: Geological Society of America Bulletin, v. 70, no. 12, Part 2, p. 1778–1779.

Hayden, R.J., and Wehrenberg, J.P., 1960, A^{40}-K^{40} dating of igneous and metamorphic rocks in western Montana: Journal of Geology, v. 68, no. 1, p. 94–97.

Jacob, L.J., 1994, A Geochemical and geochronological characterization of the Indian Creek Metamorphic Suite in the Noble Lake area, south-central Tobacco Root Mountains, southwestern Montana [B.A. thesis]: Northampton, Massachusetts, Smith College, 73 p.

Jaeger, J.C., 1964, Thermal effects of intrusions: Reviews of Geophysics, v. 2, p. 443–466.

James, H.L., 1990, Precambrian geology and bedded iron deposits of the southwestern Ruby Range, Montana: U.S. Geological Survey Professional Paper 1495, p. 39.

James, H.L., and Hedge, C.E., 1980, Age of the basement rocks of southwest Montana: Geological Society of America Bulletin, v. 91, p. 11–15.

King, J.T., 1994, A structural analysis of Archean gneisses in the Tobacco Root Mountains, southwestern Montana [B.A. thesis]: Amherst, Massachusetts, Amherst College, 44 p.

Kovaric, D.N., 1996, ^{40}Ar/^{39}Ar dating and Archean to Cretaceous thermal history of the Tobacco Root Mountains, southwestern Montana [B.A. thesis]: Northampton, Massachusetts, Smith College, 67 p.

Kovaric, D.N., Brady, J.B., Cheney, J.T., Grove, M., Jacob, L.J., and King, J.T., 1996, ^{40}Ar/ ^{39}Ar evidence for reheating events affecting basement rocks in the Tobacco Root, Ruby, and Highland Mountains, SW Montana: Geological Society of America Abstracts with Programs, v. 28, no. 7, p. 493.

Krogh, T.E., Kamo, S., and Hess, D.F., 1997, Wyoming province 3300+ Ma gneiss with 2400 Ma metamorphism, northwestern Tobacco Root Mountains, Madison County, Montana: Geological Society of America Abstracts with Programs, v. 29, no. 6, p. 408.

Larson, A.C., 1991, Talc Formation in the Regal-Keystone Mine of the Ruby Range, Dillon, MT [B.A. thesis]: Northampton, Massachusetts, Smith College, 57 p.

Lux, D.R., Gibson, D., and Hamilton, P.J., 1989, Geochronology of the Songo Pluton, western Maine, *in* Tucker, R.D., and Marvinney, R.G., eds., Igneous and metamorphic geology: Studies in Maine geology: Augusta, Maine Geological Survey, p. 101–114.

Marvin, R.F., and Dobson, S.W., 1979, Radiometric ages; compilation B, U.S. Geological Survey: Isochron/West, v. 26, p. 3–30.

McDougall, I., and Harrison, T.M., 1988, Geochronology and thermochronology by the ^{40}Ar/^{39}Ar method: Oxford Monographs on Geology and Geophysics, v. 9, p. 212.

Mogk, D.W., Burger, H.R., Mueller, P.A., D'Arcy, K., Heatherington, A.L., Wooden, J.L., Abeyta, R.L., Martin, J., and Jacob, L.J., 2004, Geochemistry of quartzofeldspathic gneisses and metamorphic mafic rocks of the Indian Creek and Pony–Middle Mountain Metamorphic Suites, Tobacco Root Mountains, Montana, *in* Brady, J.B., et al., eds., Precambrian geology of the Tobacco Root Mountains, Montana: Boulder, Colorado, Geological Society of America Special Paper 377, p. 15–46 (this volume).

Mueller, P.A., and Cordua, W.S., 1976, Rb-Sr whole rock age of gneisses from the Horse Creek area, Tobacco Root Mountains, Montana: Isochron/West, v. 16, p. 33–36.

Mueller, P.A., Wooden, J.L., Nutman, A.P., and Mogk, D.W., 1998, Early Archean crust in the northern Wyoming province: Evidence from U-Pb ages of detrital zircons: Precambrian Research, v. 91, no. 3–4, p. 295–307.

Mueller, P.A., Burger, H.R., Wooden, J.L., Heatherington, A.L., Mogk, D.W., and D'Arcy, K., 2004, Age and evolution of the Precambrian crust of the Tobacco Root Mountains, Montana, *in* Brady, J.B., et al., eds., Precambrian geology of the Tobacco Root Mountains, Montana: Boulder, Colorado, Geological Society of America Special Paper 377, p. 179–200 (this volume).

O'Neill, J.M., Duncan, M.S., and Zartman, R.E., 1988a, An early Proterozoic gneiss dome in the Highland Mountains, southwestern Montana, *in* Lewis, S.E., and Berg, R.B., eds., Precambrian and Mesozoic plate margins; Montana, Idaho, and Wyoming: Montana Bureau of Mines and Geology Special Publication 96, p. 81–88.

O'Neill, J.M., Duncan, M.S., and Zartman, R.E., 1988b, An Early Proterozoic gneiss dome, Highland Mountains, Montana: Geological Society of America Abstracts with Programs, v. 20, no. 6, p. 460–461.

Quidelleur, X., Grove, M., Lovera, O.M., Harrison, T.M., Yin, A., and Ryerson, F.J., 1997, Thermal evolution and slip history of the Renbu Zedong Thrust, southeastern Tibet: Journal of Geophysical Research, B, Solid Earth and Planets, v. 102, no. 2, p. 2659–2679.

Roberts, H., Dahl, P., Kelley, S., and Frei, R., 2002, New ^{207}Pb-^{206}Pb and ^{40}Ar-^{39}Ar ages from SW Montana, USA: Constraints on the Proterozoic and Archaean tectonic and depositional history of the Wyoming province: Precambrian Research, v. 117, no. 1–2, p. 119–143.

Tierney, K.A., 1994, The origin and evolution of the Spuhler Peak Formation along the western ridge of Thompson Peak, Tobacco Root Mountains, Montana [B.A. thesis]: Amherst, Massachusetts, Amherst College, 87 p.

Vitaliano, C.J., Burger, H.R., Cordua, W.S., Hanley, T.B., Hess, D.F., and Root, F.K., 1979a, Geologic map of southern Tobacco Root Mountains, Madison County, Montana: Geological Society of America Map and Chart Series MC31, scale 1:62,500, 1 sheet, 8 p. text.

Vitaliano, C.J., Cordua, W.S., Burger, H.R., Hanley, T.B., Hess, D.F., and Root, F.K., 1979b, Geology and structure of the southern part of the Tobacco Root Mountains, southwestern Montana; map summary: Geological Society of America Bulletin, v. 90, p. 1712–1715.

Vitaliano, C.J., Kish, S., and Towell, D.G., 1980, Potassium-argon dates and strontium isotopic values for rocks of the Tobacco Root batholith, southwestern Montana: Isochron/West, v. 28, p. 13–15.

Wooden, J.L., Vitaliano, C.J., Koehler, S.W., and Ragland, P.C., 1978, The Late Precambrian mafic dikes of the southern Tobacco Root Mountains, Montana: Geochemistry, Rb-Sr geochronology and relationship to Belt tectonics: Canadian Journal of Earth Sciences, v. 15, no. 4, p. 467–479.

Manuscript Accepted by the Society September 12, 2003

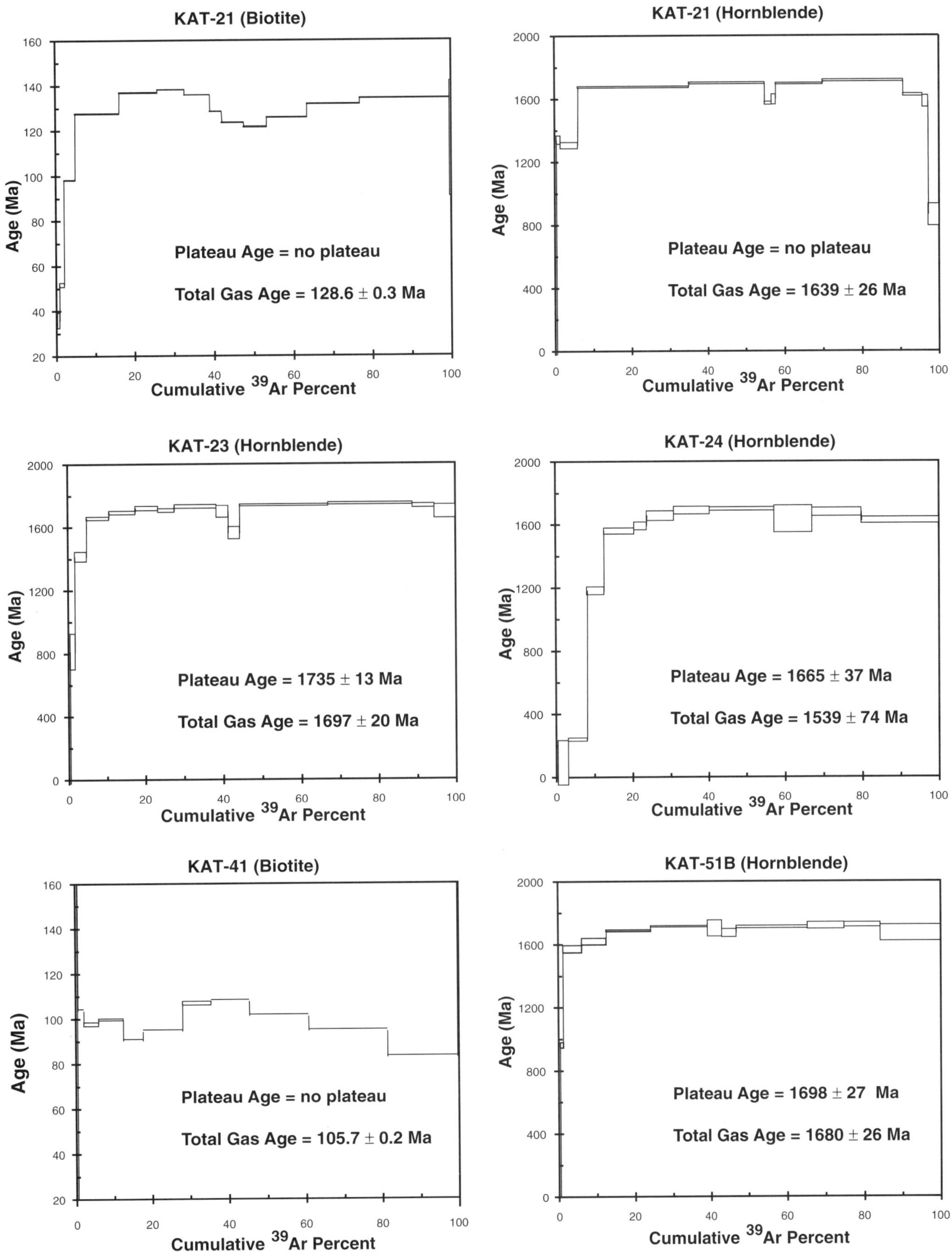

Figure A1. $^{40}Ar/^{39}Ar$ step-heating age spectra determined at University of California at Los Angeles.

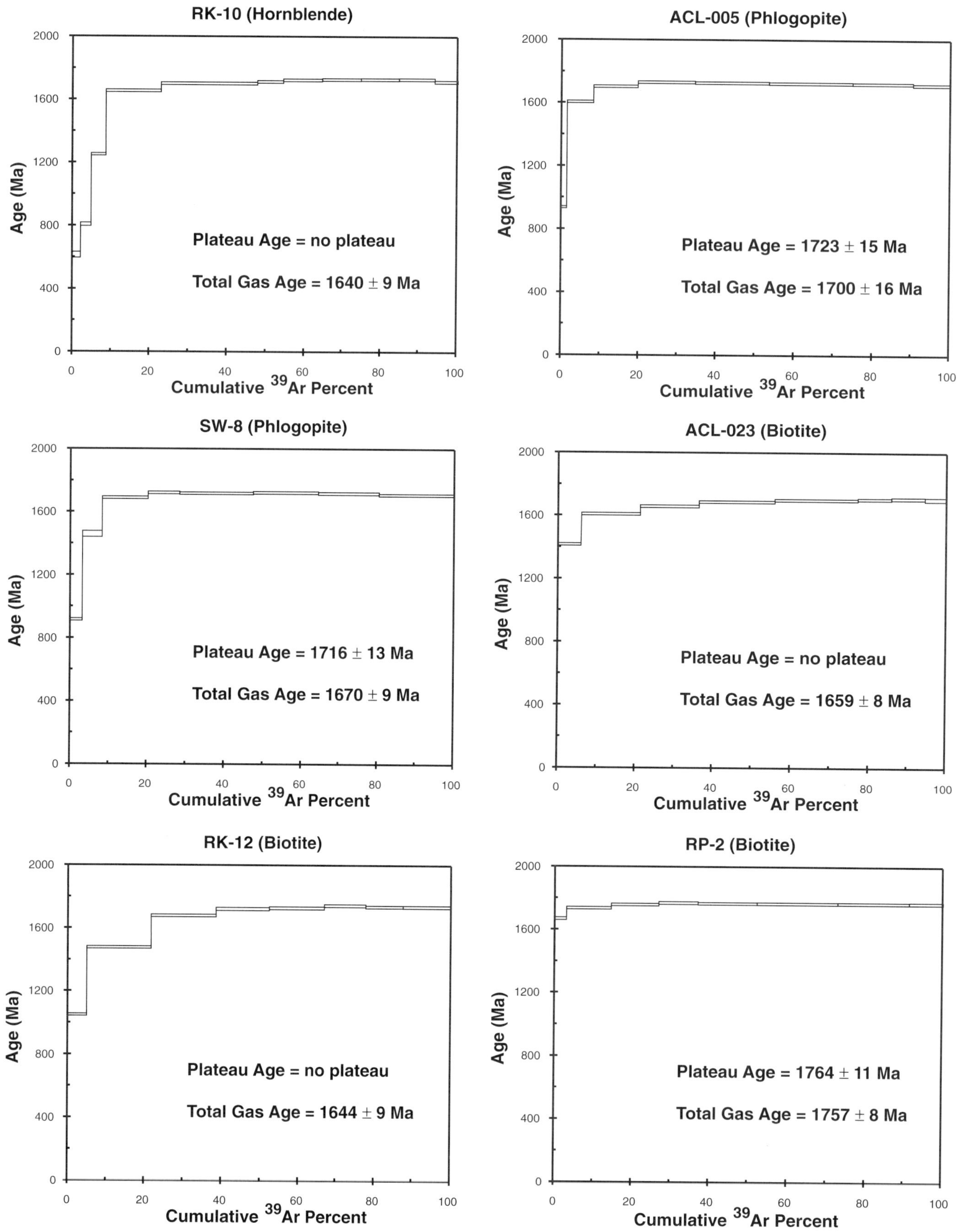

Figure A2. $^{40}Ar/^{39}Ar$ step-heating age spectra determined at the University of Maine.

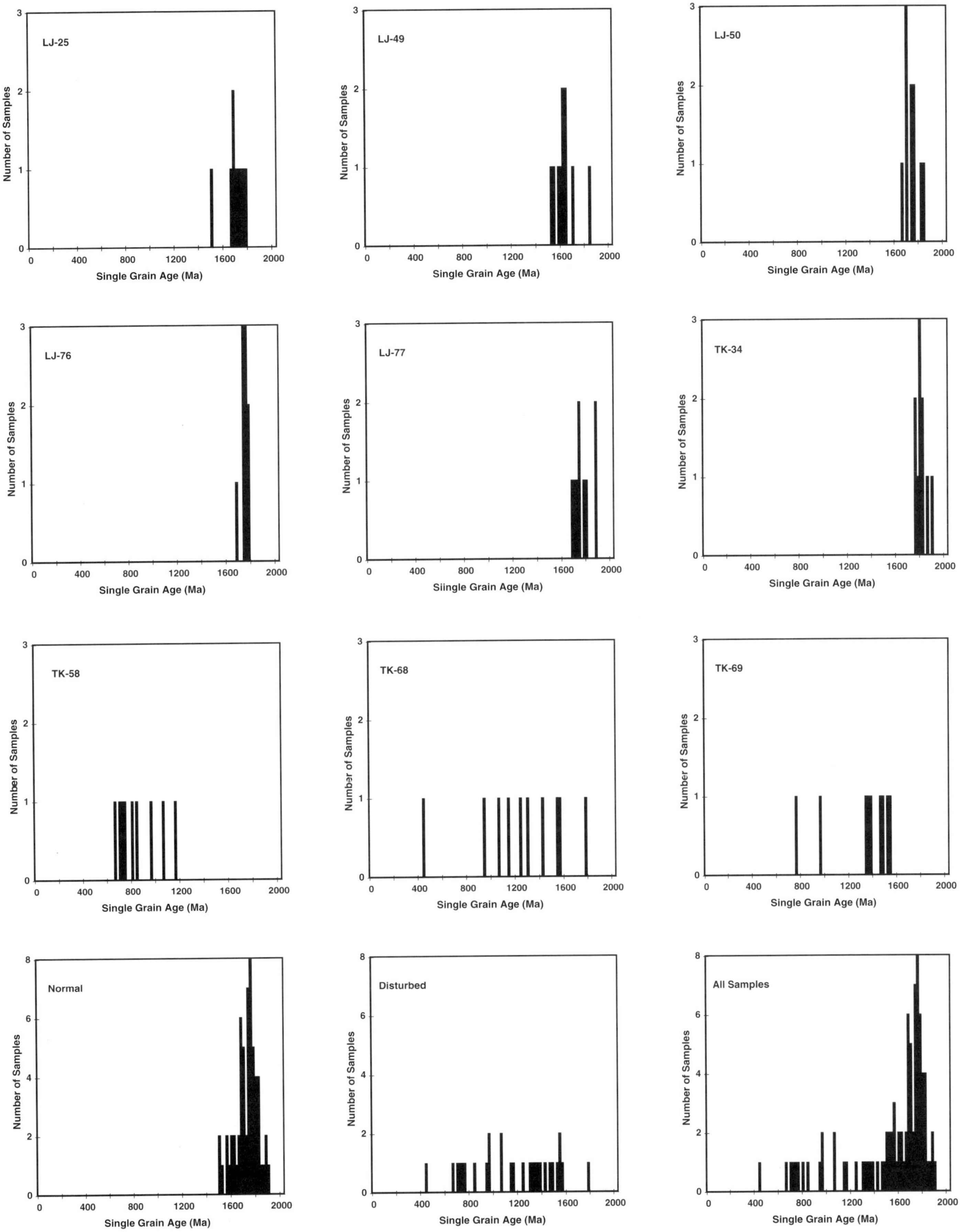

Figure A3. Single-grain UV laser microprobe $^{40}Ar/^{39}Ar$ total gas ages determined at Massachusetts Institute of Technology.

TABLE A1. UCLA Ar DATA

T (°C)	$^{40}Ar/^{39}Ar$	$^{40}Ar^*/^{39}Ar_K$	$^{37}Ar/^{39}Ar$	$^{36}Ar/^{39}Ar$	K/Ca	$\%^{39}Ar$	$\%^{40}Ar^*$	Age (Ma)	± 1σ
KAT-21 (Biotite, 19.8 mg) J = 0.006614 (± 0.000023)[†]									
500	21.82	3.140	5.43E-02	6.31E-02	9.1	0.6	14.4	37.1	3.9
600	12.70	4.485	3.49E-02	2.77E-02	14.2	1.8	35.3	52.7	1.0
680	12.66	8.490	8.40E-03	1.40E-02	58.9	4.7	67.1	98.5	0.5
750	12.20	11.07	6.86E-03	3.74E-03	72.1	16.0	90.7	127.4	0.7
800	12.31	11.90	7.05E-03	1.32E-03	70.2	25.8	96.6	136.5	0.7
840	12.33	12.01	6.21E-03	1.01E-03	79.7	32.6	97.4	137.8	0.7
880	12.22	11.81	6.85E-03	1.29E-03	72.2	38.9	96.7	135.6	0.7
920	11.58	11.15	5.76E-03	1.37E-03	85.9	41.9	96.3	128.3	0.6
960	11.27	10.71	6.35E-03	1.80E-03	77.9	47.4	95.1	123.4	0.6
990	11.03	10.54	5.71E-03	1.58E-03	86.7	53.1	95.6	121.5	0.6
1020	11.28	10.91	5.29E-03	1.17E-03	93.6	63.4	96.7	125.6	0.6
1000	11.64	11.44	7.66E-03	5.89E-04	64.6	76.8	98.3	131.5	0.6
1150	11.79	11.67	2.12E-02	3.45E-04	23.3	99.6	98.9	134.0	0.7
1350	165.1	10.07	7.19E-01	5.25E-01	0.7	100.0	6.10	116.2	29.5
Total Gas		11.18						128.6	0.3
Plateau								(no plateau)	
KAT-21 (Hornblende, 25.2 mg) J = 0.006613 (± 0.000023)									
800	195.0	167.0	8.32E-01	9.52E-02	0.588	1.1	85.6	1342.4	36.3
950	166.0	160.9	4.19E+00	1.99E-02	0.118	5.6	96.6	1307.3	28.4
1000	230.4	230.5	5.38E+00	3.59E-03	0.092	34.8	99.7	1670.8	8.8
1020	235.7	235.7	5.36E+00	4.30E-03	0.092	54.7	99.6	1695.2	9.3
1040	224.8	209.8	5.01E+00	5.43E-02	0.098	56.5	93.0	1569.9	13.7
1060	236.4	214.6	5.39E+00	7.78E-02	0.092	57.6	90.4	1593.9	44.0
1100	237.7	234.7	5.82E+00	1.47E-02	0.085	69.8	98.3	1690.2	9.0
1150	240.7	238.9	5.73E+00	1.07E-02	0.086	90.6	98.8	1709.9	9.2
1200	233.8	220.0	5.29E+00	5.06E-02	0.093	95.7	93.8	1620.3	13.5
1250	305.1	211.6	5.14E+00	3.20E-01	0.096	97.2	69.1	1579.0	50.7
1350	425.0	93.7	2.44E+00	1.12E+00	0.204	99.5	22.0	870.1	86.0
1450	5429	-86.4	2.03E+00	1.87E+01	0.244	100.0	−1.6	−1527.5	17131
Total Gas		223.8						1639	26
Plateau								(no plateau)	
KAT-23A (Hornblende, 22.5 mg) J = 0.006612 (± 0.000017)									
800	198.0	87.25	2.49E+00	3.76E-01	0.200	1.3	44.0	821.8	146.7
950	195.8	180.1	9.23E+00	5.94E-02	0.053	4.4	91.4	1414.8	43.0
980	232.7	226.6	1.15E+01	2.97E-02	0.043	10.3	96.6	1651.8	14.7
1000	241.3	234.1	1.17E+01	3.38E-02	0.042	17.3	96.2	1687.3	16.6
1010	249.3	239.6	1.21E+01	4.23E-02	0.041	23.2	95.3	1713.2	20.8
1020	245.7	237.0	1.20E+01	3.90E-02	0.041	27.4	95.7	1700.8	15.5
1040	245.3	242.6	1.24E+01	1.94E-02	0.040	38.2	98.0	1726.6	17.8
1060	253.6	235.3	1.24E+01	7.19E-02	0.040	41.2	92.0	1692.9	52.4
1100	232.9	207.9	1.11E+01	9.25E-02	0.044	44.1	88.6	1560.5	54.3
1150	245.8	244.0	1.27E+01	1.65E-02	0.038	66.9	98.4	1733.1	12.4
1200	251.7	246.0	1.26E+01	2.97E-02	0.039	88.8	96.9	1742.3	11.4
1250	300.1	242.7	1.24E+01	2.04E-01	0.040	94.6	80.2	1727.2	19.6
1350	396.5	234.6	1.21E+01	5.57E-01	0.041	100.0	58.7	1689.9	57.3
Total Gas		236.1						1697	20
Plateau						50.5		1735	13
KAT-24A (Hornblende, 19.9 mg) J = 0.006612 (± 0.000017)									
800	89.05	9.8	2.03E+00	2.68E-01	0.244	2.7	11.0	113.8	170.5
950	40.10	23.1	3.38E+00	5.85E-02	0.147	7.8	57.4	256.2	13.1
980	154.9	140.4	1.78E+01	5.94E-02	0.027	12.2	89.5	1184.4	33.4
1000	219.5	207.2	2.37E+01	5.88E-02	0.021	20.1	92.9	1556.8	26.5
1015	240.4	213.4	2.39E+01	1.09E-01	0.020	23.6	87.3	1587.9	34.5
1030	249.3	226.2	2.53E+01	9.79E-02	0.019	30.6	89.1	1649.9	42.9
1045	263.2	233.3	2.65E+01	1.22E-01	0.018	40.1	87.0	1683.8	33.6
1080	260.4	235.2	2.71E+01	1.07E-01	0.018	57.1	88.6	1692.6	17.8
1120	914.0	74.5	1.29E+01	2.85E+00	0.038	58.4	8.1	722.4	4698.8
1160	529.3	222.1	2.81E+01	1.06E+00	0.017	66.9	41.1	1630.0	108.6
1200	413.9	231.0	2.69E+01	6.40E-01	0.018	79.8	54.8	1672.6	34.9
1350	467.7	220.2	2.47E+01	8.56E-01	0.020	100.0	46.3	1621.2	25.6
Total Gas		203.7						1539	74
Plateau						76.4		1665	37

(continued)

TABLE A1. UCLA Ar DATA (*continued*)

T (°C)	$^{40}Ar/^{39}Ar$	$^{40}Ar^{*}/^{39}Ar_K$	$^{37}Ar/^{39}Ar$	$^{36}Ar/^{39}Ar$	K/Ca	$\%^{39}Ar$	$\%^{40}Ar^{*}$	Age (Ma)	± 1σ
KAT-41 (Biotite, 20.3 mg) J = 0.006608 (± 0.000017)									
500	170.4	100.51	1.62E-01	2.37E-01	3.1	0.1	59.0	918.9	26.2
680	15.97	9.594	1.19E-02	2.15E-02	41.5	1.7	60.1	110.9	0.7
750	11.20	9.101	4.65E-03	7.03E-03	106.4	5.6	81.2	105.4	0.5
800	10.25	9.311	2.65E-03	3.11E-03	186.9	12.1	90.8	107.7	0.5
840	9.175	8.562	1.96E-03	1.99E-03	251.9	17.3	93.3	99.3	0.5
880	9.483	8.881	2.11E-03	1.95E-03	234.7	27.7	93.7	102.9	0.5
920	10.37	9.861	1.91E-03	1.66E-03	258.4	35.2	95.0	113.9	0.6
960	10.54	9.996	1.71E-03	1.75E-03	289.9	45.1	94.9	115.4	0.6
1020	9.868	9.460	2.14E-03	1.30E-03	231.5	60.5	95.9	109.4	0.5
1060	9.083	8.837	2.30E-03	7.51E-04	215.1	81.1	97.3	102.4	0.5
1120	8.096	7.904	3.41E-03	5.67E-04	144.9	95.9	97.6	91.8	0.5
1200	8.202	7.871	4.02E-03	1.04E-03	123.2	99.6	96.0	91.5	0.5
1350	17.15	14.77	6.88E-02	8.00E-03	7.2	100.0	86.1	168.0	2.4
Total Gas		9.13						105.7	0.2
Plateau								(no plateau)	
KAT-51 (Hornblende, 19.5 mg) J = 0.006603 (± 0.000017)									
800	194.5	107.9	4.08E+00	2.95E-01	0.120	0.8	55.3	970.3	20.8
950	215.7	210.0	1.41E+01	2.95E-02	0.035	5.7	96.4	1569.4	31.9
980	223.0	219.5	1.28E+01	2.13E-02	0.038	12.1	97.6	1616.2	28.8
1000	234.0	233.4	1.33E+01	1.25E-02	0.037	23.9	98.8	1682.7	8.5
1020	239.0	239.0	1.36E+01	1.10E-02	0.036	38.9	99.0	1708.4	8.3
1040	245.0	236.4	1.35E+01	4.01E-02	0.036	42.8	95.6	1696.4	69.8
1080	240.2	230.1	1.38E+01	4.50E-02	0.035	46.7	94.9	1666.9	34.7
1120	239.8	238.0	1.37E+01	1.72E-02	0.036	65.3	98.3	1703.9	13.0
1160	248.5	239.8	1.34E+01	4.01E-02	0.037	74.7	95.6	1712.4	30.1
1200	258.0	241.1	1.34E+01	6.81E-02	0.036	84.1	92.6	1718.2	21.0
1350	697.1	230.0	1.28E+01	1.59E+00	0.038	100.0	32.7	1666.6	63.5
Total Gas		232.8						1680	26
Plateau						87.9		1698	27
JBB-AC-22 (Muscovite, 1.0 mg) J = 0.006598 (± 0.000049)									
500	155.6	148.2	7.42E-02	2.49E-02	6.7	6.1	95.3	1225.4	16.3
550	169.7	167.9	1.70E-02	6.05E-03	29.2	15.6	98.9	1339.6	12.3
600	168.9	167.3	1.66E-02	5.34E-03	29.9	46.1	99.1	1336.3	9.4
650	177.5	176.5	1.11E-02	3.33E-03	44.8	66.4	99.4	1387.5	12.1
700	177.1	176.1	4.31E-03	3.03E-03	114.7	79.8	99.5	1385.6	9.8
800	176.8	174.8	1.10E-02	6.49E-03	45.0	92.6	98.9	1378.3	12.0
900	133.8	130.1	3.35E-02	1.25E-02	14.7	97.2	97.2	1113.4	18.0
1100	92.99	85.80	5.02E-01	2.45E-02	1.0	100.0	92.2	805.5	16.0
Total Gas		166.2						1330	9
Plateau								(no plateau)	

†Fish Canyon sanidine standard (27.8 Ma) used for all samples.

TABLE A2. UNIVERSITY OF MAINE Ar DATA

T (°C)	$^{40}Ar/^{39}Ar$	$^{37}Ar/^{39}Ar$	$^{36}Ar/^{39}Ar$	K/Ca	$\%^{39}Ar$	$\%^{40}Ar^*$	Age (Ma)	± 1σ
RK-10 (Hornblende) J = 0.00847 †								
730	126.84	2.5840	0.2702	0.19	2.2	37.2	607.6	19
850	115.40	3.8860	0.1678	0.13	2.7	57.3	804.0	11
940	139.76	11.2900	0.0809	0.04	3.8	83.6	1247.7	8
1030	180.77	20.0740	0.0263	0.02	14.3	96.7	1653.4	9
1080	186.62	21.1150	0.0207	0.02	24.8	97.7	1700.7	9
1110	187.01	21.5850	0.0163	0.02	6.7	98.4	1711.5	8
1140	188.82	21.5630	0.0165	0.02	10.3	98.4	1722.1	8
1170	189.87	21.5020	0.0181	0.02	10.1	98.2	1725.5	8
1200	190.21	21.5170	0.0185	0.02	10.0	98.1	1726.8	8
1230	190.00	21.4560	0.0173	0.02	9.3	98.3	1727.5	8
Fused	189.49	21.2260	0.0232	0.02	6.0	97.4	1713.7	9
Total Gas					100.0		1640	9
Plateau							(no plateau)	
ACL-005 (Phlogopite) J = 0.00833								
700	87.99	0.0909	0.0240	44.75	1.5	91.9	929.2	7
820	172.86	0.0099	0.0055	49.59	6.8	99.0	1599.0	8
940	188.40	0.0093	0.0014	52.96	11.4	99.8	1700.0	8
1020	192.86	0.0112	0.0004	43.67	14.8	99.9	1727.5	8
1100	192.96	0.0171	0.0001	28.66	19.0	100.0	1724.0	8
1160	191.79	0.0110	0.0002	44.68	21.3	100.0	1721.8	8
1220	191.59	0.0098	0.0001	49.96	15.6	100.0	1720.7	8
Fused	190.58	0.0254	0.0014	19.30	9.6	99.8	1712.6	8
Total Gas					100.0		1700	16
Plateau					70.7		1724	15
SW-8 (Phlogopite) J = 0.00830								
700	85.29	0.0341	0.0215	14.39	3.2	92.5	908.9	8
820	150.74	0.0289	0.0046	16.96	5.1	99.1	1454.9	19
940	187.19	0.0162	0.0033	30.32	11.9	99.5	1685.8	8
1020	191.98	0.0151	0.0010	32.44	8.2	99.8	1717.6	8
1100	191.13	0.0176	0.0004	27.84	19.0	99.9	1713.7	8
1160	191.87	0.0362	0.0004	13.54	17.1	99.9	1718.0	8
1220	191.11	0.0236	0.0003	20.77	15.9	99.9	1713.8	9
Fused	189.63	0.0241	0.0004	20.34	19.7	99.9	1704.9	8
Total Gas					100.0		1670	9
Plateau					60.2		1716	13

(continued)

TABLE A2. UNIVERSITY OF MAINE Ar DATA (*continued*)

T (°C)	$^{40}Ar/^{39}Ar$	$^{37}Ar/^{39}Ar$	$^{36}Ar/^{39}Ar$	K/Ca	$\%^{39}Ar$	$\%^{40}Ar^*$	Age (Ma)	± 1σ
ACL-0023 (Biotite) J = 0.00841								
700	144.04	0.0311	0.0087	15.74	5.9	98.2	1413.6	8
820	171.53	0.0271	0.0030	18.10	15.2	99.5	1605.1	8
940	178.82	0.0203	0.0003	24.16	15.2	99.9	1654.9	8
1020	183.56	0.0174	0.0003	28.14	19.5	99.9	1683.4	8
1100	185.76	0.0222	0.0006	22.12	21.2	99.9	1695.9	8
1160	186.82	0.0334	0.0010	14.69	8.7	99.8	1701.5	8
1220	187.43	0.0268	0.0006	18.26	8.7	99.9	1705.7	9
Fused	188.62	0.0466	0.0009	10.51	5.6	99.8	1712.3	13
Total Gas					100.0		1659	8
Plateau							(no plateau)	
RK-12 (Biotite) J = 0.00836								
700	96.76	0.0608	0.0102	8.06	4.9	96.9	1044.0	7
820	152.50	0.0498	0.0046	9.83	16.8	99.1	1473.8	8
940	183.62	0.0340	0.0016	14.42	16.8	99.7	1675.5	8
1020	190.55	0.0310	0.0004	15.78	13.8	99.9	1718.3	10
1100	191.58	0.0190	0.0004	25.82	14.6	99.9	1724.3	9
1160	194.30	0.0212	0.0004	23.16	10.8	99.9	1740.0	9
1220	193.20	0.0218	0.0005	22.46	9.9	99.9	1733.5	8
Fused	193.16	0.0338	0.0009	14.49	12.5	99.8	1732.5	9
Total Gas					100.0		1644	9
Plateau							(no plateau)	
RP-2 (Biotite) J = 0.00843								
700	183.93	0.0243	0.0131	20.18	3.0	97.9	1665.8	9
820	192.28	0.0243	0.0030	20.18	11.6	99.5	1733.1	8
940	195.40	0.0491	0.0004	9.98	12.3	99.9	1755.5	8
1020	197.55	0.0359	0.0010	13.65	10.1	99.8	1766.8	9
1100	196.77	0.0421	0.0003	11.64	15.2	99.9	1763.6	8
1160	196.74	0.0470	0.0003	10.43	20.6	99.9	1763.4	8
1220	196.84	0.0413	0.0003	11.86	18.5	99.9	1764.0	8
Fused	196.88	0.0939	0.0003	5.22	8.8	99.9	1764.3	9
Total Gas					100.0		1757	8
Plateau					64.4		1764	11

[†]Muscovite SBG-7 standard (240.9 Ma), calibrated to MMhb-1(520.4 Ma), used for all samples.

TABLE A3. MIT SINGLE-GRAIN Ar DATA†

Sample	$^{40}Ar/^{39}Ar$	$^{40}Ar^{*}/^{39}Ar_{K}$	$^{37}Ar/^{39}Ar$	$^{36}Ar/^{39}Ar$	K/Ca	$\%^{39}Ar_{K}$	$\%^{40}Ar^{*}$	Age (Ma)	± 1σ
LJ-25	114.8	114.4	4.06E-05	1.135E-03	0.1265	0.192	99.71	1725.3	2.4
LJ-25	114.2	113.7	5.01E-05	1.135E-03	0.1032	0.326	99.71	1719.6	6.0
LJ-25	91.9	91.2	7.83E-05	1.771E-03	0.1281	0.545	99.43	1486.1	7.0
LJ-25	117.1	116.5	1.13E-04	2.472E-03	0.0975	0.769	99.38	1744.3	1.9
LJ-25	102.2	102.0	5.58E-05	1.129E-03	0.1030	0.927	99.67	1599.4	0.1
LJ-25	109.5	108.8	9.47E-05	2.579E-03	0.1292	1.092	99.30	1669.2	2.3
LJ-25	121.4	119.4	2.89E-04	6.049E-03	0.0897	1.161	98.53	1774.4	6.9
LJ-25	111.6	110.8	2.99E-04	3.465E-03	0.0540	1.258	99.08	1687.4	7.3
LJ-25	109.1	108.3	8.13E-05	2.018E-03	0.1182	1.373	99.45	1666.5	5.3
LJ-50	116.8	116.7	3.52E-05	6.178E-04	0.0780	1.565	99.84	1746.6	6.5
LJ-50	111.9	111.9	1.73E-05	3.119E-04	0.0838	1.600	99.92	1699.5	7.5
LJ-50	107.2	105.9	2.32E-04	4.025E-03	0.0843	1.704	98.89	1641.8	15.1
LJ-50	116.2	115.9	−2.93E-05	−5.907E-04	0.0903	1.766	100.15	1743.7	13.1
LJ-50	125.0	125.2	−3.65E-05	−6.897E-04	0.0787	1.821	100.16	1826.4	16.2
LJ-50	115.0	115.9	−8.55E-05	−1.451E-03	0.0768	1.890	100.37	1734.7	13.1
LJ-50	114.0	114.1	−1.94E-05	−3.354E-04	0.0790	1.932	100.09	1722.0	7.7
LJ-50	122.4	122.7	−1.29E-05	−2.388E-04	0.0788	2.078	100.06	1801.7	6.5
LJ-50	111.4	111.4	3.02E-05	5.073E-04	0.0784	2.222	99.87	1693.9	3.0
LJ-50	110.8	110.5	6.17E-05	1.070E-03	0.0813	2.398	99.71	1686.7	4.2
LJ-76	117.6	117.3	2.11E-05	6.325E-04	0.1323	2.592	99.84	1753.9	6.3
LJ-76	109.9	109.2	8.70E-05	2.272E-03	0.1236	2.726	99.39	1673.5	7.3
LJ-76	115.0	114.5	6.77E-05	1.940E-03	0.1294	2.910	99.50	1725.6	10.2
LJ-76	117.9	117.1	6.77E-05	1.980E-03	0.1291	3.179	99.50	1753.1	4.0
LJ-76	115.8	115.6	1.37E-05	4.096E-04	0.1340	3.367	99.90	1737.2	4.5
LJ-76	115.9	115.5	3.55E-05	1.114E-03	0.1408	3.666	99.72	1736.0	2.4
LJ-76	117.9	117.6	3.89E-05	1.137E-03	0.1288	3.877	99.72	1755.7	1.7
LJ-76	118.4	118.3	1.88E-05	5.433E-04	0.1271	4.118	99.86	1761.7	1.9
LJ-76	118.9	118.4	5.29E-05	1.470E-03	0.1215	4.306	99.63	1763.5	12.4
LJ-77	116.1	115.6	9.15E-05	1.809E-03	0.0885	4.553	99.54	1736.4	5.8
LJ-77	130.5	130.2	3.06E-05	7.076E-04	0.0920	4.855	99.84	1872.8	12.1
LJ-77	111.7	110.7	1.65E-04	3.159E-03	0.0891	5.057	99.16	1689.6	10.6
LJ-77	120.9	120.1	1.34E-04	2.903E-03	0.0931	5.206	99.29	1778.4	8.4
LJ-77	109.2	109.2	4.11E-05	6.641E-04	0.0769	5.380	99.82	1672.1	3.2
LJ-77	112.2	111.9	4.00E-05	7.239E-04	0.0838	5.573	99.81	1701.5	56.0
LJ-77	114.7	114.2	6.30E-05	1.159E-03	0.0834	5.718	99.70	1724.9	0.3
LJ-77	131.0	130.3	8.60E-05	1.756E-03	0.0811	5.851	99.60	1874.5	4.2
LJ-77	121.1	120.7	8.34E-05	1.582E-03	0.0815	6.097	99.61	1784.7	0.5
LJ-49	98.1	96.7	5.50E-04	5.522E-03	0.0532	6.222	98.34	1542.0	6.5
LJ-49	124.9	123.3	4.17E-04	4.580E-03	0.0458	6.313	98.92	1811.2	0.3
LJ-49	105.3	103.1	6.22E-04	7.330E-03	0.0582	6.408	97.94	1612.5	17.3
LJ-49	98.7	97.9	2.48E-04	2.593E-03	0.0551	6.512	99.22	1557.5	26.6
LJ-49	110.6	109.1	4.86E-04	5.080E-03	0.0491	6.576	98.64	1672.6	10.8
LJ-49	94.5	92.0	7.94E-04	8.399E-03	0.0582	6.637	97.37	1493.3	7.5
LJ-49	104.7	103.0	5.14E-04	5.590E-03	0.0540	6.678	98.42	1611.4	31.1
LJ-49	95.2	92.6	8.06E-04	7.891E-03	0.0535	6.750	97.55	1503.1	19.7
LJ-49	115.8	101.3	6.70E-04	6.235E-03	0.0470	6.841	98.41	1593.7	4.3
LJ-49	99.4	99.0	1.29E-04	2.536E-03	0.0577	6.911	99.25	1565.3	7.0

(continued)

TABLE A3. MIT SINGLE-GRAIN Ar DATA (*continued*)

Sample	$^{40}Ar/^{39}Ar$	$^{40}Ar^*/^{39}Ar_K$	$^{37}Ar/^{39}Ar$	$^{36}Ar/^{39}Ar$	K/Ca	$\%^{39}Ar_K$	$\%^{40}Ar^*$	Age (Ma)	± 1σ
TK-68	122.2	118.2	2.02E-03	1.313E-02	0.0276	99.156	96.82	1762.3	19.0
TK-68	148.6	83.9	2.64E-03	1.770E-02	0.0401	99.200	96.48	1401.7	3.6
TK-68	98.2	97.1	6.57E-04	4.631E-03	0.0373	99.279	98.61	1545.2	0.5
TK-68	52.4	48.6	2.32E-03	1.284E-02	0.0550	99.319	92.75	935.9	10.6
TK-68	77.4	74.6	1.48E-03	9.237E-03	0.0420	99.387	96.47	1290.3	15.5
TK-68	63.0	61.8	5.75E-04	3.654E-03	0.0524	99.550	98.29	1126.5	1.8
TK-68	19.7	19.3	2.17E-04	1.569E-03	0.1906	99.925	97.65	431.0	7.1
TK-68	63.2	55.7	6.13E-03	2.533E-02	0.0340	99.943	88.16	1040.3	31.2
TK-68	70.4	69.3	6.42E-04	3.975E-03	0.0457	99.982	98.33	1222.7	7.1
TK-68	100.7	94.5	2.97E-03	2.038E-02	0.0354	100.000	94.02	1522.1	38.8
TK-58	42.6	39.9	2.14E-03	9.573E-03	0.0545	0.000	93.36	798.0	14.7
TK-58	39.3	34.3	4.99E-03	1.707E-02	0.0453	4.077	87.16	706.4	9.2
TK-58	72.1	64.4	5.24E-03	2.602E-02	0.0358	5.240	89.34	1159.8	23.3
TK-58	39.1	33.1	8.00E-03	2.030E-02	0.0337	6.977	84.66	687.1	18.5
TK-58	52.9	49.8	1.56E-03	1.040E-02	0.0655	7.824	94.19	954.8	35.4
TK-58	32.9	30.9	1.73E-03	6.459E-03	0.0589	10.120	94.21	650.8	14.9
TK-58	39.0	35.1	2.31E-03	1.321E-02	0.0762	14.724	89.99	721.2	6.4
TK-58	58.9	55.9	1.28E-03	1.021E-02	0.0706	19.211	94.88	1043.2	9.5
TK-58	44.0	41.7	1.13E-03	7.848E-03	0.0823	24.219	94.72	828.5	4.8
TK-69	80.7	79.2	5.09E-04	4.895E-03	0.0620	26.832	98.21	1346.1	8.6
TK-69	49.7	49.1	1.67E-04	2.286E-03	0.1432	35.915	98.64	943.4	1.0
TK-69	88.5	87.6	2.06E-04	2.516E-03	0.0718	38.851	99.16	1446.2	9.9
TK-69	95.5	94.5	1.65E-04	3.077E-03	0.1018	44.295	99.05	1521.7	0.1
TK-69	36.6	36.2	8.56E-05	1.228E-03	0.2037	54.171	99.01	740.6	2.8
TK-69	91.1	90.5	1.95E-04	2.610E-03	0.0764	57.572	99.15	1474.6	9.4
TK-69	94.7	94.2	9.48E-05	1.374E-03	0.0796	63.103	99.57	1518.2	2.6
TK-69	78.6	77.4	3.42E-04	3.809E-03	0.0738	67.420	98.57	1325.5	3.9
TK-69	81.4	80.5	2.26E-04	2.999E-03	0.0850	73.365	98.91	1361.5	2.0
TK-34	123.6	121.0	1.01E-03	8.801E-03	0.0368	75.604	97.90	1788.1	11.7
TK-34	120.7	117.6	1.52E-03	1.217E-02	0.0346	77.939	97.02	1750.6	12.9
TK-34	122.8	122.1	2.84E-04	2.412E-03	0.0360	83.393	99.42	1797.7	3.7
TK-34	124.5	123.2	4.18E-04	3.861E-03	0.0386	86.435	99.08	1809.5	1.1
TK-34	125.0	122.8	9.46E-04	8.610E-03	0.0378	89.159	97.96	1801.2	12.5
TK-34	128.8	127.6	6.43E-04	5.646E-03	0.0355	91.925	98.70	1844.6	2.9
TK-34	124.2	120.8	1.27E-03	1.159E-02	0.0382	93.160	97.24	1786.1	20.3
TK-34	124.3	120.0	1.62E-03	1.462E-02	0.0377	94.717	96.53	1778.6	17.7
TK-34	122.0	116.6	2.03E-03	1.797E-02	0.0377	97.665	95.65	1747.3	21.6

[†]Hornblende, J = 0.014, Standard MMHB-1 = 520.4 Ma.

Geological Society of America
Special Paper 377
2004

In situ ion microprobe $^{207}Pb/^{206}Pb$ dating of monazite from Precambrian metamorphic suites, Tobacco Root Mountains, Montana

John T. Cheney
A. Alexander G. Webb
Department of Geology, Amherst College, Amherst, Massachusetts 01002, USA

Chris D. Coath
Kevin D. McKeegan
Department of Earth and Space Sciences, University of California, Los Angeles, California 90095, USA

ABSTRACT

The Precambrian rocks of the Tobacco Root Mountains have been separated into three suites: the Indian Creek Metamorphic Suite, the Pony–Middle Mountain Metamorphic Suite, and the Spuhler Peak Metamorphic Suite. The Indian Creek and Pony–Middle Mountain Metamorphic Suites are quartzofeldspathic gneiss suites that contain variable amounts of metasupracrustal rocks. The Spuhler Peak Metamorphic Suite contains primarily mafic rocks and is possibly ocean crust. Metamorphosed mafic dikes and sills that intruded the Indian Creek and Pony–Middle Mountain Metamorphic Suites, but not the Spuhler Peak Metamorphic Suite, indicate juxtaposition of the Spuhler Peak Metamorphic Suite with the other two suites after intrusion of the dikes at ca. 2060 Ma. All rocks have been deformed and metamorphosed together, initially at pressures greater than 1.0 GPa and temperatures greater than 750 °C, followed by differential reequilibration at ~0.6 GPa on a clockwise pressure-temperature path.

Two-hundred-seventy-two $^{207}Pb/^{206}Pb$ spot ages of monazite grains from seventeen Spuhler Peak Metamorphic Suite, five Pony–Middle Mountain Metamorphic Suite, and eight Indian Creek Metamorphic Suite rocks have been obtained from the University of California at Los Angeles ion microprobe. Based on the distribution of the ages, the samples can be divided into three groups. (1) All seventeen Spuhler Peak Metamorphic Suite, one Pony–Middle Mountain Metamorphic Suite, and two Indian Creek Metamorphic Suite samples have relatively homogeneous spot age populations that vary from ca. 1720 to ca. 1780 Ma. (2) A group of seven Indian Creek and Pony–Middle Mountain Metamorphic Suite samples has spot ages from monazite grains that form an array near 2450 Ma. (3) A group of four Indian Creek and Pony–Middle Mountain Metamorphic Suite samples are bimodal in that they contain spot ages from both the 1720–1780 Ma group and the 2450 Ma array. There are younger and a few older spot ages in these samples that likely represent mixed age domains, the former between the 1720–1780 Ma and the 2450 Ma age domains and the latter between older detrital grain cores and the 2450 Ma array.

Monazite grains in the matrix have similar ages to those that occur as inclusions in garnet and kyanite. Thus, the monazite in these rocks, as well as the peak metamorphic minerals, either grew or re-equilibrated during the higher-pressure

Cheney, J.T., Webb, A.A.G., Coath, C.D., and McKeegan, K.D., 2004, In situ ion microprobe $^{207}Pb/^{206}Pb$ dating of monazite from Precambrian metamorphic suites, Tobacco Root Mountains, Montana, *in* Brady, J.B., Burger, H.R., Cheney, J.T., and Harms, T.A., eds., Precambrian geology of the Tobacco Root Mountains, Montana: Boulder, Colorado, Geological Society of America Special Paper 377, p. 151–179. For permission to copy, contact editing@geosociety.org.

(>1.0 GPa) metamorphism. The near absence of $^{207}Pb/^{206}Pb$ ages older than 1780 Ma in the Spuhler Peak Metamorphic Suite and the common occurrence of older ages in monazite from the Indian Creek and Pony–Middle Mountain Metamorphic Suites are consistent with assembly of the Tobacco Root terrane during a prolonged (60 m.y. long) collision event, the Big Sky orogeny, beginning at ca. 1780 Ma and culminating at ca. 1720 Ma. The Early Proterozoic Big Sky orogeny significantly overprinted the effects of an earlier ca. 2450 Ma orogeny in both the Pony–Middle Mountain and Indian Creek Metamorphic Suites. This older event modified pre-existing Archean rocks. However, of the 272 spot ages on monazite grains reported here, only six are significantly older than 2450 Ma and only one of these is older than 2600 Ma—a 2988 Ma spot age from a monazite inclusion in a garnet from a Pony–Middle Mountain Metamorphic Suite sample. There is no evidence of widespread Archean events recorded in the monazite grains of the Tobacco Root Mountains.

Keywords: U-Th-Pb, monazite ages, Precambrian, Tobacco Root Mountains, Montana.

INTRODUCTION

The Tobacco Root Mountains of Montana (Fig. 1) have been the subject of close geologic scrutiny for over 50 years. The remarkable petrologic overview of Reid (1963) was followed by a large body of fieldwork incorporated in a number of theses from Indiana University. Vitaliano et al. (1979) summarized this work in the form of a geologic map and accompanying report (see reprinted map and text accompanying this volume). In the summer of 1993, under the leadership H. Robert Burger, one of the first Indiana University Ph.D. mappers, the attention of student geologists was once again focused on the Tobacco Root Mountains. This time the students were undergraduates working on senior projects supported by the Keck Geology Consortium. The fieldwork for some 31 undergraduate theses was completed during the summers 1993, 1995, and 1997. These projects were far ranging in scope but generally focused upon the tectonic evolution of the Tobacco Root Mountains. The resulting body of work, as developed in this book, has provided a detailed petrologic and structural framework for a complex polygenetic evolution of the Precambrian rocks in the area. Geochronology has been added to the growing body of knowledge most recently in the form of $^{40}Ar/^{39}Ar$ dating of hornblende and biotite (Brady et al., 2004, this volume, Chapter 7) and U-Pb dating of zircon (Mueller et al., 2004, this volume, Chapter 9). This dating has provided constraints on the age of protoliths and on the timing of cooling but has added only limited information on the timing of the orogenic events. This chapter addresses issues of the timing of metamorphism and of the tectonic assembly of the region by providing the results of a U-Pb-Th study of monazite in rocks from the major rock suites composing the Precambrian core of the Tobacco Root Mountains. Monazite is particularly well suited for constraining tectonic evolution because it can grow repeatedly during an orogenic cycle (Harrison et al., 2002). The purpose of this chapter is to integrate the results of our collaborative geologic work with 272 $^{207}Pb/^{206}Pb$ monazite spot ages acquired on the ion microprobe at the Keck Ion Microprobe Laboratory at the University of California at Los Angeles (UCLA).

BACKGROUND

Previous mapping has established the presence of three major rock units in the Tobacco Root Mountains (Fig. 1; Fig. 1 in Burger, 2004, this volume, Chapter 1). In our work, as summarized in Burger (2004, this volume, Chapter 1), the Precambrian rocks of the Tobacco Root Mountains have been divided into three packages: the Indian Creek Metamorphic Suite, the Pony–Middle Mountain Metamorphic Suite, and the Spuhler Peak Metamorphic Suite. The Spuhler Peak Metamorphic Suite contains primarily mafic igneous rocks and may be ocean crust (Burger et al., 2004, this volume, Chapter 1). The Indian Creek and Pony–Middle Mountain Metamorphic Suites are quartzofeldspathic gneiss suites that contain variable amounts of metasupracrustal rocks. The Indian Creek Metamorphic Suite is composed principally of quartzofeldspathic gneisses and amphibolite with lesser, pelitic schist, marble, quartzite, and banded iron formation. The Pony–Middle Mountain Metamorphic Suite is dominated by quartzofeldspathic gneisses with lesser amounts of hornblende + plagioclase gneiss and amphibolite, but has only minor amounts of marble, quartzite, and aluminous schists and gneisses. Geochemical analysis of quartzofeldspathic gneisses and structural observations in the Indian Creek and Pony–Middle Mountain Metamorphic Suites suggest that the gneisses in the two suites may be of similar parentage (Mogk et al., 2004, this volume, Chapter 2). One anomaly is the gedrite-bearing rocks at Cataract Mountain within the Pony–Middle Mountain Metamorphic Suite (JBB95-39a, Fig. 1) that occur as part of a Spuhler Peak Metamorphic Suite–like sequence of aluminous schist, quartzite, quartzofeldspathic gneiss and amphibolite. They were considered by Hanley (1975, 1976) as a possible tectonic slice of Spuhler Peak Metamorphic Suite rocks in the Pony–Middle Mountain Metamorphic Suite, an interpretation that we support in this paper.

Metamorphism

Both the Indian Creek and Spuhler Peak Metamorphic Suites have been metamorphosed, initially at pressures exceeding

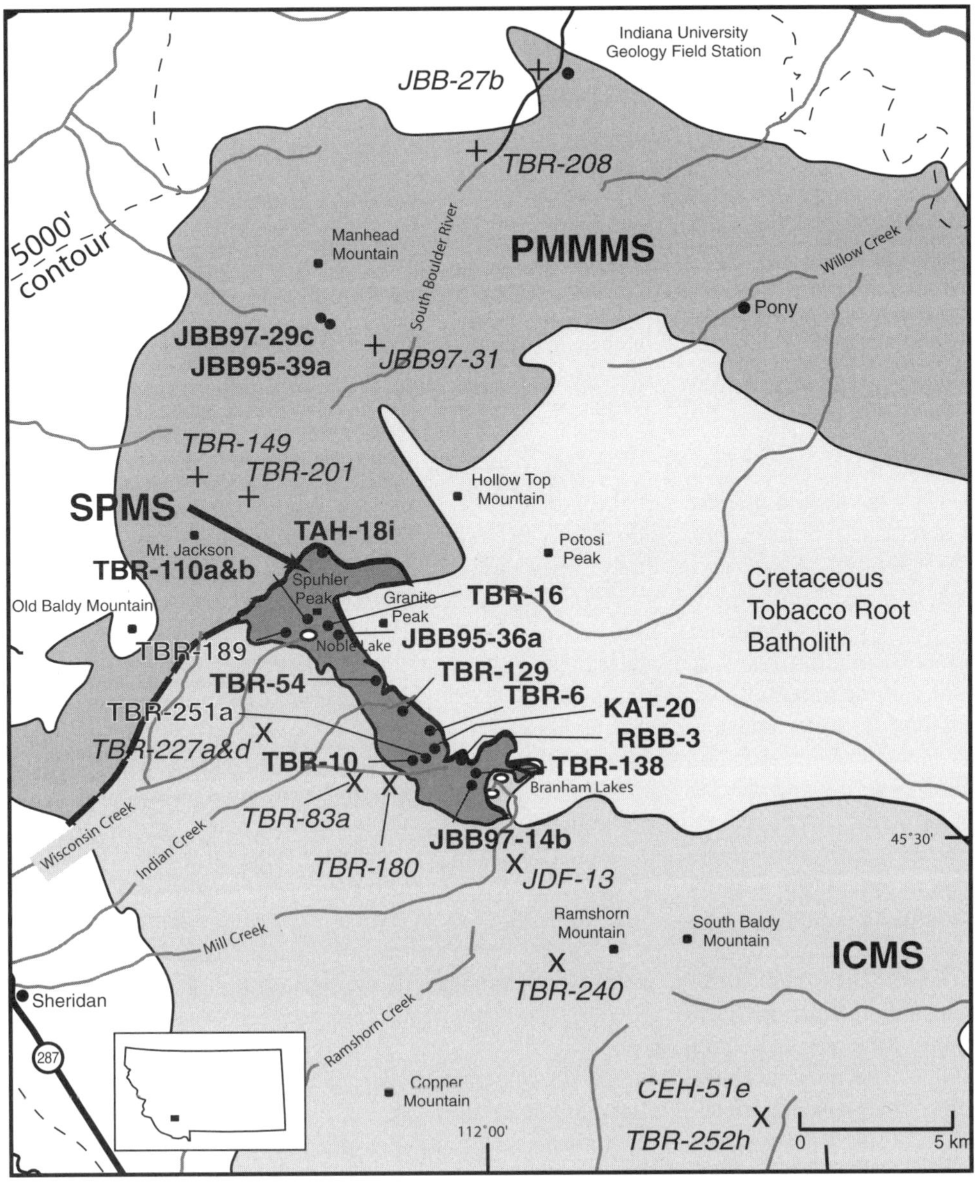

Figure 1. Generalized geologic map of the Tobacco Root Mountains (modified from Burger, 2004, this volume, Chapter 1) showing the distribution of dated samples and the metamorphic rock suite of each dated sample. The quartzofeldspathic gneiss unit to the north and east of the Spuhler Peak Metamorphic Suite (SPMS) has been designated the Pony–Middle Mountain Metamorphic Suite (PMMMS), whereas the rocks to the south of the SPMS are referred to as the Indian Creek Metamorphic Suite (ICMS). The PMMMS–ICMS boundary is marked with a thick dashed line. The geographic outline of the Tobacco Root Mountains is shown by the 5000-foot contour.

1.0 GPa (referred to as M1) and later at lower pressures of ≤0.6 GPa (referred to as M2). No evidence of a pre-M1 event has been found in the Spuhler Peak Metamorphic Suite rocks. The Indian Creek/Pony–Middle Mountain/Spuhler Peak Metamorphic Suite terrane was assembled prior to or at the time of the last ductile deformation of all the rocks because the contact is parallel to foliation within the Spuhler Peak Metamorphic Suite and in the Indian Creek/Pony–Middle Mountain Metamorphic Suite. The resulting structures are defined by minerals in both groups that are consistent with at least upper amphibolite facies conditions of ~650 °C at 0.6 GPa (Harms et al., 2004a, this volume, Chapter 10). Because the rocks all were deformed together and both the Indian Creek and Spuhler Peak Metamorphic Suites preserve evidence of higher-pressure metamorphism overprinted by lower-pressure metamorphism, assembly of the terrane may have occurred before or during the higher-pressure, M1 event. The metamorphosed mafic dikes and sills (MMDS) that intrude only the Indian Creek and Pony–Middle Mountain Metamorphic Suites have been metamorphosed at the higher pressures of M1 (Brady et al., 1998). This is consistent with the 2050 Ma intrusion age and 1763 Ma metamorphic age for the MMDS as reported by Burger et al. (1999). The higher-pressure (M1) assemblages recrystallized at the relatively low pressure (≤0.6 GPa) of M2. Calculated pressures and temperatures from near-rim compositions of quartz-bearing amphibolites from all three suites and the MMDS are similar. These metamorphic conditions of ~650–750 °C and 0.6–0.8 GPa do not vary systematically with rock suite or geographically (Cheney et al., 2004, this volume,

Chapter 6). The textures, mineral compositions, and mineral assemblages provide constraints on the evolution of the terrane that are consistent with late tectonic denudation. This involves isobaric cooling at 800 °C followed by isothermal decompression at 700 °C as shown on Figure 21 of Cheney et al. (2004, this volume, Chapter 6). $^{40}Ar/^{39}Ar$ spectra from amphiboles of all units yield ages of ca. 1710 ± 20 Ma (Brady et al., 2004, this volume, Chapter 3). This requires that amphibole in all of the units cooled through the ~500 °C isotherm at ca. 1710 Ma.

The three different rock suites have distinctive lithologic characteristics and all three rock suites contain some rock types with similar mineral assemblages (see Figure 1 in Cheney et al., 2004, this volume, Chapter 6). In particular, all three rock suites have aluminous gneisses and schists that contain varying amounts of biotite, sillimanite and/or kyanite, and garnet in addition to quartz, K-feldspar, and/or muscovite and plagioclase. These rocks typically contain abundant monazite, are most common in the Spuhler Peak Metamorphic Suite, and are least common in the Pony–Middle Mountain Metamorphic Suite. A second important rock type that typically contains abundant monazite is the orthoamphibole (gedrite) gneiss. These rocks are common in the Spuhler Peak Metamorphic Suite, rare in the Indian Creek Metamorphic Suite, and not found in the Pony–Middle Mountain Metamorphic Suite. In addition to gedrite, these rocks can contain varying amounts of sillimanite and/or kyanite, garnet, and cordierite.

Two orthoamphibole gneiss occurrences in the Indian Creek Metamorphic Suite were studied: Quartz Creek (TBR-83a of Fig. 1) and Granite Creek (TBR-252h of Fig. 1). These gneisses are similar to gedrite-bearing rocks that are characteristic of the Spuhler Peak Metamorphic Suite, but the Indian Creek Metamorphic Suite gedrite rocks are intercalated with banded iron formation (BIF) and typical Indian Creek Metamorphic Suite hornblende amphibolites at both localities. The rocks at these two localities are texturally complex and contain remnant minerals from an early coarse-grained assemblage and a later, finer-grained reaction assemblage. The polymetamorphic nature of the rocks is indicated by the following observations:

- Kyanite occurs as inclusions in garnet in some orthopyroxene + orthoamphibole gneisses.
- Staurolite occurs as inclusions in kyanite and garnet.
- Coarse-grained kyanite occurs in the gedrite matrix of some orthopyroxene-bearing rocks.
- Cordierite + orthopyroxene symplectite has partially replaced garnet.
- Sapphirine + spinel + cordierite symplectite has replaced aluminosilicate in silica-undersaturated regions of the rocks.
- Garnet rims some orthopyroxene porphyroblasts.

Previous Geochronology

Early geochronological studies of the metamorphic rocks of Tobacco Root Mountains and adjacent ranges were based principally on K/Ar studies of micas or Rb/Sr studies of micas and whole rocks (Hayden and Wehrenberg, 1959, 1960; Giletti and Gast, 1961; Giletti, 1966, 1968). These investigators consistently found ages of 1.6–1.7 Ga, which were interpreted by Giletti (1966) as evidence of a regional metamorphism of that age. Subsequent whole-rock Rb/Sr studies by Mueller and Cordua (1976) and James and Hedge (1980) yielded ages of ca. 2.7 Ga for the rocks, consistent with similar ages from elsewhere in the Wyoming province. Both Mueller and Cordua (1976) and James and Hedge (1980) argued that they were dating the major, upper amphibolite facies metamorphism and that the earlier K/Ar work dated a less significant, greenschist facies thermal event. This view was apparently not correct, as evidence for a 2.7 Ga metamorphism has not been verified by subsequent isotopic studies in the Tobacco Root Mountains.

Recent U-Pb zircon ages from the quartzofeldspathic gneisses of the Indian Creek and Pony–Middle Mountain Metamorphic Suites have been interpreted to indicate that their igneous protoliths formed between 3.2 and 3.4 Ga (Mueller et al., 2004, this volume, Chapter 9). This provides an upper age limit on the Indian Creek and Pony–Middle Mountain Metamorphic Suites metasediments deposited upon these rocks (Harms et al., 2004b, this volume, Chapter 11). U-Pb ages from detrital zircons in a Spuhler Peak Metamorphic Suite quartzite range in age from 3.8 down to 3.0 Ga (Mueller et al., 2004, this volume, Chapter 9) and provide an upper age limit for the suite. Mueller et al. (2004, this volume, Chapter 9) report one 2.45 Ga age from a single subhedral zircon in this quartzite. They argue that this grain is not clearly detrital and that its subhedral shape results from metamorphic overgrowth at ca. 2.4 Ga. Krogh et al. (1997) reported metamorphic growth of zircon at ca. 2.40 Ga in a Pony–Middle Mountain Metamorphic Suite felsic gneiss that yields a primary igneous zircon age of 3.37 Ga. Roberts et al. (2002) also have reported a 2.47 Ga age based on $^{207}Pb/^{206}Pb$ step leaching of a monazite-rich garnet. They have interpreted this age as a possible metamorphic event.

Evidence for a second and younger Proterozoic metamorphic event comes from the 1.77 Ga U-Pb age of zircon grains that crystallized from a melt pocket in a Spuhler Peak Metamorphic Suite aluminous schist (Burger et al., 1999; Mueller et al., 2004, this volume, Chapter 9). The 1.77 Ga date is therefore taken as the near-peak age of the metamorphic event. Preliminary studies of monazite, including $^{207}Pb/^{206}Pb$ secondary ion mass spectrometry (SIMS) ages (Webb et al., 2001), Th-Pb SIMS ages (Cheney et al., 1999), Th-U-Pb chemical ages from the electron microprobe (Dahl et al., 1999, 2002), and $^{207}Pb/^{206}Pb$ step-leaching ages of garnet (Roberts et al., 2002), have yielded monazite ages between ca. 1.65 and 1.85 Ga. These results provide ages of metamorphic recrystallization and give further support for the existence of a younger Proterozoic metamorphic event ca. 1.77 Ga in age.

$^{40}Ar/^{39}Ar$ ages of 1.71 Ga of both hornblende and biotite from the Tobacco Root Mountains and adjacent Ruby Range are consistent with rapid cooling following the younger Proterozoic event (Brady et al., 2004, this volume, Chapter 7). Roberts et al. (2002) have also recently provided $^{40}Ar/^{39}Ar$ ages from a selected subset of biotite ages that are somewhat older and average 1.76 Ga for the same area. This discrepancy is unresolved, but Brady et al. (2004, this volume, Chapter 7) appeal to the data selection and/or inter-lab errors as possible solutions.

IN SITU ION PROBE AGES

Sample Selection

Several hundred samples from the Tobacco Root Mountains have been examined with a petrographic microscope and with the backscattered electron detector of a scanning electron microscope, both to identify the presence of monazite grains and to characterize the textural and spatial relationships of these grains. Samples were selected for dating that have monazite grains in a variety of habits such as inclusions in different metamorphic minerals (e.g., garnet, kyanite, sillimanite, and cordierite), in reaction textures typically involving the production of cordierite at the expense of garnet, and in the matrix. Samples were also selected from the same outcrops to check for consistency of results. Other samples were selected for dating to provide geographic and lithologic coverage across the terrane.

SIMS Analyses

Ion microprobe analyses of monazite were obtained from the IMS 1270 ion microprobe at the Keck National Ion Microprobe facility at UCLA. The data were acquired during three seven-day working sessions. The analytical procedures described by Harrison et al. (1995, 1997) were followed for the August 1998 session. A similar procedure was followed in August of 1999 and December of 1999, but a different standard monazite was required in order to fully evaluate the U-Th-Pb characteristics of Tobacco Root Mountains monazite. Monazite "UCLA76" was used as standard because it has sufficient concentration of U to allow measurement of ^{238}U and $^{254}UO/^{238}U$, in addition to the normally determined masses of ^{232}Th, $^{264}ThO_2$, ^{238}U, $^{254}UO/^{238}U$, ^{204}Pb, ^{206}Pb, ^{207}Pb, and ^{208}Pb. Although the normal protocol is to mount standard monazite grains in the same mount as the unknown monazite, the limited quantity of UCLA76 required the use of a separate standard mount. This in turn required alternately swapping the standard and unknown mounts in the ion microprobe sample chamber. Most of the grains analyzed during the August 1998 session were reanalyzed during August of 1999. In general, two calibration curves of $^{264}ThO_2/^{232}Th$ versus $^{208}Pb/^{232}Th$ and $^{254}UO/^{238}U$ versus $^{206}Pb/^{238}U$ were obtained by repeated analyses of the standard using natural instrument drift as a means of spreading out the analyses along the standard curve. During all three sessions the standard curves were built up over a seven-day session. Sample charging or instrument drift are necessary to produce spread along the standard curve and to constrain the slope of the standard curve. The calibrated values of $^{264}ThO_2/^{232}Th$, for example, for the standard should bracket those of the unknowns. The ThO_2/Th ratios do not in all cases fall within the brackets, thus producing possible extrapolation errors on the resulting ages. A third standard curve for ($^{251}UO/^{235}U$ vs. $^{207}Pb/^{235}U$) was generated by assuming that the ratio $^{238}U/^{235}U$ is 137. In this way, three independent age determinations were obtained on each spot by measurement of the three Pb isotopes ($^{208}Pb/^{232}Th$, $^{206}Pb/^{238}U$, and $^{207}Pb/^{235}U$). In addition $^{207}Pb/^{206}Pb$ ages were calculated.

Results of monazite analyses are summarized in Table 1. As discussed below, nearly all spot analyses are discordant on $^{206}Pb/^{238}U$ versus $^{207}Pb/^{235}U$ concordia diagrams. Normal discordance is commonly attributed to Pb loss or mixing of ages. As the normal discordance in many cases is manifest in a line from concordia through the origin, analytic error associated with the standardization is the likely cause for this behavior (Parrish, 1990). Pb loss cannot be excluded in individual cases, especially for those ages younger than 1700 Ma. These may reflect disturbance related to the Cretaceous Tobacco Root batholith. In the case of either modern Pb loss or error, the $^{207}Pb/^{206}Pb$ dates are the better indicator of crystallization age.

The cause of reverse discordance is typically attributed (e.g., Parrish, 1990) to the ^{206}Pb derived from radioactive decay of ^{230}Th that was incorporated into the monazite during its growth or so-called "unsupported" ^{206}Pb. However, F. Spear (2003, personal commun.) interprets the reverse discordance observed in many analyses as an analytical effect. This interpretation is based on the observation that the magnitude of the discordance is impossible to obtain from ^{230}Th decay given the small quantities of ^{230}Th that were likely available during monazite growth. Accordingly he argues that the most likely cause is related to sample charging on the unknown monazite that is different from that on the standard monazite. Charging that produces oxides in the plasma that are not accounted for in the standard curve produces incorrect ages. The common practice in $^{206}Pb/^{238}U$ versus $^{207}Pb/^{235}U$ studies (e.g., Wing et al., 2003) of building a standard curve over several days, from the calibration curves derived from the same standard in several different sample mounts, appears to suffer from this problem. This problem is possibly accentuated by the sample-standard swapping used in this study. Similar calibration problems may affect the $^{232}Th/^{208}Pb$ system. Because of the possible discordance in the $^{208}Pb/^{232}Th$, $^{206}Pb/^{238}U$, and $^{207}Pb/^{235}U$ systems, we use only directly measured $^{207}Pb/^{206}Pb$ ages in this paper. $^{207}Pb/^{206}Pb$ ages are independent of standard curves and, therefore, are clearly the most accurate crystallization ages for the monazite considered here. Monazite ages discussed throughout the paper are $^{207}Pb/^{206}Pb$ ages reported with 1σ (one standard error) uncertainty based on the data given in Table 1.

Results

Results of SIMS analyses are presented in Table 1 and illustrated by the concordia diagrams in Figure 2. Histograms of age distributions for each of the three rock suites are given individually in Figure 3 and collectively in Figure 4. The location of samples is shown on the map of the Tobacco Root Mountains in Figure 1.

TABLE 1. ION MICROPROBE AGES OF MONAZITES FROM THE TOBACCO ROOT MOUNTAINS, MONTANA

Sample/Grain/Spot#	Date	Grain size (microns)	Grain shape*	Textural setting	X-ray-map (X = map)	Age (Ma)† $^{207}Pb/^{206}Pb$	1 s.e.§ $^{207}Pb/^{206}Pb$	Age (Ma) $^{208}Pb/^{232}Th$	1 s.e.§ $^{208}Pb/^{232}Th$	Age (Ma) $^{206}Pb/^{238}U$	1 s.e.§ $^{206}Pb/^{238}U$	Age (Ma) $^{207}Pb/^{235}U$	1 s.e.§ $^{207}Pb/^{235}U$	Th O_2/Th	% Radio-genic ^{208}Pb	% Radio-genic ^{207}Pb	% Radio-genic ^{206}Pb
ICMS																	
TBR-252	**Granite Creek**	**ICMS**	**Garnet-Orthoamphibole Gneiss**														
TBR252H_g1s1	1999-08-16Aug	150	E/A	in garnet	X	1766	8	1821	10	1941	15	1858	8	6.26	99.86	99.14	99.9
TBR252H_g1s2	1999-08-16Aug					1826	25	1976	265	2011	242	1922	127	5.58	99.73	98.21	99.79
TBR252H_g1s3	1999-08-16Aug					1797	35	1739	44	1765	60	1780	30	6.50	99.43	96.45	99.58
TBR252H_g1s4	1999-08-16Aug					1827	17	1847	18	1922	20	1877	14	6.62	99.79	98.34	99.8
TBR252H_g1s5	1999-08-16Aug					1745	17	1769	32	1769	38	1758	21	6.32	99.04	94.3	99.32
TBR252H_g1s6	1999-08-16Aug					1810	37	1643	34	1872	48	1843	25	7.06	99.54	96.41	99.57
TBR252H_g1s7	1999-08-16Aug					1777	17	1683	22	1835	33	1808	20	6.24	99.93	99.53	99.95
TBR-252h_Mon1_pt1	98_08_28aug			same as grain 1		1747	3	1631	7					4.02			
TBR-252h_Mon1_pt2	98_08_28aug					1760	3	1703	25					3.19			
TBR-252h_Mon1_pt3	98_08_28aug					1757	6	1292	24					3.21			
TBR252H_g2s2	1999-08-16Aug	75	E	in garnet	X	1743	5	1825	13	1799	15	1773	8	5.98	99.94	99.79	99.98
TBR252H_g2s3	1999-08-16Aug					1746	14	1715	22	1737	27	1741	15	5.50	99.95	99.69	99.96
TBR252H_g2s4	1999-08-16Aug					1755	17	1662	39	1755	40	1755	22	4.90	99.89	99.46	99.94
TBR252H_g2s5	1999-08-16Aug					1741	12	1786	26	1858	24	1804	12	5.54	99.82	99.29	99.92
TBR252H_g2s6	1999-08-16Aug					1747	11	1729	16	1857	19	1806	8	5.77	99.87	99.4	99.93
TBR-252h_Mon2_pt1	98_08_28aug			same as grain 2		1765	4	1740	12					3.67			
TBR-252h_Mon2_pt3	98_08_28aug					1766	5	1773	39					3.00			
TBR-252h_Mon2_pt2	98_08_28aug					1759	3	1704	32					2.87			
TBR-252h_Mon2_pt4	98_08_28aug					1762	4	1421	61					2.37			
						1754	**10**										
CEH-51	**Granite Creek**	**ICMS**	**Garnet-Orthoamphibole Gneiss**														
CEH-51_g1s1	1999-08-17Aug	250	P	in garnet	X	1736	10	1674	8	1772	9	1755	7	6.94	99.87	99.22	99.91
CEH-51_g1s2	1999-08-17Aug					1732	7	1707	9	1782	17	1759	8	7.83	99.83	99.67	99.96
CEH-51_g1s3	1999-08-17Aug					1750	7	1694	9	1769	12	1760	8	7.54	99.85	99.72	99.97
CEH-51_g1s4	1999-08-17Aug					1759	6	1714	9	1783	12	1772	7	7.27	99.79	99.59	99.95
CEH-51_g1s5	1999-08-17Aug					1760	8	1698	8	1782	12	1772	6	7.14	99.89	99.29	99.92
CEH-51b_sp1	98_08_26Aug\			same as grain 1		1749	9	1638	14					4.87			
CEH-51b_sp2	98_08_26Aug\					1739	4	1642	13					4.92			
CEH-51b_sp3	98_08_26Aug\					1738	6	1653	13					4.85			
CEH-51b_sp4	98_08_26Aug\					1735	11	1667	12					4.67			
CEH-51b_sp5	98_08_26Aug\					1744	11	1590	10					4.47			
						1744	**10**										
CEH-51_g2s1	1999-08-17Aug	160	E/A	in Gedrite	X	1750	10	1580	10	1662	16	1701	10	5.77	99.91	99.05	99.89
CEH-51_g2s2	1999-08-17Aug					1762	5	1667	12	1722	14	1740	8	5.60	99.86	99.53	99.95
CEH-51_g2s3	1999-08-17Aug					1738	7	1711	16	1792	19	1767	10	5.22	99.87	98.64	99.85
CEH-51_g2s4	1999-08-17Aug					1766	10	1594	15	1725	22	1743	12	5.43	99.88	99.31	99.92
CEH-51_g2s5	1999-08-17Aug					1749	8	1656	15	1752	21	1751	12	5.34	99.93	99.28	99.92
CEH-51_g2s6	1999-08-17Aug					1728	5	1607	22	1720	25	1724	14	4.71	99.88	99.23	99.91
						1749	**14**										
CEH-51e_sp1	98_08_26Aug\	?		?		1815	47	1260	47					5.49			
CEH-51e_sp1@1	98_08_26Aug\					1743	16	1764	33					3.71			

(continued)

TABLE 1. ION MICROPROBE AGES OF MONAZITES FROM THE TOBACCO ROOT MOUNTAINS, MONTANA (*continued*)

Sample/Grain/Spot#	Date	Grain size (microns)	Grain shape*	Textural setting	X-ray-map (X = map)	Age (Ma)† $^{207}Pb/^{206}Pb$	1 s.e.§ $^{207}Pb/^{206}Pb$	Age (Ma) $^{208}Pb/^{232}Th$	1 s.e.§ $^{208}Pb/^{232}Th$	Age (Ma) $^{206}Pb/^{238}U$	1 s.e.§ $^{206}Pb/^{238}U$	Age (Ma) $^{207}Pb/^{235}U$	1 s.e.§ $^{207}Pb/^{235}U$	Th O_2/Th	% Radio-genic ^{208}Pb	% Radio-genic ^{207}Pb	% Radio-genic ^{206}Pb
TBR-83	**Quartz Creek Ridge**	**ICMS**	**Garnet-Orthoamphibole Gneiss**														
TBR-83g1sp1	1999-08-18Aug	80	P/A	matrix	X	2414	8	2295	14	2408	20	2411	10	6.93	99.8	99.14	99.86
TBR-83g1sp2	1999-08-18Aug					1924	12	1818	9	1900	13	1912	9	6.73	99.88	98.81	99.85
TBR-83g1sp3	1999-08-18Aug					2154	6	1839	8	1957	10	2055	6	6.64	99.9	99.4	99.91
TBR-83g1sp4	1999-08-18Aug					2294	10	1999	14	2197	25	2248	12	7.57	99.9	99.41	99.91
TBR-83g1sp5	1999-08-18Aug					1987	8	1933	11	1982	12	1985	6	6.18	99.81	98.57	99.81
TBR-83g1sp6	1999-08-18Aug					1765	13	1756	17	1881	27	1826	14	5.77	99.75	97.05	99.66
TBR-83a_gr1_sp1	98_08_27Aug\			same as grain 1		1879	4	1788	14					3.68			
TBR-83a_gr1_sp2	98_08_27Aug\					1978	7	1737	50					2.93			
TBR-83a_gr1_sp3	98_08_27Aug\					2226	7	1832	114					2.34			
TBR-83a_gr1_sp4	98_08_27Aug\					2031	6	1850	31					3.28			
TBR-83g3sp1	1999-08-18Aug	25	E	matrix	X	2084	10	1888	13	2015	16	2050	8	5.85	99.92	99.34	99.91
TBR-83g3sp2	1999-08-18Aug					2219	11	1899	8	2027	13	2124	8	6.52	99.96	99.7	99.96
TBR-83g3sp3	1999-08-18Aug					2327	16	2267	21	2284	28	2307	14	5.32	99.84	98.62	99.78
TBR-83a_gr2_sp1	98_08_27Aug\	50	E	in cordierite	X	1722	17	1744	64					2.77			
TBR-83a_gr2_sp2	98_08_27Aug\					1948	7	1823	37					3.17			
JDF-13	**Mill Creek**	**ICMS**	**Aluminous Gneiss**														
JDF-13_g1	1999-08-21Aug	15	E	in garnet	X	2474	5	2578	27	2735	29	2587	12	3.39	99.64	94.7	99.23
JDF-13_g2	1999-08-21Aug	50	E	matrix	X	2485	24	2331	49	2458	61	2473	28	4.12	99.98	99.11	99.85
JDF-13_g2@1	1999-08-21Aug					2471	20	2433	38	2545	66	2504	29	4.52	99.97	98.67	99.77
JDF-13_g2@2	1999-08-21Aug					2455	15	2622	93	2914	116	2651	47	3.34	99.97	99.07	99.84
JDF-13_g2@3	1999-08-21Aug					2427	7	2447	15	2600	21	2504	10	5.07	99.6	98.59	99.76
						2451	**22**										
JDF-13_g4	1999-08-21Aug	50	E/A	matrix	X	2068	74	2327	90	2582	122	2305	73	5.45	99.97	99.7	99.95
JDF-13_g5	1999-08-21Aug	25	E/A	matrix	X	2407	6	2414	19	2601	24	2494	10	5.72	99.97	99.46	99.91
TBR-240	**Bivens Creek ridge**	**ICMS**	**Aluminous Gneiss**														
TBR-240_gc1	1999-08-21Aug	150	P/A	matrix	X	2270	9	2170	10	2260	13	2265	7	6.36	99.96	99.77	99.96
TBR-240_gc1@1	1999-08-21Aug					1770	8	1713	7	1793	13	1783	8	7.10	99.93	99.43	99.94
TBR-240_gc1@2	1999-08-21Aug					1759	6	1685	8	1746	10	1752	6	7.47	99.96	99.63	99.96
TBR-240_gc1@3	1999-08-21Aug					1955	9	1844	10	1955	11	1955	7	6.66	99.92	99.43	99.93
TBR-240_gc3	1999-08-21Aug	100	P	matrix	X	2439	5	2450	11	2497	15	2465	7	6.47	99.97	99.76	99.96
TBR-240_gc3@1	1999-08-21Aug	50	P	in quartz	X	1977	5	1858	10	1892	13	1933	7	5.96	99.94	99.63	99.95
TBR-180	**Quartz Creek Ridge**	**ICMS**	**Aluminous Gneiss**														
TBR-180AgA1	1999-08-22Aug	200	P/A	matrix	X	1771	7	1601	12	1970	15	1875	8	5.65	99.94	99.26	99.92
TBR-180AgA1@1	1999-08-22Aug					1745	9	1565	12	1742	16	1743	9	5.64	99.89	97.43	99.7
TBR-180AgA2	1999-08-22Aug	100		matrix		1760	9	1565	9	1821	14	1792	8	6.10	99.93	99.28	99.92
TBR-180AgA2@1	1999-08-22Aug					1740	7	1609	12	1784	16	1764	9	5.72	99.96	99.26	99.92
TBR-180AgA2@2	1999-08-22Aug					1753	9	1560	11	1771	15	1763	9	6.05	99.8	96.7	99.62
TBR-180AgB3	1999-08-22Aug	30	E/A	in garnet	X	2576	11	2475	15	2335	17	2466	8	6.63	99.73	98.38	99.7
TBR-180AgB3@1	1999-08-22Aug					2459	6	2339	14	2423	20	2443	9	5.97	99.85	99.22	99.87

(continued)

TABLE 1. ION MICROPROBE AGES OF MONAZITES FROM THE TOBACCO ROOT MOUNTAINS, MONTANA (*continued*)

Sample/Grain/Spot#	Date	Grain size (microns)	Grain shape*	Textural setting	X-ray-map (X = map)	Age (Ma)† $^{207}Pb/^{206}Pb$	1 s.e.§ $^{207}Pb/^{206}Pb$	Age (Ma) $^{208}Pb/^{232}Th$	1 s.e.§ $^{208}Pb/^{232}Th$	Age (Ma) $^{206}Pb/^{238}U$	1 s.e.§ $^{206}Pb/^{238}U$	Age (Ma) $^{207}Pb/^{235}U$	1 s.e.§ $^{207}Pb/^{235}U$	Th O_2/Th	% Radio-genic ^{208}Pb	% Radio-genic ^{207}Pb	% Radio-genic ^{206}Pb
TBR-227d	**Skihi Ridge**	**ICMS**	**Aluminous Gneiss**														
TBR-227d_g1sp1	1999-12-07Dec	500	P	matrix	X	2464	8	2247	45	2388	32	2429	15	4.741	99.97	99.65	99.94
TBR-227d_g1sp2	1999-12-07Dec					2445	13	2127	46	2245	44	2351	20	4.712	99.96	99.47	99.91
TBR-227d_g1sp3	1999-12-07Dec					2444	10	2191	43	2322	35	2387	17	4.884	99.97	99.07	99.84
TBR-227d_g1sp4	1999-12-07Dec					2381	10	2034	40	2180	31	2286	15	4.815	99.97	99.55	99.93
TBR-227d_g1sp5	1999-12-07Dec					2447	12	2173	35	2321	36	2389	16	4.963	99.97	99.27	99.88
TBR-227d_g1sp6	1999-12-08Dec					2451	6	2242	15	2399	21	2427	10	4.645	99.97	99.61	99.93
TBR-227d_g1sp7	1999-12-08Dec					2463	9	2165	19	2301	19	2388	9	4.862	99.97	99.52	99.92
TBR-227d_g1sp8	1999-12-08Dec					2444	12	2199	23	2302	30	2378	13	4.922	99.98	99.37	99.89
TBR-227d_g1sp9	1999-12-08Dec					2445	10	2214	19	2320	17	2387	8	4.855	99.97	99.55	99.92
TBR-227d_g1sp10	1999-12-08Dec					2410	10	2077	15	2239	20	2330	9	4.723	99.95	99.37	99.9
						2439	**25**										
TBR-227a	**Skihi Ridge**	**ICMS**	**Aluminous Gneiss**														
TBR-227a_g1sp1	1999-12-08Dec	150	E	matrix	X	2453	7	2089	11	2186	19	2327	9	4.85	99.96	99.39	99.9
TBR-227a_g1sp2	1999-12-08Dec					2432	11	2115	18	2280	29	2361	14	4.593	99.96	99.29	99.88
TBR-227a_g1sp3	1999-12-08Dec					2457	11	2137	18	2300	30	2384	14	4.598	99.98	99.61	99.93
TBR-227a_g1sp4	1999-12-08Dec					2440	14	2227	28	2307	30	2378	14	5.367	99.96	99.3	99.88
TBR-227a_g1sp5	1999-12-08Dec					2454	11	2076	22	2198	23	2334	13	4.612	99.96	99.32	99.89
						2447	**11**										
JBB-27b	**IU Field Camp**	**PMMMS**	**Aluminous Gneiss**														
JBB-27b_g1sp1	1999-08-19Aug	250	P	matrix biotite	X	1765	4	1717	16	1839	18	1804	10	5.14	99.95	99.72	99.97
JBB-27b_g1sp2	1999-08-19Aug					1740	6	1721	6	1802	9	1773	6	6.70	99.98	99.86	99.98
JBB-27b_g1sp3	1999-08-19Aug					1724	6	1717	6	1773	10	1751	6	6.56	99.97	99.79	99.98
JBB-27b_g1sp4	1999-08-19Aug					1748	3	1692	9	1762	12	1756	6	5.84	99.95	99.71	99.97
JBB-27b_g1sp5	1999-08-19Aug					1769	3	1673	9	1778	11	1774	6	5.83	99.97	99.85	99.98
JBB-27b_g1sp6	1999-08-19Aug					1783	4	1767	7	1823	9	1804	5	6.51	99.97	99.78	99.97
JBB-27b_g1sp7	1999-08-19Aug					1770	5	1666	11	1765	14	1767	7	5.61	99.98	99.87	99.98
JBB-27b	98_08_27Aug			same as grain 1		1768	5	1733	12					3.79			
JBB-27b_g3sp1	1999-08-19Aug	60	P	in kyanite	X	1778	5	1719	7	1802	10	1791	6	6.59	99.98	99.83	99.98
JBB-27b_g3sp2	1999-08-19Aug					1779	5	1746	17	1925	22	1856	11	5.22	99.93	99.48	99.94
JBB-27b_g3sp3	1999-08-19Aug					1769	10	1559	19	1638	21	1696	12	5.03	99.98	99.81	99.98
						1775	**6**										
JBB-27b_gr2_sp1@1	98_08_27Aug\	50	E/A	in matrix quartz		1778	6	1769	30					3.28			
JBB-27b_gr2_sp2	98_08_27Aug\					1769	31	1701	24					3.41			
JBB-27b_gr2_sp3	98_08_27Aug\					1759	5	1725	19					3.51			
JBB-27b_gr2_sp4	98_08_27Aug\					1763	5	1671	14					3.68			
JBB-27b_gr2_sp5	98_08_27Aug\					1766	7	1781	24					3.42			
						1767	**7**										

(continued)

TABLE 1. ION MICROPROBE AGES OF MONAZITES FROM THE TOBACCO ROOT MOUNTAINS, MONTANA (*continued*)

Sample/Grain/Spot#	Date	Grain size (microns)	Grain shape*	Textural setting	X-ray-map (X = map)	Age (Ma)† $^{207}Pb/^{206}Pb$	1 s.e.§ $^{207}Pb/^{206}Pb$	Age (Ma) $^{208}Pb/^{232}Th$	1 s.e.§ $^{208}Pb/^{232}Th$	Age (Ma) $^{206}Pb/^{238}U$	1 s.e.§ $^{206}Pb/^{238}U$	Age (Ma) $^{207}Pb/^{235}U$	1 s.e.§ $^{207}Pb/^{235}U$	Th O_2/Th	% Radio-genic ^{208}Pb	% Radio-genic ^{207}Pb	% Radio-genic ^{206}Pb
TBR-208	**IU Field Camp**	**PMMMS**	**Aluminous Gneiss**														
TBR-208gAsp1	1999-08-20Aug	150	E	matrix	X	2509	6	2327	10	2425	20	2471	9	7.22	99.96	99.7	99.95
TBR-208gAsp2	1999-08-20Aug					2449	6	2335	12	2373	18	2414	8	7.75	99.97	99.57	99.93
TBR-208gAsp3	1999-08-20Aug					2433	6	2248	10	2368	13	2403	7	7.21	99.97	99.58	99.93
TBR-208gBsp1	1999-08-20Aug	125	E	matrix	X	2438	6	2327	10	2474	15	2454	7	6.90	99.96	99.46	99.91
TBR-208gBsp2	1999-08-20Aug					1767	17	1671	8	1739	11	1752	10	6.93	99.28	90.97	98.88
TBR-208gCsp1	1999-08-20Aug	200	E	matrix	X	2472	4	2414	12	2545	15	2504	6	6.09	99.96	99.69	99.95
TBR-208gCsp2	1999-08-20Aug					1866	6	1853	12	1949	13	1909	7	5.87	99.54	94.68	99.33
TBR-208gCsp3	1999-08-20Aug					2417	7	2353	11	2501	15	2455	6	6.45	99.96	99.66	99.94
TBR-208gCsp4	1999-08-20Aug					2455	5	2345	12	2474	15	2464	7	6.57	99.95	99.75	99.96
TBR-208gCsp5	1999-08-20Aug					2460	5	2397	12	2518	19	2486	8	7.49	99.94	99.66	99.94
TBR-208gCsp6	1999-08-20Aug					2373	5	2194	10	2364	13	2369	6	6.63	99.95	99.44	99.91
TBR-149	**Boulder Lakes**	**PMMMS**	**Aluminous Gneiss**														
TBR-149_gA1	1999-08-21Aug	150	P	matrix	X	2447	5	2337	12	2320	15	2388	7	6.25	99.96	99.7	99.95
TBR-149_gA1@1	1999-08-21Aug					2480	7	2236	13	2289	19	2392	8	6.00	99.94	99.36	99.89
TBR-149_gA1@2	1999-08-21Aug					2448	6	2280	18	2396	24	2424	12	5.42	99.96	99.66	99.94
TBR-149_gA1@3	1999-08-21Aug					2398	19	2159	28	2264	46	2335	24	4.85	99.51	91.47	98.5
						2443	**34**										
TBR-149_gA2	1999-08-21Aug	75	P	matrix	X	2426	5	2338	13	2530	17	2473	8	5.89	99.95	99.64	99.94
TBR-149_gA2@1	1999-08-21Aug					2425	6	2199	13	2406	15	2416	7	5.92	99.97	99.67	99.95
						2426	**1**										
TBR-201	**Lake Shore Peak**	**PMMMS**	**Aluminous Gneiss**														
TBR-201ga1	1999-08-22Aug	125	E	in garnet	X	2449	7	2367	12	2495	19	2470	8	6.25	99.98	99.66	99.94
TBR-201ga1@1	1999-08-22Aug					2442	5	2260	14	2458	16	2450	8	6.46	99.97	99.65	99.94
TBR-201ga1@2	1999-08-22Aug					2440	7	2271	18	2437	27	2438	13	5.59	99.97	99.59	99.93
TBR-201ga1@3	1999-08-22Aug					2473	9	2405	34	2513	38	2491	17	4.60	99.92	99.08	99.84
						2451	**15**										
TBR-201gb1	1999-08-22Aug	100	E/A	in garnet	X	2988	5	2371	17	2653	21	2848	9	5.57	99.89	99.63	99.91
TBR-201gb1@1	1999-08-22Aug					2586	18	1951	40	2198	45	2406	22	4.38	99.94	99.85	99.97
TBR-201gb2	1999-08-22Aug	150	P/A	on garnet edge	X	2309	6	2004	15	2141	20	2228	11	5.55	99.95	98.75	99.8
TBR-201gb2@1	1999-08-22Aug					2329	6	1992	18	2187	22	2261	11	5.22	99.97	99.2	99.87
TBR-201gb2@2	1999-08-22Aug					2365	8	2094	16	2290	25	2330	11	5.45	99.97	99.19	99.87
TBR-201gb2@3	1999-08-22Aug					2432	5	2225	14	2365	17	2401	7	5.69	99.97	99.58	99.93
JBB97-31g	**Cataract Mtn.**	**PMMMS**	**Aluminous Gneiss**														
JBB97_31gBsp1	1999-12-08Dec	200	P/A	matrix	X	2448	9	2024	22	2188	24	2325	13	4.454	99.97	99.78	99.96
JBB97_31gBsp2	1999-12-08Dec					2457	12	2085	23	2203	30	2337	14	4.627	99.97	99.8	99.97
JBB97_31gBsp3	1999-12-08Dec					2447	16	2074	35	2216	50	2338	24	4.698	99.97	99.74	99.96
JBB97_31gBsp4	1999-12-08Dec					2434	12	1981	30	2163	30	2305	15	4.436	99.97	99.76	99.96
JBB97_31gBsp5	1999-12-08Dec					2459	11	1970	23	2080	29	2277	15	4.582	99.97	99.79	99.96
						2449	**10**										
JBB97_31g1so1	1999-12-09Dec	100	P/A	matrix	X	2457	6	2338	13	2351	17	2408	8	5.118	99.97	99.75	99.96
JBB97_31g1sp2	1999-12-09Dec					2522	8	2290	14	2332	16	2435	6	5.266	99.95	99.79	99.96
JBB97_31g1sp3	1999-12-09Dec					2524	7	2381	18	2413	18	2474	8	5.421	99.94	99.79	99.96
JBB97_31g1sp4	1999-12-09Dec					2463	8	2366	15	2476	19	2469	8	5.103	99.96	99.69	99.95
JBB97_31g2sp6	1999-12-09Dec					2454	7	2087	37	2098	34	2284	17	4.192	99.97	99.79	99.96

(*continued*)

TABLE 1. ION MICROPROBE AGES OF MONAZITES FROM THE TOBACCO ROOT MOUNTAINS, MONTANA *(continued)*

Sample/Grain/Spot#	Date	Grain size (microns)	Grain shape*	Textural setting	X-ray-map (X = map)	Age (Ma)† $^{207}Pb/^{206}Pb$	1 s.e.§ $^{207}Pb/^{206}Pb$	Age (Ma) $^{208}Pb/^{232}Th$	1 s.e.§ $^{208}Pb/^{232}Th$	Age (Ma) $^{206}Pb/^{238}U$	1 s.e.§ $^{206}Pb/^{238}U$	Age (Ma) $^{207}Pb/^{235}U$	1 s.e.§ $^{207}Pb/^{235}U$	Th O_2/Th	% Radio-genic ^{208}Pb	% Radio-genic ^{207}Pb	% Radio-genic ^{206}Pb
TBR-129	**Leggat Ridge**	**SPMS**	**Aluminous Gneiss**														
TBR-129g2s1	1999-08-16Aug	75	E/A	in biotite	X	1746	61	1696	124	1879	154	1817	83				
TBR-129_g2s6	1999-08-17Aug					1738	38	1421	41	1533	59	1621	34	4.73	99.98	98.76	99.86
TBR-129_g2s7	1999-08-17Aug					1920	31	1588	60	1680	89	1790	53	3.65	99.96	97.14	99.64
TBR-129_g2s8	1999-08-17Aug					1713	18	1437	25	1510	33	1597	22	6.79	99.94	99.55	99.95
TBR-129_g2s9	1999-08-17Aug					1738	19	1553	18	1683	29	1708	16	6.48	99.69	97.51	99.71
TBR-129_g2s10	1999-08-17Aug					1716	18	1491	32	1563	40	1629	21	5.72	99.98	99.7	99.97
TBR-129_g2s11	1999-08-17Aug					1756	43	1598	67	1896	99	1830	52	3.36	99.96	97.66	99.73
TBR-129_g2s12	1999-08-17Aug					1740	25	1586	41	1718	63	1728	35	4.86	99.98	98.48	99.83
TBR-129_Mon2_sp1	98_08_28aug\			same as grain 2		1753	3	1606	7					4.35			
TBR-129_g8s2	1999-08-17Aug	60	E	in cordierite	X	1765	27	1680	38	1624	73	1686	43	6.32	99.92	98.54	99.83
TBR-129_g9s1	1999-08-17Aug	110	P/A	Crd-Grt-Sill mass	X	1713	19	1593	21	1618	30	1659	20	6.61	99.95	99.7	99.97
TBR-129_g9s2	1999-08-17Aug					1763	16	1681	16	1837	28	1802	14	6.57	99.94	98.91	99.88
TBR-129_g9s3	1999-08-17Aug					1747	20	1584	23	1580	29	1653	20	6.98	99.91	99.15	99.9
TBR-129_g9s4	1999-08-17Aug					1753	6	1680	10	1703	14	1726	8	6.20	99.95	99.68	99.96
TBR-129_g9s5	1999-08-17Aug					1776	9	1652	10	1652	14	1708	8	6.57	99.92	99.45	99.94
TBR-129_g9s7	1999-08-17Aug					1771	7	1623	12	1691	15	1727	8	5.72	99.97	99.72	99.97
TBR-129_g9s8	1999-08-17Aug					1767	11	1561	8	1741	16	1753	10	6.75	99.86	98.58	99.84
TBR-129_Mon1_sp1	98_08_28aug \		P/A	same as grain 9		1747	13	1671	41					3.35			
TBR-129_Mon3_sp1	98_08_28aug \	85	P/A	matrix		1772	9	1593	8					4.19			
TBR-129_Mon3_sp2	98_08_28aug \					1765	4	1496	9					3.76			
TBR-16	**Noble Lake Ridge**	**SPMS**	**Garnet-Orthoamphibole Gneiss**														
TBR-16_g1sp1	1999-08-18Aug	100	E/A	matrix	X	1740	9	1676	8	1709	15	1723	8	6.86	99.92	98.78	99.86
TBR-16_g1sp2	1999-08-18Aug					1785	9	1660	9	1756	14	1769	8	7.16	99.96	99.45	99.94
TBR-16_g1sp3	1999-08-18Aug					1756	9	1669	9	1738	13	1746	8	7.22	99.96	99.33	99.92
TBR-16_g1sp4	1999-08-18Aug					1764	9	10	0	1806	18	1787	10	7.09	92.54	98.7	99.85
TBR-16_gr1_sp1	98_08_27Aug\			same as grain 1		1748	6	1528	9					3.64			
TBR-16_gr1_sp2	98_08_27Aug\					1772	5	1626	9.99					3.62			
TBR-16_gr1_sp3	98_08_27Aug\					1749	7	1656	18					3.54			
						1759	**16**										
TBR-16_g2sp1	1999-08-18Aug	100	P/A	matrix	X	1771	9		0	1909	14	1844	9	6.48	91.24	99.06	99.89
TBR-16_g2sp2	1999-08-18Aug					1758	11	1650	13	1739	23	1748	13	5.50	99.98	99.24	99.91
TBR-16_gr2_sp1	98_08_27Aug\			same as grain 2		1751	7	1601	10					3.89			
TBR-16_gr2_sp2	98_08_27Aug\					1756	26	1662	15					3.77			
						1759	**9**										
TBR-16_g3sp2	1999-08-18Aug	100	E/A	matrix	X	1757	13	1678	7	1738	17	1746	10	7.04	99.95	98.49	99.83
TBR-16_g3sp3	1999-08-18Aug					1759	11	1687	7	1733	16	1744	10	7.23	99.94	98.25	99.8
TBR-16g3	1999-08-18Aug					1738	21	1680	47	1722	65	1729	35	7.33	99.82	95.34	99.45

(continued)

TABLE 1. ION MICROPROBE AGES OF MONAZITES FROM THE TOBACCO ROOT MOUNTAINS, MONTANA (*continued*)

Sample/Grain/Spot#	Date	Grain size (microns)	Grain shape*	Textural setting	X-ray-map (X = map)	Age (Ma)† $^{207}Pb/^{206}Pb$	1 s.e.§ $^{207}Pb/^{206}Pb$	Age (Ma) $^{208}Pb/^{232}Th$	1 s.e.§ $^{208}Pb/^{232}Th$	Age (Ma) $^{206}Pb/^{238}U$	1 s.e.§ $^{206}Pb/^{238}U$	Age (Ma) $^{207}Pb/^{235}U$	1 s.e.§ $^{207}Pb/^{235}U$	Th O_2/Th	% Radio-genic ^{208}Pb	% Radio-genic ^{207}Pb	% Radio-genic ^{206}Pb
TAH-18i	**Sunrise Lake**	**SPMS**	**Aluminous Gneiss**														
TAH-18i_g1sp1	1999-08-19Aug	120	P/A	in matrix/ garnet rim	X	1715	15	1629	7	1659	16	1684	10	7.20	99.95	92.05	99.05
TAH-18i_g1sp2	1999-08-19Aug					1734	37	1377	24	1595	31	1656	22	5.34	99.91	90.75	98.87
TAH-18i_g1sp3	1999-08-19Aug					1730	13	1641	9	1744	20	1738	13	6.19	99.96	93.27	99.2
TAH-18i_g1sp4	1999-08-19Aug					1752	9	1723	7	1793	15	1774	8	7.49	99.97	98.11	99.78
TAH-18i_gra_sp1	98_08_27Aug\			**?? same as grain 1**		1775	5	1655	13					3.95			
TAH-18i_gra_sp2	98_08_27Aug\			looks like grain 6		1774	7	1725	17					3.58			
TAH-18i_gra_sp2@1	98_08_27Aug\					1771	11	1767	16					3.67			
TAH-18i_g2sp2	1999-08-19Aug	50	E/A	matrix	X	1752	15	1699	8	1821	19	1789	12	6.46	99.94	94.86	99.39
TAH-18i_g6sp2	1999-08-19Aug					1742	5	1694	8	1820	9	1784	5	6.30	99.94	99.21	99.91
TAH-18i_gr2_sp1	98_08_27Aug\			same as grain 2		1585	148	1777	16					4.16			
TAH-18i_gr2_sp2	98_08_27Aug\					1728	21	1803	12					3.91			
TAH-18i_gr2_sp3	98_08_27Aug\					1727	20	1666	13					3.79			
TAH-18i_gr3_sp1	98_08_27Aug\	50	E/A	in matrix w/ kyanite	X	1720	24	1623	15					3.64			
TAH-18i_gr3_sp2	98_08_27Aug\					1740	57	1738	17					3.94			
TAH-18i_gr3_sp3	98_08_27Aug\					1739	43	1641	24					3.46			
TAH-18i_gr4_sp1	98_08_27Aug\	70	P/A	in kyanite		1753	8	1709	14					3.69			
TAH-18i_gr4_sp2	98_08_27Aug\					1760	13	1462	13					3.69			
TAH-18i_g6sp1	1999-08-19Aug	50	E/A	in garnet		1771	8	1738	8	1828	17	1801	8	6.63	99.93	99.16	99.9
TBR-251a	**Thompson Peak Ridge**	**SPMS**	**Aluminous Gneiss**														
TBR-251a_g1sp1	1999-08-19Aug	60	E/A	matrix	X	1786	6	1765	7	1873	10	1832	5	6.44	99.99	99.89	99.99
TBR-251a_g1sp2	1999-08-19Aug					1772	5	1721	7	1779	10	1776	6	6.68	99.98	99.77	99.97
TBR-251a_g1sp3	1999-08-19Aug					1761	5	1697	8	1776	9	1769	5	6.14	99.98	99.78	99.98
TBR-251a_g1sp4	1999-08-19Aug					1773	20	1524	22	1617	23	1686	13	5.74	99.97	99.8	99.98
TBR-251a_g1sp5	1999-08-19Aug					1773	4	1669	16	1740	17	1755	9	5.26	99.99	99.87	99.99
						1773	**9**										
TBR-110A	**Noble Lake**	**SPMS**	**Aluminous Gneiss**														
TBR-110AgBsp1	1999-08-20Aug	75	E	matrix		1782	9	1651	10	1773	12	1777	7	7.25	99.93	99.32	99.92
TBR-110AgBsp2	1999-08-20Aug					1768	11	1634	9	1573	15	1658	10	6.98	99.88	98.38	99.81
TBR-110AgBsp3	1999-08-20Aug					1757	11	1690	8	1651	18	1698	10	6.65	99.83	97.98	99.77
						1769	**13**										
TBR-110B	**Noble Lake**	**SPMS**	**Aluminous Gneiss**														
TBR-110BgB1	1999-08-22Aug	250	P	matrix		1757	7	1775	12	1840	16	1801	8	5.71	99.93	99.64	99.96
TBR-110BgB1@1	1999-08-22Aug					1721	6	1782	14	1846	15	1788	9	5.55	99.84	99.14	99.9
TBR-110BgB1@3	1999-08-22Aug					1769	6	1753	8	1828	9	1800	5	6.25	99.96	99.69	99.96
TBR-110BgB2	1999-08-22Aug	30	E	matrix	X	1838	8	1809	13	1813	17	1824	10	5.56	99.93	99.49	99.94
TBR-110BgB2@1	1999-08-22Aug					1755	7	1778	12	1861	16	1811	8	5.77	99.92	99.56	99.95

(*continued*)

TABLE 1. ION MICROPROBE AGES OF MONAZITES FROM THE TOBACCO ROOT MOUNTAINS, MONTANA (*continued*)

Sample/Grain/Spot#	Date	Grain size (microns)	Grain shape*	Textural setting	X-ray-map (X = map)	Age (Ma)† $^{207}Pb/^{206}Pb$	1 s.e.§ $^{207}Pb/^{206}Pb$	Age (Ma) $^{208}Pb/^{232}Th$	1 s.e.§ $^{208}Pb/^{232}Th$	Age (Ma) $^{206}Pb/^{238}U$	1 s.e.§ $^{206}Pb/^{238}U$	Age (Ma) $^{207}Pb/^{235}U$	1 s.e.§ $^{207}Pb/^{235}U$	Th O_2/Th	% Radio-genic ^{208}Pb	% Radio-genic ^{207}Pb	% Radio-genic ^{206}Pb
TBR-189	**Noble Lake**	**SPMS**	**Aluminous Gneiss**														
TBR-189gAsp1	1999-08-20Aug	100	E/A	matrix		1783	5	1693	11	1749	12	1765	6	5.81	99.94	99.49	99.94
TBR-189gAsp2	1999-08-20Aug					1781	7	1690	7	1697	9	1735	6	7.04	99.88	99.51	99.94
						1782	**1**										
TBR-189gbsp1	1999-08-20Aug	50	E	matrix		1781	6	1745	11	1686	11	1729	6	6.16	99.78	98.89	99.87
TBR-189gbsp2	1999-08-20Aug					1765	10	1503	15	1492	16	1609	10	5.80	99.85	99.42	99.93
						1773	**11**										
TBR-189gCsp1	1999-08-20Aug	60	E	matrix		1749	14	1692	8	1680	20	1711	11	6.84	99.94	96.59	99.6
TBR-189gCsp2	1999-08-20Aug					1730	15	1677	10	1688	20	1707	11	6.10	99.97	97.98	99.77
						1740	**13**										
TBR-138	**Leggat Ridge**	**SPMS**	**Garnet-Orthoamphibole Gneiss**														
TBR-138_g1	1999-08-21Aug	50	E	matrix		1777	13	1687	8	1726	20	1749	12	6.99	99.89	97.64	99.72
TBR-138g5sp1@1	1999-08-21Aug	50	E/A	matrix		1785	12	1696	12	1654	26	1712	15	7.37	99.81	99.02	99.89
TBR-138_g7sp1	1999-08-21Aug	100	E	matrix		1771	8	1644	8	1646	11	1702	7	7.06	99.91	99.43	99.93
TBR-138_g7sp1@1	1999-08-21Aug			matrix		1774	11	1644	19	1723	24	1746	12	6.09	98.56	97.97	99.76
TBR-138_g8	1999-08-21Aug	50	P	in crd after grt		1776	12	1689	11	1736	15	1754	10	5.97	99.51	97.32	99.68
						1777	**5**										
TBR-54	**Granite Peak ridge**	**SPMS**	**Garnet-Orthoamphibole Gneiss**														
TBR-54gai	1999-08-22Aug	150	E	matrix	X	1866	9	1696	13	1556	16	1692	9	7.01	99.96	99.55	99.95
TBR-54gai@1	1999-08-22Aug					1785	11	1633	8	1722	17	1750	7	6.55	99.95	99.57	99.95
TBR-54gai@2	1999-08-22Aug					1741	13	1486	11	1602	20	1663	10	7.04	99.96	98.64	99.85
TBR-54gai@3	1999-08-22Aug					1749	31	1647	22	1739	32	1744	15	6.90	99.91	96.56	99.6
						1785	**57**										
TBR-10	**Thompson Peak Ridge**	**SPMS**	**Garnet-Orthoamphibole Gneiss**														
TBR-10_4s1g1	1999-12-06Dec	175	P	in crd after grt		1786	7	1698	14	1776	17	1781	10	4.80	99.95	99.60	99.95
TBR-10_4g1sp2	1999-12-06Dec					1795	7	1701	13	1711	19	1749	11	4.98	99.94	99.42	99.93
TBR-10_4g1sp3	1999-12-06Dec					1823	9	1604	9	1578	12	1686	8	5.29	99.97	99.40	99.93
TBR-10_4gr1sp4	1999-12-06Dec					1753	11	1603	15	1732	23	1741	14	4.73	99.96	98.96	99.88
TBR-10_4gr1sp5	1999-12-06Dec					1748	10	1619	20	1793	26	1772	14	4.50	99.97	99.16	99.90
TBR-10_4gr1sp6	1999-12-06Dec					1770	7	1662	9	1560	10	1651	5	5.32	99.96	99.68	99.96
TBR-10_4gr1sp7	1999-12-06Dec					1765	8	1598	16	1602	22	1674	13	4.69	99.95	99.29	99.92
TBR-10_4gr1sp8	1999-12-06Dec					1768	7	1552	15	1596	16	1671	10	4.74	99.95	99.27	99.92
TBR-10_4gr1sp8	1999-12-06Dec					1769	8	1555	13	1596	18	1672	11	5.06	99.97	99.40	99.93
TBR-10_4gr1sp10	1999-12-06Dec					1767	7	1597	21	1632	23	1692	13	4.56	99.97	99.27	99.92
TBR-10_gr2bsp1	1999-12-06Dec	110	E	in crd after grt		1762	12	1492	9	1533	17	1632	10	5.52	99.96	99.65	99.96
TBR-10_gr2bsp2	1999-12-06Dec					1768	10	1502	9	1575	10	1659	6	5.54	99.97	99.70	99.97
TBR-10_gr2bsp3	1999-12-06Dec					1771	9	1593	10	1559	13	1652	8	5.87	99.96	99.63	99.96
TBR-10_gr2bsp4	1999-12-06Dec					1772	9	1605	16	1837	19	1807	10	4.91	99.95	99.72	99.97
TBR-10_gr2bsp5	1999-12-06Dec					1766	10	1403	21	1713	26	1737	14	4.95	99.97	99.40	99.93
TBR-10_gr2bsp6	1999-12-06Dec					1763	9	1369	16	1622	22	1685	12	4.73	99.97	99.58	99.95
						1767	**4**										

(*continued*)

TABLE 1. ION MICROPROBE AGES OF MONAZITES FROM THE TOBACCO ROOT MOUNTAINS, MONTANA (*continued*)

Sample/Grain/Spot#	Date	Grain size (microns)	Grain shape*	Textural setting	X-ray-map (X = map)	Age (Ma)† $^{207}Pb/^{206}Pb$	1 s.e.§ $^{207}Pb/^{206}Pb$	Age (Ma) $^{208}Pb/^{232}Th$	1 s.e.§ $^{208}Pb/^{232}Th$	Age (Ma) $^{206}Pb/^{238}U$	1 s.e.§ $^{206}Pb/^{238}U$	Age (Ma) $^{207}Pb/^{235}U$	1 s.e.§ $^{207}Pb/^{235}U$	Th O_2/Th	% Radio-genic ^{208}Pb	% Radio-genic ^{207}Pb	% Radio-genic ^{206}Pb
KAT-20	**Thompson Peak Ridge**	**SPMS**	**Garnet-Orthoamphibole Gneiss**														
KAT-20_g1sp1	1999-12-07Dec	150	E/A	matrix		1765	9	1510	19	1594	21	1669	11	5.19	99.81	98.07	99.78
KAT-20_g1sp2	1999-12-07Dec					1770	10	1581	26	1695	29	1729	17	5.11	99.89	99.00	99.89
KAT-20_g1sp3	1999-12-07Dec					1770	9	1320	27	1452	21	1587	13	4.78	99.91	99.31	99.92
KAT-20_g1sp4	1999-12-07Dec					1760	12	1505	11	1521	13	1624	8	5.63	99.74	97.79	99.74
						1766	**5**										
KAT-20_g2sp1	1999-12-07Dec	100	E/A	matrix		1773	7	1576	22	1659	19	1710	11	5.03	99.97	99.61	99.96
KAT-20_g2sp2	1999-12-07Dec					1764	9	1534	30	1632	26	1691	14	4.79	99.97	99.61	99.96
KAT-20_g2sp3	1999-12-07Dec					1720	8	1602	12	1614	15	1660	9	5.43	99.85	99.58	99.95
KAT-20_g2sp4	1999-12-07Dec					1734	9	1572	24	1704	21	1717	12	4.96	99.77	97.80	99.75
TBR-6	**Thompson Peak Ridge**	**SPMS**	**Garnet-Orthoamphibole Gneiss**														
TBR-6_g1sp1	1999-12-07Dec	125	P	in biotite		1765	10	1557	14	1643	17	1697	10	5.31	99.96	99.25	99.91
TBR-6_g1sp2	1999-12-07Dec					1766	10	1508	10	1589	19	1667	11	5.45	99.97	99.00	99.89
TBR-6_g1sp3	1999-12-07Dec					1764	15	1483	10	1522	21	1626	13	5.51	99.97	99.01	99.89
						1765	**1**										
TBR-6_g2sp1	1999-12-07Dec	100	E	in biotite		1774	10	1624	10	1719	13	1744	7	5.48	99.92	99.45	99.94
TBR-6_g2sp2	1999-12-07Dec					1750	13	1631	19	1665	30	1702	17	5.65	99.96	98.98	99.88
TBR-6_g2sp3	1999-12-07Dec					1759	12	1498	13	1577	17	1656	10	5.30	99.96	99.14	99.90
						1761	**12**										
TBR-6_g3sp1	1999-12-07Dec	50	E	in biotite		1774	12	1605	17	1690	18	1728	12	5.34	99.94	99.08	99.89
TBR-6_g3sp12	1999-12-07Dec					1748	13	1548	15	1650	22	1693	12	5.28	99.96	99.17	99.91
JBB95-36a	**Rossitter Lake**	**SPMS**	**Aluminous Gneiss**														
JBB95_36a_g1Bsp1	1999-12-08Dec	50	E/A	in Ilm in Sil		1752	14	1678	18	1560	21	1644	13	6.436	99.67	95.45	99.46
JBB95_36a_g1Bsp2	1999-12-08Dec					1761	13	1698	15	1780	24	1771	13	5.159	99.93	99.05	99.89
JBB95_36a_g1Bsp3	1999-12-08Dec					1776	16	1623	16	1634	23	1697	11	4.994	99.9	**98.5**	99.83
						1763	**12**										
JBB97-14b	**Leggat Ridge**	**SPMS**	**Aluminous Gneiss**														
JBB97_14b_g1asp1	1999-12-08Dec	300	P	matrix		1716	29	1492	19	1566	21	1631	16	4.632	99.95	**93.64**	99.25
JBB97_14b_g1asp2	1999-12-08Dec					1736	70	1492	11	1552	32	1632	37	4.708	99.78	**87.23**	98.37
JBB97_14b_g1asp3	1999-12-08Dec					1669	115	1082	13	1350	27	1479	53	4.24	99.45	**59.54**	93.12
JBB97_14b_g1asp4	1999-12-08Dec					1773	23	1401	13	1452	33	1588	21	4.513	99.97	**97.11**	99.66
JBB97_14b_g1asp5	1999-12-08Dec					1765	52	1422	11	1484	30	1604	32	4.679	99.79	**87.81**	98.44
JBB97_14b_gMsp1	1999-12-08Dec	125	E/A	matrix		1726	28	1563	9	1639	37	1678	23	5.423	99.96	94.41	99.34
JBB97_14b_gMsp2	1999-12-08Dec					1713	31	1548	10	1677	31	1693	17	5.574	99.91	89.88	98.77
JBB97_14b_gMsp3	1999-12-08Dec					1737	27	1516	9	1633	48	1679	28	5.046	99.96	94.7	99.38
JBB97_14b_gMsp4	1999-12-08Dec					1717	27	1516	21	1593	36	1647	21	5.381	99.97	95.88	99.52
						1723	**11**										
JBB97-29c	**Cataract Mtn.**	**SPMS**	**Garnet-Orthoamphibole Gneiss**														
JBB97_29c_g2sp1	1999-12-08Dec	150	E	matrix		1759	10	1607	12	1606	14	1673	9	5.247	99.96	98.76	99.86
JBB97_29c_g2sp2	1999-12-08Dec					1728	15	1539	10	1540	15	1621	11	5.19	99.91	96.64	99.61
JBB97_29c_g2sp3	1999-12-08Dec					1746	15	1520	9	1528	21	1622	15	5.603	99.89	96.13	99.55
JBB97_29c_g2sp4	1999-12-08Dec					1727	15	1545	11	1532	17	1616	9	5.429	99.94	98.13	99.79
						1740	**15**										

(*continued*)

TABLE 1. ION MICROPROBE AGES OF MONAZITES FROM THE TOBACCO ROOT MOUNTAINS, MONTANA (*continued*)

Sample/Grain/Spot#	Date	Grain size (microns)	Grain shape*	Textural setting	X-ray-map (X = map)	Age (Ma)† $^{207}Pb/^{206}Pb$	1 s.e.§ $^{207}Pb/^{206}Pb$	Age (Ma) $^{208}Pb/^{232}Th$	1 s.e.§ $^{208}Pb/^{232}Th$	Age (Ma) $^{206}Pb/^{238}U$	1 s.e.§ $^{206}Pb/^{238}U$	Age (Ma) $^{207}Pb/^{235}U$	1 s.e.§ $^{207}Pb/^{235}U$	Th O_2/Th	% Radio-genic ^{208}Pb	% Radio-genic ^{207}Pb	% Radio-genic ^{206}Pb
JBB95-39A	**Cataract Mtn.**	**SPMS**	**Quartzofeldspathic Gneiss**														
JBB95_39Ag2sp1	1999-12-09Dec	80	E	matrix		1717	12	1435	12	1463	14	1570	10	4.804	99.84	95.69	99.5
JBB95_39Ag2sp2	1999-12-09Dec					1737	11	1541	12	1550	12	1631	8	5.06	99.96	98.97	99.88
JBB95_39Ag2sp3	1999-12-09Dec					1739	17	1326	31	1275	30	1459	20	3.948	99.97	99.11	99.9
						1731	**12**										
RBB-3	**Leggat Saddle**	**SPMS**	**Quartzite**														
RBB-3_g2sp1	1999-12-09Dec	100		matrix		469	417	79	1	77	1	91	17	4.462	87.98	14.27	73.68
RBB-3_g2sp3	1999-12-09Dec		E/A	cluster of 4 grains		0	0	74	1	83	2	46	24	4.436	94.97	8.998	78.5
RBB-3_g2sp4	1999-12-09Dec					0	0	76	1	72	2	11	23	4.469	84.45	1.189	61.97
RBB-3_g2sp4@1	1999-12-09Dec					1750	52	1420	15	1703	43	1724	34	4.804	98.38	57.11	92.17

* Grain shape codes: P—prismatic; E—equant; A—irregular or anhedral shape; ICMS—Indian Creek Metamorphic Suite; PMMMS—Pony–Middle Mountain Metamorphic Suite; SPMS—Spuhler Peak Metamorphic Suite.

†Boxes outline spot ages that are averaged, and the average is shown in the outlined boxes. The number to the right of the average, in the outlined boxes, is the standard deviation of the average.

§These are one sigma standard errors.

All results are reported in Table 1, but as discussed above, the $^{207}Pb/^{206}Pb$ ages are believed to be more consistent indicators of monazite growth ages and are used exclusively in this study.

As shown by the histograms in Figures 3 and 4, all three rock suites contain ages between ca. 1720 and ca. 1780 Ma, but the monazite age populations for the Indian Creek and Pony–Middle Mountain Metamorphic Suites are bimodal. Both the Indian Creek and Pony–Middle Mountain Metamorphic Suites contain a younger group of ages varying from 1720 to 1780 Ma and a second group of ages that form an array ca. 2450 Ma.

The 272 spot ages reported in Table 1 were obtained from 70 different monazite grains in four different rock types among the three metamorphic suites. The most common rock types studied were aluminous gneiss, orthoamphibole gneiss, quartzofeldspathic gneiss, and quartzite. The monazite grains selected for dating in these rocks tend to appear as equant or elongate grains reflecting end section and prismatic orientations, respectively; within a given sample the prismatic grains tend to be larger (Table 1). These grains range in size from less than 25 microns up to 500 microns across, but most are 100 microns across or less. The selected monazite grains vary from anhedral to euhedral, but there are no systematic trends in shape or size with regard to rock type or rock suite. Although exceptions occur, matrix grains tend to be larger than those monazite grains that occur as inclusions in other minerals.

Relationship of Ages to Chemical Zonation of Monazite Grains

Spot age locations on most monazite grains and existing X-ray maps are provided in Webb (2001). These materials are available from the senior author and include X-ray maps for nearly all (30 of 33) of the dated Indian Creek and Pony–Middle Mountain Metamorphic Suite monazite grains (see Table 1) as well as back-scattered electron images showing the locations and values of spot ages from most of the samples reported here. Because this study evolved with the development of SIMS techniques for monazite dating, our approach also evolved. During our initial session in June of 1998 we did not have X-ray maps of all the monazite grains prior to dating the monazite. Our initial age results and initial X-ray maps of Spuhler Peak Metamorphic Suite monazite grains indicated that they are relatively homogeneous, and we concentrated our efforts on the more heterogeneous monazite grains from the Indian Creek and Pony–Middle Mountain Metamorphic Suite samples. However, it has become apparent that X-ray maps are essential for interpreting the ages of all monazite grains (e.g., see Williams et al., 1999; Williams and Jercinovic, 2002). Nonetheless, as shown in Table 1, only 12 of 37 Spuhler Peak Metamorphic Suite dated monazite grains are accompanied by X-ray maps.

The available X-ray maps of the monazite grains are diverse and have varying and complex relationships among metamorphic rock suites, lithologies, and/or textural settings of the monazite grains. In some cases similar zonation patterns occur in several grains of the same sample. The distribution of Th and Y and the

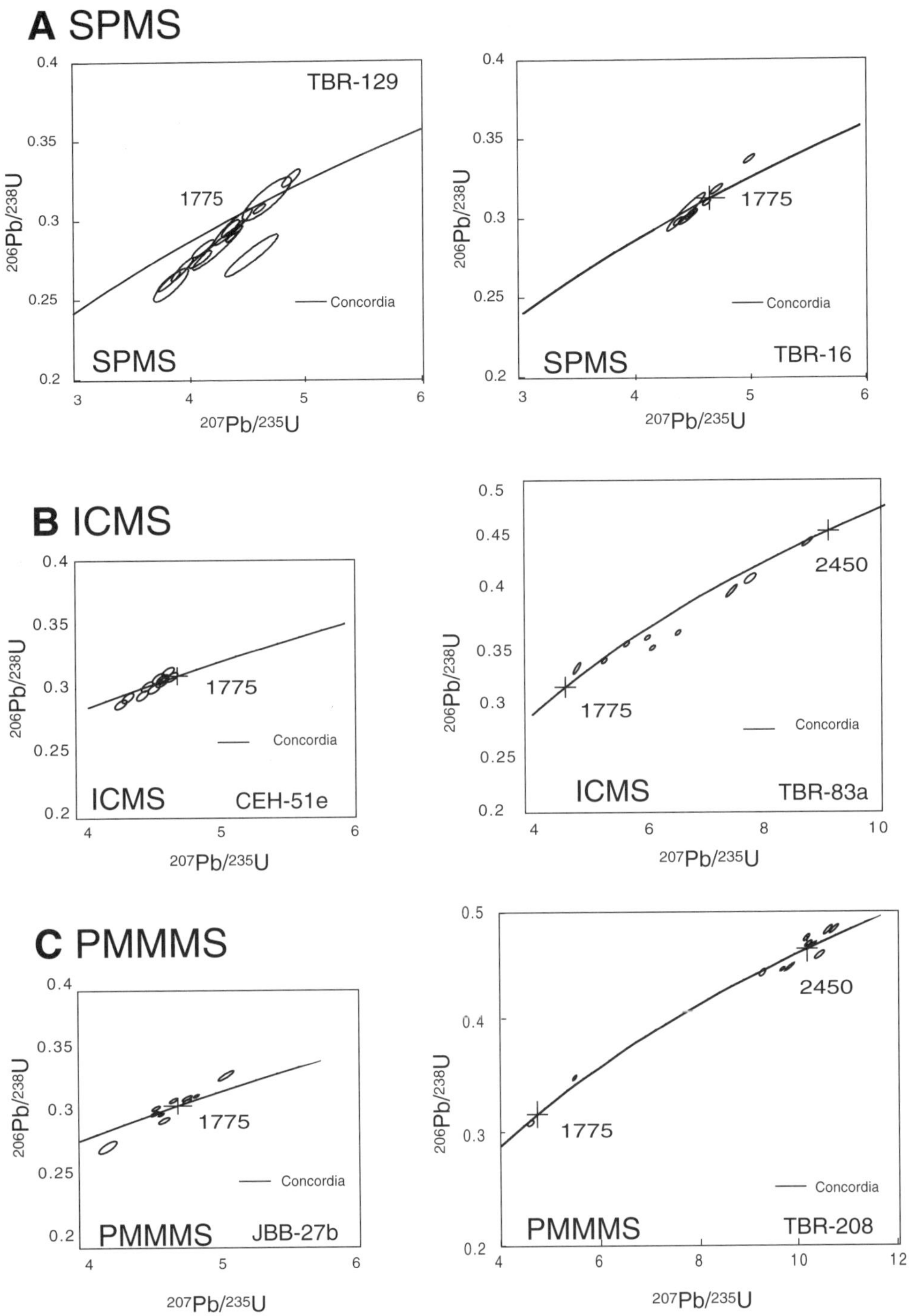

Figure 2. Representative concordia plots for spot ages by metamorphic rock suite. A: Spuhler Peak Metamorphic Suite (SPMS). B: Indian Creek Metamorphic Suite (ICMS). C: Pony–Middle Mountain Metamorphic Suite (PMMMS). Error ellipses are 1σ.

geometry of their compositional variations are shown for a number of samples in Figures 5 through 11. Most of the grains are compositionally zoned, although homogenous grains exist (Fig. 10A). Other grains, particularly in the Indian Creek and Pony–Middle Mountain Metamorphic Suites, have irregular or patchy rims (Figs. 8D and 9B) or cores (Figs. 9B and 10D) that are especially pronounced in the Y maps. In some cases, the Y variation is inversely related to the changes in Th-U and Pb (Fig. 11). The variation in Pb or U (Fig. 9) is less distinct and the elements tend to be uniformly distributed. Nd as a proxy for the rare earth elements in monazite is also evenly distributed in most samples, but there are exceptions. These features are less common in Spuhler Peak Metamorphic Suite samples, and the suite's monazite tends to have simpler zonation, illustrated principally by Th variation.

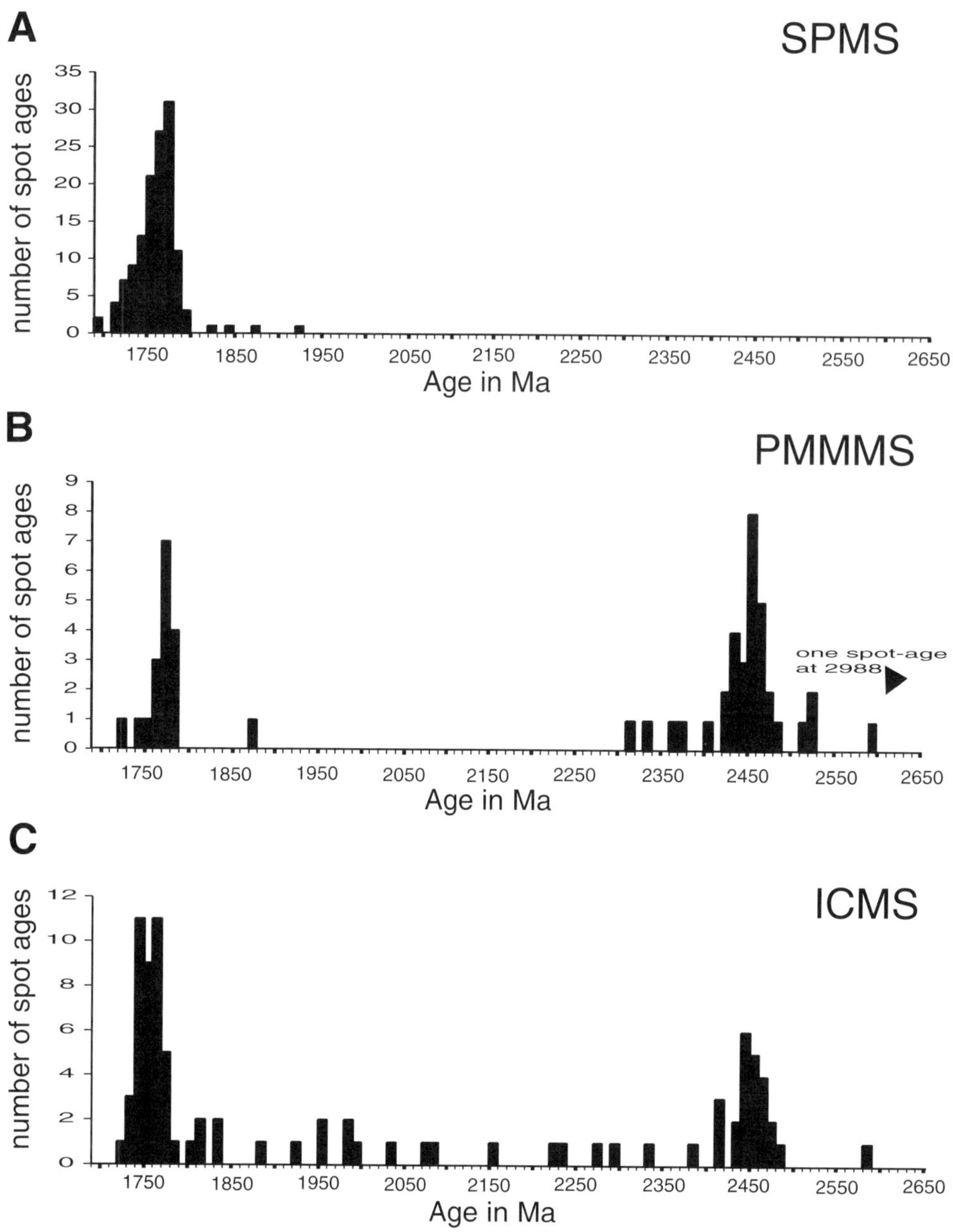

Figure 3. Histograms showing $^{207}Pb/^{206}Pb$ spot age distribution by metamorphic rock suite. A: Spuhler Peak Metamorphic Suite (SPMS). B: Pony–Middle Mountain Metamorphic Suite (PMMMS). C: Indian Creek Metamorphic Suite (ICMS). The bin size on all three diagrams is 10 m.y.

Some Spuhler Peak Metamorphic Suite monazite grains have well-defined high-Th rims (Fig. 5), whereas others have irregular, partial high-Th rims. This relationship can be reversed, even in the same sample as shown by Figure 6. The weak low-Th rims in the Spuhler Peak Metamorphic Suite aluminous gneiss of Figure 6B, the Spuhler Peak Metamorphic Suite orthoamphibolite in Figure 7C, and the Pony–Middle Mountain Metamorphic Suite aluminous gneiss in Figure 9A, coupled with the irregular high-Y cores in the Indian Creek Metamorphic Suite aluminous gneiss of Figure 8B or with the orthoamphibole-bearing rock in Figure 10D, illustrate that similar monazite zoning patterns can occur in rocks of differing composition and metamorphic suite.

Discerning the relationship between chemical zoning and age distribution can be difficult when the data available are limited because (1) X-ray maps are not available for each grain; (2) some chemically zoned grains are not age zoned, and some homogenous grains have significant, but not systematic, age variation (e.g., Figs. 8B and 10A); (3) we did not sample all areas of zoned grains with the ion probe beam; (4) the scale of the age variation and the chemical zoning is commonly smaller than the 20–30 micron beam of the ion microprobe and thus beyond the spatial resolution obtainable. In the context of point 3, Webb (2001) presents some high-resolution X-ray maps and exploratory electron microprobe ages provided by Mike Jercinovic at the University

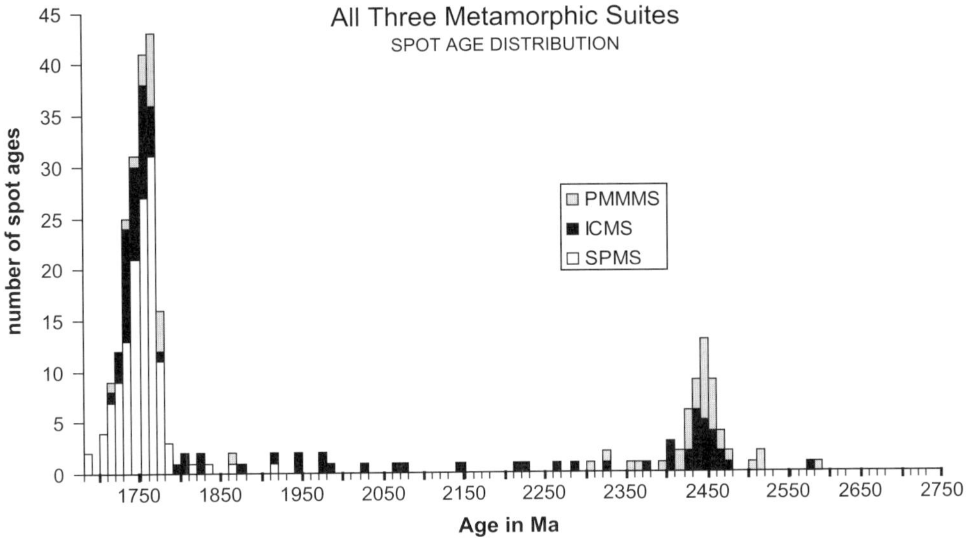

Figure 4. Histogram showing the distribution of spot ages from all three metamorphic rock suites. PMMMS—Pony–Middle Mountain Metamorphic Suite, ICMS—Indian Creek Metamorphic Suite, SPMS—Spuhler Peak Metamorphic Suite. The bin size is 10 m.y.

of Massachusetts. The higher spatial resolution of the electron microprobe permits evaluation of the small-scale age variations that can occur in complex geometric configurations, as reported by several workers (e.g., Williams et al., 1999; Williams and Jercinovic, 2002; Pyle et al., 2002; Spear and Pyle, 2002). These preliminary data coupled with existing X-ray maps that have ion microprobe spot ages plotted on them as in Figure 5B indicate that the ion microprobe may be sampling different age domains in some samples. This problem is particularly exacerbated if the younger age is from relatively thin rims, as suggested in many of the X-ray maps (e.g., Figs. 5A, 9A–C, 11A).

Spuhler Peak Metamorphic Suite Monazite Ages

With six exceptions, all 137 spot ages on 37 grains from 17 Spuhler Peak Metamorphic Suite samples are, within the standard error of each spot age (Table 1), between 1720 and 1780 Ma. The six exceptions include two younger ages that have uncharacteristically large errors, in excess of 100 m.y., and five older ages from three different samples.

The single-grain spot ages vary within the standard error of each analysis in about half of the analyzed grains (Table 1). Typically, the spot ages of grains that exceed the spot age standard error can be related to spot location and/or compositional zonation (Figs. 5, 6, 7). Although there are exceptions, the older ages match up with the grain interiors whereas the younger grains, especially those younger than 1730 Ma, are confined to the compositionally defined grain edges. Intermediate ages likely result from the mixing of age population in the relatively large ion probe beam (see Fig. 11A).

Most of the Spuhler Peak Metamorphic Suite data are from matrix monazite grains, which have spot ages that vary from 1715 ± 15 to 1786 ± 6 Ma. A very limited number of the spot ages, as indicated in Table 1, are from monazite inclusions in garnet, kyanite, or biotite (Fig. 6D); other grains occur in cordierite ± sillimanite intergrowths that have partially replaced garnet (Fig. 5C). The inclusion ages, although very limited in number, do place upper age limits on the growth of some important minerals in these samples. With the exception of one spot age (1713 ± 18 Ma) from a monazite inclusion in biotite, all inclusion ages are older than ca. 1750 Ma. The four biotite inclusions yield ages that are older than 1753 ± 3 Ma. Likewise the spot ages from one inclusion in garnet are older than 1771 ± 8 Ma, and those from one inclusion in kyanite are older than 1753 ± 8 Ma. The monazite grains that occur as inclusions in cordierite, or as part of polycrystalline cordierite ± sillimanite pods that have partially replaced garnet, range in age from 1747 ± 13 to 1786 ± 7 Ma. Thus, spot ages younger than ca. 1750 Ma are restricted to matrix monazite grains in the Spuhler Peak Metamorphic Suite.

The oldest age is 1920 Ma and occurs in the core of an aluminous gneiss monazite (Fig. 6). The other four ages between 1780 and 1920 Ma may also reflect preservation of detrital cores. The 60 m.y. spread (1720–1780 Ma) in Spuhler Peak Metamorphic Suite ages (Fig. 3A) is from two to six times the standard error of most of the analyses. Moreover, the single-grain age variation exceeds the standard error in about half of the grains studied. Much, but not all, of this age variation can be related to compositional zoning and to the mixing of age populations in the ion probe beam. This suggests that these grains grew during multiple growth stages, or that they grew continuously over an extended period, or both. Accordingly, the spread in Spuhler Peak Metamorphic Suite ages from 1720 to 1780 Ma represents different age populations mixing in the ion beam. But we do not know the identity or even the total number of these ages. There are a significant number of homogenous grains with a spot age near 1750 Ma. Although this 1750 Ma remains problematic, the 1720–1780 span of ages represents the metamorphic growth of monazite over an extended orogenic period of some 60 m.y.

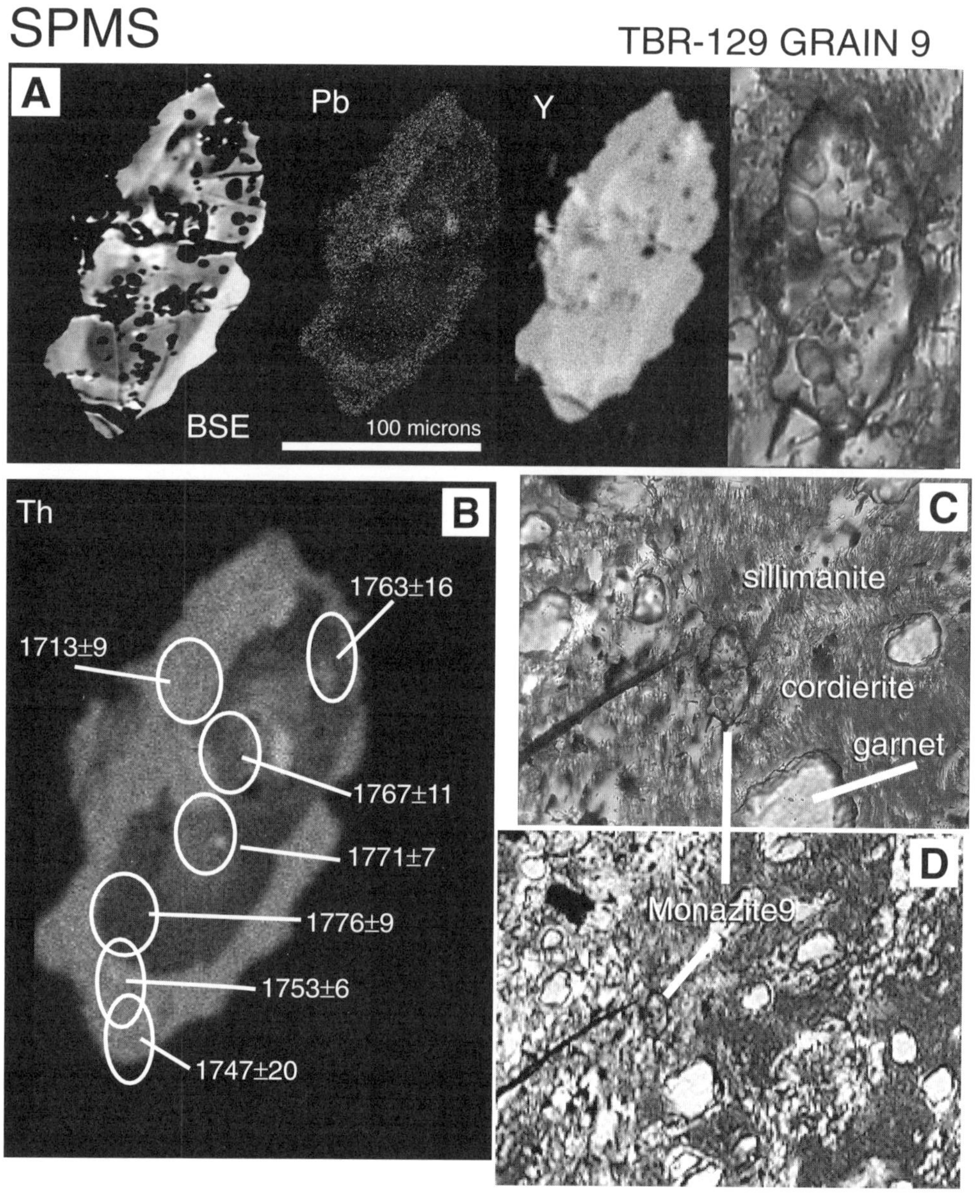

Figure 5. Monazite images from grain 9 of TBR-129, an aluminous gneiss from the Spuhler Peak Metamorphic Suite (SPMS). Backscattered electron image (A) and X-ray maps illustrating systematic Th (B) and Pb (A) and weak Y compositional zoning of monazite grains. B: Th X-ray map showing ion microprobe spot locations and $^{207}Pb/^{206}Pb$ ages for grain 9. C and D: Photomicrographs showing textural setting of grain 9 in sillimanite fibers in a moat of cordierite (darker gray) after garnet reaction zone.

Two samples (JBB95-39a, JBB97-29c) from Cataract Mountain give only ages in the 1720–1780 Ma range. Although Cataract Mountain is surrounded by Pony–Middle Mountain Metamorphic Suite gneisses, these rocks were identified by Hanley (1975) as very similar to the Spuhler Peak Metamorphic Suite rocks of Gillmeister (1971). The absence of monazite ages older than the 1720–1780 Ma group is consistent with Hanley's (1975) interpretation that Cataract Mountain is a possible tectonic slice of Spuhler Peak Metamorphic Suite. We agree with this idea and have grouped the Cataract Mountain samples with the Spuhler Peak Metamorphic Suite samples.

Finally, it is worth pointing out that the one Spuhler Peak Metamorphic Suite sample collected from closest to the Tobacco Root batholith contains an interesting mixed suite of ages that include a 1750 ± 52 Ma and three $^{232}Th/^{208}Pb$ ages of 79, 74, and 76 Ma, all ± 1 Ma. These young monazite grains provide a date of intrusion of the Tobacco Root batholith. The Th-Pb ages are used here for two reasons. First, the error in the ^{207}Pb in such young rocks makes the U-Pb system unreliable. Secondly, even with a very large error of ±5% in Th-Pb due to calibration errors, the resulting error in the age would be within ±3 Ma of the intrusion age, here taken to be 76 ± 2.5 Ma—the average of the analyses in Table 1.

Pony–Middle Mountain Metamorphic Suite Monazite Ages

The bimodal age distribution shown on Figure 3B is from 48 spot ages of 13 monazite grains in five Pony–Middle Mountain Metamorphic Suite samples. There are three different types of

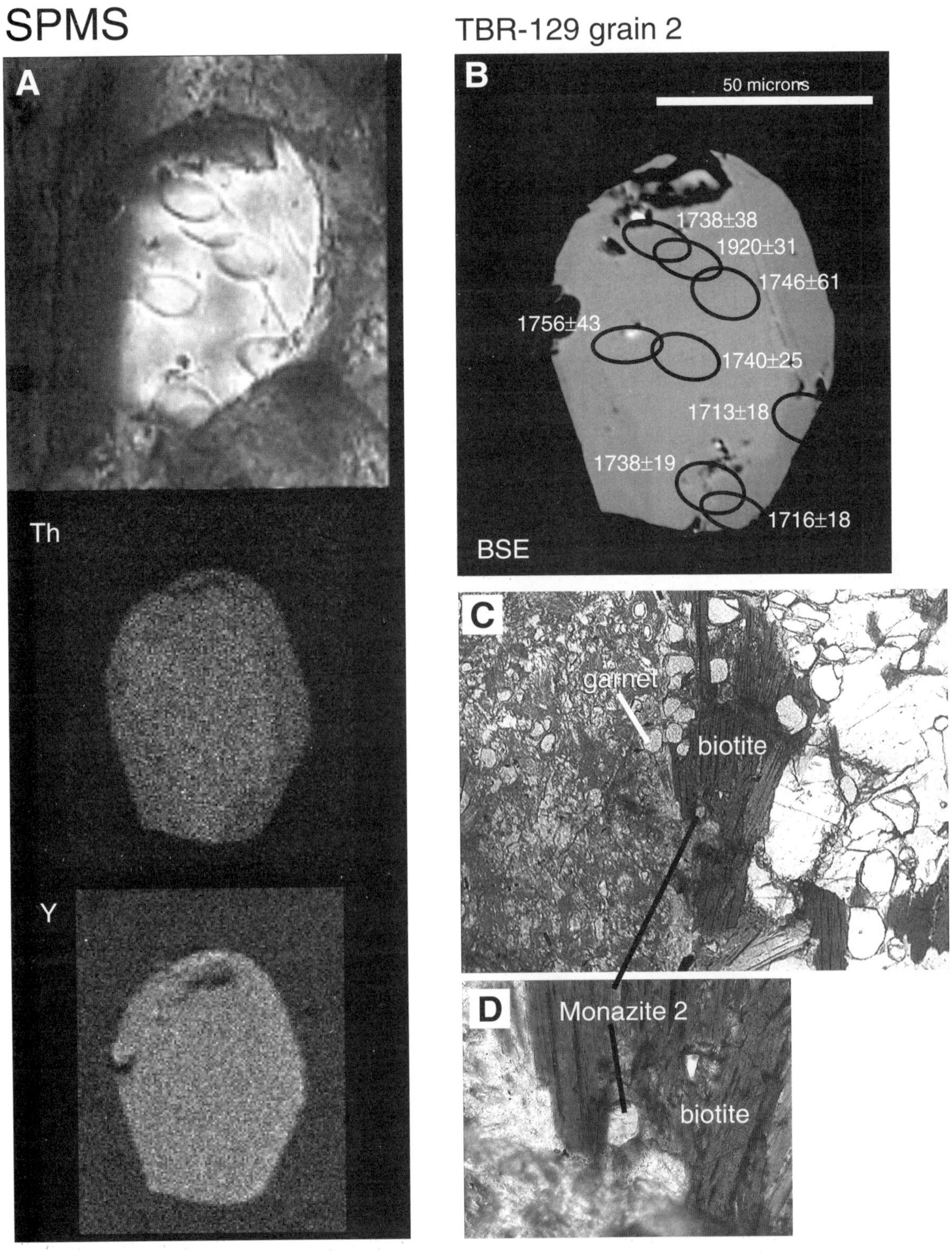

Figure 6. Monazite images from grain 2 of sample TBR-129, an aluminous gneiss from the Spuhler Peak Metamorphic Suite (SPMS). A: Backscattered electron (BSE) image and X-ray maps showing very weak inverse Y and Th compositional zoning. B: BSE image showing ion microprobe spot locations and $^{207}Pb/^{206}Pb$ ages for grain 2. C and D: Photomicrographs showing textural setting of grain 2 as inclusion in coarse-grained biotite that grew from back reaction of melt during cooling. Note the many garnet fragments enclosed in the biotite.

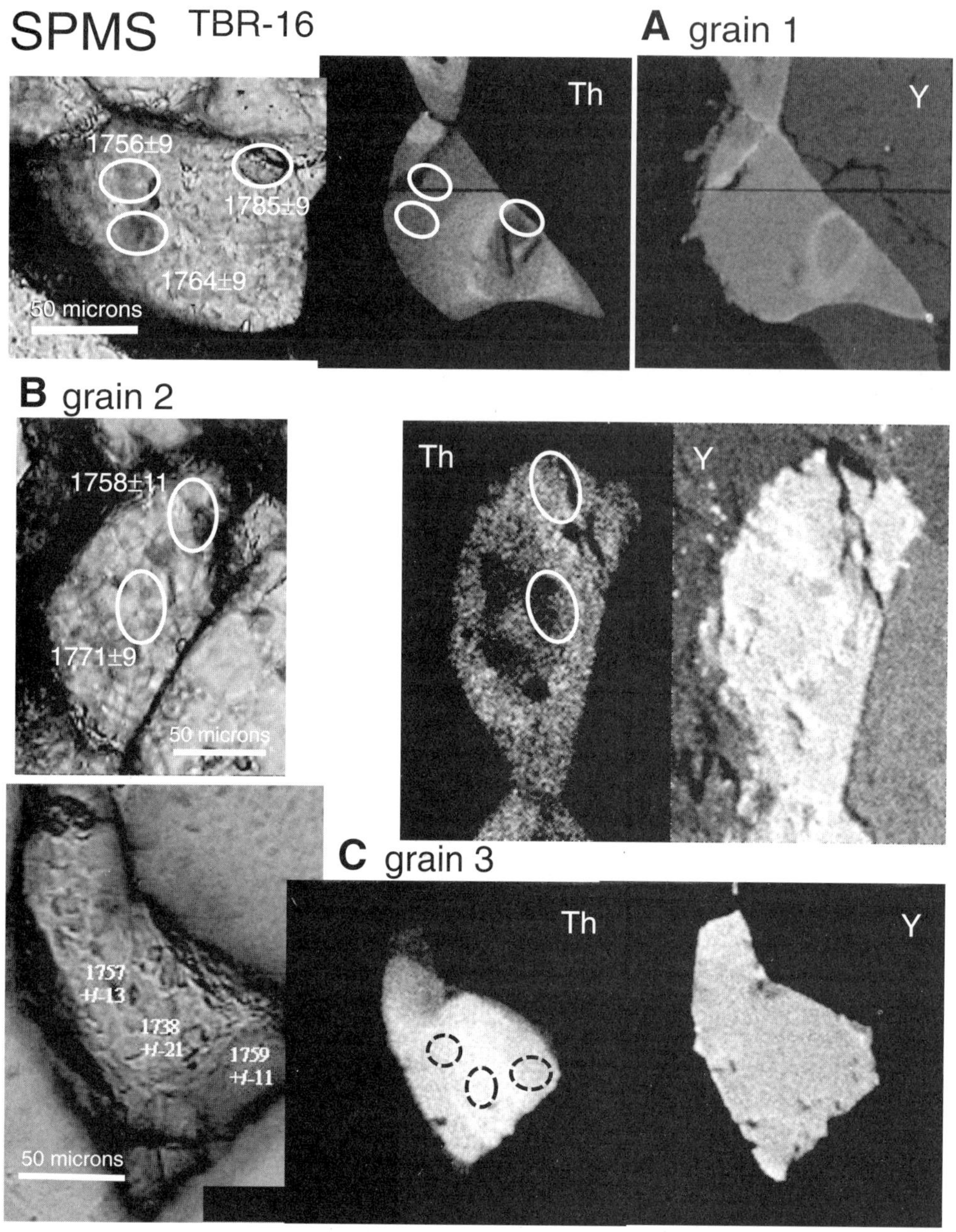

Figure 7. Monazite images from sample TBR-16, an orthoamphibole gneiss from the Spuhler Peak Metamorphic Suite (SPMS). A, B, and C: $^{207}Pb/^{206}Pb$ spot age locations and X-ray maps for grains 1, 2, and 3 respectively.

monazite age distributions from these five Pony–Middle Mountain Metamorphic Suite samples. One sample (JBB97-27b, see Fig. 1) contains monazite with ages only between 1724 ± 6 and 1783 ± 4 Ma, analogous to Spuhler Peak Metamorphic Suite monazite. Each of the three grains in this sample is from a different textural setting (Table 1; Fig. 8). Grain 1, an inclusion in a large biotite, has an irregular high-Y core and possibly a very narrow low-Y rim (Fig. 8B). Moreover, with one unresolved exception, all of the core ages are older than 1765 ± 4 Ma. There is also an anomalous 1783 ± 4 Ma spot on the grain edge. In contrast, grain 3 is an inclusion in kyanite, but it is zoned in the opposite sense. This smaller grain (Fig. 8D) has a low-Y core and a higher-Y rim, but the three spot ages are ca. 1775 Ma within error of each other. Thus, it is tempting to equate the monazite enclosed in kyanite with at least part of the center of the monazite enclosed in the biotite grain. We have no X-ray map for monazite grain 2 from this sample. It is also relatively small (60 microns) and occurs as an inclusion in a matrix quartz grain. The five spot ages range from 1759 ± 5 to 1778 ± 6 Ma (Table 1), similar to the core of grain 1, the inclusion in biotite.

A second group of three samples includes 26 spot analyses from seven grains that are older than 2309 Ma. This group is

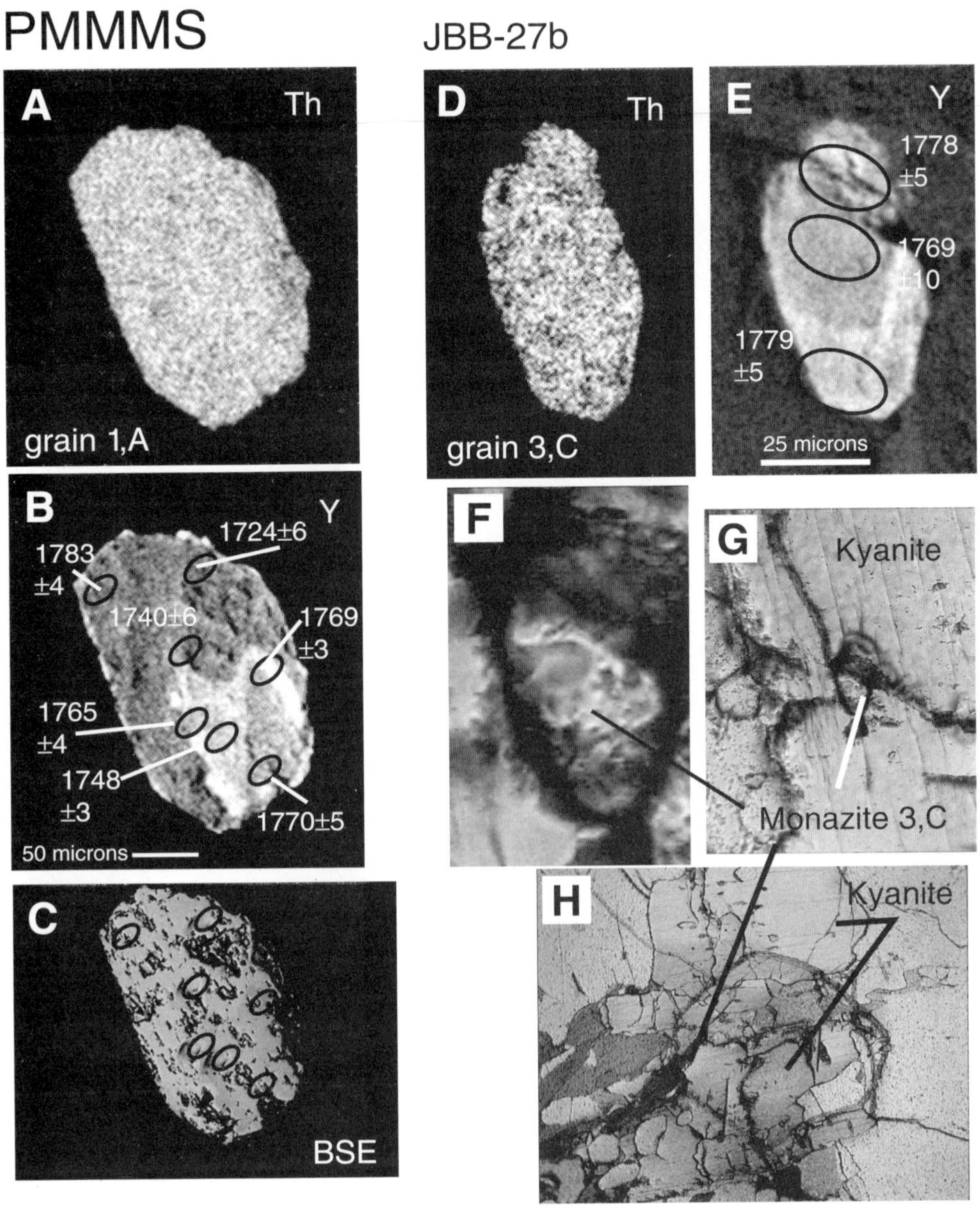

Figure 8. Monazite images from sample JBB-27b, an aluminous gneiss from the Pony–Middle Mountain Metamorphic Suite (PMMMS). A: Th X-ray map. B: Y X-ray map shows the spot age locations and Y zoning of grain 1. C: The ion probe pits and the textural setting of grain 1 in the matrix enclosed by coarse biotite laths. D, E, and F: $^{207}Pb/^{206}Pb$ spot age locations and X-ray maps for grain 3, an inclusion in kyanite as shown in G and H. Note that the high-Y rim is the opposite of the zoning for grain 1, which has a high-Y core.

somewhat heterogeneous with 17 of these older ages ranging continuously between 2425 and 2480 Ma. These ages cluster about 2450 Ma, as shown on Figure 3B. Five other spot ages range up in age from 2522 to 2988 Ma and are likely mixed ages between the 2450 Ma age and older remnant core ages, possibly from detrital grains. The four slightly younger spot ages that range from 2398 Ma down to 2309 Ma may also result from mixing of the 2420 Ma age, but with younger 1780 Ma ages. Most of the monazite grains from these samples are zoned, particularly in Y (similar to that in Fig. 9) as is common in the Pony–Middle Mountain Metamorphic Suite (described above). Unfortunately, much of this compositional variation is at a scale smaller than the ion probe beam size, resulting in the mixed spot ages.

The last type of Pony–Middle Mountain Metamorphic Suite sample contains both the younger 1720–1780 Ma and older 2450 Ma ages. In TBR-208 (see Fig. 1), one matrix monazite grain has only the older spot ages, whereas the other two monazite grains contain a mixture of the younger and older ages. All three grains from this sample have complexly zoned cores and very narrow (~20 microns), low-Th rims (Fig. 9). The low-Th rims are the source of the young ages (Figs. 9B and 9C). The resulting age variation in each grain reflects this zoning, especially the partial overlap between the cores and the narrow rims

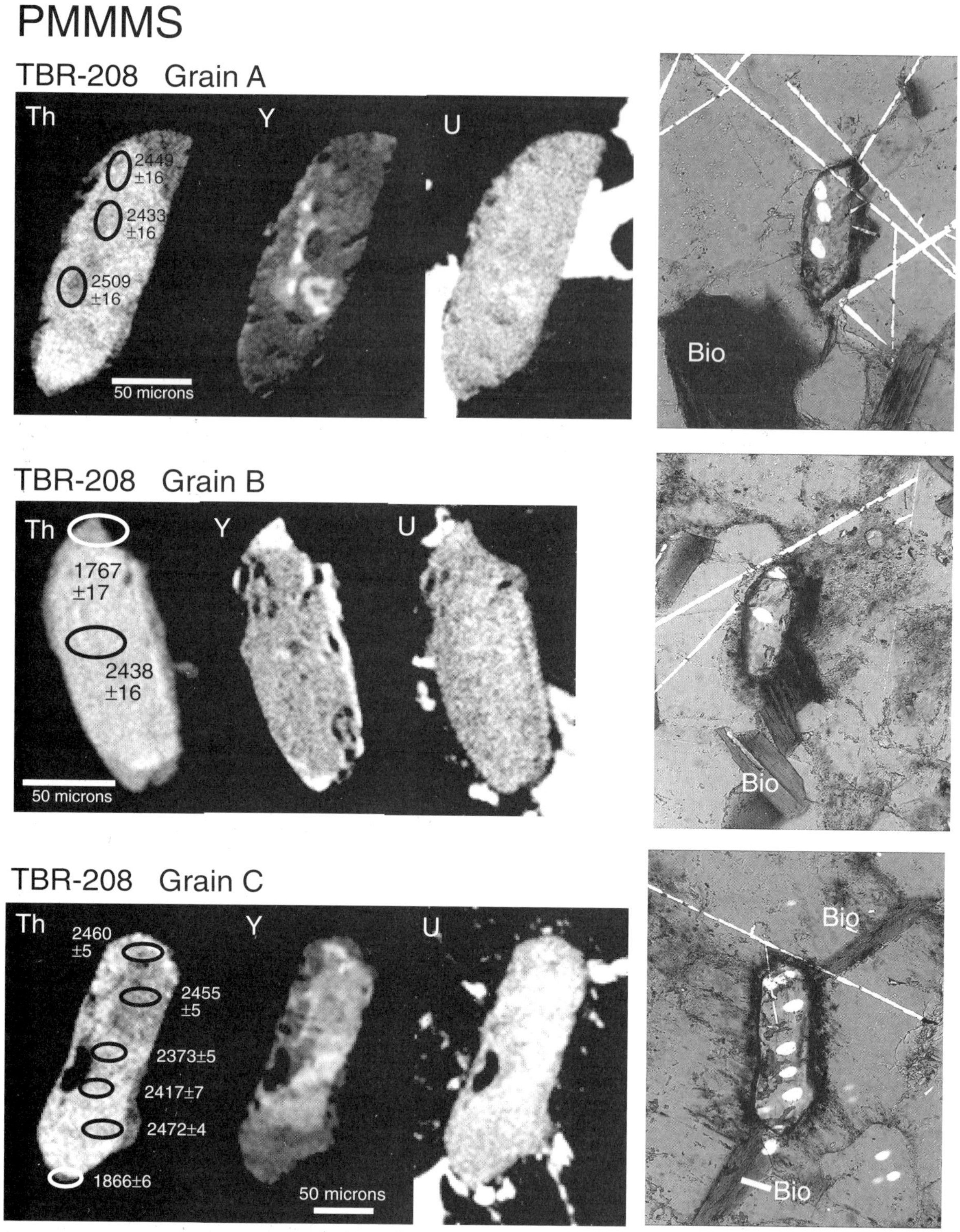

Figure 9. Monazite images from sample TBR-208, also an aluminous gneiss from the Pony–Middle Mountain Metamorphic Suite (PMMMS). A, B, and C: Spot age locations and X-ray maps for grains A, B, and C, respectively. The accompanying photomicrographs show ion probe pits and the setting of the three grains in the matrix. The straight white stripes in the photomicrographs are scratches in the gold coating. Note the thin low-Th and high-Y discontinuous rims on these grains and the younger ages that are from this rim in B and C. The younger ages are from ion probe spots that seem to overlap the chemical zoning.

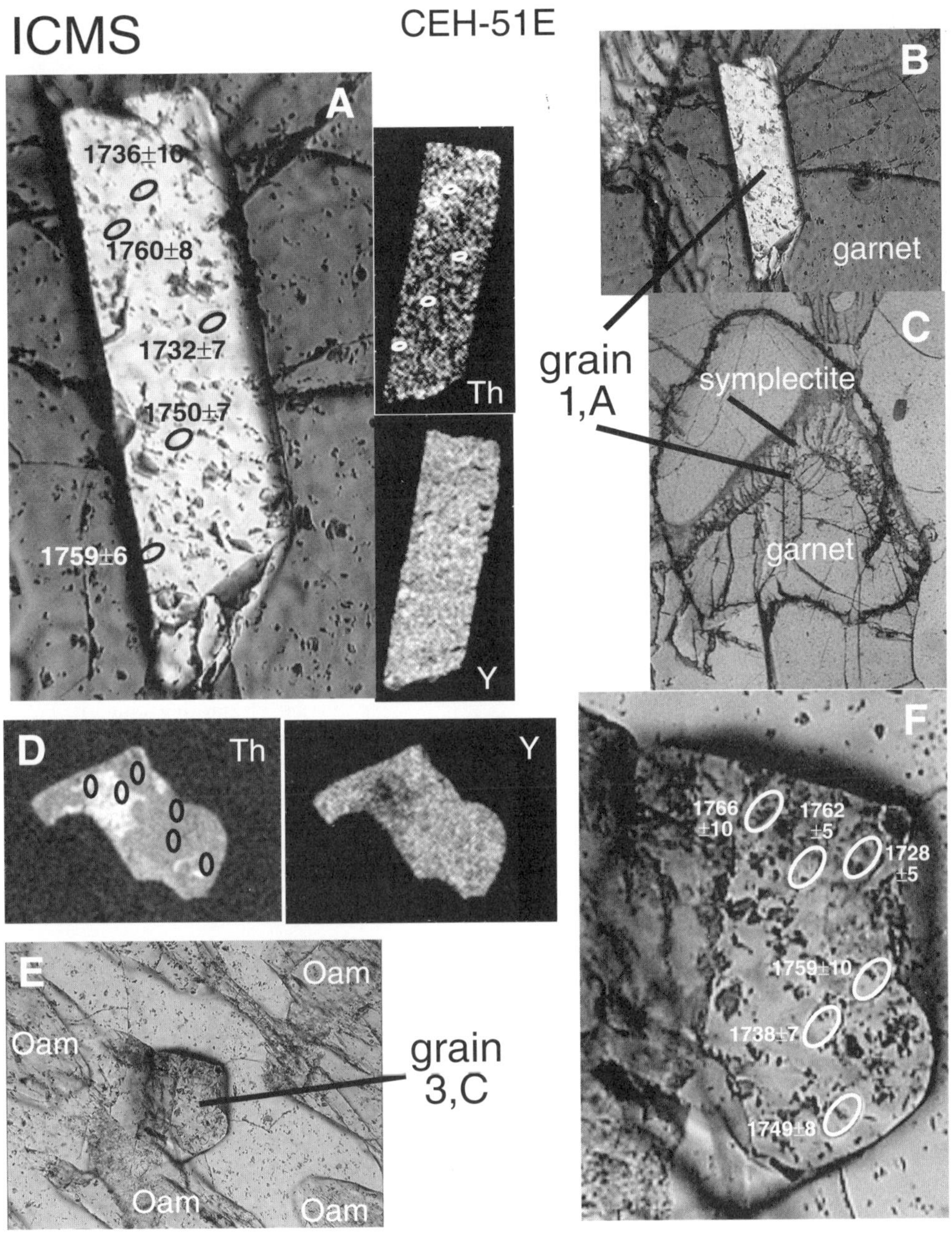

Figure 10. Monazite images from sample CEH-51e, an orthoamphibole gneiss from the Indian Creek Metamorphic Suite (ICMS). A: $^{207}Pb/^{206}Pb$ spot ages. Insets show the compositional homogeneity of grain 1. The small white circles show the analyses locations on the Th X-ray map. B and C: The euhedral monazite as an inclusion in garnet. A cordierite + orthopyroxene symplectite has developed during decompression on the edge of the garnet. D: An irregular core with high Th and low Y in grain 3, a matrix monazite that is largely enclosed in Oam (gedrite) as shown in E. F: The spot ages from grain 3.

by the ~20 micron beam (e.g., the 1866 ± 6 Ma rim age in Fig. 9C). Again the oldest spot age in this sample, 2509 ± 16 Ma, is from a spot that partially overlaps an indistinct Y high and Th low in the grain center, possibly a detrital remnant.

Indian Creek Metamorphic Suite Monazite Ages

Table 1 contains 87 spot ages from 20 different grains in eight Indian Creek Metamorphic Suite samples—three gedrite gneisses and five aluminous gneisses. As shown on Figure 3C, the age distribution is similar to that of the Pony–Middle Mountain Metamorphic Suite monazite, but there are a greater number of ages intermediate between the two peaks.

As in the Pony–Middle Mountain Metamorphic Suite, monazite grains from two Indian Creek Metamorphic Suite samples (two gedrite-bearing rocks, from the same location) contain only the 1720–1780 Ma ages, three samples (two from the same outcrop) contain only the older spot ages, and three samples have mixed age populations. This age distribution is independent of the textural occurrence of the analyzed monazite grains because matrix grains and grains included in garnet occur in all three groups of samples. Twelve of the dated Indian Creek

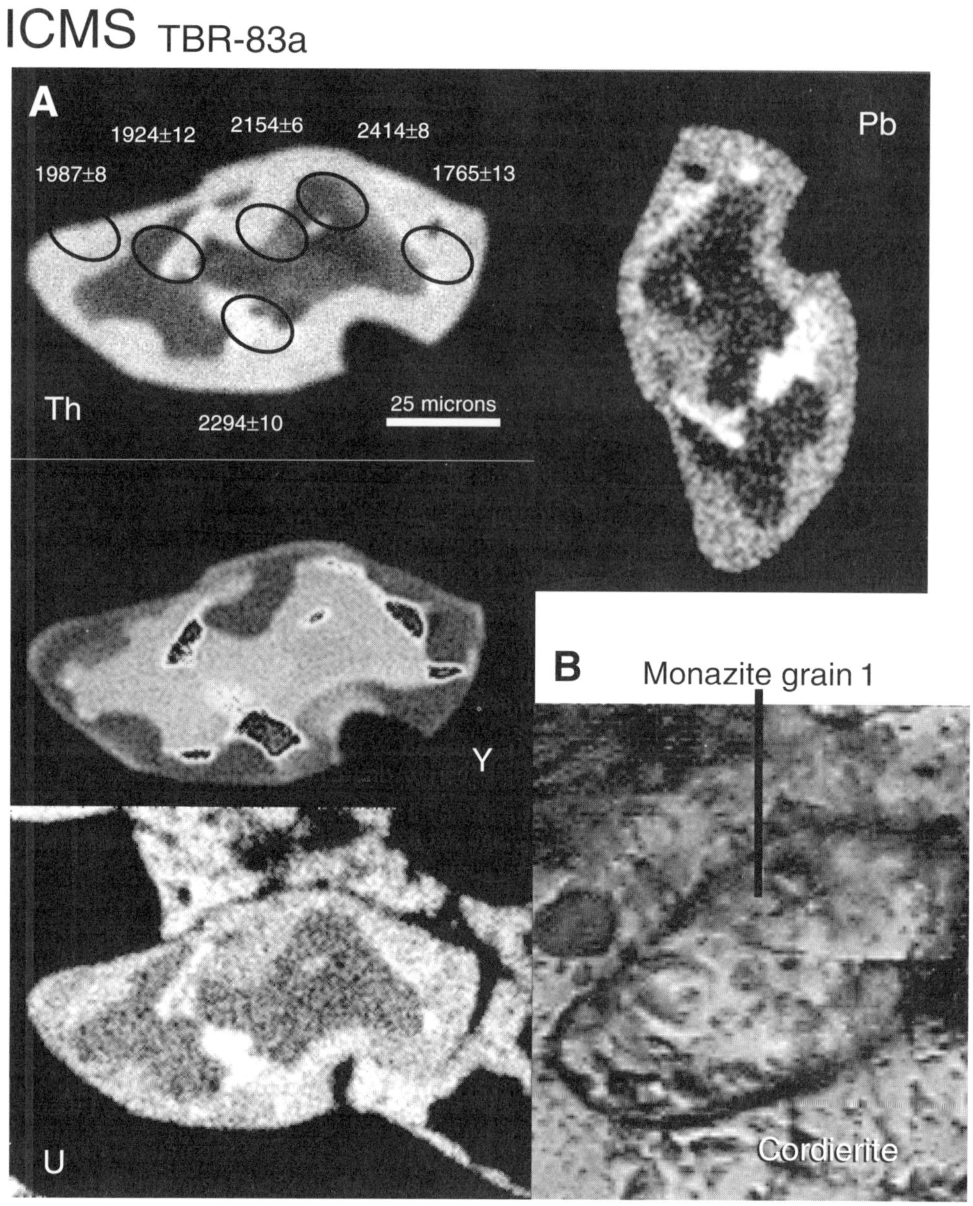

Figure 11. Monazite images from sample TBR-83a, an orthoamphibole gneiss from the Indian Creek Metamorphic Suite (ICMS). A: The systematic relationship between spot ages and the inverse variation of Th and Y. This set of figures again illustrates the potential overlap of some ion microprobe spots and the chemical zoning. B: The textural setting of this grain as an inclusion in cordierite.

Metamorphic Suite grains are from the matrix, whereas eight occur as inclusions. The youngest ages recorded by the inclusions are 1732 ± 7 Ma from the three grains in garnet from a gedrite amphibolite, 2459 ± 6 Ma from the two grains in garnet from aluminous gneiss, 1728 ± 5 Ma from the single monazite in gedrite, and 1977 ± 5 Ma from one inclusion in matrix quartz from an aluminous gneiss. For comparison, matrix monazite grains range down in age to 1740 ± 7 Ma. Spot ages from two Indian Creek Metamorphic Suite rocks, CEH-51e (only 1720–1780 Ma ages) and TBR-83a (bimodal) are shown on Th X-ray maps in Figures 10 and 11, respectively. Both are cordierite-garnet-gedrite rocks that contain kyanite. The monazite in Figure 11 occurs as an inclusion in cordierite. This cordierite is texturally unique in that it occurs as coarse (1 mm) subhedral grains. The Th-Y inverse compositional zoning correlates with the bimodal spot age distribution in that the high-Th/low-Y rims are younger (ca. 1765 ± 13 Ma) than the cores (up to 2414 ± 8 Ma). Most of the ion probe spots are relatively large, and they overlap the compositional zones. Cordierite from other analyzed samples is much finer grained, anhedral, and confined to garnet reaction zones or symplectic intergrowths, as the case for the cordierite in sample CEH-51e (Fig. 10C). Although the spot age variation in both grains from this sample exceeded the error of the individual analyses, only one is apparently compositionally zoned. The unzoned grain is an inclusion in garnet (Figs. 10A–C). The spot ages vary from 1732 ± 7 to 1760 ± 8 Ma (Fig. 10A) for reasons that remain unclear. The zoned monazite has an irregular core defined by high Y, and it occurs inside a large (>5 mm), inclusion-riddled gedrite. The core contains spot ages that exceed 1762 ± 5 Ma whereas the rim ages range from the overlap age of 1759 ± 10 Ma down to 1728 ± 5 Ma (Fig. 10F), about the same age span as for the monazite included in garnet.

This age distribution of Indian Creek and Pony–Middle Mountain Metamorphic Suite samples results in part from age zoning of individual monazite grains and partly from the occurrence of only younger or older spot ages in some Indian Creek and Pony–Middle Mountain Metamorphic Suite samples. The spot ages between 1790 Ma and 2450 Ma likely reflect mixing of these two or more domains in both of these suites.

DISCUSSION

There are two distinct populations of monazite spot ages from the three Tobacco Root Mountain metamorphic rock suites. This bimodal age distribution is described by a wide "peak" at 1720–1780 Ma consisting of all Spuhler Peak Metamorphic Suite and some Indian Creek and Pony–Middle Mountain Metamorphic Suite spot ages and a second age peak near ca. 2450 Ma (Fig. 4) composed only of Indian Creek and Pony–Middle Mountain Metamorphic Suite spot ages. The monazite age populations for the Indian Creek and Pony–Middle Mountain Metamorphic Suite monazite grains are similar in that both are significantly more heterogeneous than the Spuhler Peak Metamorphic Suite. The heterogeneity results in part from age zoning of individual monazite grains and partly from the restricted occurrence of some Indian Creek and Pony–Middle Mountain Metamorphic Suite monazite grains to either the 1720–1780 Ma array or the ca. 2450 Ma array. There are younger and a few older spot ages in some Indian Creek and Pony–Middle Mountain Metamorphic Suite samples that represent mixed age domains. The spot ages between 1790 Ma and 2450 Ma are from the mixing of two or more of these domains. The ages older than 2450 Ma likely represent mixtures between older remnants of detrital Archean grains and the 2450 Ma material.

The near absence of $^{207}Pb/^{206}Pb$ ages older than 1780 Ma in the Spuhler Peak Metamorphic Suite and the common occurrence of these ages in monazite from the Indian Creek and Pony–Middle Mountain Metamorphic Suites are consistent with assembly of the Tobacco Root Terrane during an orogenic event that began as a collision at ca. 1780 Ma and continued for some 60 Ma until ca. 1720 Ma. We call this event the Big Sky orogeny (Fig. 12; see also Harms et al., 2004b, this volume, Chapter 11). We believe that this assembly involved the juxtaposition of previously unmetamorphosed Spuhler Peak Metamorphic Suite and a 2450 Ma terrane consisting of the Pony–Middle Mountain Metamorphic Suite, the Indian Creek Metamorphic Suite, and the 2060 Ma mafic dikes and sills. These results corroborate, extend, and clarify the preliminary findings of Webb et al. (2001), Cheney et al. (1999), Dahl et al. (1999, 2002), and Roberts et al. (2002).

The distribution of $^{207}Pb/^{206}Pb$ monazite ages for the Big Sky orogeny (Fig. 12) results from matrix grains, grains associated with cordierite-garnet reaction textures, and monazite grains that occur as inclusion in biotite, gedrite, quartz, garnet, and kyanite. These relationships are summarized in Figure 13. Many of the grains in these different textural settings are compositionally zoned. The matrix monazite grains vary in age from ca. 1786 Ma to ca. 1713 Ma. The rims of the zoned matrix grains tend to yield those ages younger than ca. 1730 Ma. Spot ages younger than ca. 1735 Ma from monazite inclusions in all minerals are uncommon. Taken together these observations indicate that most of the significant mineral growth in these rocks occurred before ca. 1735 Ma, but that rims grew on matrix monazite grains between ca. 1735 and ca. 1715 Ma. Based on the spot ages from monazite inclusions in kyanite, the rocks were in the kyanite stability field and at the higher pressure of M1 until ca. 1750 Ma. The few Spuhler Peak and Pony–Middle Mountain Metamorphic Suite monazite inclusions in garnet from aluminous gneisses are also older than ca. 1750 Ma (Fig. 13). All of Indian Creek Metamorphic Suite monazite inclusions in garnet are from the kyanite orthoamphibolites (see location CEH-51e of Fig. 1). The spot ages of the monazite inclusions in these garnet grains are as young as 1732 Ma. Garnet from the aluminous gneisses likely formed during M1, but in the kyanite orthoamphibolites, garnet can grow with decreasing pressure via reactions like kyanite + gedrite = garnet + cordierite (Cheney et al., 2004, this volume, Chapter 6). The end of garnet growth nearly coincides with the apparent end of mineral growth in other rocks of the terrane at ca. 1735 Ma.

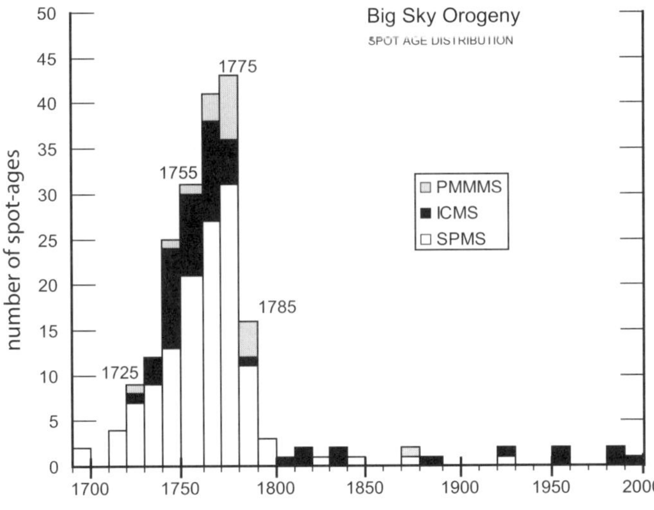

Figure 12. Enlarged histogram showing the distribution of Big Sky $^{207}Pb/^{206}Pb$ spot ages for all three metamorphic rock suites. PMMMS—Pony–Middle Mountain Metamorphic Suite, ICMS—Indian Creek Metamorphic Suite, SPMS—Spuhler Peak Metamorphic Suite. The bin size is 10 m.y.

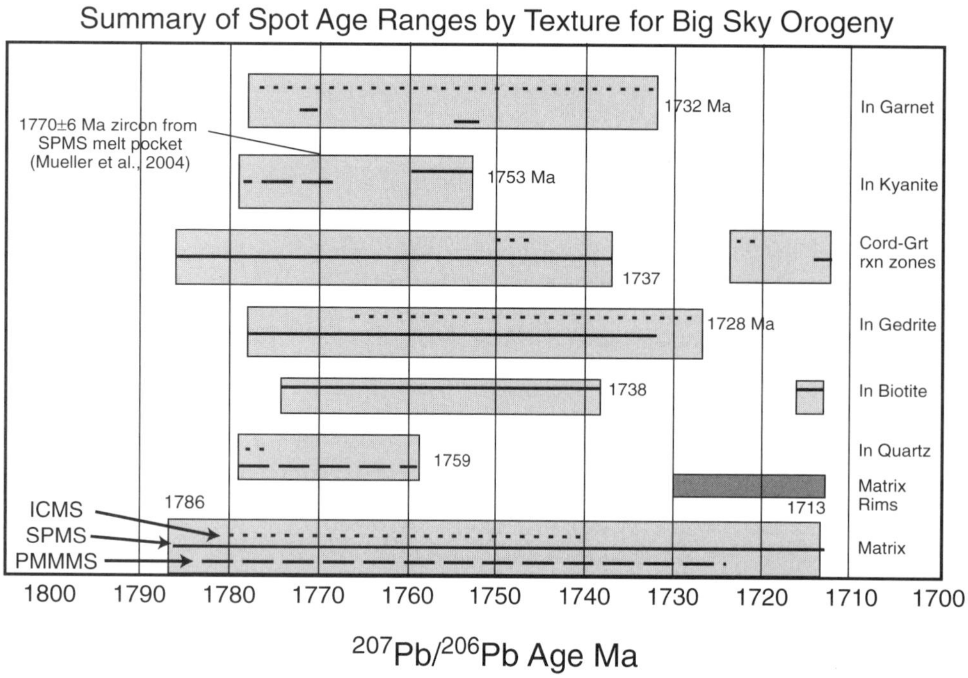

Figure 13. Summary of $^{207}Pb/^{206}Pb$ spot ages from monazite grains by textural setting. The gray boxes outline the range of ages for each of the monazite textures as (1) inclusions in garnet, kyanite, gedrite, biotite, or quartz, (2) in cordierite-garnet reaction zones, and (3) in the matrix. The range of ages from monazite matrix rims is also shown. Within each box, the range of ages is indicated for samples from each metamorphic rock suite, Spuhler Peak Metamorphic Suite (SPMS) by the solid line, Indian Creek Metamorphic Suite (ICMS) by the short dashes, and Pony–Middle Mountain Metamorphic Suite (PMMMS) by the longer dashes.

The Big Sky orogeny produces partial melts and zircon growth in those melts at 1770 ± 6 Ma (Mueller et al., 2004, this volume, Chapter 9). The age of zircon crystallization and the occurrence of migmatites in aluminous gneisses are consistent with the formation of the leucosomes by muscovite dehydration melting in the kyanite stability field during M1 (Cheney et al., 2004, this volume, Chapter 6). Initial monazite growth in aluminous rocks from many terranes is commonly coincident with the growth of garnet (Spear and Pyle, 2002). The oldest monazite ages are 1786 Ma and precede the growth of magmatic zircon by only 15 Ma; this requires fairly rapid heating of the terrane.

Because the kyanite is demonstrably early in nearly all rocks from this terrane, the age of monazite inclusions in the kyanite significantly constrains the timing of the early high-pressure part of the Big Sky *P-T*-t path of Cheney et al. (2004, this volume, Chapter 6), shown here in Figure 14. This clockwise path involves higher-pressure metamorphism, initially in the kyanite + orthopyroxene stability field, followed by melting at 1770 Ma in the kyanite + K-feldspar field, then nearly isobaric cooling prior to isothermal decompression. The end of coarse-grained mineral growth at 1735 Ma coincides with isobaric cooling and the crystallization of any remaining migmatite (see Fig. 14). The crystallizing melts would facilitate retrograde hydration reactions and could produce cordierite + biotite after garnet (Fig. 6D). Isobaric decompression further consumes garnet in many rocks, producing strong zoning in the garnet rims. The matrix monazite rims are possibly related to this decompression, which ends by 1713 Ma with further cooling.

Especially provocative is the abrupt termination of the Big Sky orogeny as recorded by monazite ages. The lower limit of $^{207}Pb/^{206}Pb$ ages end relatively sharply in all rock suites at ca. 1720 Ma (Fig. 12). This is virtually the same age as the $^{40}Ar/^{39}Ar$ cooling dates on both hornblende and biotite, at ca. 1710 Ma in these same rocks (Brady et al., 2004, this volume, Chapter 7), indicative of rapid cooling of the terrane. In turn, this is remarkably consistent with tectonic denudation suggested by the *P-T*-t path in Figure 14. The coincidence of the two dating techniques both confirms the time and constrains the rapid rate of this cooling, thereby fixing the age of the end of the Big Sky *P-T*-t path.

The nature of the 2450 Ma orogeny is not well constrained, but there are several additional indications of a significant event in the area at this time. These include the aforementioned 2450 Ma overgrowths on a 3370 Ma magmatic zircon from a Pony–Middle Mountain Metamorphic Suite quartzofeldspathic gneiss (Krogh et al., 1997) and the occurrence of ca. 2450 Ma ages from inherited zircon grains in the Cretaceous Tobacco Root batholith (Fig. 1). The 2450 Ma zircon grains have been interpreted as magmatic grains that formed in some of the material melted to make the Cretaceous magma (Mueller et al., 1998a, 1998b). We have not identified specific minerals—other than monazite—or textures that are unequivocally related to the 2450 Ma event. However, the MMDS crosscut a gneissic fabric in the Pony–Middle Mountain and Indian Creek Metamorphic Suites during their intrusion at ca. 2060 Ma. Thus, as developed by Harms et al. (2004b, this volume, Chapter 11), there is a pre–Big Sky fabric in the terrain, and this fabric could be a remnant of the Early Proterozoic orogeny at ca. 2450 Ma.

The Archean age of the quartzofeldspathic basement gneisses of the Pony–Middle Mountain and Indian Creek Metamorphic Suites was established from the ages of magmatic

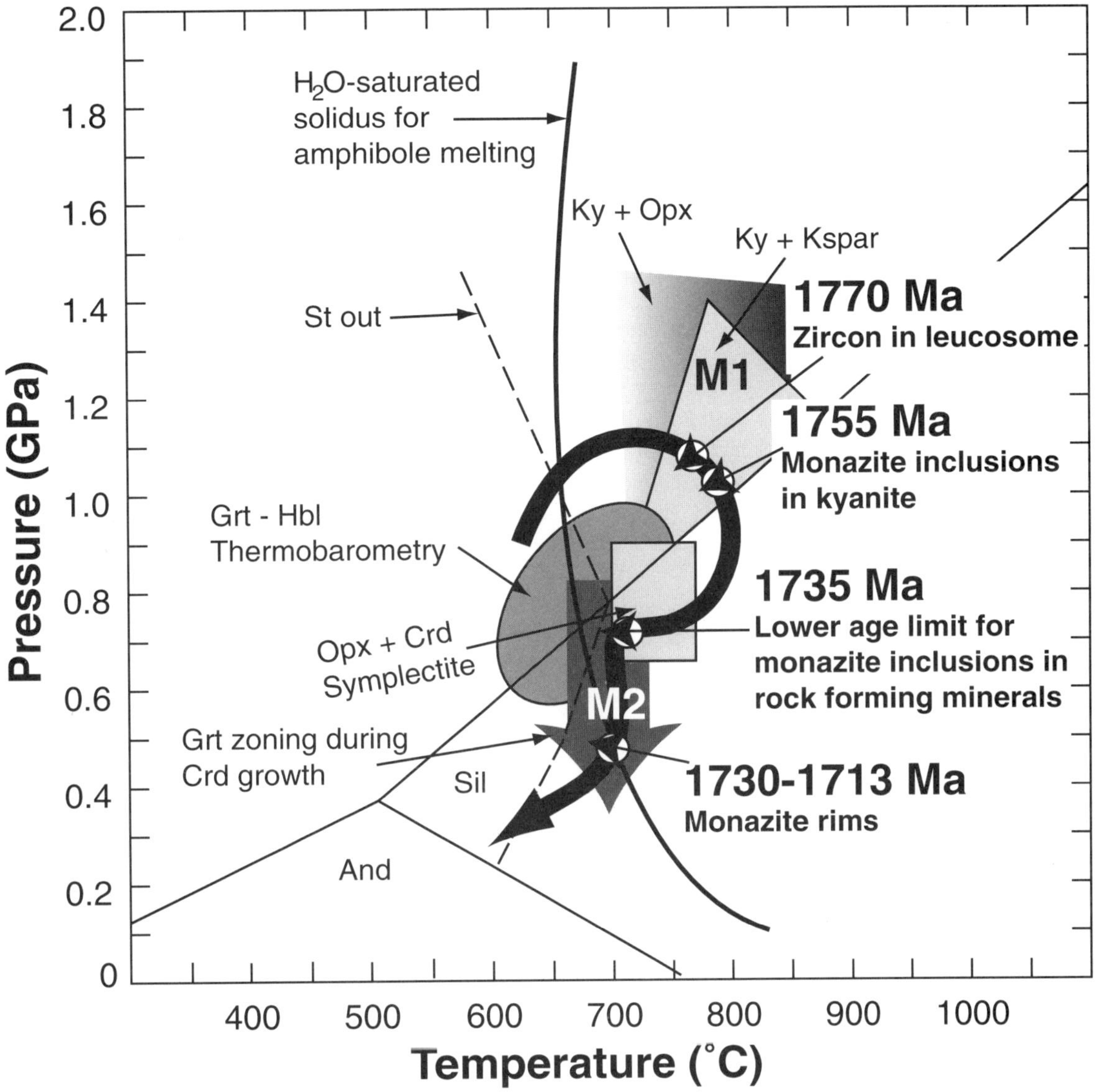

Figure 14. Summary diagram, modified from Figure 21 of Cheney et al. (2004, this volume, Chapter 6), showing timing constraints on the *P-T*-t evolution of the metamorphic rocks from the Tobacco Root Mountains during the 1.78–1.72 Ga Big Sky orogeny. The *P-T* path is shown by the heavy black line. Specific fields outlined on the diagram are (1) the lower part of the stability field of kyanite + orthopyroxene, (2) the lower part of the stability field of kyanite + K-feldspar produced by dehydration melting from F. Spear (1999; personal commun.), (3) the range of temperatures and pressures (dark oval) calculated from the garnet-hornblende-plagioclase-quartz thermobarometer, (4) calculated pressures and temperatures from garnet + orthopyroxene equilibria field (rectangular box) from the orthopyroxene + cordierite symplectite, (5) a calculated decompression path (wide arrow) resulting from cordierite growth based on observed Mn zoning of garnet, (6) the upper stability of staurolite + quartz (dashed line), and (7) a dehydration melting curve for amphibolite (solid line). The details of these constraints are provided in Figure 21 of Cheney et al. (2004, this volume, Chapter 6).

zircons from these rocks as discussed in Mueller et al. (2004, this volume, Chapter 9). These Archean basement rocks were significantly overprinted initially during the Early Proterozoic at ca. 2450 Ma and subsequently during the Big Sky orogeny at 1720–1780 Ma. However, there is a limited record of Archean-age monazite in the Tobacco Root Mountains. The few Archean spot ages reported in Table 1 are from monazite grains in metasediments of the Pony–Middle Mountain and Indian Creek Metamorphic Suites, the oldest being at least ca. 2990 Ma, a spot age from a monazite inclusion in a garnet from a Pony–Middle Mountain Metamorphic Suite sample. Accordingly, there is no evidence of widespread Archean events recorded in the monazite grains of the Tobacco Root Mountain region of the northern Wyoming province.

ACKNOWLEDGMENTS

We much appreciated the use of the University of California at Los Angeles National Ion Microprobe Facility. Mark Harrison provided invaluable expertise and guidance on monazite dating. We gratefully acknowledge the remarkable hospitality and help of Marty Grove, George Jarzebinski, Liz Catlos, Jessica D'Andrea, and Steve Mojzsis. Frank Spear served as a willing lab partner, and he has provided continuous and useful advice. Joe Pyle of the Rensselaer Polytechnic Institute Electron Microprobe Facility generously provided most of the X-ray maps used in this project. Mike Jercinovic of the University of Massachusetts at Amherst Electron Microprobe Facility kindly provided additional X-ray maps on short notice. Funding for this work was provided by an Amherst College Research Award Program grant to JTC. Particular thanks go to the W.M. Keck Foundation for its continued support of the Keck Geology Consortium and undergraduate research. Special gratitude is extended to the students and their faculty advisors who participated in the Montana Keck project. JTC is especially pleased to acknowledge the hard work and support of Montana project director H. Robert Burger, and project colleagues John B. Brady and Tekla A. Harms. The manuscript has benefited immensely from excellent reviews by Frank Spear and Michael Williams and the help of John Brady.

REFERENCES CITED

Brady, J.B., Carmichael, S., Burger, H.R., Harris, C., Cheney, J.T., and Harms, T.A., 1998, Thermobarometry and geochemistry of metamorphosed mafic dikes and sills (MMDS) of the Tobacco Root Mountains, Montana: Geological Society of America Abstracts with Programs, v. 30, no. 7, p. A-97.

Brady, J.B., Kovaric, D.N., Cheney, J.T., Jacob, L.J., and King, J.T., 2004, $^{40}Ar/^{39}Ar$ ages of metamorphic rocks from the Tobacco Root Mountains region, Montana, *in* Brady, J.B., et al., eds., Precambrian geology of the Tobacco Root Mountains, Montana: Boulder, Colorado, Geological Society of America Special Paper 377, p. 131–149 (this volume).

Burger, H.R., 2004, General geology and tectonic setting of the Tobacco Root Mountains, Montana, *in* Brady, J.B., et al., eds., Precambrian geology of the Tobacco Root Mountains, Montana: Boulder, Colorado, Geological Society of America Special Paper 377, p. 1–14 (this volume).

Burger, H.R., Brady, J.B., Cheney, J.T., Harms, T.A., Mueller, P., Heatherington, A., and Wooden, J., 1999, Evidence for a major, Early Proterozoic orogenic event in the Tobacco Root Mountains of southwestern Montana: Geological Society of America Abstracts with Programs, v. 31, no. 7, p. 177–178.

Burger, H.R., Peck, W.H., Johnson, K.E., Tierney, K.A., Poulsen, C.J., Cady, P., Lowell, J., MacFarlane, W.A., Sincock, M.J., Archuleta, L.L., Pufall, A., and Cox, M.J., 2004, Geology and geochemistry of the Spuhler Peak Metamorphic Suite, *in* Brady, J.B., et al., eds., Precambrian geology of the Tobacco Root Mountains, Montana: Boulder, Colorado, Geological Society of America Special Paper 377, p. 47–70 (this volume).

Cheney, J.T., Harms, T.A., Brady, J.B., and Burger, H.R., 1999, Early Proterozoic $^{208}Pb/^{232}Th$ in situ ion probe dating of monazite from Archean metamorphic suites, Tobacco Root Mountains, Montana: Geological Society of America Abstracts with Programs, v. 31, no. 7, p. A-178.

Cheney, J.T., Brady, J.B., Tierney, K.A., DeGraff, K.A., Mohlman, H.K., Frisch, J.D., Hatch, C.E., Steiner, M.L., Carmichael, S.K., Fisher, R.G.M., Tuit, C.B., Steffen, K.J., Cady, P., Lowell, J., Archuleta, L., Hirst, J., Wegmann, K.W., and Monteleone, B., 2004, Proterozoic metamorphism of the Tobacco Root Mountains, Montana, *in* Brady, J.B., et al., eds., Precambrian geology of the Tobacco Root Mountains, Montana: Boulder, Colorado, Geological Society of America Special Paper 377, p. 105–129 (this volume).

Dahl, P.S., Dorais, M.J., Roberts, H.J., Kelley, S.P., and Frei, R., 1999, Electron microprobe geochronometry of age-zoned monazite crystals in Archean metapelites from the Wyoming province: The nature of Pb rejuvenation and implications for regional tectonism: Geological Society of America Abstracts with Programs, v. 31, no. 7, p. A39.

Dahl, P.S., Hamilton, M.A., Terry, M.P., Roberts, H.J., Kelly, S.P., Frei, R., Jercinovic, M.J., and Williams, M.L., 2002, Comparative ion and electron microprobe dating of Wyoming province monazite, with tectonic and analytical implications: Geological Society of America Abstracts with Programs, v. 34, no. 1, p. A9.

Giletti, B.J., 1966, Isotopic ages from southwestern Montana: Journal of Geophysical Research, v. 71, p. 4029–4036.

Giletti, B.J., 1968, Isotopic geochronology of Montana and Wyoming, *in* Hamilton, E.I., and Farquhar, R.M., eds., Radiometric dating for geologists: London, Interscience Publishers, p. 111–146.

Giletti, B.J., and Gast, P.W., 1961, Absolute age of Pre-Cambrian rocks in Wyoming and Montana: Annals of the New York Academy of Sciences, v. 91, p. 454–458.

Gillmeister, N.M., 1971, Petrology of Precambrian rocks in the central Tobacco Root Mountains, Madison County, Montana [Ph.D. thesis]: Cambridge, Harvard University, 210 p.

Hanley, T.B., 1975, Structure and petrology of the northwestern Tobacco Root Mountains, Madison County, Montana [Ph.D. thesis]: Bloomington, Indiana University, 289 p.

Hanley, T.B., 1976, Stratigraphy and structure of the Central Fault Block, northwestern Tobacco Root Mountains, Madison County, Montana, *in* Anonymous, ed., Guidebook, The Tobacco Root Geological Society 1976 Field Conference: Butte, Montana, Montana Bureau of Mines and Geology Special Publication 73, p. 7–14.

Harms, T.A., Burger, H.R., Blednick, D.G., Cooper, J.M., King, J.T., Owen, D.R., Lowell, J., Sincock, M.J., Kranenburg, S.R., Pufall, A., and Picornell, C.M., 2004a, Character and origin of Precambrian fabrics and structures in the Tobacco Root Mountains, Montana, *in* Brady, J.B., et al., eds., Precambrian geology of the Tobacco Root Mountains, Montana: Boulder, Colorado, Geological Society of America Special Paper 377, p. 203–226 (this volume).

Harms, T.A., Brady, J.B., Burger, H.R., and Cheney, J.T., 2004b, Advances in the geology of the Tobacco Root Mountains, Montana, and their implications for the history of the northern Wyoming province, *in* Brady, J.B., et al., eds., Precambrian geology of the Tobacco Root Mountains, Montana: Boulder, Colorado, Geological Society of America Special Paper 377, p. 227–243 (this volume).

Harrison, T.M., McKeegan, K.D., and Le Fort, P., 1995, Detection of inherited monazite in the Manaslu leucogranite by $^{208}Pb/^{232}Th$ ion microprobe dating: Crystallization age and tectonic implications: Earth and Planetary Science Letters, v. 133, p. 271–282.

Harrison, T.M., Ryerson, F.J., Le Fort, P., Yin, A., Lovera, O.M., and Catlos, E.J., 1997, A Late Miocene-Pliocene origin for the Central Himalayan inverted metamorphism: Earth and Planetary Science Letters, v. 146, p. 1–7.

Harrison, T.M., Catlos, E.J., and Montel, J.M., 2002, U-Th-Pb dating of phosphate minerals, *in* Kohn, M.J., et al., eds., Phosphates: Geochemical, geobiological, and materials importance: Washington, D.C., Mineralogical Society of America, Reviews in Mineralogy and Geochemistry 48, p. 337–362.

Hayden, R.J., and Wehrenberg, J.P., 1959, Potassium-argon dating in western Montana: Geological Society of America Bulletin, v. 70, no. 12, Part 2, p. 1778–1779.

Hayden, R.J., and Wehrenberg, J.P., 1960, A^{40}-K^{40} dating of igneous metamorphic rocks in western Montana: Journal of Geology, v. 68, no. 1, p. 94–97.

James, H.L., and Hedge, C.E., 1980, Age of the basement rocks of southwest Montana: Geological Society of America Bulletin, v. 91, p. 11–15.

Krogh, T.E., Kamo, S., and Hess, D.F., 1997, Wyoming province 3300+ gneiss with 2400 Ma metamorphism, northwestern Tobacco Root Mountains, Madison County, Montana: Geological Society of America Abstracts with Programs, v. 29, no. 6, p. A-408.

Mogk, D.W., Burger, H.R., Mueller, P.A., D'Arcy, K., Heatherington, A.L., Wooden, J.L., Abeyta, R.L., Martin, J., and Jacob, L.J., 2004, Geochemistry of quartzofeldspathic gneisses and metamorphic mafic rocks of the Indian Creek and Pony–Middle Mountain Metamorphic Suites, Tobacco Root Mountains, Montana, *in* Brady, J.B., et al., eds., Precambrian geology of the Tobacco Root Mountains, Montana: Boulder, Colorado, Geological Society of America Special Paper 377, p. 15–46 (this volume).

Mueller, P.A., and Cordua, W.S., 1976, Rb-Sr whole rock age of gneisses from the Horse Creek area, Tobacco Root Mountains, Montana: Isochron/West, v. 16, p. 33–36.

Mueller, P., Heatherington, A., Wooden, J., Mogk, D., and Bowes, D., 1998a, The Wyoming province and its place in the evolution of Precambrian North America: Geological Society of America Abstracts with Programs, v. 30, no. 7, p. A-46.

Mueller, P.A., Wooden, J.L., Nutman, A.P., and Mogk, D.W., 1998b, Early Archean crust in the northern Wyoming province: Evidence from U-Pb ages of detrital zircons: Precambrian Research, v. 91, p. 295–307.

Mueller, P.A., Burger, H.R., Wooden, J.L., Heatherington, A.L., Mogk, D.W., and D'Arcy, K., 2004, Age and evolution of the Precambrian crust of the Tobacco Root Mountains, Montana, *in* Brady, J.B., et al., eds., Precambrian geology of the Tobacco Root Mountains, Montana: Boulder, Colorado, Geological Society of America Special Paper 377, p. 181–202 (this volume).

Parrish, R., 1990, U-Pb dating of monazite and its application to geological problems: Canadian Journal of Earth Sciences, v. 27, p. 1431–1450.

Pyle, J.M., Spear, F.S., and Wark, D.A., 2002, Electron microprobe analysis of REE in apatite, monazite, and xenotime: Protocols and pitfalls, *in* Kohn, M.J., et al., eds., Phosphates: geochemical, geobiological, and materials importance: Washington, D.C., Mineralogical Society of America, Reviews in Mineralogy and Geochemistry 48, p. 337–362.

Reid, R.R., 1963, Metamorphic rocks of the northern Tobacco Root Mountains, Madison County: Geological Society of America Bulletin, v. 74, p. 293–306.

Roberts, H.J., Kelley, S.P., Dahl, P.S., and Frei, R., 2002, New ^{207}Pb-^{206}Pb and ^{40}Ar-^{39}Ar ages from SW Montana, USA: Constraints on the Proterozoic and Archean tectonic and depositional history of the Wyoming province: Precambrian Research, v. 117, p. 119–143.

Spear, F.S., and Pyle, J.M., 2002, Apatite, monazite, and xenotime in metamorphic rocks, *in* Kohn, M.J., et al., eds., Phosphates: Geochemical, geobiological, and materials importance: Washington, D.C., Mineralogical Society of America, Reviews in Mineralogy and Geochemistry, 48, p. 293–335.

Vitaliano, C.J., Burger, H.R., Cordua, W.S., Hanley, T.B., Hess, D.F., and Root, F.K., 1979, Geologic map of southern Tobacco Root Mountains, Madison County, Montana: Geological Society of America Map and Chart Series MC31, scale 1:62,500, 1 sheet, 8 p. text.

Webb, A.A., 2001, Ion microprobe dating of monazite from Precambrian metamorphic suites, Tobacco Root Mountains, Montana [B.A. thesis]: Amherst, Massachusetts, Amherst College, 99 p.

Webb, A.A., Cheney, J.T., Harms, T.A., and Coath, C.D., 2001, ^{207}Pb/^{206}Pb ion microprobe dating of monazite from Precambrian metamorphic suites, Tobacco Root Mountains, southwestern Montana: Geological Society of America Abstracts with Programs, v. 33, no. 1, p. A10.

Williams, M.L., and Jercinovic, M.J., 2002, Microprobe monazite geochronology: Putting absolute time into microstructural analysis: Journal of Structural Geology, v. 24, p. 1013–1028.

Williams, M.L., Jercinovic, M.J., and Terry, M.P., 1999, Age mapping and dating of monazite on the electron microprobe: Deconvoluting multistage tectonic histories: Geology, v. 27, p. 1023–1026.

Wing, B.A., Ferry, J.M., and Harrison, T.M., 2003, Prograde destruction and formation of monazite and allanite during contact and regional metamorphism of pelites: Petrology and geochronology: Contributions to Mineralogy and Petrology, v. 145, p. 228–250.

Manuscript Accepted by the Society September 12, 2003

Printed in the USA

Geological Society of America
Special Paper 377
2004

Age and evolution of the Precambrian crust of the Tobacco Root Mountains, Montana

Paul A. Mueller
Department of Geological Sciences, University of Florida, Gainesville, Florida 32611, USA

H. Robert Burger
Department of Geology, Smith College, Northampton, Massachusetts 01063, USA

Joseph L. Wooden
U.S. Geological Survey, 345 Middlefield Road, Menlo Park, California 94025, USA

Ann L. Heatherington
Department of Geological Sciences, University of Florida, Gainesville, Florida 32611, USA

David W. Mogk
Department of Earth Sciences, Montana State University, Bozeman, Montana 59717, USA

Kimberly D'Arcy
Department of Geological Sciences, University of Florida, Gainesville, Florida 32611, USA

ABSTRACT

U-Pb analyses of zircons from gneisses, anatectic leucosome, metasedimentary rocks, and a younger (metamorphosed) mafic dike from the Tobacco Root Mountains of southwestern Montana document a Precambrian history that extends from at least 3.90–1.77 Ga. The oldest U-Pb age reported here (3.8 Ga) is from a detrital zircon from a quartzite within the Spuhler Peak Metamorphic Suite, although younger ages of clearly detrital grains suggest the protolith was deposited subsequent to 3.2 Ga. Alternatively, a Pb-Pb age of ca. 2.45 Ga from a single subhedral zircon from this quartzite suggests the quartzite, and perhaps other Spuhler Peak Metamorphic Suite lithologies, may have formed in the Proterozoic. An Archean age, however, seems most compatible with the Archean Sm-Nd model ages of mafic and metasedimentary components of the Spuhler Peak Metamorphic Suite and the age distribution of zircons from the quartzite, which is very similar to the age distribution present in Archean quartzites in the region.

The Spuhler Peak Metamorphic Suite lies in tectonic contact with volumetrically dominant, Archean, quartzofeldspathic gneisses and intercalated metasedimentary rocks. The protoliths of these gneisses were apparently emplaced 3.2–3.4 Ga, and are interpreted to be the basement upon which the intercalated (meta)sedimentary rocks were deposited. U-Pb analyses of zircons from anatectic leucosome near the boundary between the gneisses and the Spuhler Peak Metamorphic Suite, however, yield a significant population of 1.77 Ga grains, which are interpreted to have crystallized from the leucosome. All other grains are Archean (to 3.48 Ga) and interpreted to derive from the metasedimentary source of the leucosome. In addition, U-Pb analyses of zircons

Mueller, P.A., Burger, H.R., Wooden, J.L., Heatherington, A.L., Mogk, D.W., and D'Arcy, K., 2004, Age and evolution of the Precambrian crust of the Tobacco Root Mountains, Montana, *in* Brady, J.B., Burger, H.R., Cheney, J.T., and Harms, T.A., eds., Precambrian geology of the Tobacco Root Mountains, Montana: Boulder, Colorado, Geological Society of America Special Paper 377, p. 181–202. For permission to copy, contact editing@geosociety.org.

extracted from a granulite facies mafic dike that cuts across Archean gneissic banding indicate the dike was intruded at 2.06 Ga, but reached granulite facies at 1.76 Ga. Structural, petrologic, and geochronologic data suggest all lithologies experienced granulite facies metamorphism at ca. 1.77 Ga and that the Spuhler Peak Metamorphic Suite was tectonically emplaced after 2.06 Ga, but before 1.77 Ga. This Paleoproterozoic tectonic activity is most likely a result of burial during terrane collision (e.g., the juxtaposition of the Wyoming and Hearne provinces) and/or to postcollisional mafic underplating.

Keywords: Precambrian, tectonics, geochronology, Montana, isotopes (Nd), metamorphism.

INTRODUCTION

Precambrian basement exposed in the Tobacco Root Mountains of southwestern Montana is composed primarily of Archean quartzofeldspathic gneisses (Vitaliano et al., 1979a, 1979b; see reprinted map and text accompanying this volume) and is part of the Montana metasedimentary subprovince of the Wyoming province (Mogk et al., 1992). These gneisses are intercalated with a variety of lithologies, including quartzite-bearing metasupracrustal assemblages, and host a variety of later mafic rocks that cut across distinct structural features. A major lithologic, temporal, and structural discontinuity occurs in the north-central part of the range where the distinct mafic to ultramafic rocks of the Spuhler Peak Metamorphic Suite occur in tectonic contact with the Archean gneisses of the Indian Creek and Pony–Middle Mountain Metamorphic Suites (Burger, 2004, this volume, Chapter 1; Fig. 1).

Despite numerous geochronologic studies of the Precambrian rocks in the Tobacco Root Mountains over the past ~40 years, critical age relations among the various metamorphic suites (i.e., Spuhler Peak, Indian Creek, and Pony–Middle Mountain), as well as between distinct lithologies within these metamorphic suites, remain unclear (Burger, 2004, this volume, Chapter 1). The early history of Precambrian crust in the Tobacco Root Mountains and adjacent areas can be generally constrained by the age distribution of detrital zircons from metapsammitic rocks. Mueller et al. (1998) reported ages as old as ca. 3.9 Ga in a suite of detrital zircons from a quartzite in the Indian Creek Metamorphic Suite. The mode of the distribution, however, was in the range of 3.2–3.4 Ga, similar to other suites of detrital zircons from Precambrian quartzites from the northern Wyoming province (Mueller et al., 1998). The youngest, clearly detrital grain reported from the Tobacco Root sample gave an age of ca. 3.13 Ga, which provides a maximum age for deposition of this unit. The minimum age of deposition cannot be specified, but two observations suggest that this quartzite was deposited in the Archean: (1) the lack of grains representative of the widely distributed 2.75–2.85 Ga gneisses of the northern Wyoming province and (2) the lack of any 2.4–3.0 Ga grains, which constitute the inherited zircon suite found in the Cretaceous Tobacco Root batholith (Mueller et al., 1996a). Together, these data suggest that deposition occurred in the Late Archean, between ca. 2.8 and ca. 3.1 Ga.

The age(s) of the quartzofeldspathic gneisses of the Tobacco Root Mountains were originally estimated to be in the range of 2.6–2.7 Ga, based on whole-rock Rb-Sr measurements (Mueller and Cordua, 1976; James and Hedge, 1980). Subsequent U-Pb zircon analyses have shown, however, that gneisses from the central part of the range have protolith crystallization ages of 3.2–3.4 Ga (Mueller et al., 1996a; Krogh et al., 1997). U-Pb analyses of zircons reported here also suggest that the Tobacco Root Mountains experienced a metamorphic event ca. 2.8 Ga. The 2.6–2.7 Ga Rb-Sr whole-rock ages, therefore, may represent the time of isotopic equilibration of the Rb-Sr system in the Tobacco Root and proximal ranges subsequent to this event.

The extent and distribution of the older Archean crust (older than 3.4 Ga) is difficult to specify, but the existence of 3.4–4.0 Ga detrital zircons and numerous Sm-Nd depleted mantle model ages older than 3.4 Ga throughout the northern Wyoming province suggest that 3.4–4.0 Ga crust was present (e.g., Mueller et al., 1993, 1996c, 1998). The ages of inherited zircons from the Tobacco Root batholith, however, also suggest the possible presence of Paleoproterozoic crust (2.2–2.4 Ga) in the Tobacco Root basement (Mueller et al., 1996a). Although mineral ages indicative of ca. 2.4 Ga metamorphism and/or magmatism have been reported from the Tobacco Root Mountains (e.g., zircon: Mueller et al., 1996a; Krogh et al., 1997; monazite: Cheney et al., 2004b, this volume, Chapter 8; Roberts et al., 2002) and elsewhere in southwestern Montana (Mueller et al., 1994, 1996b), no rocks of this age have been documented in the Tobacco Root Mountains.

The early work of Hayden and Wehrenberg (1959), Giletti and Gast (1961), and Giletti (1966) first demonstrated that Proterozoic thermal disturbances had been recorded by Archean rocks throughout southwestern Montana. Subsequent K-Ar, Rb-Sr, and U-Pb mineral studies (e.g., O'Neill et al., 1988; Erslev and Sutter, 1990; Harlan et al., 1996; Brady et al., 1998; Roberts et al., 2002) also were interpreted to indicate Proterozoic thermal activity in this area as summarized in Burger (2004, this volume, Chapter 1) and discussed in Cheney et al. (2004a, this volume, Chapter 6); Brady et al. (2004b, this volume, Chapter 7); and Cheney et al. (2004b, this volume, Chapter 8). The exact timing and duration of the thermal episode(s) remain problematic, although data reported in this volume shed considerable light on the Proterozoic history of the Archean gneisses of the Tobacco Root Mountains. In this chapter, we address this Proterozoic history using the results of U-Pb determinations of zircons from an anatectic melt and a (metamorphosed) mafic dike that can be directly tied to the structural evolution of the Archean rocks. In

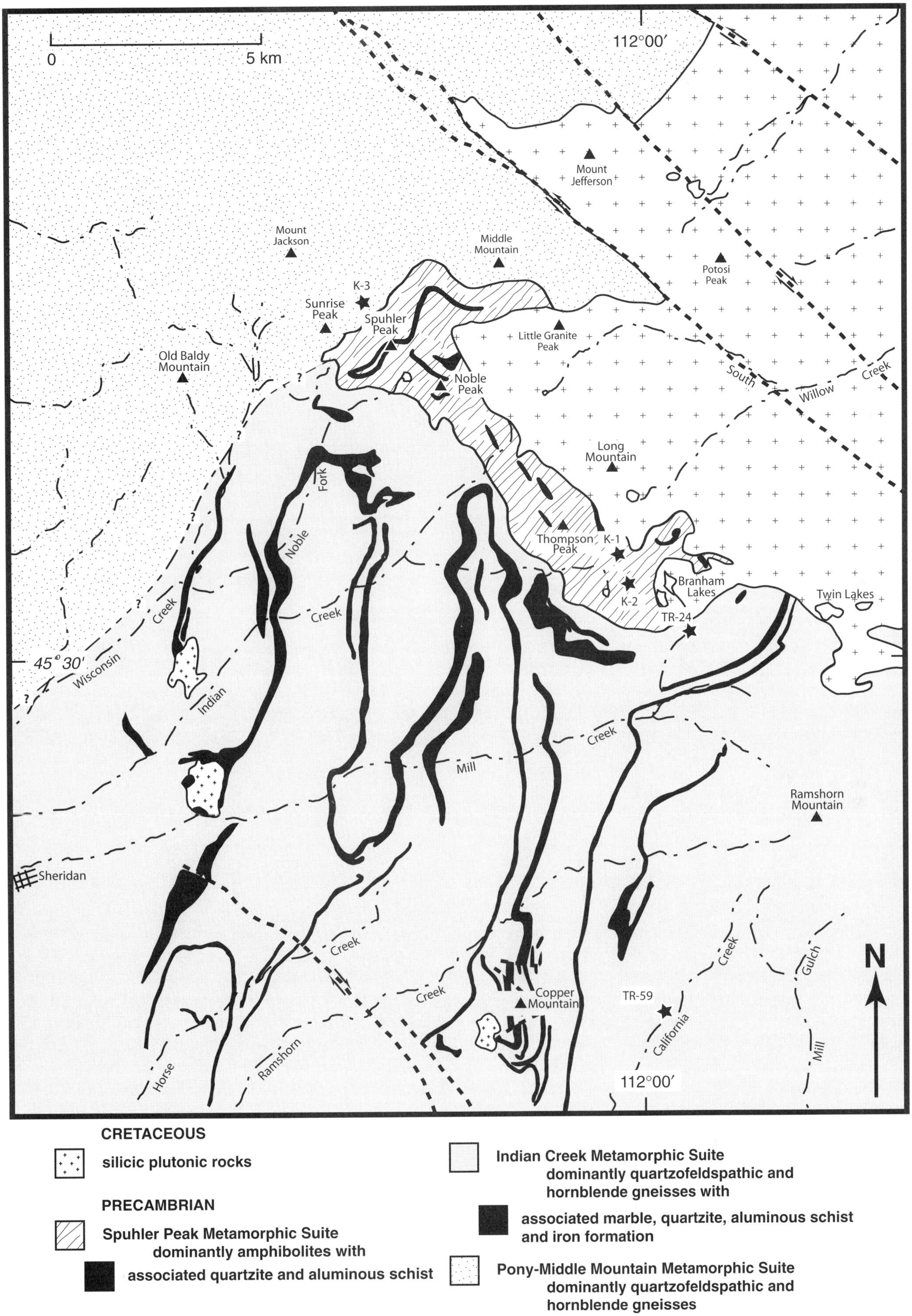

Figure 1. Generalized geologic map showing location of dated samples and basement lithologies in the Tobacco Root Mountains.

addition, we provide U-Pb and Sm-Nd data that bear on the age and origin of the metamorphosed mafic, ultramafic, and sedimentary rocks of the Spuhler Peak Metamorphic Suite, the Archean quartzofeldspathic gneisses, and a distinctive group of metamorphosed mafic dikes and sills (MMDS).

SAMPLE DESCRIPTIONS AND FIELD RELATIONS

Samples TR-24 and TR-59 are from quartzofeldspathic gneisses of tonalitic-trondhjemitic composition typical of the Archean gneisses of the Indian Creek Metamorphic Suite (Fig. 1). Sample TR-24 was collected from blasted outcrops along the road ~50 m below the spillway of lower Branham Lake; sample TR-59 was collected along California Creek at the forest boundary below the Bins mine.

Sample K-1 was collected from an ~15 m thick quartzite bed that lies within a sequence typical of the Spuhler Peak Metamorphic Suite 0.5 km due east of Thompson Peak (Fig. 1). Field relations suggest the quartzites within the Spuhler Peak Metamorphic Suite were originally cherty and arenitic sediments deposited in a shallow marine environment in association with other shallow marine sediments and extensive basaltic and komatiitic lava flows (Burger et al., 2004, this volume, Chapter 3).

Sample K-2 was collected from a crystallized felsic melt derived at least in part from a schist (aluminous paragneiss) within the Spuhler Peak Metamorphic Suite (Fig. 1). The sample was collected within a zone of intense deformation within the Spuhler Peak Metamorphic Suite and near the contact between the Archean gneisses and the suite. The sample location is in a small cirque ~0.6 km west of lower Branham Lake at the head of the Mill Creek drainage. The geological significance of this sample stems from its relationship to structures within the zone of deformation. The sample was collected between two amphibolite boudins (Fig. 2A), and it appears that the melt was generated

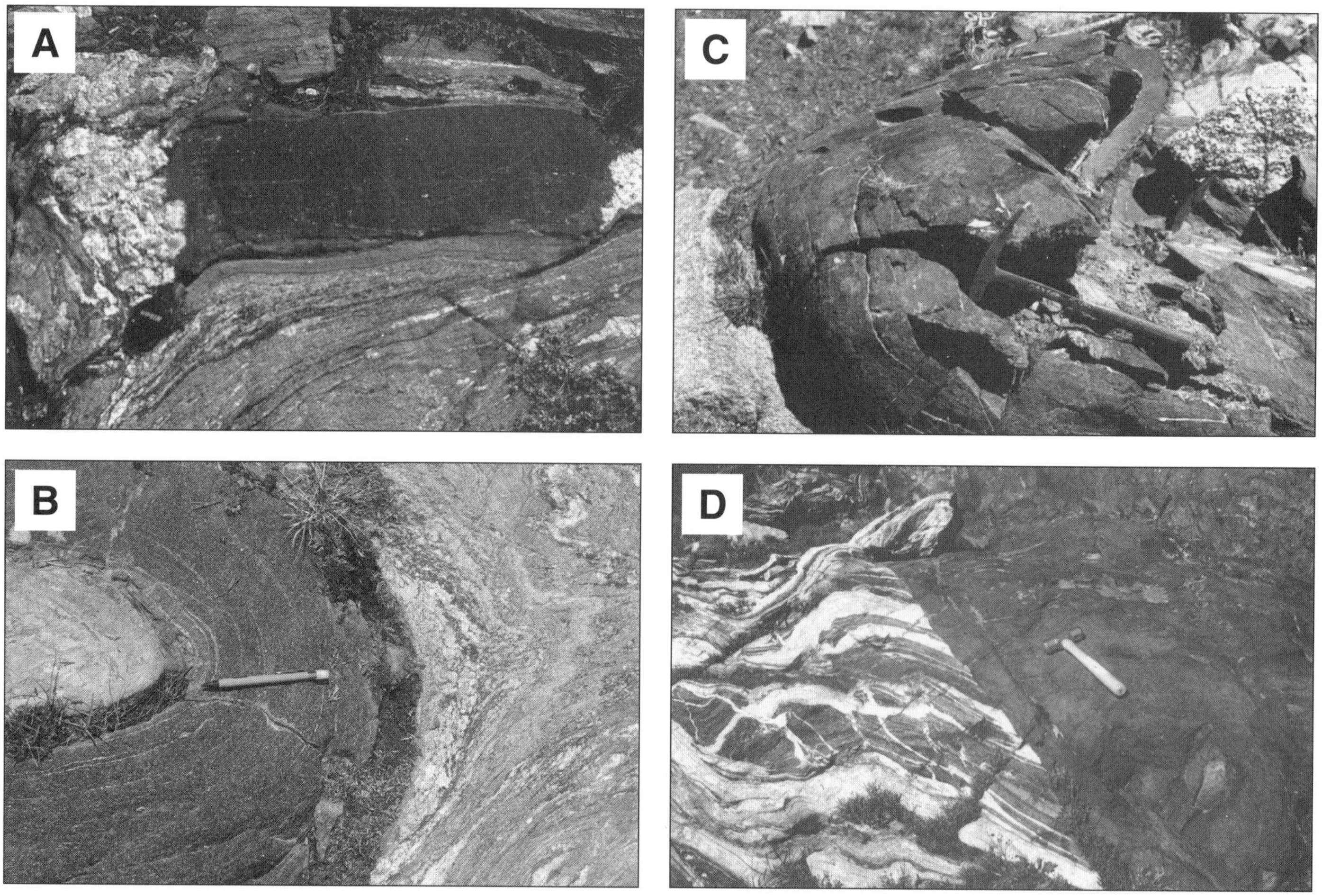

Figure 2. A: Melt zone in boudinaged limb of fold illustrated in Figure 2B. Lens cap for scale = 5.1 cm diameter. Sample K-2 was collected from this zone. B: Hinge zone of fold noted in Figure 2A. Minor melt in hinge zone is identical to that sampled for K-2. Pencil for scale = 14.5 cm. C: Sheath fold in amphibolite located in same zone of intense deformation as fold illustrated in Figure 2B. Hammer for scale; handle length 30.0 cm. D: Dike of metamorphic mafic dikes and sills crosscutting gneissic banding in Indian Creek Metamorphic Suite. Sample K-3 was collected at this site. Hammer for scale; handle length 35.5 cm.

from melting of the immediately adjacent aluminous gneiss or schist (quartz, sillimanite, garnet, biotite, plagioclase, and cordierite) during extension of the amphibolite. The boudinaged amphibolite is located on one limb of a tight isoclinal fold. The hinge of this fold is located ~2.5 m from the melt zone. Examination of the hinge area of the fold (Fig. 2B) reveals a small quantity of solidified melt within the aluminous gneiss unit. The total geometry of fold hinge, minor structures within the hinge area, and limb extension suggest the melt formed during generation of these structures. Thus, zircon ages derived from the sample date the time of formation of these structures. The Indian Creek Metamorphic Suite–Spuhler Peak Metamorphic Suite contact typically marks a zone of high strain and intense deformation, within which are often developed magnificent sheath folds (Fig. 2C) that are related to the same intense regional simple shear event that produced the structures illustrated in Figures 2A and 2B (Harms et al., 2004a, this volume, Chapter 10).

Sample K-3 is from one of many MMDS of tholeiitic composition commonly found in the Precambrian rocks of the Tobacco Root Mountains (Brady et al., 2004a, this volume, Chapter 5; Wooden et al., 1978). They are especially abundant in the Pony–Middle Mountain Metamorphic Suite near its contact with the Spuhler Peak Metamorphic Suite. Sample K-3 was collected from an ~6 m thick basaltic dike of the MMDS located ~100 m northwest of Sunrise Lake at the head of the Wisconsin Creek drainage (Fig. 1). The MMDS cut across the gneissic banding of the Archean gneisses (Fig. 2D) and, therefore, postdate the tectonism and metamorphism that produced that banding. The MMDS, however, have been metamorphosed to produce the mineral assemblage Grt + Pl + Cpx + Hbl + Qtz + Ilm ± Opx and exhibit a weak foliation that is generally parallel to their well-defined, intrusive margins. In addition, the MMDS has been intensely folded with fold axes parallel to the fold axes in the intruded gneisses. Structural analysis of folds in the Spuhler Peak Metamorphic Suite adjoining Archean gneisses suggests a consistent movement pattern (see Harms et al., 2004a, this volume, Chapter 10), thereby demonstrating that an intense deformation occurred subsequent to the intrusion of the MMDS. The Grt-Hbl-Pl-Qtz thermobarometer of Spear et al. (1991) yields pressures of 6–10 kbar and temperatures of 600–700 °C. These results are similar to those obtained from the same mineral assemblage in amphibolites within the host Archean gneisses and from Spuhler Peak Metamorphic Suite amphibolites (Cheney et al., 2004a, this volume, Chapter 6; Brady et al., 2004b, this volume, Chapter 7; Cheney et al., 2004b, this volume, Chapter 8). Crystal size in the MMDS is small (0.2 mm), especially near their edges, but can be up to 1 mm in the middle of large (up to 30 m wide) sills. These textures may reflect chilled margins of the original intrusions, which is consistent with injection of these melts into relatively cool crust. Additional evidence for intrusion of the MMDS into relatively cool crust includes sharp contacts with host rocks and sharp, nonfaulted offsets of the contacts, suggesting availability of brittle fractures during intrusion.

RESULTS

Zircons from four samples were analyzed by sensitive high resolution ion microprobe methods (SHRIMP) I at the Australian National University and SHRIMP RG (reverse geometry) at the Stanford–U.S. Geological Survey Microanalytical Center (Table 1; Compston et al., 1984; Compston and Williams, 1992). Ages are reported at ±2 sigma; data reduction utilized the SQUID and ISOPLOT routines of Ludwig (1999). Whole-rock Sm-Nd isotopic analyses of 36 samples (Table 2) were completed at the University of Florida using two-stage chromatographic separations followed by thermal ionization mass spectrometric analysis (Heatherington and Mueller, 1991). Major and trace element analyses for these samples are reported in Mogk et al. (2004, this volume, Chapter 2).

Geochronology

Sample TR-24 (quartzofeldspathic gneiss) yielded abundant brown to pink, transparent zircons that in many cases exhibit narrow rims (too narrow to accurately analyze by ion probe). Analyses from the central parts of 11 grains were obtained using the SHRIMP I, and yielded a range of $^{207}Pb/^{206}Pb$ ages from 2.53 to 3.33 Ga (Table 1; Mueller et al., 1996c). Of these 11 grains, eight yield $^{207}Pb/^{206}Pb$ ages of ca. 3.3 Ga. Regression of these data yields a concordia intercept age of 3.305 ± 0.030 Ga (Fig. 3). Grains that yield generally younger, less concordant, low Th/U (<0.16) analyses are interpreted to result from lead loss from zircon grown in contact with hydrothermal solutions circulating during two distinct metamorphic episodes at ca. 2.8 Ga and ca. 2.4 Ga. The ages from these low-Th/U grains are compatible with previous reports of activity in this region at ca. 2.8 Ga and ca. 2.4 Ga (e.g., Mueller and Cordua, 1976; James and Hedge, 1980; Mueller et al., 1994, 1996a, 1996b; Krogh et al., 1997; Dahl et al., 2002; Roberts et al., 2002; Cheney et al., 2004b, this volume, Chapter 8).

Sample TR-59 yielded brown, well-formed grains, many with optically and analytically distinguishable rims. Regression of data obtained from analyses of distinct cores of six grains suggests that the protolith of this gneiss crystallized at ca. 3.217 ± 0.016 Ga (Fig. 4). In addition, one whole grain and rims on three grains yield analyses that are <10% discordant and define younger ages (ca. 2.8 Ga and ca. 2.4 Ga). The regions analyzed are not only optically distinct (visible light and cathode luminescence [CL]), but have low Th/U ratios (0.01–0.14) characteristic of growth from hydrothermal solutions. These data provide additional support for both the Late Archean and Early Proterozoic metamorphic events evident in data from sample TR-24.

Results from another sample of quartzofeldspathic gneiss from the Indian Creek Metamorphic Suite reported by Krogh et al. (1997) suggest a crystallization age of ca. 3.37 Ga and metamorphism at 2.4 Ga. Collectively, these data suggest that the tonalitic to trondhjemitic protoliths of the Archean gneisses were emplaced over a period of time from at least 3.2–3.4 Ga. These

TABLE 1. U-PB DATA

Labels	U/ppm	Th/ ppm	Th/U	Pb*/ ppm	^{204}Pb/ ppb	$^{204}Pb/^{206}Pb$	± 4/6	$^{208}Pb/^{206}Pb$	± 8/6	$^{208}Pb/^{232}Th$	± 8/32	$^{206}Pb/^{238}U$	± 6/38	$^{207}Pb/^{235}U$	± 7/35
K-1-1.1	222	33	0.15	52	6	0.00015	0.00004	0.09680	0.00212	0.13917	0.00375	0.21052	0.00274	5.689	0.0925
K-1-2.1	109	38	0.35	76	3	0.00005	0.00003	0.09419	0.00196	0.15918	0.00536	0.58622	0.01333	23.072	0.5649
K-1-3.1	839	81	0.10	84	2	0.00003	0.00001	0.03999	0.00067	0.03874	0.00079	0.09296	0.00102	2.666	0.0330
K-1-4.1	1767	46	0.03	29	71	0.00261	0.00042	0.00692	0.01669	0.00454	0.01095	0.01716	0.00028	0.232	0.0193
K-1-5.1	774	341	0.44	408	1	0.00000	0.00000	0.12094	0.00081	0.13167	0.00174	0.47895	0.00526	10.515	0.1290
K-1-6.1	176	120	0.68	135	2	0.00002	0.00001	0.18515	0.00244	0.16892	0.00436	0.62015	0.01223	21.680	0.5353
K-1-7.1	391	217	0.55	246	4	0.00002	0.00001	0.14914	0.00186	0.14265	0.00271	0.53006	0.00736	16.845	0.3013
K-1-8.1	494	64	0.13	67	5	0.00010	0.00005	0.16468	0.00230	0.14486	0.00398	0.11332	0.00237	3.404	0.0765
K-1-9.1	786	121	0.15	42	4	0.00010	0.00007	0.04424	0.00263	0.01475	0.00090	0.05150	0.00062	1.106	0.0268
K-1-10.1	291	17	0.06	42	5	0.00014	0.00005	0.01824	0.00232	0.04642	0.00658	0.14806	0.00830	2.618	0.1660
K-1-11.1	394	139	0.35	192	0	0.00000	0.00000	0.12122	0.00182	0.14156	0.00343	0.41304	0.00695	14.170	0.3039
K-1A-1.1	268	88	0.33	81	209	0.00374	0.00044	0.28251	0.02321	0.19612	0.01645	0.22826	0.00366	7.936	0.3615
K-1A-2.1	163	76	0.47	103	8	0.00011	0.00003	0.16803	0.00202	0.18509	0.00417	0.51550	0.00858	18.216	0.3678
K-1A-3.1	98	69	0.71	64	3	0.00006	0.00004	0.19756	0.00235	0.14770	0.00331	0.52689	0.00895	17.776	0.3313
K-1A-4.1	4159	238	0.06	49	203	0.00406	0.00041	0.01946	0.01787	0.00440	0.00404	0.01294	0.00018	0.088	0.0125
K-1A-5.1	423	151	0.36	154	4	0.00004	0.00001	0.10384	0.00090	0.09151	0.00141	0.31530	0.00375	10.288	0.1448
K-1A-6.1	668	402	0.60	205	3	0.00002	0.00001	0.18886	0.00120	0.07916	0.00169	0.25234	0.00465	7.856	0.1571
K-1A-7.1	297	141	0.47	227	2	0.00001	0.00000	0.12726	0.00095	0.17106	0.00253	0.63719	0.00779	23.439	0.3356
K-1A-8.1	267	133	0.50	253	3	0.00002	0.00001	0.13311	0.00126	0.19302	0.00314	0.72298	0.00902	38.535	0.5303
K-1A-9.1	201	54	0.27	107	9	0.00011	0.00002	0.09661	0.00182	0.16579	0.00514	0.45793	0.01010	15.652	0.4410
K-1A-10.1	250	403	1.61	149	13	0.00015	0.00006	0.52458	0.03548	0.12840	0.01048	0.39445	0.01780	12.404	0.8997
K-1A-11.1	431	96	0.22	156	3	0.00003	0.00002	0.10027	0.00130	0.14147	0.00248	0.31311	0.00358	10.422	0.1340
K-1A-12.1	226	172	0.76	157	1	0.00001	0.00001	0.19279	0.00204	0.14074	0.00285	0.55615	0.00884	19.810	0.4174
K-1A-13.1	1329	51	0.04	29	39	0.00144	0.00013	0.00565	0.00533	0.00340	0.00321	0.02299	0.00039	0.264	0.0118
K-1A-14.1	276	258	0.94	160	6	0.00005	0.00001	0.23945	0.00293	0.11535	0.00414	0.45101	0.00762	15.448	0.3446
K-1A-15.1	48	30	0.61	33	0	0.00002	0.00004	0.17359	0.00365	0.15771	0.00462	0.55548	0.00992	19.598	0.4539
K-1A-16.1	403	72	0.18	168	5	0.00004	0.00001	0.06477	0.00181	0.13347	0.00463	0.37061	0.00746	12.488	0.3611
K-1A-17.1	205	186	0.91	163	1	0.00001	0.00000	0.24555	0.00303	0.16544	0.00298	0.61180	0.00750	21.927	0.3304
K-1A-18.1	162	120	0.75	128	1	0.00001	0.00001	0.20422	0.00156	0.17242	0.00290	0.62921	0.00879	22.227	0.3560
K-1A-19.1	588	909	1.55	72	13	0.00027	0.00010	0.27696	0.00539	0.01709	0.00068	0.09537	0.00151	2.866	0.0702
K-1A-20.1	372	212	0.57	336	1	0.00000	0.00000	0.14982	0.00170	0.18931	0.00340	0.72071	0.00937	30.398	0.4455
K-1A-21.1	369	85	0.23	167	3	0.00003	0.00001	0.07589	0.00097	0.13023	0.00235	0.39790	0.00477	13.383	0.1730
K-1A-22.1	447	169	0.38	121	35	0.00043	0.00005	0.20795	0.00220	0.11757	0.00233	0.21406	0.00318	7.709	0.1264

(continued)

TABLE 1. U-PB DATA (*continued*)

Labels	$^{207}Pb/^{206}Pb$	± 7/6	$^{238}U/^{206}Pb$	± 38/6	$^{235}U/^{207}Pb$	± 35/7	$^{206}Pb/^{207}Pb$	± 6/7	age 8/32	± age 8/32	age 6/38	± age 6/38	age 7/35	± age 7/35	age 7/6	± age 7/6	% conc
K-1-1.1	0.19599	0.00162	4.7502	0.06185	0.17578	0.00286	5.10233	0.04213	2634	67	1232	15	1930	14	2793	14	44
K-1-2.1	0.28545	0.00191	1.7059	0.03879	0.04334	0.00106	3.50329	0.02339	2986	94	2974	54	3230	24	3393	10	88
K-1-3.1	0.20798	0.00094	10.7568	0.11787	0.37511	0.00464	4.80809	0.02183	768	15	573	6	1319	9	2890	7	20
K-1-4.1	0.09788	0.00782	58.2781	0.95194	4.31833	0.36063	10.21672	0.81613	92	219	110	2	211	16	1584	157	7
K-1-5.1	0.15922	0.00068	2.0879	0.02291	0.09510	0.00117	6.28048	0.02687	2500	31	2523	23	2481	11	2447	7	103
K-1-6.1	0.25354	0.00319	1.6125	0.03181	0.04613	0.00114	3.94409	0.04967	3155	76	3111	49	3169	24	3207	20	97
K-1-7.1	0.23049	0.00223	1.8866	0.02619	0.05936	0.00106	4.33864	0.04197	2695	48	2742	31	2926	17	3055	16	90
K-1-8.1	0.21787	0.00133	8.8244	0.18422	0.29376	0.00660	4.58996	0.02796	2734	70	692	14	1505	18	2965	10	23
K-1-9.1	0.15574	0.00305	19.4169	0.23311	0.90420	0.02192	6.42077	0.12593	296	18	324	4	756	13	2410	34	13
K-1-10.1	0.12823	0.00301	6.7539	0.37866	0.38200	0.02423	7.79854	0.18288	917	128	890	47	1306	48	2074	42	43
K-1-11.1	0.24882	0.00283	2.4211	0.04072	0.07057	0.00151	4.01897	0.04570	2676	61	2229	32	2761	21	3177	18	70
K-1A-1.1	0.25214	0.01026	4.3809	0.07027	0.12602	0.00574	3.96600	0.16131	3620	280	1325	19	2224	42	3198	66	41
K-1A-2.1	0.25629	0.00245	1.9399	0.03228	0.05490	0.00111	3.90190	0.03725	3432	71	2680	37	3001	20	3224	15	83
K-1A-3.1	0.24469	0.00143	1.8979	0.03222	0.05626	0.00105	4.08690	0.02390	2784	58	2728	38	2978	18	3151	9	87
K-1A-4.1	0.04951	0.00690	77.2530	1.08790	11.31600	1.60560	20.19600	2.81620	89	81	83	1	86	12	172	297	48
K-1A-5.1	0.23664	0.00147	3.1715	0.03768	0.09720	0.00137	4.22580	0.02622	1770	26	1767	18	2461	13	3098	10	57
K-1A-6.1	0.22580	0.00130	3.9628	0.07307	0.12729	0.00255	4.42870	0.02556	1540	32	1451	24	2215	18	3023	9	48
K-1A-7.1	0.26678	0.00162	1.5694	0.01918	0.04266	0.00061	3.74830	0.02282	3192	44	3178	31	3245	14	3287	10	97
K-1A-8.1	0.38657	0.00172	1.3832	0.01726	0.02595	0.00036	2.58680	0.01151	3567	53	3507	34	3734	14	3858	7	91
K-1A-9.1	0.24790	0.00372	2.1837	0.04814	0.06389	0.00180	4.03400	0.06054	3101	89	2430	45	2856	27	3171	24	77
K-1A-10.1	0.22807	0.01173	2.5352	0.11443	0.08062	0.00585	4.38470	0.22546	2442	189	2143	83	2635	71	3039	85	71
K-1A-11.1	0.24140	0.00112	3.1937	0.03647	0.09595	0.00123	4.14240	0.01928	2674	44	1756	18	2473	12	3129	7	56
K-1A-12.1	0.25834	0.00310	1.7981	0.02857	0.05048	0.00106	3.87090	0.04642	2662	51	2851	37	3082	21	3237	19	88
K-1A-13.1	0.08339	0.00328	43.5030	0.74372	3.78370	0.16951	11.99200	0.47163	69	65	147	2	238	10	1278	79	12
K-1A-14.1	0.24841	0.00313	2.2172	0.03744	0.06473	0.00144	4.02560	0.05078	2206	75	2400	34	2843	21	3175	20	76
K-1A-15.1	0.25588	0.00324	1.8002	0.03215	0.05103	0.00118	3.90810	0.04954	2960	81	2848	41	3072	23	3222	20	88
K-1A-16.1	0.24439	0.00450	2.6983	0.05428	0.08007	0.00232	4.09180	0.07534	2532	83	2032	35	2642	28	3149	30	65
K-1A-17.1	0.25994	0.00192	1.6345	0.02003	0.04561	0.00069	3.84710	0.02835	3094	52	3077	30	3181	15	3246	12	95
K-1A-18.1	0.25620	0.00161	1.5893	0.02219	0.04499	0.00072	3.90330	0.02459	3215	50	3146	35	3194	16	3223	10	98
K-1A-19.1	0.21795	0.00367	10.4850	0.16576	0.34893	0.00854	4.58830	0.07721	342	13	587	9	1373	19	2966	27	20
K-1A-20.1	0.30590	0.00163	1.3875	0.01805	0.03290	0.00048	3.26900	0.01740	3504	58	3499	35	3500	15	3500	8	100
K-1A-21.1	0.24394	0.00088	2.5132	0.03011	0.07472	0.00097	4.09930	0.01471	2474	42	2159	22	2707	12	3146	6	69
K-1A-22.1	0.26118	0.00138	4.6715	0.06947	0.12972	0.00213	3.82880	0.02024	2247	42	1250	17	2198	15	3254	8	38

(*continued*)

TABLE 1. U-PB DATA (*continued*)

Labels	U/ppm	Th/ ppm	Th/U	Pb*/ ppm	^{204}Pb/ ppb	^{204}Pb/^{206}Pb	± 4/6	^{208}Pb/^{206}Pb	± 8/6	^{208}Pb/^{232}Th	± 8/32	^{206}Pb/^{238}U	± 6/38	^{207}Pb/^{235}U	± 7/35
K-2-1.1	385	21	0.05	113	2	0.00002	0.00001	0.01457	0.00034	0.08184	0.00217	0.30332	0.00352	4.561	0.0647
K-2-2.1	644	16	0.02	183	1	0.00001	0.00000	0.00697	0.00014	0.08551	0.00206	0.29620	0.00332	4.394	0.0545
K-2-3.1	532	22	0.04	164	1	0.00001	0.00000	0.01204	0.00028	0.09297	0.00248	0.32003	0.00359	4.798	0.0608
K-2-4.1	406	16	0.04	121	1	0.00001	0.00000	0.01148	0.00027	0.09065	0.00246	0.31006	0.00375	4.592	0.0624
K-2-5.1	322	14	0.04	93	0	0.00001	0.00000	0.01273	0.00042	0.08968	0.00320	0.29980	0.00349	4.499	0.0674
K-2-6.1	299	36	0.12	92	0	0.00001	0.00000	0.03414	0.00043	0.08907	0.00157	0.31344	0.00360	4.650	0.0617
K-2-7.1	452	19	0.04	139	2	0.00001	0.00001	0.01186	0.00046	0.09221	0.00378	0.32029	0.00388	4.802	0.0649
K-2-8.1	446	85	0.19	148	3	0.00002	0.00002	0.07569	0.00144	0.12280	0.00373	0.31057	0.00636	7.062	0.1725
K-2-9.1	456	42	0.09	152	–1	–0.00001	0.00001	0.03606	0.00057	0.12842	0.00259	0.32925	0.00385	6.298	0.0788
K-2-10.1	458	15	0.03	142	2	0.00002	0.00001	0.00939	0.00028	0.09430	0.00319	0.32324	0.00437	4.798	0.0762
K-2-11.1	649	132	0.20	233	3123	0.01503	0.00139	0.28305	0.07099	0.39661	0.10092	0.28438	0.01136	7.079	1.2139
K-2-12.1	584	38	0.07	185	1	0.00001	0.00000	0.01876	0.00024	0.09461	0.00163	0.32684	0.00345	4.914	0.0586
K-2-13.1	438	23	0.05	133	0	0.00000	0.00000	0.01510	0.00049	0.09258	0.00325	0.31516	0.00375	4.703	0.0664
K-2-14.1	361	44	0.12	100	1	0.00002	0.00001	0.03550	0.00071	0.08241	0.00196	0.28223	0.00335	4.197	0.0625
K-2-15.1	522	227	0.44	160	3	0.00003	0.00002	0.12105	0.00122	0.07650	0.00119	0.27478	0.00307	6.719	0.0820
K-2-16.1	186	191	1.03	164	–1	0.00000	0.00000	0.28259	0.00116	0.18216	0.00273	0.66110	0.00874	24.078	0.3477
K-2-17.1	381	54	0.14	129	1	0.00000	0.00000	0.04917	0.00088	0.11348	0.00273	0.32831	0.00454	6.799	0.1144
K-2-18.1	511	28	0.05	319	–1	0.00000	0.00001	0.01546	0.00030	0.15685	0.00831	0.55053	0.02393	22.937	1.0920
K-2-19.1	386	105	0.27	163	3	0.00003	0.00001	0.10984	0.00105	0.15346	0.00242	0.37841	0.00451	9.567	0.1265
K-2-20.1	302	121	0.40	161	2	0.00001	0.00001	0.14935	0.00085	0.16908	0.00300	0.45470	0.00683	13.268	0.2126
K-2-21.1	452	19	0.04	101	1	0.00001	0.00001	0.01552	0.00051	0.08609	0.00305	0.23197	0.00269	3.464	0.0472

(continued)

TABLE 1. U-PB DATA (*continued*)

Labels	^{207}Pb/^{206}Pb	± 7/6	^{238}U/^{206}Pb	± 38/6	^{235}U/^{207}Pb	± 35/7	^{206}Pb/^{207}Pb	± 6/7	age 8/32	± age 8/32	age 6/38	± age 6/38	age 7/35	± age 7/35	age 7/6	± age 7/6	% conc
K-2-1.1	0.10907	0.00074	3.2969	0.03825	0.21923	0.00311	9.16860	0.06253	1590	41	1708	17	1742	12	1784	12	96
K-2-2.1	0.10759	0.00044	3.3761	0.03785	0.22759	0.00282	9.29480	0.03808	1658	38	1672	17	1711	10	1759	8	95
K-2-3.1	0.10874	0.00051	3.1247	0.03501	0.20840	0.00264	9.19590	0.04294	1797	46	1790	18	1785	11	1778	9	101
K-2-4.1	0.10741	0.00052	3.2252	0.03902	0.21777	0.00296	9.30970	0.04536	1754	46	1741	18	1748	11	1756	9	99
K-2-5.1	0.10883	0.00088	3.3356	0.03881	0.22229	0.00333	9.18850	0.07436	1736	59	1690	17	1731	13	1780	15	95
K-2-6.1	0.10759	0.00058	3.1904	0.03659	0.21506	0.00285	9.29440	0.05015	1725	29	1758	18	1758	11	1759	10	100
K-2-7.1	0.10874	0.00050	3.1222	0.03783	0.20825	0.00281	9.19630	0.04265	1783	70	1791	19	1785	11	1778	8	101
K-2-8.1	0.16492	0.00181	3.2199	0.06598	0.14160	0.00346	6.06360	0.06662	2341	67	1744	31	2119	22	2507	19	70
K-2-9.1	0.13872	0.00045	3.0372	0.03552	0.15879	0.00199	7.20860	0.02331	2442	46	1835	19	2018	11	2211	6	83
K-2-10.1	0.10766	0.00073	3.0937	0.04186	0.20841	0.00331	9.28820	0.06318	1821	59	1806	21	1785	13	1760	12	103
K-2-11.1	0.18054	0.02922	3.5164	0.14041	0.14126	0.02422	5.53900	0.89649	6752	1516	1613	57	2121	165	2658	297	61
K-2-12.1	0.10905	0.00048	3.0596	0.03230	0.20348	0.00243	9.17010	0.04019	1827	30	1823	17	1805	10	1784	8	102
K-2-13.1	0.10823	0.00068	3.1730	0.03775	0.21262	0.00300	9.23920	0.05775	1790	60	1766	18	1768	12	1770	11	100
K-2-14.1	0.10787	0.00083	3.5433	0.04203	0.23824	0.00355	9.27080	0.07093	1601	37	1603	17	1674	12	1764	14	91
K-2-15.1	0.17735	0.00066	3.6392	0.04066	0.14882	0.00182	5.63850	0.02097	1490	22	1565	16	2075	11	2628	6	60
K-2-16.1	0.26415	0.00116	1.5126	0.02000	0.04153	0.00060	3.78570	0.01656	3382	47	3271	34	3272	14	3272	7	100
K-2-17.1	0.15019	0.00120	3.0459	0.04215	0.14709	0.00248	6.65840	0.05326	2173	50	1830	22	2086	15	2348	14	78
K-2-18.1	0.30218	0.00445	1.8164	0.07895	0.04360	0.00208	3.30930	0.04877	2945	146	2827	100	3224	47	3481	23	81
K-2-19.1	0.18337	0.00081	2.6426	0.03148	0.10452	0.00138	5.45360	0.02418	2886	42	2069	21	2394	12	2684	7	77
K-2-20.1	0.21162	0.00085	2.1993	0.03304	0.07537	0.00121	4.72530	0.01895	3158	52	2416	30	2699	15	2918	7	83
K-2-21.1	0.10831	0.00063	4.3108	0.05003	0.28865	0.00393	9.23240	0.05380	1669	57	1345	14	1519	11	1771	11	76

(continued)

TABLE 1. U-PB DATA (*continued*)

Labels	U/ppm	Th/ ppm	Th/U	Pb*/ ppm	^{204}Pb/ ppb	$^{204}Pb/^{206}Pb$	± 4/6	$^{208}Pb/^{206}Pb$	± 8/6	$^{208}Pb/^{232}Th$	± 8/32	$^{206}Pb/^{238}U$	± 6/38	$^{207}Pb/^{235}U$	± 7/35
K-3-1	429	247	0.58	165	5	0.00004	0.00001	0.16458	0.00107	0.09906	0.00163	0.34609	0.00481	5.916	0.0930
K-3-2	278	114	0.41	117	5	0.00005	0.00005	0.12869	0.00240	0.12287	0.00292	0.39093	0.00516	6.551	0.1072
K-3-3	174	107	0.62	68	7	0.00013	0.00003	0.17849	0.00126	0.10109	0.00156	0.34989	0.00448	6.111	0.0890
K-3-4	341	79	0.23	109	18	0.00019	0.00004	0.06497	0.00175	0.08936	0.00276	0.31850	0.00432	4.681	0.0767
K-3-5	365	235	0.64	148	7	0.00007	0.00002	0.18552	0.00157	0.10341	0.00573	0.35889	0.01619	6.340	0.2959
K-3-6	194	77	0.40	81	10	0.00016	0.00005	0.11844	0.00187	0.11609	0.00267	0.38944	0.00586	6.782	0.1233
K-3-7	402	69	0.17	123	8	0.00007	0.00002	0.04857	0.00082	0.08698	0.00177	0.30747	0.00336	4.602	0.0572
K-3-8	353	14	0.04	108	3	0.00003	0.00002	0.01210	0.00081	0.09438	0.00651	0.31757	0.00421	4.741	0.0702
Keck3-1.1	290	168	0.58	128	4	0.00004	0.00002	0.16489	0.00151	0.11303	0.00184	0.39718	0.00443	6.983	0.1254
Keck3-2.1	458	134	0.29	152	3	0.00002	0.00001	0.08493	0.00102	0.09344	0.00149	0.32329	0.00251	4.817	0.0688
Keck3-3.1	528	416	0.79	237	1	0.00000	0.00001	0.22729	0.00128	0.11087	0.00150	0.38481	0.00394	6.797	0.0797
Keck3-4.1	414	84	0.20	131	4	0.00003	0.00001	0.05944	0.00118	0.09269	0.00200	0.31550	0.00224	4.667	0.0470
Keck3-5.1	366	61	0.17	114	1	0.00001	0.00001	0.04685	0.00077	0.08848	0.00181	0.31444	0.00326	4.678	0.0636
Keck3-6.1	423	100	0.24	135	2	0.00002	0.00001	0.06877	0.00088	0.09212	0.00201	0.31547	0.00470	4.653	0.0770
Keck3-7.1	393	128	0.33	115	113	0.00112	0.00011	0.05711	0.00401	0.05197	0.00373	0.29613	0.00360	3.716	0.0897
Keck3-8.1	278	163	0.59	116	0	0.00000	0.00000	0.17575	0.00152	0.11144	0.00229	0.37183	0.00569	6.637	0.1304
Keck3-9.1	571	491	0.86	249	9	0.00005	0.00002	0.23719	0.00473	0.10228	0.00304	0.37032	0.00560	6.525	0.1238
K3-10.1	267	136	0.51	97	21	0.00027	0.00008	0.14631	0.00396	0.09521	0.00299	0.33195	0.00454	5.708	0.1145
K3-11.1	452	323	0.71	200	2	0.00001	0.00001	0.21119	0.00316	0.11350	0.00236	0.38376	0.00459	6.830	0.0966
K3-12.1	363	90	0.25	115	2	0.00002	0.00001	0.07663	0.00158	0.09535	0.00281	0.31017	0.00467	4.660	0.0943
K3-13	294	132	0.45	108	1	0.00001	0.00001	0.13342	0.00184	0.10050	0.00249	0.33919	0.00577	5.859	0.1301
K3-14	661	603	0.91	295	5	0.00002	0.00001	0.26562	0.00130	0.10816	0.00103	0.37157	0.00255	6.538	0.0565
K3-15.1	355	207	0.58	149	2	0.00002	0.00004	0.16862	0.00194	0.10891	0.00195	0.37603	0.00436	6.678	0.1030
K3-16.1	636	526	0.83	250	7	0.00004	0.00001	0.23174	0.00136	0.09396	0.00127	0.33567	0.00334	5.877	0.0709
K3-17.1	319	193	0.61	120	2	0.00003	0.00004	0.17876	0.00213	0.09939	0.00174	0.33720	0.00377	5.569	0.0817
K3-18.1	262	157	0.60	113	2	0.00003	0.00003	0.18058	0.00336	0.11556	0.00274	0.38304	0.00492	6.737	0.1269
K3-19.1	508	389	0.77	211	19	0.00012	0.00006	0.22357	0.00267	0.10455	0.00167	0.35786	0.00322	6.259	0.0812
K3-20.1	385	108	0.28	117	39	0.00039	0.00006	0.07053	0.00323	0.07602	0.00372	0.30128	0.00417	4.249	0.0840
K3-21.1	394	70	0.18	122	5	0.00005	0.00002	0.04970	0.00109	0.08745	0.00216	0.31249	0.00259	4.636	0.0519
K3-22.1	355	101	0.29	116	23	0.00023	0.00007	0.07394	0.00328	0.08374	0.00529	0.32307	0.01155	4.801	0.2197
K3-23.1	359	68	0.19	110	–1	–0.00001	0.00001	0.05438	0.00115	0.08830	0.00219	0.30614	0.00301	4.624	0.0659
K3-24.1	340	88	0.26	106	14	0.00016	0.00010	0.07054	0.00395	0.08340	0.00483	0.30722	0.00364	4.472	0.0955
K3-25.1	257	46	0.18	75	24	0.00036	0.00006	0.03923	0.00224	0.06426	0.00377	0.29577	0.00280	4.233	0.0679

(*continued*)

TABLE 1. U-PB DATA (*continued*)

Labels	$^{207}Pb/^{206}Pb$	± 7/6	$^{238}U/^{206}Pb$	± 38/6	$^{235}U/^{207}Pb$	± 35/7	$^{206}Pb/^{207}Pb$	± 6/7	age 8/32	± age 8/32	age 6/38	± age 6/38	age 7/35	± age 7/35	age 7/6	± age 7/6	% conc
K-3-1	0.12397	0.00072	2.8894	0.04018	0.16904	0.00266	8.06670	0.04692	1909	30	1916	23	1964	14	2014	10	94
K-3-2	0.12154	0.00099	2.5580	0.03374	0.15265	0.00250	8.22810	0.06722	2342	53	2127	24	2053	15	1979	15	106
K-3-3	0.12668	0.00070	2.8581	0.03656	0.16364	0.00238	7.89420	0.04367	1946	29	1934	21	1992	13	2052	10	93
K-3-4	0.10660	0.00081	3.1397	0.04260	0.21362	0.00350	9.38120	0.07159	1730	51	1782	21	1764	14	1742	14	101
K-3-5	0.12812	0.00097	2.7864	0.12573	0.15774	0.00736	7.80550	0.05893	1989	105	1977	77	2024	42	2072	13	94
K-3-6	0.12630	0.00107	2.5678	0.03864	0.14745	0.00268	7.91770	0.06734	2220	48	2120	27	2083	16	2047	15	102
K-3-7	0.10855	0.00051	3.2524	0.03557	0.21731	0.00270	9.21250	0.04308	1686	33	1728	17	1750	10	1775	9	96
K-3-8	0.10827	0.00056	3.1489	0.04176	0.21093	0.00313	9.23590	0.04760	1823	121	1778	21	1774	13	1771	9	99
Keck3-1.1	0.12752	0.00162	2.5177	0.02808	0.14320	0.00257	7.84190	0.09991	2165	34	2156	20	2109	16	2064	23	105
Keck3-2.1	0.10807	0.00120	3.0932	0.02398	0.20759	0.00296	9.25340	0.10234	1806	28	1806	12	1788	12	1767	20	102
Keck3-3.1	0.12810	0.00059	2.5987	0.02660	0.14713	0.00172	7.80620	0.03568	2125	27	2099	18	2085	10	2072	8	101
Keck3-4.1	0.10728	0.00068	3.1696	0.02251	0.21427	0.00216	9.32110	0.05872	1792	37	1768	11	1761	8	1754	12	101
Keck3-5.1	0.10789	0.00082	3.1802	0.03292	0.21378	0.00291	9.26840	0.07061	1714	34	1763	16	1763	11	1764	14	100
Keck3-6.1	0.10698	0.00060	3.1698	0.04718	0.21490	0.00356	9.34760	0.05240	1781	37	1768	23	1759	14	1749	10	101
Keck3-7.1	0.09102	0.00176	3.3768	0.04109	0.26908	0.00650	10.98700	0.21297	1024	72	1672	18	1575	20	1447	37	116
Keck3-8.1	0.12945	0.00137	2.6894	0.04115	0.15068	0.00296	7.72510	0.08165	2136	42	2038	27	2064	17	2091	19	98
Keck3-9.1	0.12779	0.00124	2.7004	0.04080	0.15326	0.00291	7.82540	0.07615	1968	56	2031	26	2049	17	2068	17	98
K3-10.1	0.12472	0.00163	3.0125	0.04120	0.17519	0.00352	8.01820	0.10481	1838	55	1848	22	1933	17	2025	23	91
K3-11.1	0.12909	0.00080	2.6058	0.03114	0.14641	0.00207	7.74680	0.04822	2173	43	2094	21	2090	13	2086	11	100
K3-12.1	0.10896	0.00129	3.2240	0.04850	0.21461	0.00434	9.17780	0.10829	1841	52	1742	23	1760	17	1782	22	98
K3-13	0.12528	0.00154	2.9482	0.05016	0.17068	0.00379	7.98200	0.09819	1936	46	1883	28	1955	19	2033	22	93
K3-14	0.12762	0.00057	2.6913	0.01850	0.15295	0.00132	7.83590	0.03486	2076	19	2037	12	2051	8	2065	8	99
K3-15.1	0.12879	0.00114	2.6594	0.03085	0.14976	0.00231	7.76440	0.06848	2090	36	2058	20	2070	14	2082	16	99
K3-16.1	0.12698	0.00072	2.9791	0.02962	0.17016	0.00205	7.87550	0.04483	1815	24	1866	16	1958	11	2057	10	91
K3-17.1	0.11978	0.00099	2.9656	0.03311	0.17957	0.00264	8.34880	0.06869	1915	32	1873	18	1911	13	1953	15	96
K3-18.1	0.12755	0.00157	2.6107	0.03350	0.14844	0.00280	7.83990	0.09620	2210	50	2091	23	2077	17	2065	22	101
K3-19.1	0.12685	0.00105	2.7944	0.02518	0.15976	0.00207	7.88300	0.06529	2010	31	1972	15	2013	11	2055	15	96
K3-20.1	0.10229	0.00128	3.3191	0.04591	0.23535	0.00465	9.77660	0.12217	1481	70	1698	21	1684	16	1666	23	102
K3-21.1	0.10759	0.00071	3.2001	0.02648	0.21572	0.00241	9.29480	0.06112	1695	40	1753	13	1756	9	1759	12	100
K3-22.1	0.10777	0.00263	3.0953	0.11069	0.20830	0.00953	9.27890	0.22685	1626	99	1805	57	1785	39	1762	45	102
K3-23.1	0.10956	0.00100	3.2664	0.03208	0.21624	0.00308	9.12780	0.08369	1710	41	1722	15	1754	12	1792	17	96
K3-24.1	0.10556	0.00173	3.2550	0.03854	0.22364	0.00478	9.47330	0.15502	1619	90	1727	18	1726	18	1724	30	100
K3-25.1	0.10379	0.00123	3.3810	0.03197	0.23625	0.00379	9.63450	0.11399	1259	72	1670	14	1680	13	1693	22	99

(*continued*)

TABLE 1 U-PB DATA (*continued*)

Labels	U/ppm	Th/ ppm	Th/U	Pb*/ ppm	^{204}Pb/ ppb	^{204}Pb/^{206}Pb	± 4/6	^{208}Pb/^{206}Pb	± 8/6	^{208}Pb/^{232}Th	± 8/32	^{206}Pb/^{238}U	± 6/38	^{207}Pb/^{235}U	± 7/35
TR59-1.1	1773	5037	2.84	1467	62	0.00009	0.00001	0.74556	0.00287	0.12596	0.00085	0.48003	0.00235	16.730	0.1087
TR59-2.1	582	128	0.22	350	7	0.00003	0.00001	0.01518	0.00047	0.03815	0.00129	0.55204	0.00558	19.200	0.2157
TR59-2.2T	205	29	0.14	97	8	0.00010	0.00003	0.03822	0.00130	0.12214	0.00454	0.45882	0.00562	9.970	0.1861
TR59-3.1	508	5	0.01	286	2	0.00001	0.00001	0.00214	0.00031	0.13119	0.01933	0.54529	0.00677	14.904	0.2178
TR59-3.2T	172	4	0.02	90	−1	−0.00002	0.00001	0.01024	0.00085	0.21594	0.01884	0.50346	0.00868	14.231	0.3061
TR59-4.1C	142	154	1.08	119	1	0.00001	0.00001	0.28545	0.00413	0.16624	0.00463	0.62930	0.01295	22.464	0.5253
TR59-5.1C	80	52	0.65	60	−1	−0.00003	0.00002	0.19024	0.00302	0.17494	0.00583	0.59916	0.01501	20.670	0.6712
TR59-6.1T	578	5	0.01	306	0	0.00000	0.00000	0.00183	0.00010	0.10583	0.00624	0.51480	0.00624	13.818	0.1787
TR59-7.1C	386	195	0.51	300	2	0.00001	0.00001	0.13871	0.00098	0.17780	0.00434	0.64817	0.01271	22.862	0.4970
TR59-8.1C	612	67	0.11	368	1	0.00000	0.00000	0.02968	0.00036	0.15175	0.00261	0.55577	0.00483	17.346	0.2046
TR59-9.1C	170	144	0.85	130	4	0.00004	0.00002	0.23793	0.00255	0.16705	0.00416	0.59453	0.01099	20.933	0.4125
TR-24-1.1	29	10	0.33	24	0	0.00000	0.00000	0.09108	0.00216	0.19347	0.01369	0.70534	0.04027	26.179	1.5616
TR-24-2.1	172	3	0.02	90	5	0.00006	0.00005	0.00099	0.00164	0.02644	0.04395	0.50418	0.04755	14.473	1.4084
TR-24-3.1	79	27	0.34	53	2	0.00005	0.00006	0.09399	0.00416	0.15892	0.00982	0.57768	0.02243	21.105	1.0025
TR-24-4.1	192	30	0.16	103	0	0.00000	0.00000	0.05095	0.00296	0.16112	0.01175	0.50220	0.02059	13.028	0.7029
TR-24-5.1	46	18	0.38	37	1	0.00004	0.00002	0.10155	0.00333	0.18183	0.00899	0.68158	0.02036	25.691	0.8222
TR-24-6.1	217	121	0.56	158	0	0.00000	0.00000	0.15166	0.00151	0.16287	0.00785	0.59863	0.02349	21.561	0.8729
TR-24-7.1	124	54	0.43	94	1	0.00001	0.00003	0.11605	0.00555	0.17267	0.01244	0.64299	0.03230	23.287	1.4773
TR-24-8.1	156	65	0.42	121	1	0.00002	0.00003	0.11196	0.00127	0.17671	0.00474	0.65631	0.01520	24.330	0.6377
TR-24-9.1	243	19	0.08	133	3	0.00003	0.00003	0.02396	0.00198	0.15252	0.01609	0.50556	0.02912	16.012	1.0421
TR-24-9.2	1582	37	0.02	409	8	0.00002	0.00001	0.00870	0.00036	0.09634	0.00480	0.25598	0.00672	5.892	0.1574
TR-24-10.1	41	10	0.23	30	2	0.00008	0.00010	0.06323	0.00421	0.17572	0.01551	0.65258	0.03312	23.234	1.3082
TR-24-11.1	92	33	0.36	70	0	0.00000	0.00000	0.09775	0.00271	0.17548	0.01086	0.64726	0.03069	24.221	1.2371

(*continued*)

TABLE 1. U-PB DATA (*continued*)

Labels	$^{207}Pb/^{206}Pb$	± 7/6	$^{238}U/^{206}Pb$	± 38/6	$^{235}U/^{207}Pb$	± 35/7	$^{206}Pb/^{207}Pb$	± 6/7	age 8/32	± age 8/32	age 6/38	± age 6/38	age 7/35	± age 7/35	age 7/6	± age 7/6	% conc
TR59-1.1	0.25277	0.00093	2.0832	0.01020	0.05977	0.00039	3.95611	0.01463	2398	15	2527	10	2920	6	3202	6	79
TR59-2.1	0.25225	0.00096	1.8115	0.01832	0.05208	0.00059	3.96434	0.01504	757	25	2834	23	3052	11	3199	6	89
TR59-2.2T	0.15760	0.00199	2.1795	0.02671	0.10030	0.00187	6.34507	0.08012	2329	82	2434	25	2432	17	2430	22	100
TR59-3.1	0.19823	0.00125	1.8339	0.02275	0.06710	0.00098	5.04474	0.03189	2491	348	2806	28	2809	14	2812	10	100
TR59-3.2T	0.20500	0.00223	1.9863	0.03424	0.07027	0.00151	4.87796	0.05309	3952	316	2629	37	2765	21	2866	18	92
TR59-4.1C	0.25889	0.00229	1.5891	0.03269	0.04452	0.00104	3.86258	0.03415	3108	80	3147	51	3204	23	3240	14	97
TR59-5.1C	0.25020	0.00444	1.6690	0.04182	0.04838	0.00157	3.99676	0.07099	3258	100	3026	61	3123	32	3186	28	95
TR59-6.1T	0.19467	0.00063	1.9425	0.02355	0.07237	0.00094	5.13683	0.01658	2033	114	2677	27	2737	12	2782	5	96
TR59-7.1C	0.25582	0.00185	1.5428	0.03026	0.04374	0.00095	3.90905	0.02831	3308	75	3221	50	3221	21	3221	11	100
TR59-8.1C	0.22635	0.00158	1.7993	0.01563	0.05765	0.00068	4.41786	0.03080	2856	46	2849	20	2954	11	3026	11	94
TR59-9.1C	0.25537	0.00124	1.6820	0.03111	0.04777	0.00094	3.91594	0.01909	3122	72	3008	45	3136	19	3218	8	94
TR-24-1.1	0.26919	0.00311	1.4178	0.08094	0.03820	0.00228	3.7149	0.04294	3575	233	3441	154	3353	60	3301	18	104
TR-24-2.1	0.20820	0.00311	1.9834	0.18706	0.06909	0.00672	4.8030	0.07167	527	847	2632	207	2781	97	2892	24	91
TR-24-3.1	0.26497	0.00608	1.7311	0.06721	0.04738	0.00225	3.7740	0.08658	2981	172	2939	92	3143	47	3277	37	90
TR-24-4.1	0.18815	0.00571	1.9913	0.08162	0.07676	0.00414	5.3149	0.16117	3019	206	2623	89	2682	52	2726	51	96
TR-24-5.1	0.27338	0.00228	1.4672	0.04382	0.03892	0.00125	3.6580	0.03056	3377	154	3350	79	3335	32	3325	13	101
TR-24-6.1	0.26122	0.00162	1.6705	0.06555	0.04638	0.00188	3.8282	0.02368	3050	137	3024	95	3164	40	3254	10	93
TR-24-7.1	0.26267	0.00866	1.5552	0.07813	0.04294	0.00272	3.8071	0.12551	3220	216	3201	128	3239	64	3263	53	98
TR-24-8.1	0.26887	0.00261	1.5237	0.03530	0.04110	0.00108	3.7193	0.03609	3289	82	3253	59	3282	26	3299	15	99
TR-24-9.1	0.22970	0.00550	1.9780	0.11395	0.06246	0.00406	4.3535	0.10418	2869	284	2638	126	2877	64	3050	39	87
TR-24-9.2	0.16694	0.00046	3.9065	0.10248	0.16972	0.00453	5.9902	0.01633	1859	89	1469	35	1960	23	2527	5	58
TR-24-10.1	0.25822	0.00487	1.5324	0.07777	0.04304	0.00242	3.8726	0.07300	3272	268	3238	131	3237	56	3236	30	100
TR-24-11.1	0.27140	0.00379	1.5450	0.07326	0.04129	0.00211	3.6846	0.05143	3268	188	3218	121	3277	51	3314	22	97

Note: Sample designations that end in "C" indicate the core of the grain was analyzed; those that end in T indicate that a tip or rim of the grain was analyzed. K-3 and Keck-3 are the same sample but are from different probe mounts. All ratios are atomic. % conc—percent of concordance.

TABLE 2. Sm-Nd DATA

Sample	Rock	Sm	Nd	$^{147}Sm/^{144}Nd$	$^{143}Sm/^{144}Nd$	ε(0)	T_{chur}	T_{dm}	t(Ga)	ε(t)
ICMS/PMMMS										
TR-2	Qfg	00.16	01.02	0.0949	0.51041	–43.4	3.31	3.43	3.3	–0.1
TR-4	Qfg	03.69	19.88	0.1123	0.51096	–32.7	3.01	3.18	3.3	3.3
TR-10	Qfg	01.17	07.97	0.0888	0.51041	–43.4	3.12	3.26	3.3	2.5
TR-51	MS	07.91	39.48	0.1212	0.51106	–30.7	3.16	3.32	3.3	1.4
TR-56	Qfg	05.70	29.29	0.1177	0.51084	–35.0	3.43	3.57	3.3	–1.4
TR-59	Qfg	07.40	50.60	0.0884	0.51029	–45.8	3.28	3.4	3.3	0.3
KD-009	HG	03.53	12.41	0.1720	0.51208	–10.9	3.44	3.73	3.3	–0.4
KD-010	Qfg	04.68	29.28	0.0967	0.51057	–40.4	3.13	3.27	3.3	2.2
KD-014	MS	12.65	51.80	0.1477	0.51148	–22.7	3.59	3.75	3.3	–1.8
KD-022	Qfg	06.45	32.67	0.1194	0.51092	–33.5	3.36	3.5	3.3	–0.6
KD-030	Qfg	04.21	23.03	0.1106	0.51054	–40.9	3.68	3.79	3.3	–4.3
KD-036	Qfg	08.28	48.72	0.1028	0.51082	–35.4	2.93	3.09	3.3	4.6
RC-2	A	00.13	08.43	0.0093	0.51201	–12.3	0.51	0.78	3.3	68.0
RC-9	Qfg	04.79	18.01	0.1609	0.51172	–17.9	3.87	4.02	3.3	–2.7
RC-10	Qfg	01.67	10.71	0.0943	0.51052	–41.4	3.13	3.27	3.3	2.2
RC-11	Qfg	00.51	03.23	0.0955	0.51068	–38.3	2.93	3.09	3.3	4.9
RC-13	Qfg	05.97	24.86	0.1452	0.51150	–22.1	3.33	3.53	3.3	–0.2
RC-24	A	06.61	26.00	0.1538	0.51249	–02.8	0.51	1.39	3.3	15.6
RP-25	Qfg	08.73	35.32	0.1495	0.51160	–20.2	3.31	3.52	3.3	–0.1
RP-26	Qfg	05.72	24.19	0.1430	0.51140	–24.2	3.50	3.66	3.3	–1.4
RP-35	Qfg	00.69	02.95	0.1415	0.51126	–27.0	3.78	3.91	3.3	–3.5
WC-3	Qfg	02.58	21.04	0.0742	0.51020	–47.6	3.01	3.14	3.3	4.6
K-2	M	00.52	03.69	0.0852	0.51081	–35.7	2.49	2.67	1.77	–10.4
SPMS										
TR-12	A	00.78	02.69	0.1754	0.51223	–7.9	2.89	3.37	N.D.	N.D.
TR-13	A	01.19	06.13	0.1174	0.51081	–35.6	3.48	3.61	N.D.	N.D.
TR-30	A	11.44	53.10	0.1303	0.51131	–25.9	3.03	3.22	N.D.	N.D.
TR-31	MS	03.90	15.80	0.1493	0.51179	–16.5	2.71	3.02	N.D.	N.D.
TR-32	A	04.11	19.90	0.1249	0.51126	–26.9	2.91	3.11	N.D.	N.D.
TR-34	A	16.94	115.93	0.0884	0.51091	–33.6	2.41	2.61	N.D.	N.D.
TR-36	A	02.43	10.90	0.1348	0.51127	–26.7	3.34	3.51	N.D.	N.D.
TR-43	MS	00.40	01.99	0.1216	0.51091	–33.7	3.48	3.62	N.D.	N.D.
MMDS										
K-3	D	02.16	06.36	0.2054	0.51275	2.2	1.96	3.78	2.06	–0.1
K-4	D	03.08	11.60	0.1606	0.51195	–13.4	2.89	3.22	2.06	–3.9
K-5	D	02.59	08.65	0.1811	0.51231	–6.4	3.18	3.65	2.06	–2.3
K-6	D	01.87	06.80	0.1663	0.51227	–7.2	1.84	2.51	2.06	0.9
K-7	D	02.07	07.22	0.1734	0.51220	–8.5	2.85	3.31	2.06	–2.4

Note: chur—Sm-Nd chondritic model age; dm—Sm-Nd depleted mantle model age; A—amphibolite; HG—hornblende gneiss; Qfg—quartzofeldspathic gneiss; MS—metasedimentary rock; D—dike; M—anatectite. ε(0)—epsilon today; ε(t)—epsilon at estimated age of crystallization, is calculated for the value listed as t(Ga), which is the estimated age of crystallization of protolith. ICMS—Indian Creek Metamorphic Suite; PMMMS—Pony–Middle Mountain Metamorphic Suite; SPMS—Spuhler Peak Metamorphic Suite; MMDS—metamorphosed mafic dikes and sills.

N.D.—no data.

gneisses, therefore, are generally coeval with the protoliths of tonalite-trondhjemite-granodiorite suite gneisses found throughout the Montana metasedimentary province (e.g., Mueller et al., 1993, 1996a, 1996c).

Sample K-1 (quartzite) yielded abundant, mostly blocky, clear to cloudy grains with pronounced, cloudy rims on some grains. Analyses of cloudy material as whole grains and rims reveal low Th/U ratios and strongly discordant data typical of ancient, hydrothermally grown zircon. The grain with the youngest $^{207}Pb/^{206}Pb$ age (0.17 Ga, 3% discordant) is probably an older grain that suffered Pb loss at the time of emplacement of the Tobacco Root batholith (Cretaceous), based on its $^{206}Pb/^{238}U$ age of 0.086 Ga. The oldest grain (3.86 Ga, 6% discordant) falls within the range of ages reported for detrital zircons for the northern Wyoming province, including the Tobacco Root Mountains (Mueller et al., 1998).

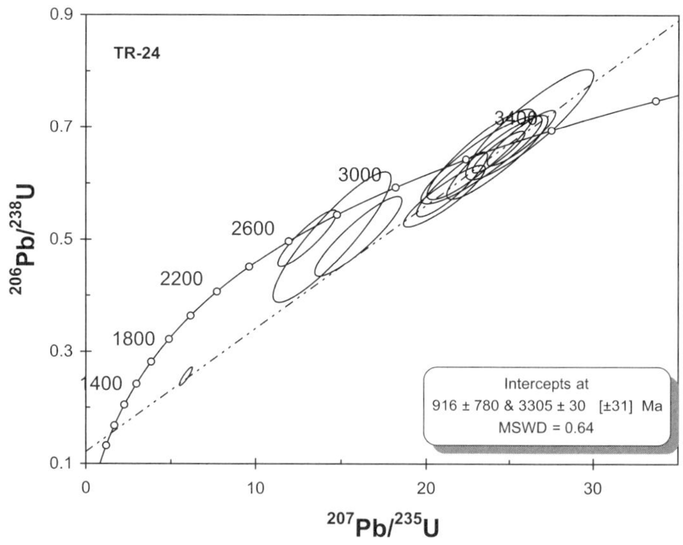

Figure 3. U-Pb concordia plot for sample TR-24 (gneiss). Upper intercept interpreted as crystallization age of protolith. Cluster at ca. 2.8 Ga interpreted to result from later metamorphism. Imprecise lower intercept may not have geologic meaning. MSWD—mean square of weighted deviates.

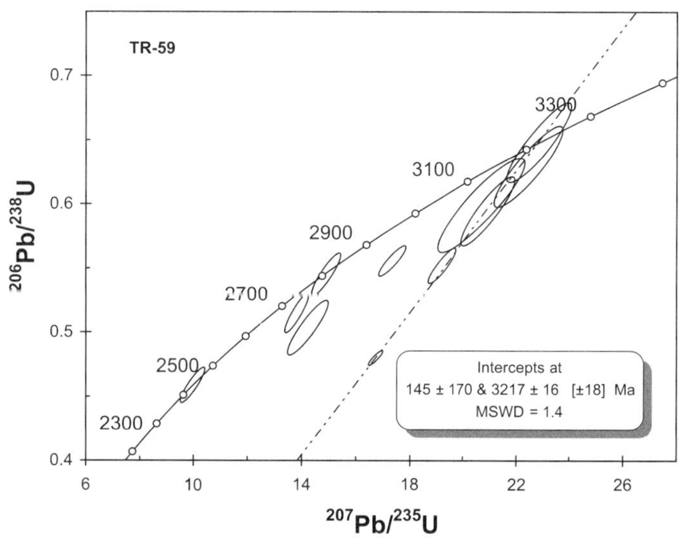

Figure 4. U-Pb concordia for sample TR-59 (gneiss). Upper intercept interpreted to indicate age of crystallization of protolith. Younger ages reflect later metamorphism (see text). MSWD—mean square of weighted deviates.

A distinctive characteristic of sample K-1, however, is the significant discordance in all material analyzed (i.e., only 8 of 33 analyses are within 10% of concordia). Of these eight analyses, four are in a restricted range of 3.2–3.3 Ga. If all data are considered, 23 of 33 analyses yield $^{207}Pb/^{206}Pb$ ages between 3.02 and 3.39 Ga. The concentration of ages in the range of 3.0–3.4 Ga strongly mimics the age pattern from Archean quartzites throughout the northern Wyoming province, including a sample from the Indian Creek Metamorphic Suite (TR-1) reported by Mueller et al. (1998). The paucity of rounded grains and the limited range of $^{207}Pb/^{206}Pb$ ages in sample K-1 suggest that the protolith for this quartzite was not a regional, sheet-type sand with a temporally diverse provenance. A more plausible model may be a sedimentary rock deposited in a restricted continental setting (e.g., rift basin), which contains a significant volcaniclastic component as proposed by Mueller et al. (1998) for other quartzites from the northern Wyoming province. The more pronounced discordance in sample K-1, compared to sample TR-1, may be related to greater interaction with fluids, which may have been more prevalent near the fault separating the Spuhler Peak Metamorphic Suite from the Archean gneisses.

The similarity in overall age distribution of K-1 to TR-1 and other quartzites in the region (Fig. 5B; Mueller et al., 1998), including the lack of 2.7–2.8 Ga detrital grains, suggests that this unit was deposited in the Late Archean from a temporally restricted provenance. Specifically, the age of deposition of this quartzite is clearly younger than 3.21 Ga, the age of the youngest clearly detrital grain in the sample that is less than 5% discordant. This observation strongly suggests that the metasupracrustal rocks are younger than the enclosing gneisses, but places no minimum age constraint on the time of deposition. One nearly concordant grain (3% reverse discordance), however, yields a $^{207}Pb/^{206}Pb$ age of 2.45 ± 0.01 Ga. If this grain is unaltered and detrital, it would place the age of deposition at younger than 2.45 Ga. Unfortunately, this grain is not rounded and its age is coincident with the strong ca. 2.40 Ga metamorphic event that affected the region (Mueller et al., 1994, 1996a; Cheney et al., 2004b, this volume, Chapter 8; Krogh et al., 1997; Dahl et al., 2002; Roberts et al., 2002). It does not, however, have the low Th/U ratio characteristic of hydrothermal zircon. If this unit was deposited in the Proterozoic, it must have formed prior to the 1.77 Ga tectonothermal event, which appears to have deformed both the Spuhler Peak Metamorphic Suite and underlying gneisses. In addition, the lack of 2.7–2.9 Ga zircons in a Proterozoic sandstone in this area would be highly unlikely, unless the source of the sediment was outside of the northern Wyoming province and/or had a very special provenance.

Sample K-2 (anatectic melt) yielded zircons in a range of sizes, colors, shapes, and ages, as would be expected for an anatectic melt of much older sedimentary rock. Of the 21 grains analyzed, the oldest grain yielded a $^{207}Pb/^{206}Pb$ age of 3.48 Ga. Only 11 of the 21 grains yielded data within 10% of concordia. Of these 11 grains, 10 grains are characterized by the low Th/U ratios (<0.15) suggestive of growth from a water-rich fluid. These 10 low-Th/U grains are chronologically coherent and yield an error-weighted mean $^{207}Pb/^{206}Pb$ age of 1.770 ± 0.006 Ga. Regression of these data yields an indistinguishable age (1.772 ± 0.008 Ga; Fig. 6). This date (1.77 Ga) is taken to represent the time at which the melt crystallized and provides a likely time for the development of peak metamorphic temperatures and

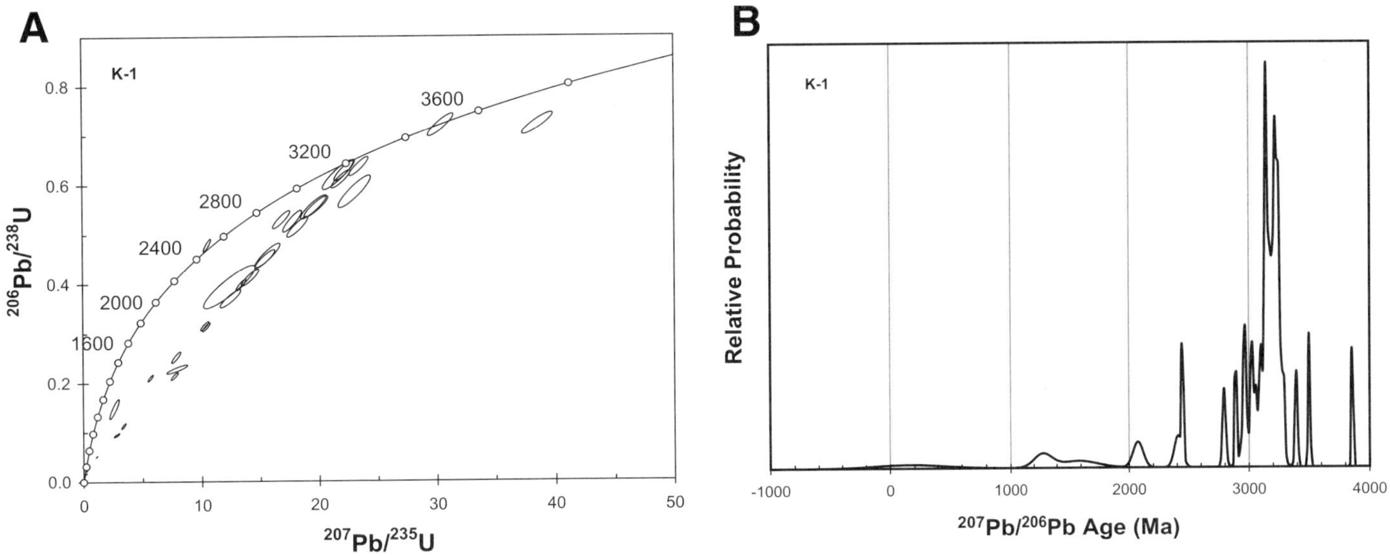

Figure 5. U-Pb concordia (A) and probability (B) plots ($^{207}Pb/^{206}Pb$ ages) for sample K-1 (quartzite). Concentration of concordant and discordant age at ca. 3.2–3.3 Ga suggests limited provenance for the protolith.

a minimum age for the Spuhler Peak Metamorphic Suite. As suggested by the Sm-Nd data (T_{dm} = 2.6 Ga, Table 2), the melt was likely produced from a mixture of melt derived from older metasedimentary rocks and younger materials that were probably introduced during the 1.77 Ga event.

Sample K-3 (metamorphosed mafic dike of MMDS) yielded zircons of two distinct morphologies: very small, prismatic grains and very small, generally spherical grains. Data from 16 of 18 prismatic grains analyzed were <10% discordant and older than 2.0 Ga. They yielded an error-weighted mean $^{207}Pb/^{206}Pb$ age of 2.060 ± 0.006 Ga. Twelve of 13 analyses of the smaller, rounder grains yielded ages <10% discordant and an error-weighted mean $^{207}Pb/^{206}Pb$ age of 1.763 ± 0.008 Ga (Table 1); regression of these data yields an analytically indistinguishable age (1.759 ± 0.026 Ga; Fig. 7). This date (1.76 Ga) is interpreted to be the time that the dike was subjected to granulite facies metamorphism.

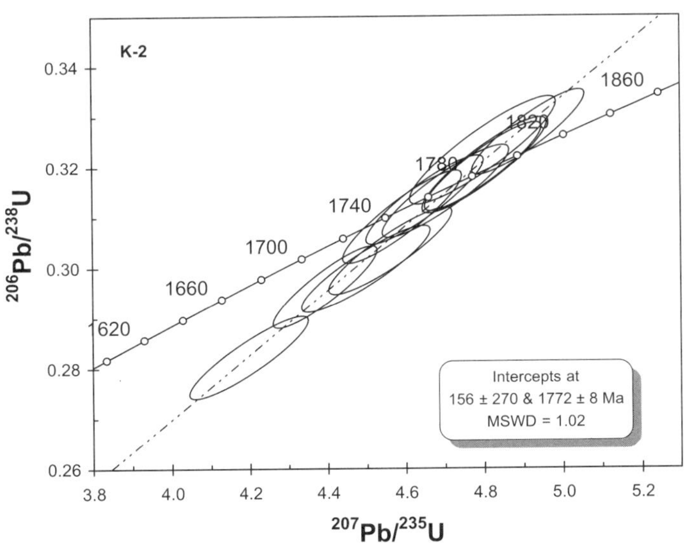

Figure 6. U-Pb concordia for sample K-2 (leucosome). Upper intercept interpreted as crystallization age of leucosome. MSWD—mean square of weighted deviates.

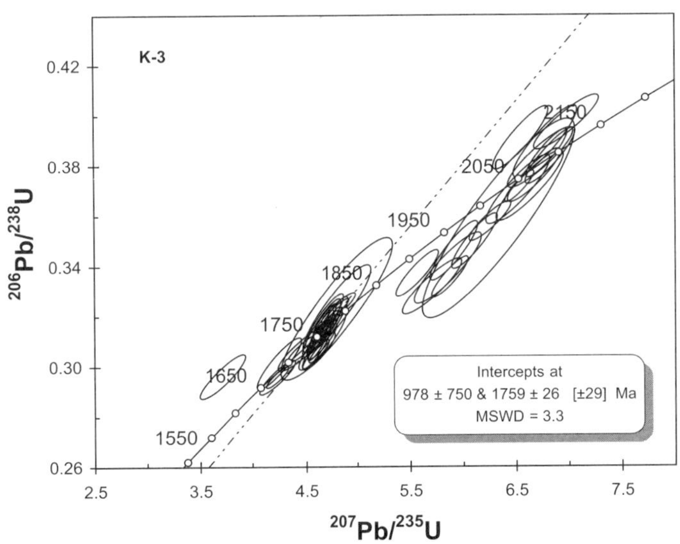

Figure 7. U-Pb concordia for sample K-3 (metamorphosed mafic dike). Intercept age of 1.76 a interpreted as time of metamorphism of dike originally intruded at 2.06 Ga. MSWD—mean square of weighted deviates.

Although these grains also appear to have grown at ca. 1.76 Ga, none have Th/U ratios as low as those present in the ca. 1.77 Ga grains from the anatectic melt pod (leucosome).

The relatively lower, but distinct, Th/U ratios of the metamorphic (vs. older and/or igneous) zircons in each sample suggest that they formed in separate environments. For example, the average Th/U ratio of the 1.77 Ga grains from the leucosome (0.05) is much lower (~6×) than the average Th/U of the older grains (0.31). Although metamorphic zircons from the mafic dike also exhibit lower Th/U than the coexisting igneous grains (averages of 0.21 vs. 0.64), their Th/U ratios are distinctly higher than the Th/U ratios of zircons that formed in the leucosome (average of 0.05). The lower Th/U ratios of the metamorphic zircons in each case suggest they formed in the presence of a fluid capable of preferentially mobilizing U. It appears unlikely, however, that the grains grew from a common fluid.

Isotope Systematics

Whole-rock Sm-Nd isotopic analyses were performed on 36 samples from the Tobacco Root Mountains, including samples of quartzofeldspathic gneiss, amphibolites, meta-ultramafites, metasedimentary rocks, and metamorphosed mafic dikes (Table 2). Epsilon Nd values (ε) and depleted mantle model ages (T_{dm}) range from 2.2 to −45.8 and from 0.78 to 4.02 Ga, respectively. Interpretations of the initial ε values are complicated by the fact that accurate crystallization ages are not available for each of these samples, and that these samples have been subjected to a variety of igneous, sedimentary, and metamorphic processes that may have altered their Sm-Nd systematics. In addition, some of the metamorphosed mafic and sedimentary rocks have relatively high Sm/Nd ratios (i.e., approaching chondritic values), which warrant more cautious interpretations than samples with lower, more crust-like Sm/Nd ratios. Despite these limitations, Sm-Nd systematics can provide insight into two important aspects of crustal evolution.

Age and Origin of Basement Gneisses

Sm-Nd systematics provides significant insight into the origin of the quartzofeldspathic gneisses and the extent of crustal recycling, especially when compared to U-Pb ages of detrital zircon suites. For example, a comparison of the T_{dm} of an individual sample and its crystallization age (e.g., U-Pb zircon age) can be used to distinguish samples that contain significant amounts of older crust (i.e., those with T_{dm} > crystallization age are likely to contain recycled crust and/or lithosphere). Alternatively, a sample that is 100% juvenile relative to the depleted mantle model will yield T_{dm} equal to its crystallization age. In a similar but less rigorous fashion, the T_{dm} of gneisses on a regional scale can be compared with detrital zircon populations from the same region to gain at least a first-order appreciation of the degree of recycling in that region and the extent of ancient crust. In the case of the Tobacco Root rocks, the distribution of T_{dm} corresponds generally with the distribution of detrital zircon ages from quartzites (e.g., Mueller et al., 1998; Fig. 8). If large degrees of crustal recycling had occurred, then the mode and range of detrital zircon ages and T_{dm} would be expected to differ. The extent of the discrepancies would likely increase with time and the extent of recycling because the crustal Nd reservoir tends to be very conservative in general, while the U-Pb age of an individual zircon is more likely to be preserved in a sedimentary system than an igneous system (e.g., zircon ages are more likely to be at least partly degraded with each new melting/crystallization cycle).

In the case of the Tobacco Root gneisses and quartzites, the oldest detrital zircon is 3.93 Ga and the oldest T_{dm} is 4.02 Ga. In the age distribution of the detrital zircons, there is a clear peak around 3.2 Ga. Similarly, the T_{dm}s show a concentration between 3.2 and 3.5 Ga (50% of all analyses). These similarities (Fig. 8) suggest that a major crust-forming event(s) of largely juvenile origin occurred ca. 3.2–3.3 Ga, but that some older crust (at least 3.7–4.0 Ga) was involved. Similar relationships exist for other uplifts of Precambrian rocks in the Montana metasedimentary province and were interpreted to indicate Late Archean deposition of the quartzites (Mueller et al., 1998, 1993, 1996c).

Spuhler Peak Association

Whole-rock Sm-Nd systematics also provides insight into the origin of the Spuhler Peak Metamorphic Suite. Though composed mostly of metamorphosed mafic and ultramafic rocks, metasedimentary rocks are also present. As noted above, a quartzite from the Spuhler Peak Metamorphic Suite (K-1) appears to contain dominantly Archean zircons. When compared to zircons from sample TR-1 (Mueller et al., 1998), however, zircons from sample K-1 are generally more disturbed (only 24% within 10% of concordia vs. 60% for TR-1). Although the disturbed nature of the K-1 zircons may obscure some details, it seems clear that the age distribution from sample K-1 generally follows that of TR-1 and other detrital suites from the Montana metasedimentary province (Mueller et al., 1998). In addition, two whole-rock Sm-Nd analyses of metasedimentary rocks of the Spuhler Peak Metamorphic Suite yield T_{dm}s of 3.0 and 3.6 Ga. Sources for sedimentary rocks deposited during the Paleoproterozoic in the northern Wyoming province would be characterized by both >3.0 Ga material and abundant Late Archean (2.6–2.9 Ga) zircons. The lack of such Late Archean zircons in sample K-1 makes it probable that the protolith of K-1 was deposited subsequent to 3.21 Ga (youngest clearly detrital zircon), but before ca. 2.8 Ga. If the single 2.45 Ga zircon analyzed is detrital, the depositional age would have to be younger than 2.45 Ga, and the source of the sediment would most likely have been outside of the northern Wyoming province, which may be feasible in light of Spuhler Peak Metamorphic Suite's structural history.

If the protoliths to Spuhler Peak Metamorphic Suite metasedimentary rocks were originally derived from the Wyoming province, then an Archean depositional age would seem most probable and generally compatible with whole-rock Sm-Nd data from the more abundant mafic and ultramafic amphibolites (Burger et al., 2004, this volume, Chapter 3; Table 2; Fig. 5). This agreement is somewhat problematical, however, because

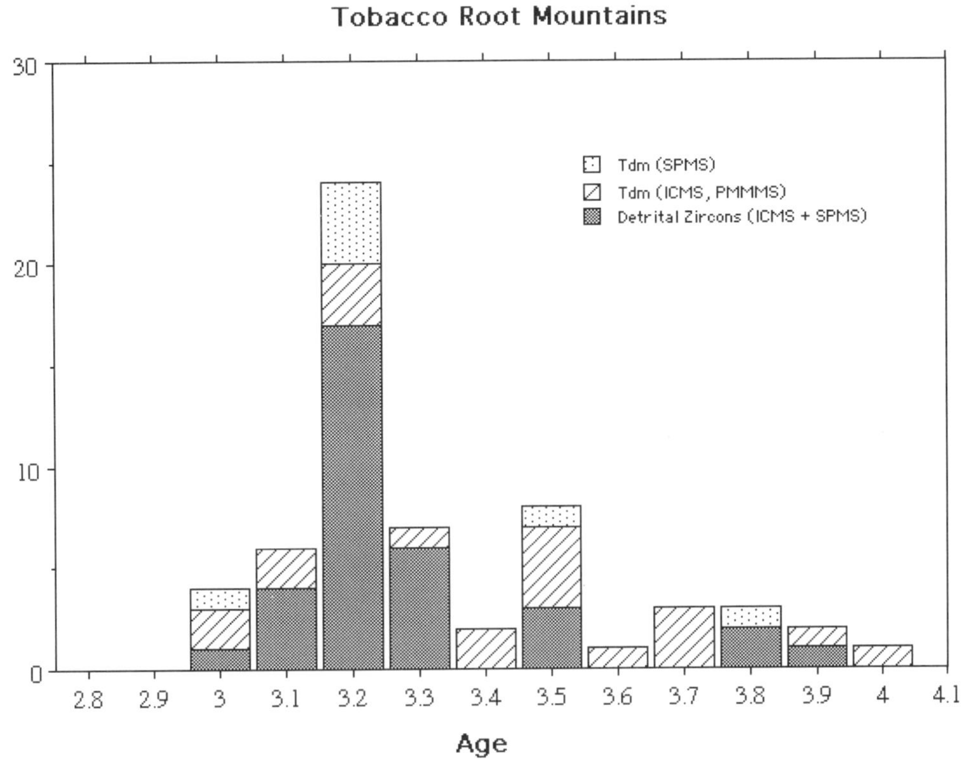

Figure 8. Composite histogram of (1) detrital zircon ages from K-1 (SPMS) and TR-1 (Mueller et al., 1998) that are less than 10% discordant (shaded), (2) T_{dm} of SPMS meta-igneous and metasedimentary rocks (dotted), and (3) T_{dm} for Indian Creek Metamorphic Suite and Pony–Middle Mountain Metamorphic Suite gneisses and amphibolites (hachured). SPMS—Spuhler Peak Metamorphic Suite; ICMS—Indian Creek Metamorphic Suite; PMMMS—Pony–Middle Mountain Metamorphic Suite.

these amphibolites, in at least some cases, have experienced significant hydrothermal alteration (Burger et al., 2004, this volume, Chapter 3). This alteration is clearly evident in some samples in the form of relatively high Nd contents (e.g., TR-34 = 116 ppm), which exceed typical values for oceanic tholeiites or average continental crust (Taylor and McLennan, 1985). In addition, the $^{147}Sm/^{144}Nd$ ratios of the high-Nd samples are more crust-like and yield generally lower model ages. These samples have most likely experienced post-crystallization alteration of their Sm-Nd systems, and their model ages are not likely to be reliable. The fact that the sample with the highest Nd content (TR-34) still yields an Archean model age, however, suggests that these rocks are likely to have originally formed in the Archean, or interacted primarily with fluids derived from Archean crust.

For samples with more typical Sm and Nd contents (TR-12, -13, and -36), T_{dm}s are also Archean (3.4–3.6 Ga) over a broad range of $^{147}Sm/^{144}Nd$ (0.11–0.17). The model ages are comparable to those of the surrounding gneisses and younger mafic rocks (see below), and suggestive of an Archean origin for these rocks as well. Alternative interpretations (e.g., these are Proterozoic igneous rocks contaminated by Archean crust/lithosphere) cannot be ruled out with the present database; however, the low-Nd samples do not exhibit evidence of mixing, e.g., they lack a regular relationship between model age and Nd content. Consequently, these data do not require the mafic rocks of the Spuhler Peak Metamorphic Suite to be Archean, but are compatible with such an interpretation. Ultimately, a more comprehensive isotopic and geochronological investigation of the rocks of the Spuhler Peak Metamorphic Suite will be required to confirm its origin. The emplacement of these rocks (Spuhler Peak Metamorphic Suite) into their present structural position, however, appears well constrained between ca. 1.77 and 2.06 Ga based on the age relations discussed above.

Metamorphosed Mafic Dikes and Sills (MMDS)

The Sm-Nd data for the mafic dikes of the MMDS (Table 2) show a wide range of T_{dm}s (2.51–3.78 Ga) in which all ages exceed the intrusion age determined for sample K-3 (2.06 Ga). For samples K-5 and K-3, the T_{dm}s are greater than the age of the country rock. None of the T_{dm}s, however, must be viewed as having any direct chronologic relation to the time of intrusion or the age of the source from which they were extracted. In part, this is because these rocks are characterized by Sm/Nd ratios (0.16–0.21) that largely overlap the values used in chondritic and depleted mantle evolution models. Consequently, small errors (geologic or analytical) in measured Sm/Nd ratios (or Nd isotopic composition) relative to the true Sm/Nd that characterized the sample at the time of crystallization can lead to large errors in the model age and/or estimates of initial ε values.

In particular, lack of age control (only sample K-3 has been dated) may be partly responsible for the range in initial ε values calculated for these rocks (0.9 to –3.9 at 2.06 Ga). For example, if sample K-7 were 2.5 Ga, its initial ε value would be –1 rather than –2.4. To the extent that the age and Sm/Nd systematics are correct, the lack of correlation between T_{dm} and Nd contents argues against a simple model of crustal contamination of asthenospherically derived mafic melts to explain the generally negative initial ε values. It seems more likely that the Sm-Nd systematics of these rocks is related to derivation from, or interaction with, an even more ancient lithosphere. This argument is plausible in light of the fact that isotopic systematics of Phanerozoic igneous rocks suggests that Archean subcontinental lithosphere exists beneath the Wyoming craton today (e.g., Wooden and Mueller, 1988; Carlson and Irving, 1994).

DISCUSSION

Age Relationships

Since the initial report of Paleoproterozoic ages within what is now recognized as the Montana metasedimentary province and/or Great Falls tectonic zone in southwestern Montana (e.g., Hayden and Wehrenberg, 1959; Giletti, 1966; O'Neill and Lopez, 1985; Mogk et al., 1992), thermochronologic studies of reset, recrystallized, and new growth minerals in Archean gneisses have been characterized by significant variability (e.g., 1.6–2.4 Ga; Burger, 2004, this volume, Chapter 1; Brady et al., 2004b, this volume, Chapter 7). This variability, in part, reflects a complex Archean history (e.g., Mueller et al., 1993, 1996c), as well as the impact of at least two distinct, post-Archean, thermal maxima, one at ca. 2.4 Ga (Mueller et al., 1994, 1996b; Krogh et al., 1997; Cheney et al., 2004b, this volume, Chapter 8; Roberts et al., 2002) and one represented by the average of the two weighted mean U-Pb ages of zircons reported here (1.767 ± 0.004 Ga), $^{40}Ar/^{39}Ar$ data reported in Brady et al. (2004b, this volume, Chapter 7), and similar monazite U-Th-Pb ages (Cheney et al., 2004b, this volume, Chapter 8). The two concordant U-Pb zircon ages reported here are unique, however, because they represent growth of a mineral (zircon) during metamorphism that has a closure temperature (~900 °C; Mezger, 1990) that exceeds mineral thermometric estimates of peak metamorphic temperatures in this region (750 °C, Cheney et al., 2004a, this volume, Chapter 6). As such, these grains are not likely to have any memory of previous igneous or metamorphic events, nor are they likely to have experienced even limited degradation by subsequent events. Consequently, this age provides a critical datum for understanding the Paleoproterozoic tectonothermal evolution of the Tobacco Root Mountains, as well as other uplifts of Archean rocks in the Montana metasedimentary province and Great Falls tectonic zone.

The extent to which these new U-Pb ages can be used to better understand the Paleoproterozoic evolution of the Montana metasedimentary province and Great Falls tectonic zone and their extant mineral ages depends upon several factors. In general, the high-grade metamorphic conditions that fostered growth of the 1.77 Ga zircons suggest that these ages are likely to be representative of a time that many, if not all, uplifts throughout the Montana metasedimentary province and Great Falls tectonic zone reached thermal maxima. These data, however, do not imply that the thermal maxima in all areas were the same. In terms of the Tobacco Root Mountains, we interpret the combined metamorphic petrology, structural analysis, mineral thermometry, and geochronology to indicate that the peak, post–2.4 Ga, metamorphic temperatures (>750 °C; Cheney et al., 2004a, this volume, Chapter 6; 2004b, this volume, Chapter 8) were probably reached at 1.77 Ga. This is slightly younger than the ca. 1820 Ma age proposed by Roberts et al. (2002) for peak metamorphic conditions for the Montana metasedimentary province as a whole.

Roberts et al.'s (2002) estimate of ca. 1.82 Ga for the time of metamorphism in this region, and their proposal for a single late Paleoproterozoic metamorphic event and synchronous uplift of the region, is based on step leaching of Pb from two monazite-bearing samples of garnet (Ruby Range, 1.822 ± 0.032 Ga; Highland Mountains, 1.786 ± 0.004). The temperatures nominally associated with these step-leaching ages would be the closure temperature for the U-Pb system in monazite (estimated to be 700–750 °C; Mezger, 1990; Parrish, 1990). If all three uplifts had a similar metamorphic history and synchronous cooling histories, then one would expect all monazite ages to be less than the zircon age, i.e., to reflect cooling through the closure temperature of monazite. This does not appear to be the case for the two dates listed above, and is clearly not the case for the ca. 2.5 Ga date obtained from step leaching of a monazite-bearing sample of garnet from the Tobacco Root Mountains (Roberts et al., 2002; Dahl et al., 2002).

This inversion of U-Pb ages relative to their purported closure temperatures (e.g., 1.77 Ga zircon vs. 2.45 Ga monazite) may be related to both the nature of the experiments that extract Pb from monazite-bearing garnet by step leaching and the response of some monazite to the 1.77 Ga metamorphism. In the leaching experiments, the Pb recovered is dominantly derived from various U, Th, and Pb-bearing minerals such as monazite, zircon, etc. that are included in the garnet. In some instances the isotopic ratios of the extracted Pb can help distinguish high-Th/U sources from low-Th/U sources (e.g., distinguish monazite from zircon; Frei and Kamber, 1995). It is more difficult, however, to recognize the presence of pre-metamorphic grains of the primary U- and Pb-carrying phases, unless they are selectively leached in a distinct step. Consequently, the >1.77 Ga Pb-Pb dates derived from monazites included in garnets from the Tobacco Root Mountains (Roberts et al., 2002), as well as the Highland and Ruby uplifts (Roberts et al., 2002), may be related to the presence of small amounts of older monazite included in the younger garnets (e.g., Copeland et al., 1988; Smith and Giletti, 1997). That some monazite apparently retained isotopic systematics established before 1.77 Ga also suggests that the U-Pb monazite system in the analyzed garnets may have closure temperatures greater than the thermal maximum reached at 1.77 Ga in the specific areas sampled. This may be directly applicable to the Ruby Range because the maximum metamorphic temperature reported is ~665 °C (Roberts et al., 2002). Alternatively, the thermal history of the metamorphic event (i.e., time-temperature profile) and the grain size and host mineral of the older monazite grains may not have allowed these grains to completely reequilibrate via diffusion at 1.77 Ga (e.g., Smith and Giletti, 1997). Ultimate resolution of these discrepancies, however, must await a clear definition of the time at which rocks of the Ruby and Highland Ranges reached their peak, Paleoproterozoic metamorphic conditions.

The fact that the distribution of K-Ar–based ages from various uplifts in the region also does not correspond to a simple model of post–1.77 Ga cooling may be related to (1) the Ruby, Highland, and Tobacco Root Ranges have different cooling histories (i.e., the synchronous metamorphism/uplift model proposed by Roberts et al. [2002] does not apply); (2) younger events with peak metamorphic temperatures below the closure temperature of zircon (but above that of other mineral systems) have affected various parts of the region (e.g., Brady et al., 1998); and/or (3) most of the reported K-Ar-based data are not accompanied by sufficient analytical information to permit standardization to a common reference value. Although these factors can help explain the wide range of post–1.77 Ga K-Ar–based ages, they are unlikely to explain those ages older than 1.77 Ga. Ages older than 1.77 Ga are more likely to be the product of excess Ar or incomplete resetting of minerals in certain areas/lithologies. For the purposes of this discussion, however, all reported K-Ar–based ages are taken at face value and are compared directly to the reported U-Pb ages without attempting to correct for systematic errors between laboratories or systems.

Accepting the age of 1.77 Ga as the time of the metamorphic thermal maximum for the entire region adds complexity to the interpretations of K-Ar and Ar-Ar dates throughout the Montana metasedimentary province and Great Falls tectonic zone, particularly those older than 1.77 Ga (see Table 1 in Burger, 2004, this volume, Chapter 1). For example, Roberts et al. (2002) present the most extensive Ar-Ar database for the Montana metasedimentary province (32 biotite and amphibole separates from 29 individual samples). They interpret these data to indicate synchronous cooling through the biotite closure temperature in the Ruby, Tobacco Root, and Highland uplifts during the interval 1.78–1.74 Ga. Discounting younger dates attributed to Cretaceous igneous activity as identified by Roberts et al. (2002), Ar-Ar dates vary from 1.735 to 1.788 Ga in the Ruby Range and from 1.688 to 1.755 Ga in the Tobacco Root Range. Although Roberts et al. (2002) reported results from only one sample from the Highland Range (1.751 Ga), Harlan et al. (1996) reported Ar-Ar dates from both biotite and hornblende from a quartzofeldspathic gneiss in the Highland Range of ca. 1.82 Ga. This date extends the range of Ar-Ar biotite dates in the Highland uplift to ca. 60 m.y., comparable to the ranges reported for the Ruby and Tobacco Root uplifts (Roberts et al., 2002) and greater than variations expected from inter-laboratory calibration differences. These data (e.g., Harlan et al., 1996; Roberts et al., 2002) also emphasize another characteristic of the systematics of the K-Ar-based systems in the Great Falls tectonic zone, i.e., that mica dates and amphibole dates are often identical or even inverted relative to their closure temperatures.

Another important characteristic of the total population of dates derived from biotite, and to a lesser extent amphibole, from the three areas is the wide range of ages (>100 m.y.). This range appears too large to be compatible with the synchronous uplift hypothesis of Roberts et al. (2002), unless this terrane lingered in the partial retention zone of biotite for a significant length of time. Though such an evolution might explain the range of ages, it does not offer an explanation for the fact that the upper limits of the range of biotite dates approaches or exceeds the time of peak metamorphic conditions indicated by the U-Pb zircon ages reported here (and Pb-Pb ages reported by Roberts et al., 2002). These older biotite ages are difficult to explain, but may represent interaction with Ar-rich fluids during later metamorphic events (e.g., 1.40 Ga; Brady et al., 1998), as proposed by Roberts et al. (2002) to explain several anomalous dates from amphiboles. Even in the case of the Tobacco Root Mountains, for which the upper limit for the range of biotite dates reported by Roberts et al. (2002) is younger than 1.77 Ga, an unusually high cooling rate is indicated, i.e., on the order of 30 °C/m.y. (1500 m/Mm.y. at 20 °C/km). Such rates of cooling are generally restricted to rapidly extending environments (e.g., Foster and John, 1999). Unfortunately, there are inadequate thermochronologic data available to critically evaluate this possibility.

Tectonic Relationships

The question remains, therefore, as to the nature and cause of the orogenic episode(s) that produced the late Paleoproterozoic tectonothermal record of the Montana metasedimentary province and Great Falls tectonic zone. As can be seen in Figure 9, Archean rocks compose large cratonic areas adjacent to the Montana metasedimentary province, as well as along the northern and eastern sides of the Wyoming craton. Previous authors (e.g., O'Neill and Lopez, 1985; Hoffman, 1989; O'Neill, 1998) postulated that the Paleoproterozoic reworking of the Archean gneisses was related to the suturing of the Hearne province (and/or Medicine Hat Block) and the Wyoming province (e.g., Mueller et al., 2002). Other workers have argued against this interpretation, primarily on the basis of geophysical information derived from the extensive buried interval that currently separates surface exposures of the Hearne and Wyoming provinces (e.g., Boerner et al., 1998; Lemieux et al., 2000). These workers supported an Archean age for the Wyoming-Hearne/Medicine Hat collision and suggested that the Great Falls tectonic zone is an intracontinental shear zone reactivated in the Paleoproterozoic as a consequence of collisional activity along the Trans-Hudson orogen.

More recently, however, Gorman et al. (2002) interpret seismic data to indicate a relict slab indicative of north-dipping subduction (present coordinates) between the Medicine Hat Block and the Wyoming craton. Although these authors interpret this subduction to have occurred during the Late Archean, the only potential, temporal restrictions on the age of subduction are derived from the ages of lower crustal xenoliths recovered from the Medicine Hat Block (1.75–1.85 Ga; Davis et al., 1995). Mueller et al. (2002), however, report U-Pb zircon ages of 1.86 Ga for quartzofeldspathic gneisses and a later diorite from the Little Belt Mountains of Montana (Fig. 9). These rocks are

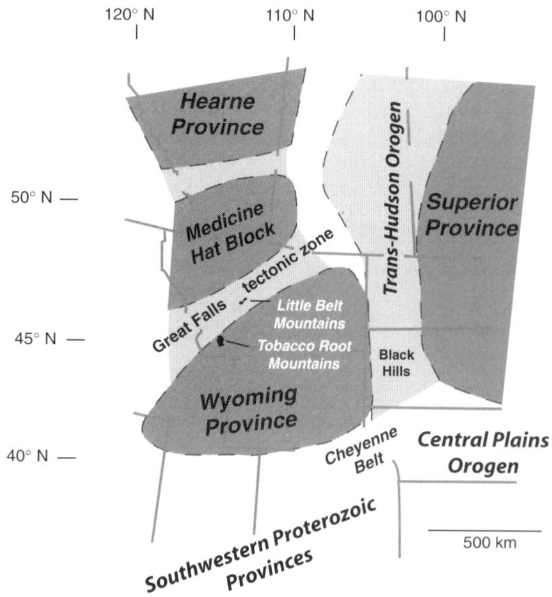

Figure 9. Schematic depiction of major Archean and Proterozoic features at ca. 1.8 Ga in western Laurentia (after Mueller et al., 2002).

part of a complex assemblage of Late Archean and Proterozoic gneisses and intrusive rocks (Vogl et al., 2002) that constitute one of only two significant exposures of Precambrian rocks between the Hearne and Wyoming provinces. The Little Belt rocks are calc-alkaline, exhibit arc-like trace element signatures, have relatively primitive Nd isotopic signatures, and are interpreted to result from consumption of oceanic lithosphere in a collisional environment (Mueller et al., 2002). These observations suggest that the subduction remnant reported by Gorman et al. (2002) between the Medicine Hat Block and Wyoming craton may have actually developed in the Paleoproterozoic.

Although Paleoproterozoic igneous rocks are relatively abundant in the Little Belt Mountains, it is important to note that no Paleoproterozoic magmatic rocks have been documented in the exposed portions of the Montana metasedimentary province. The lack of Paleoproterozoic magmatism within this region is a key observation that must be accommodated in any model invoked to explain the origin of the 1.77 Ga tectonothermal event. In general, there are two different mechanisms that could result in Paleoproterozoic heating of the Archean rocks of southwestern Montana.

One model invokes a collisional orogeny, perhaps with the Hearne province to the north or other terrane to the west. This interpretation is compatible with the seismic evidence of Gorman et al. (2002), although it would place the time of subduction at ca. 1.8–1.9 Ga rather than 2.5 Ga as suggested by Gorman et al. (2002). This model is also compatible with reports of Archean and >1.77 Ga Paleoproterozoic crust (and/or zircons inherited from such crust) within the Great Falls tectonic zone (e.g., Mueller et al., 1996a; Foster et al., 2002), and to the west (e.g., Leeman et al., 1985; Farmer and DePaolo, 1983; Aleinikoff et al., 1986; Hedge et al., 1983). Outward-directed subduction to the north, west, or northwest would account for the lack of ca. 1.77 Ga igneous rocks in the Montana metasedimentary province, and would allow for burial of the Montana metasedimentary province by the overriding plate as the causative mechanism for the development of the high pressures and temperatures of the 1.77 Ga tectonothermal event. If this mechanism is valid, then the calc-alkaline rocks of the Little Belt Mountains would represent an early stage of collision because they formed more than 90 m.y. before the time of peak metamorphic conditions in the Tobacco Root Mountains (i.e., Harms et al., 2004b, this volume, Chapter 11).

Alternatively, heating and deformation in the western Great Falls tectonic zone may be related to the emplacement of substantial volumes of mafic rocks in the lower crust during the Paleoproterozoic. Geophysical studies of the broad buried interval between the exposed limits of the Wyoming and Hearne provinces indicate the presence of a thick (15–25 km), high-velocity, lower crustal layer that extends from southern Alberta southward beneath the Medicine Hat Block, the Great Falls tectonic zone, and the northern Wyoming craton (e.g., Henstock et al., 1998; Clowes et al., 2002). U-Pb zircon ages of xenoliths of mafic granulite ascribed to this layer in southern Alberta (Medicine Hat Block) yield ages between 1.85 and 1.75 Ga (e.g., Davis et al., 1995). In addition, U-Pb ages of zircons from mantle xenoliths from the Great Falls tectonic zone north of the Montana metasedimentary province show similar ages (e.g., 1.78–1.82 Ga; Rudnick et al., 1999). To the extent that these xenoliths are younger than the collision-related, 1.86 Ga magmatism in the Little Belt Mountains and are chronologically representative of the high-velocity lower crustal layer, they suggest the presence of an extensive province of (post)orogenic mafic underplating within the Wyoming-Hearne collision zone (sensu lato). Extension of underplating on this scale beneath the Montana metasedimentary province at ca. 1.77 Ga would provide a heat source for the 1.77 Ga tectonothermal event that would be compatible with the rapid exhumation suggested by some of the extant thermochronologic data. Conversely, the existence of Paleoproterozoic rocks west of the surface limits of the Montana metasedimentary province (e.g., Foster et al., 2002) may not be compatible with complete reliance on a single episode of magmatic underplating and heating to explain all facets of Paleoproterozoic orogenesis in southwestern Montana. Distinguishing these alternatives (i.e., extensional or compressional environments at 1.77 Ga) is critical, however, for assessing models for the accretion of the Wyoming province to Laurentia because ca. 1.77 Ga is also the proposed time of convergence and juxtaposition of the Wyoming, Superior, and Colorado provinces (e.g., Dahl et al., 1999; Chamberlain, 1998).

CONCLUSIONS

U-Pb data for zircons from quartzofeldspathic gneisses and metasedimentary rocks, along with whole-rock Sm-Nd data, suggest that the majority of the felsic crust exposed in the Tobacco

Root Mountains formed 3.2 to 3.4 Ga. Much of this crust appears to be relatively juvenile, but evidence for the involvement of older crust of at least 4.0 Ga in age is evident in both Sm-Nd and U-Pb systems. Extensive metamorphic and deformational events affected these rocks at various times (e.g., 2.8 Ga, 2.4 Ga) up until ca. 2.06 Ga, when at least some members of the MMDS were intruded. At ca. 1.77 Ga, the region was affected by a major tectonothermal event that resulted in melting of the older gneisses in some locations, and development of a common penetrative fabric element that affected all earlier rocks. The emplacement of the Spuhler Peak Metamorphic Suite probably occurred coevally with this tectonothermal event. Although the meta-igneous rocks of the Spuhler Peak Metamorphic Suite are likely to be Archean, sedimentary rocks intercalated with them may have been deposited in the 1.77–2.45 Ga interval.

ACKNOWLEDGMENTS

The authors wish to extend their appreciation to Smith College, Amherst College, and the National Science Foundation (EAR-0106592) for financial support of this project. We also benefited from reviews by Steve Harlan, Dave Foster, and Robert Shuster.

REFERENCES CITED

Aleinikoff, J., Nielson, D., Hedge, C., and Evans, S., 1986, Geochronology of Precambrian Mineral Mountains, south-central Utah: U.S. Geological Survey Bulletin 1622, p. 1–12.

Boerner, D.E., Craven, J.A., Kurtz, R.D., Ross, G.M., and Jones, F.W., 1998, The Great Falls tectonic zone: Suture or intracontinental shear zone?: Canadian Journal of Earth Sciences, v. 35, p. 175–183.

Brady, J.B., Cheney, J.T., Rhodes, A.L., Vasquez, A., Green, C., Duvall, M., Kogut, A., Kaufman, L., and Kovaric, D., 1998, Isotope geochemistry of Proterozoic talc occurrences in Archean marbles of the Ruby Mountains, southwest Montana, USA: Geological Materials Research, v. 1, no. 2, p.1–41.

Brady, J.B., Mohlman, H.K., Harris, C., Carmichael, S.K., Jacob, L.K., and Chaparro, W.R., 2004a, General geology and geochemistry of metamorphosed Proterozoic mafic dikes and sills, Tobacco Root Mountains, Montana, *in* Brady, J.B., et al., eds., Precambrian geology of the Tobacco Root Mountains, Montana: Boulder, Colorado, Geological Society of America Special Paper 377, p. 89–104 (this volume).

Brady, J.B., Kovaric, D.N., Cheney, J.T., Jacob, L.J., and King, J.T., 2004b, $^{40}Ar/^{39}Ar$ ages of metamorphic rocks from the Tobacco Root Mountains region, Montana, *in* Brady, J.B., et al., eds., Precambrian geology of the Tobacco Root Mountains, Montana: Boulder, Colorado, Geological Society of America Special Paper 377, p. 131–149 (this volume).

Burger, H.R., 2004, General geology and tectonic setting of the Tobacco Root Mountains, Montana, *in* Brady, J.B., et al., eds., Precambrian geology of the Tobacco Root Mountains, Montana: Boulder, Colorado, Geological Society of America Special Paper 377, p. 1–14 (this volume).

Burger, H.R., Peck, W.H., Johnson, K.E., Tierney, K.A., Poulsen, C.J., Cady, P., Lowell, J., MacFarlane, W.A., Sincock, M.J., Archuleta, L.L., Pufall, A., and Cox, M.J., 2004, Geology and geochemistry of the Spuhler Peak Metamorphic Suite, *in* Brady, J.B., et al., eds., Precambrian geology of the Tobacco Root Mountains, Montana: Boulder, Colorado, Geological Society of America Special Paper 377, p. 47–70 (this volume).

Carlson, R., and Irving, A., 1994, Depletion and enrichment history of subcontinental lithospheric mantle: An Os, Sr, Nd and Pb isotopic study of ultramafic xenoliths from the northwestern Wyoming craton: Earth and Planetary Science Letters, v. 126, p. 457–472.

Chamberlain, K., 1998, Medicine Bow orogeny: Timing of deformation and model of crustal structure produced during continent-arc collision, ca. 1.78 Ga, southeastern Wyoming: Rocky Mountain Geologist, v. 33, p. 259–277.

Cheney, J.T., Brady, J.B., Tierney, K.A., DeGraff, K.A., Mohlman, H.K., Frisch, J.D., Hatch, C.E., Steiner, M.L., Carmichael, S.K., Fisher, R.G.M., Tuit, C.B., Steffen, K.J., Cady, P., Lowell, J., Archuleta, L., Hirst, J., Wegmann, K.W., and Monteleone, B., 2004a, Proterozoic metamorphism of the Tobacco Root Mountains, Montana, *in* Brady, J.B., et al., eds., Precambrian geology of the Tobacco Root Mountains, Montana: Boulder, Colorado, Geological Society of America Special Paper 377, p. 105–129 (this volume).

Cheney, J.T., Webb, A.A., Coath, C.D., and McKeegan, K.D., 2004b, In situ ion microprobe $^{207}Pb/^{206}Pb$ dating of monazite from Precambrian metamorphic suites, Tobacco Root Mountains, Montana, *in* Brady, J.B., et al., eds., Precambrian geology of the Tobacco Root Mountains, Montana: Boulder, Colorado, Geological Society of America Special Paper 377, p. 151–179 (this volume).

Clowes, R., Burianyk, M., Gorman, A., and Kanasewich, E., 2002, Crustal velocity structure from SAREX, the Southern Alberta Refraction Experiment: Canadian Journal of Earth Sciences, v. 39, p. 351–373.

Compston, W., and Williams, I., 1992, Ion probe ages for the British Ordovician and Silurian stratotypes, *in* Webby, B., and Laurie, J., eds., Global perspectives on ordovician geology: Rotterdam, Netherlands, Balkema, p. 59–67.

Compston, W., Williams, I.S., and Meyer, C., 1984, U-Pb geochronology of zircons from lunar breccia 73217 using a sensitive high mass-resolution ion microprobe, *in* Proceedings, Lunar and Planetary Science Conference, 14th: Journal of Geophysical Research, v. 89, p. B525–B534.

Copeland, P., Parrish, R., and Harrison, M., 1988, Identification of inherited radiogenic Pb in monazite and implications for U-Pb systematics: Nature, v. 333, p. 760–763.

Dahl, P.S, Hamilton, M., Terry, M., Roberts, H., Kelley, S., Frei, R., Jercinovic, M., and Williams, M., 2002, Comparative ion and electron microprobe dating of Wyoming province monazite, with tectonic and analytical implications: Geological Society of America Abstracts with Programs, v. 34, no. 1, p. A9.

Dahl, P.S., Holm, D.K., Gardner, E.T., Hubacher, F.A., and Foland, K.A., 1999, New constraints on the timing of Early Proterozoic tectonism in the Black Hills (South Dakota), with implications for docking of the Wyoming province with Laurentia: Geological Society of America Bulletin, v. 111, p. 1335–1349.

Davis, W.J., Berman, R., and Kyarsgaard, B., 1995, U-Pb geochronology and isotopic studies of crustal xenoliths from the Archean Medicine Hat Block, northern Montana and southern Alberta: Paleoproterozoic reworking of Archean crust, *in* Ross, G.M., ed., 1995 Alberta Basement Transects Workshop: Lithoprobe Secretariat, University of British Columbia, Lithoprobe Report 47, p. 330–335.

Erslev, E.A., and Sutter, J.F., 1990, Evidence for Proterozoic mylonitization in the northwestern Wyoming province: Geological Society of America Bulletin, v. 102, p. 1681–1694.

Farmer, G., and DePaolo, D., 1983, Origin of Mesozoic and Tertiary granite in the western United States and implications for pre-Mesozoic crustal structure, 1, Nd and Sr isotopic studies in the geocline of the northern Great Basin: Journal of Geophysical Research, v. 88, p. 3379–3401.

Foster, D., Mueller, P., Heatherington, A., Vogl, J., Meert, J., Lewis, R., and Wooden, J., 2002, Configuration of the 2.0–1.6 Ga accretionary margin of the NW Wyoming province: Implications for the Proterozoic continental reconstructions: Geological Society of America Abstracts with Programs, v. 34, no. 6, p. 559.

Foster, D., and John, B., 1999, Quantifying tectonic exhumation in an extensional orogen with thermochronology: Examples from the southern Basin and Range province, *in* Ring, U., et al., eds., Exhumation processes: Normal faulting, ductile flow and erosion: Geological Society [London] Special Publication 154, p. 343–364.

Frei, R., and Kamber, B., 1995, Single mineral Pb-Pb dating: Earth and Planetary Science Letters, v. 129, p. 261–268.

Giletti, B.J., 1966, Isotopic ages from southwestern Montana: Journal of Geophysical Research, v. 71, p. 4029–4036.

Giletti, B., and Gast, P., 1961, Absolute age of Pre-Cambrian rocks in Wyoming and Montana: Annals of New York Academy of Sciences, v. 91, p. 454–458.

Gorman, A., Clowes, R., Ellis, R., Henstock, T., Spence, G., Keller, G., Levander, A., Snelson, C., Burianyk, M., Kanasewich, E., Asudeh, I., Zoltan, H., and Miller, K., 2002, Deep Probe: Imaging the roots of western North America: Canadian Journal of Earth Sciences, v. 39, p. 375–398.

Harlan, S., Geissman, J., Snee, L., and Reynolds, R., 1996, Late Cretaceous remagnetization of Proterozoic mafic dikes, southern Highland Mountains, southwestern Montana: A paleomagnetic and $^{40}Ar/^{39}Ar$ study: Geological Society of America Bulletin, v. 108, p. 653–668.

Harms, T.A., Burger, H.R., Blednick, D.G., Cooper, J.M., King, J.T., Owen, D.R., Lowell, J., Sincock, M.J., Kranenburg, S.R., Pufall, A., and Picornell, C.M., 2004a, Character and origin of Precambrian fabrics and structures in the Tobacco Root Mountains, Montana, *in* Brady, J.B., et al., eds., Precambrian geology of the Tobacco Root Mountains, Montana: Boulder, Colorado, Geological Society of America Special Paper 377, p. 203–226 (this volume).

Harms, T.A., Brady, J.B., Burger, H.R., and Cheney, J.T., 2004b, Advances in the geology of the Tobacco Root Mountains, Montana, and their implications for the history of the northern Wyoming Province, *in* Brady, J.B., et al., eds., Precambrian geology of the Tobacco Root Mountains, Montana: Boulder, Colorado, Geological Society of America Special Paper 377, p. 227–243 (this volume).

Hayden, R.J., and Wehrenberg, J.P., 1959, Potassium-argon dating in western Montana: Geological Society of America Bulletin, v. 70, no. 12, Part 2, p. 1778–1779.

Heatherington, A.L., and Mueller, P.A., 1991, Geochemical evidence for Triassic rifting in southwestern Florida: Tectonophysics, v. 188, p. 291–302.

Hedge, C., Stacey, J., and Bryant, B., 1983, Geochronology of the Farmington Canyon Complex, Wasatch Mountains, Utah: Geological Society of America Memoir 157, p. 37–44.

Henstock, T.J., Levander, A., Snelson, C.M., Keller, G.R., Miller, K.C., Harder, S.H., Gorman, A.R., Clowes, R.M., Burianyk, M.J.A., and Humphreys, E.D., 1998, Probing the Archean and Proterozoic lithosphere of western North America: GSA Today, v. 8, no. 7, p. 1–5, 16–17.

Hoffman, P.E., 1989, Precambrian geology and tectonic history of North America, *in* Bally, A.W., and Palmer, A.R., eds., The geology of North America—An overview: Boulder, Colorado, Geological Society of America, Geology of North America, v. A, p. 447–512.

James, H.L., and Hedge, C.E., 1980, Age of the basement rocks of southwest Montana: Geological Society of America Bulletin, v. 91, p. 11–15.

Krogh, T.E., Kamo, S., and Hess, D.F., 1997, Wyoming province 3300+ gneiss with 2400 Ma metamorphism, northwestern Tobacco Root Mountains, Madison County, Montana: Geological Society of America Abstracts with Programs, v. 29, no. 6, p. A-408.

Leeman, W., Menzies, M., Matty, D., and Embree, G., 1985, Strontium, neodymium and lead isotopic compositions of deep crustal xenolith from the Snake River Plain: Evidence for Archean basement: Earth and Planetary Science Letters, v. 75, p. 354–368.

Lemieux, S., Ross, G., and Cook, F., 2000, Crustal geometry and tectonic evolution of the Archean crystalline basement beneath the southern Alberta Plains, from new seismic reflection and potential-field studies: Canadian Journal of Earth Sciences, v. 37, p. 1473–1491.

Ludwig, K., 1999, Isoplot/Ex2 version 2.06: Berkeley Geochronology Center, Special Publication 1a, www.bgc.org/klprogrammenu.html.

Mezger, K., 1990, Geochronology in granulires, *in* Vielzeuf, D., and Vidal, Ph., eds., Granulites and crustal evolution: Dordrecht, Netherlands, Kluwer Academic Publishers, p. 451–470.

Mogk, D., Mueller, P., and Wooden, J., 1992, The significance of Archean terrane boundaries: Evidence from the northern Wyoming province: Precambrian Research, v. 55, p. 155–168.

Mogk, D.W., Burger, H.R., Mueller, P.A., D'Arcy, K., Heatherington, A.L., Wooden, J.L., Abeyta, R.L., Martin, J., and Jacob, L.J., 2004, Geochemistry of quartzofeldspathic gneisses and metamorphic mafic rocks of the Indian Creek and Pony–Middle Mountain Metamorphic Suites, Tobacco Root Mountains, Montana, *in* Brady, J.B., et al., eds., Precambrian geology of the Tobacco Root Mountains, Montana: Boulder, Colorado, Geological Society of America Special Paper 377, p. 15–46 (this volume).

Mueller, P.A., and Cordua, W.S., 1976, Rb-Sr whole rock age of gneisses from the Horse Creek area, Tobacco Root Mountains, Montana: Isochron/West, v. 16, p. 33–36.

Mueller, P.A., Shuster, R.D., Wooden, J.L., Erslev, E.A., and Bowes, D.R., 1993, Age and composition of Archean crystalline rocks from the southern Madison Range, Montana: Geological Society of America Bulletin, v. 105, p. 437–446.

Mueller, P., Heatherington, A., Wooden, J., Shuster, R., Mogk, D., and Nutman, A., 1994, Paleoproterozoic evolution of the NW margin of the Wyoming craton: Evidence for 1.7 and 2.4 Ga terranes: Geological Society of America Abstracts with Programs, v. 26, no. 7, p. 34.

Mueller, P.A., Heatherington, A.L., D'Arcy, K.A., Wooden, J.L., and Nutman, A.P., 1996a, Contrasts between Sm-Nd whole-rock and U-Pb zircon systematics in the Tobacco Root batholith, Montana: Implications for the determination of crustal age provinces: Tectonophysics, v. 265, p. 169–179.

Mueller, P., Heatherington, A., Wooden, J., Mogk, D., and Nutman, A., 1996b, Proterozoic evolution of the northwestern Wyoming craton: Geological Society of America Abstracts with Programs, v. 28, no. 7, p. A-314.

Mueller, P.A., Wooden, J.L., Mogk, D.W., Nutman, A.P., and Williams, I.S., 1996c, Extended history of a 3.5 Ga trondhjemitic gneiss, Wyoming province, USA: Evidence from U-Pb systematics in zircon: Precambrian Research, v. 78, p. 41–52.

Mueller, P.A., Wooden, J.L., Nutman, A.P., and Mogk, D.W., 1998, Early Archean crust in the northern Wyoming province: Evidence from U-Pb ages of detrital zircons: Precambrian Research, v. 91, p. 295–307.

Mueller, P., Heatherington, A., Kelley, D., Wooden, J., and Mogk, D., 2002, Paleoproterozoic crust within the Great Falls tectonic zone: Implications for the assembly of southern Laurentia: Geology, v. 30, p. 127–130.

O'Neill, J.M., 1998, The Great Falls tectonic zone, Montana-Idaho: An Early Proterozoic collisional orogen beneath and south of the Belt Basin, *in* Berg, R.B., ed., Belt Symposium III—1993: Montana Bureau of Mines and Geology Special Publication 112, p. 222–228.

O'Neill, J.M., and Lopez, D.A., 1985, Character and significance of Great Falls tectonic zone, east-central Idaho and west-central Montana: American Association of Petroleum Geologists Bulletin, v. 69, p. 437–447.

O'Neill, J.M., Duncan, M.S., and Zartman, R.E., 1988, An Early Proterozoic gneiss dome in the Highland Mountains, southwestern Montana, *in* Lewis, S.E., and Berg, R.B., eds., Precambrian and Mesozoic plate margins; Montana, Idaho, and Wyoming, with field guides for the 8th International Conference on Basement Tectonics: Montana Bureau of Mines and Geology Special Publication 96, p. 81–88.

Parrish, R., 1990, U-Pb dating of monazite and its application to geological problems: Canadian Journal of Earth Science, v. 27, p. 1431–1450.

Roberts, H., Dahl, P., Kelley, S., and Frei, R., 2002, New ^{207}Pb-^{206}Pb and ^{40}Ar-^{39}Ar ages from SW Montana, USA: Constraints on the Proterozoic and Archean tectonic and depositional history of the Wyoming province: Precambrian Research, v. 117, p. 119–143.

Rudnick, R., Ireland, T., Gehrels, G., Irving, T., Chesley, J., and Hanchar, J., 1999, Dating mantle metasomatism: U-Pb geochronology of zircons in cratonic mantle xenoliths from Montana and Tanzania, *in* Gurney, J., et al., eds., Proceedings of the 7th International Kimberlite Conference, vol. I: Cape Town, South Africa, Redroof Design, p. 728–735.

Smith, H., And Giletti, B., 1997, Lead diffusion in monazite: Geochimica et Cosmochimica Acta, v. 61, p. 1047–1055.

Spear, F.S., Peacock, S.M., Kohn, M.J., Florence, F.P., and Menard, T., 1991, Computer programs for petrologic P-T-t path calculations: American Mineralogist, v. 76, p. 2009–2012, http://ees2.geo.rpi.edu/MetaPetaRen/Frame_software.html (January 2004).

Taylor, S., and McLennan, S., 1985, The continental crust: Its composition and evolution: Oxford, Blackwell Scientific Publications, p. 312.

Vitaliano, C.J., Burger, H.R., Cordua, W.S., Hanley, T.B., Hess, D.F., and Root, F.K., 1979a, Geologic map of southern Tobacco Root Mountains, Madison County, Montana: Geological Society of America Map and Chart Series MC31, scale 1:62,500, 1 sheet, 8 p. text.

Vitaliano, C.J., Burger, H.R., Cordua, W.S., Hanley, T.B., Hess, D.F., and Root, F.K., 1979b, Explanatory text to accompany geologic map of the southern Tobacco Root Mountains, Madison County, Montana: Geological Society of America Map and Chart Series MC31, scale 1:62,500, 1 sheet, 8 p. text.

Vogl, J., Mueller, P., Foster, D., Mogk, D., and Wooden, J., 2002, Character and age of Precambrian basement in the Little Belt Mountains, Montana: Implications for the role of the Great Falls Tectonic Zone in the Paleoproterozoic assembly of North America: Geological Society of America Abstracts with Programs, v. 34, no. 6, p. 273.

Wooden, J.L., and Mueller, P.A., 1988, Pb, Sr, and Nd isotopic compositions of a suite of Late Archean, igneous rocks, eastern Beartooth Mountains: Implications for crust-mantle evolution: Earth and Planetary Science Letters, v. 87, p. 59–72.

Wooden, J.L., Vitaliano, C.J., Koehler, S.W., and Ragland, P.C., 1978, The Late Precambrian mafic dikes of the southern Tobacco Root Mountains, Montana: Geochemistry, Rb-Sr geochronology, and relationship to belt tectonics: Canadian Journal of Earth Sciences, v. 15, p. 467–479.

Manuscript Accepted by the Society September 12, 2003

Printed in the USA

Geological Society of America
Special Paper 377
2004

Character and origin of Precambrian fabrics and structures in the Tobacco Root Mountains, Montana

Tekla A. Harms
Department of Geology, Amherst College, Amherst, Massachusetts 01002, USA

H. Robert Burger
Department of Geology, Smith College, Northampton, Massachusetts 01063, USA

Daniel G. Blednick
Jacob M. Cooper
J. Toby King
David R. Owen
Department of Geology, Amherst College, Amherst, Massachusetts 01002, USA

Josh Lowell
Colorado College, Colorado Springs, Colorado 80903, USA

M. Jennifer Sincock
Franklin & Marshall College, Lancaster, Pennsylvania 17604-3003, USA

Steven R. Kranenburg
Sam Houston State University, Huntsville, Texas 77341, USA

Ann Pufall
Department of Geology, Smith College, Northampton, Massachusetts 01063, USA

Carlos M. Picornell
University of New Orleans, New Orleans, Louisiana 70148, USA

ABSTRACT

A Proterozoic map-scale sheath fold is present in the crystalline core of the Tobacco Root Mountains. The fold has distinctive metamafic and associated metasedimentary rocks of the Spuhler Peak Metamorphic Suite in its core and quartzofeldspathic gneisses of the Indian Creek and Pony–Middle Mountain Metamorphic Suites on its outer flank. Fabric and structures throughout the range relate geometrically to this sheath fold and together demonstrate the character of ductile deformation during the 1.72–1.78 Ga Big Sky orogeny in the Tobacco Root Mountains. Intense, progressive, noncoaxial strain either transposed or formed compositional layering, mineral-alignment foliation, and lineation and generated outcrop-scale sheath folds, cylindrical folds with axes dispersed in the plane of foliation, and larger curtain folds. While it can be demonstrated that some gneissic fabrics and folds in the Indian Creek and Pony–Middle Mountain Metamorphic Suites formed in a much earlier phase of deformation that predates the intrusion of a suite of metamorphosed mafic dikes and

Harms, T.A., Burger, H.R., Blednick, D.G., Cooper, J.M., King, J.T., Owen, D.R., Lowell, J., Sincock, M.J., Kranenburg, S.R., Pufall, A., and Picornell, C.M., 2004, Character and origin of Precambrian fabrics and structures in the Tobacco Root Mountains, Montana, *in* Brady, J.B., Burger, H.R., Cheney, J.T., and Harms, T.A., eds., Precambrian geology of the Tobacco Root Mountains, Montana: Boulder, Colorado, Geological Society of America Special Paper 377, p. 203–226. For permission to copy, contact editing@geosociety.org.

sills, the Big Sky orogeny effectively reoriented preexisting structures and fabrics into a single Proterozoic pattern. It is most likely that tectonic juxtaposition of the Spuhler Peak Metamorphic Suite with the quartzofeldspathic gneiss suites was an integral part of this shear strain. Mesoscopic asymmetric folds record the displacement of the Spuhler Peak Metamorphic Suite down and to the north relative to the Indian Creek and Pony–Middle Mountain Metamorphic Suites during the formation of the map-scale sheath fold. "Unfolding" this sheath fold demonstrates that the Indian Creek and Pony–Middle Mountain Metamorphic Suites are lateral equivalents but leaves undetermined whether the Spuhler Peak Metamorphic Suite was juxtaposed above or below the quartzofeldspathic gneisses during Big Sky deformation.

Keywords: Tobacco Root Mountains, Precambrian, sheath fold, curtain fold, Laramide.

INTRODUCTION

The Tobacco Root Mountains of southwestern Montana lie along the northwestern limit of exposure of the Precambrian Wyoming province. Underlain by metamorphic rocks long considered Archean, we now recognize that the basement of the Tobacco Root Mountains was profoundly reworked during a major orogenic pulse in the Proterozoic, which we call the Big Sky orogeny (see Burger, 2004, this volume, Chapter 1; Cheney et al., 2004a, this volume, Chapter 6, 2004b, this volume, Chapter 8, and references therein). Metamorphic rocks across the range record peak pressures of >1.0 GPa and peak temperatures of >700 °C (Cheney et al., 2004a, this volume, Chapter 6); gneissic fabric and widespread, intrafolial folds and boudins are evidence for regional ductile deformation that is consistent with this grade of metamorphism.

Precambrian rocks of the Tobacco Root Mountains comprise three major map units (Burger, 2004, this volume, Chapter 1) (Fig. 1). Two of these, the Indian Creek Metamorphic Suite and the Pony–Middle Mountain Metamorphic Suite, are dominated by quartzofeldspathic gneiss with associated amphibolite or hornblende gneiss. They are differentiated from each other mainly by the presence of subordinate marble, quartzite, iron formation, and pelitic schist in the Indian Creek Metamorphic Suite and the near absence of these metasupracrustal rock types in the Pony–Middle Mountain Metamorphic Suite. Because the abundant quartzofeldspathic gneisses of the two suites are indistinguishable in outcrop, no formal contact has been mapped between the Indian Creek and Pony–Middle Mountain Metamorphic Suites, although one is schematically placed in the vicinity of Wisconsin Creek. The Spuhler Peak Metamorphic Suite is a lithologically distinct sequence of gedrite-garnet amphibolite, hornblende amphibolite, pelitic schist, quartzite, meta-ultramafic rocks, and minor quartzofeldspathic gneiss. A fourth Precambrian geologic unit, the metamorphosed mafic dikes and sills (MMDS), while not regionally mappable, provides a critical datum for constraining the geologic history of the Tobacco Root Mountains. These bodies intrude both the Indian Creek and Pony–Middle Mountain Metamorphic Suites but are not found in the Spuhler Peak Metamorphic Suite (Gillmeister, 1971).

The Pony–Middle Mountain and Indian Creek Metamorphic Suites appear to have originated in a distinct tectonic environment and potentially at a very different time from the Spuhler Peak Metamorphic Suite (Burger et al., 2004, this volume, Chapter 3; Mogk et al., 2004, this volume, Chapter 2; Mueller et al., 2004, this volume, Chapter 9). All three suites experienced the same *P-T* path during Big Sky metamorphism (Cheney et al., 2004a, this volume, Chapter 6) and were assembled during or before the Big Sky orogeny. The nature of their juxtaposition has potentially far-reaching implications for the evolution of the northern Wyoming province and the history of the Big Sky orogen. Yet despite previous investigations, the geometric, genetic, tectonic, and age relationships between these three units remained unresolved. The Indian Creek Metamorphic Suite, correlated with the Cherry Creek Series at the time these interpretations were made, has been considered to underlie and depositionally predate (Reid, 1963; Gillmeister, 1971, 1972; Hanley, 1975), or conversely to postdate (Tansley et al., 1933) the Pony–Middle Mountain Metamorphic Suite, to be tectonically juxtaposed against the Pony–Middle Mountain Metamorphic Suite (Gillmeister, 1971; Duncan, 1978), or, with that suite, to be part of a single sequence with internal variations (Vitaliano et al., 1979b; see reprinted map and text acccompanying this volume). The Spuhler Peak Metamorphic Suite is generally believed to overlie both the Pony–Middle Mountain and Indian Creek Metamorphic Suites across what was originally either an unconformity (Gillmeister, 1971; Duncan, 1978) or a fault (Burger, 1966, 1969), although some have placed the Spuhler Peak Metamorphic Suite between the Indian Creek and Pony–Middle Mountain Metamorphic Suites in a pre-metamorphic stratigraphic sequence, now partially truncated by faulting along contacts (Hanley, 1975; Duncan, 1978). Finally, Gillmeister (1971, 1972) proposed that the Spuhler Peak Metamorphic Suite occurs in a large open syncline.

None of these scenarios is fully consistent with the complex distribution of the three suites as presently observed in the crystalline core of the Tobacco Root Mountains (Fig. 1). The Cretaceous Tobacco Root batholith intruded in a central band across the eastern part of the range, obliterating many of the contact relationships between the three suites. Lithologic contacts within and between the three Precambrian suites are locally folded but regionally dip gently to the north (Fig. 2).

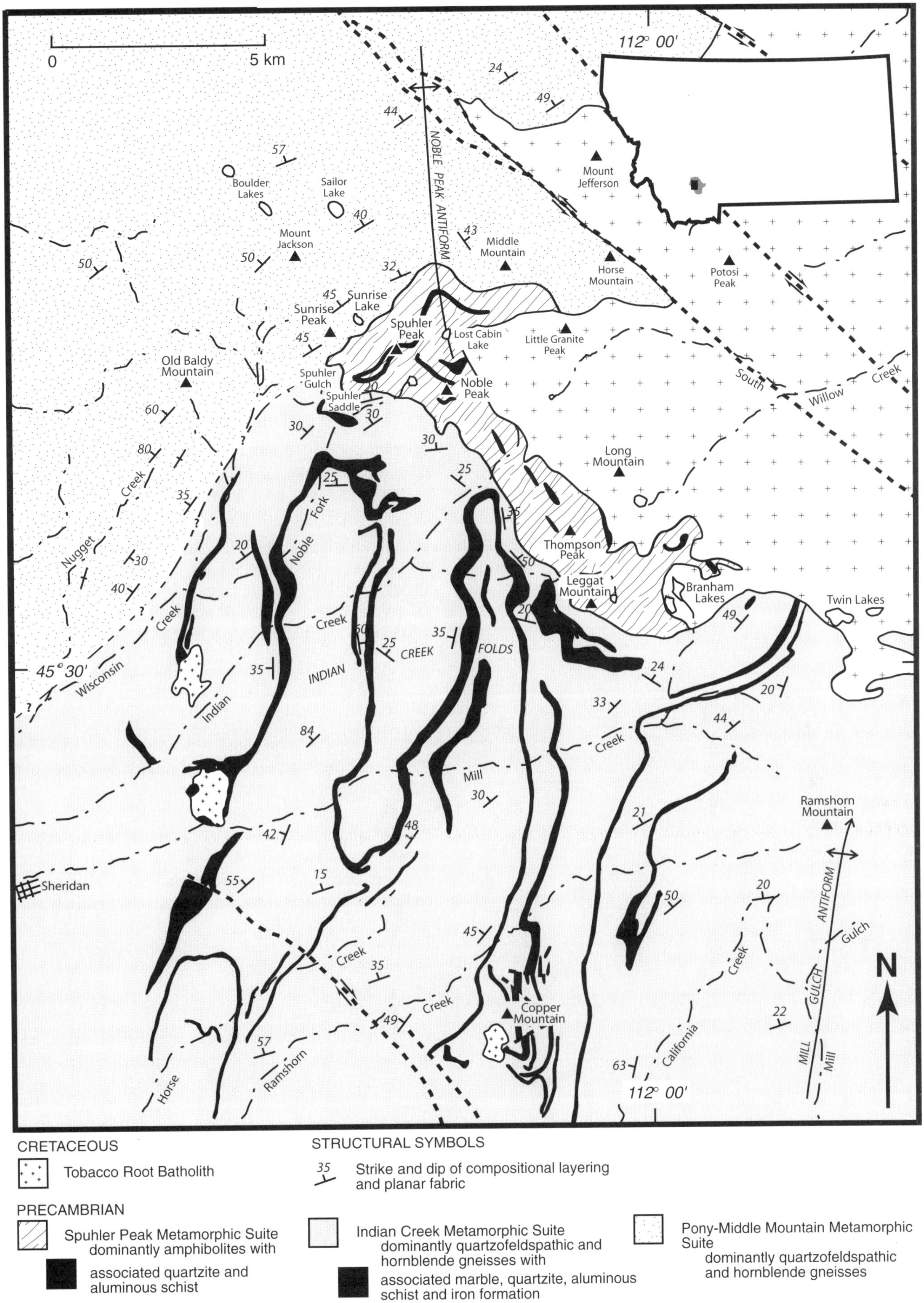

Figure 1. Schematic geologic map of the crystalline core of the Tobacco Root Mountains with representative orientations of compositional layering and foliation. Adapted from Vitaliano et al. (1979a). Inset map at upper right shows the outline of Montana with the location of the Tobacco Root Mountains (gray) and the geologic map area (black).

The Pony–Middle Mountain Metamorphic Suite is present in the northern half of the range and the Indian Creek Metamorphic Suite occupies the southern half. The Spuhler Peak Metamorphic Suite lies in a restricted area in the west center of the range, on the western flank of the Tobacco Root batholith. Thus, in the west-central part of the Tobacco Root Mountains, where the three metamorphic suites coexist, the Indian Creek Metamorphic Suite presently structurally underlies the Pony–Middle Mountain Metamorphic Suite, and the Spuhler Peak Metamorphic Suite is a wedge-shaped body between the Indian Creek and Pony–Middle Mountain Metamorphic Suites so that it structurally overlies the former and underlies the latter. A mappable aluminous schist horizon (Vitaliano et al., 1979a) appears to trace out a large-scale isoclinal fold within the tapered Spuhler Peak Metamorphic Suite.

Structural studies were undertaken along and across the contact between the quartzofeldspathic gneiss suites and the Spuhler Peak Metamorphic Suite, particularly in the area of Spuhler Gulch where all three units converge. We reconsider the relationship between the Spuhler Peak Metamorphic Suite and the quartzofeldspathic gneiss suites and between the Indian Creek and Pony–Middle Mountain Metamorphic Suites on the basis of our findings and provide constraints on the geologic history of their juxtaposition. Our structural analysis also allows us to characterize the deformation associated with the Big Sky orogeny. The results of this relatively localized study are augmented by the regional studies of Cordua (1973) in the southern and southeastern part of the Tobacco Root Mountains, Burger (1966, 1967, 1969) in the southwestern part of the range, Duncan (1978) and Gillmeister (1971) in the west-central region, Hanley (1975) in the northwest, Root (1965) in the northeast, and Hess (1967) who examined Precambrian rocks in a selvage around the Tobacco Root batholith.

The present elevation of the Tobacco Root Mountain range is attributable both to shortening on basement-involving thrust faults during the Laramide orogeny (Schmidt and Garihan, 1983; Kellogg et al., 1995) and to younger normal block faulting (Pardee, 1950). Either or both of these episodes are likely to have rotated the Precambrian crystalline basement of the range, probably about roughly north-trending axes. All attitudes and orientations referred to here will, nevertheless, be given in present coordinates. Intrusion of the Tobacco Root batholith does not appear to have reoriented host rocks along the margins of the pluton (Hess, 1967).

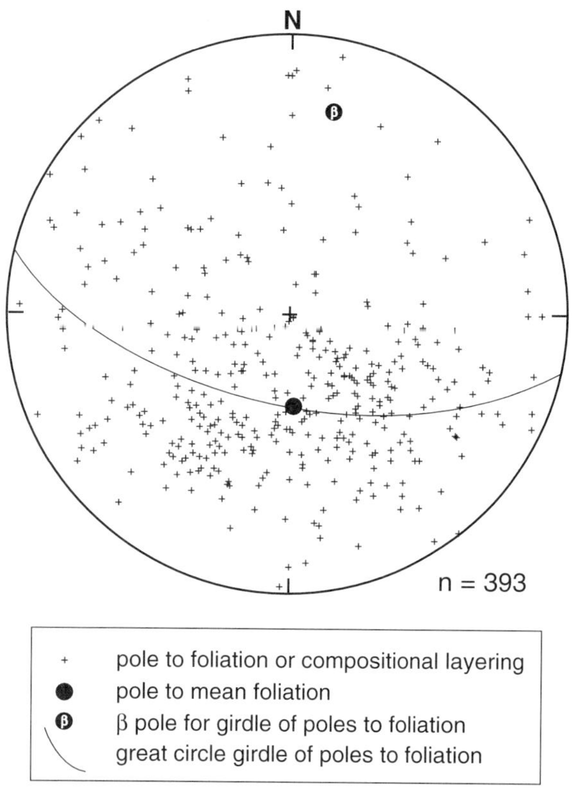

Figure 2. Equal-area, lower-hemisphere projection of poles to mineral foliations and compositional layering in all metamorphic suites within the area of the structural study. The calculated mean planar fabric lies at 268°/27°. The best-fit girdle (great circle) to the poles plotted provides a β pole of 012°/27° for the distribution of planar fabric.

FABRICS OF THE TOBACCO ROOT MOUNTAINS

Planar Fabrics

There are two planar fabrics in the Tobacco Root Mountains: compositional layering and a single mineral-alignment foliation. Compositional layering dominates the appearance of all three Precambrian metamorphic suites on all scales. With some exceptions, most notably ultramafic bodies that are podiform, the distinctive lithologies that characterize the Indian Creek, Pony–Middle Mountain, and Spuhler Peak Metamorphic Suites occur as tabular, laterally continuous bodies. At the largest scale, compositional layers occur as mappable horizons such as the marbles of the Indian Creek Metamorphic Suite or aluminous schist of the Spuhler Peak Metamorphic Suite (Fig. 1). Although too thin to be mapped, one- to several-meter-thick horizons of amphibolite within the quartzofeldspathic gneisses of the Indian Creek and Pony–Middle Mountain Metamorphic Suites, or interlayered amphibolite, gedrite-garnet schist, quartzite, and aluminous schist in the Spuhler Peak Metamorphic Suite, can be traced along strike to the limits of outcrop (Fig. 3A). Gneissic banding, typified by felsic and mafic segregations in the quartzofeldspathic units, is the finest compositional layering and occurs on the scale of a few centimeters to just millimeters (Figs. 3B, 3C, and 3D).

Foliation, as defined by shape-preferred orientation of minerals, parallels compositional layering ubiquitously. Foliation is variously produced by any combination of aligned platy (e.g., biotite), bladed (kyanite), or acicular (hornblende) minerals, flattened quartz and feldspar grains, and flattened aggregates of single mineral species.

Figure 3. Compositional layering in Precambrian rocks of the Tobacco Root Mountains. A: Mafic and felsic layers in quartzofeldspathic gneiss of the Pony–Middle Mountain Metamorphic Suite. Field of view ~6 m across. B: Thin segregations of mafic minerals define fine compositional laminae and a parallel mineral-alignment foliation in quartz-rich, garnet-bearing gneiss in the Spuhler Peak Metamorphic Suite. C: Typical gneissic banding in a hornblende gneiss of the Indian Creek Metamorphic Suite. D: Photomicrograph of ribbon quartz grains in a thin section of quartzofeldspathic gneiss in which most other grains are equidimensional. From the Indian Creek Metamorphic Suite. Photomicrograph taken with cross-polarized light. Field of view is ~12 mm across.

Compositional layering is conformable within units and with unit contacts. In general, planar fabric in the Tobacco Root Mountains dips gently to the north (Fig. 2). (Very similar stereograms from more widespread areas of the range can be found in Burger [1966], Gillmeister [1971], Cordua [1973], Hanley [1975] and Duncan [1978].) More locally, within the study area, compositional layering follows the contact between the Spuhler Peak Metamorphic Suite and the quartzofeldspathic gneiss suites (Fig. 1). Layering throughout the Tobacco Root Mountains is also folded on the map and outcrop scale, which will be taken up in detail in the next section.

Because they are uniformly parallel, unless otherwise specified, references made to layering or foliation through the remainder of this paper can be assumed to apply to both types of planar fabric. The dominance and concordance of compositional layering and the parallelism of mineral foliation with that layering are observations that have been made in all previous studies in the Tobacco Root Mountains (Burger, 1966; Root, 1965; Hess, 1967; Gillmeister, 1971; Cordua, 1973; Hanley, 1975) and are simply reconfirmed by the work of this project.

Linear Fabrics

Due to the granular texture of many lithologies and the equidimensional habit of many of the major rock-forming minerals, especially in the quartzofeldspathic gneisses, lineation is less well and less uniformly developed in the metamorphic rocks of the Tobacco Root Mountains than is planar fabric. Nevertheless, alignment of prismatic minerals, stretched quartz and feldspar grains, and rodlike monomineralic aggregates are present and were analyzed in this study. Others (Burger, 1966; Root, 1965; Cordua, 1973) have included boudins in their analyses of linear fabrics, but we found that boudins are not well enough exposed to measure accurately as a linear element. Mineral lineations have a general north-northeast trend and gentle plunge within the study area (Fig. 4). This distribution is consistent with lineation data in Burger (1966), Root (1965), Gillmeister (1971), Cordua (1973), and Hanley (1975), all of whom found that lineations cluster just east or west of north throughout the Tobacco Root Mountains.

Garbenschiefers, in which prismatic minerals are randomly distributed within the foliation plane, are a prominent exception to this pattern. Garbenschiefers are particularly well developed in parts of the Spuhler Peak Metamorphic Suite, typically in gedrite in gedrite-garnet amphibolite (see Figure 2 in Burger et al., 2004, this volume, Chapter 3) and in kyanite or sillimanite in aluminous schist. In extreme cases, the gedrite or aluminosilicate may not even lie in the plane of foliation. Garbenschiefers in the Tobacco Root Mountains are consistently associated with very coarse-grained versions of the rock type and can occur in close proximity to lineated counterparts. There is no textural evidence that garbenschiefers postdate the fabric in adjacent rocks. We can offer no explanation for this fabric at this time, but suggest that it may be the result of reaction kinetics in isolated domains that experienced elevated fluid pressure due to metamorphic dewatering or partial melting.

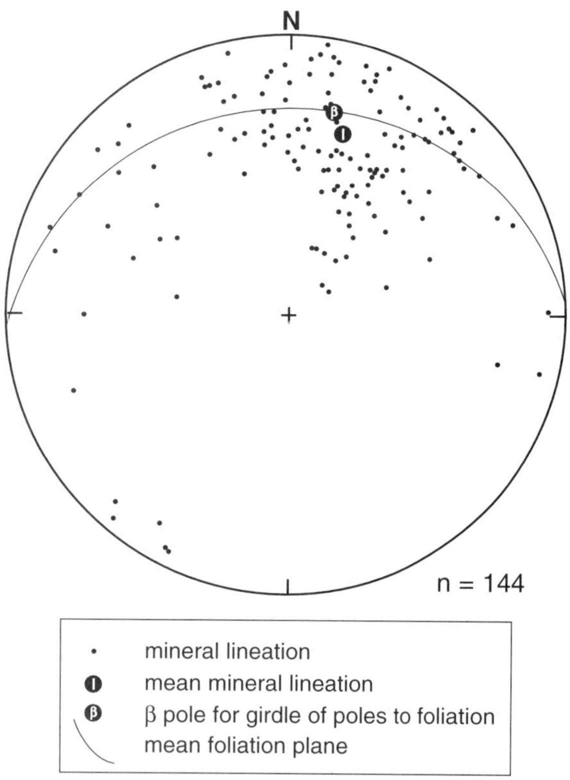

Figure 4. Equal-area, lower-hemisphere projection of the orientations of mineral lineations in all metamorphic suites within the area of the structural study. The calculated mean lineation (017°/32°) is compared to the β pole from Figure 2, and the distribution of lineations is compared to the mean foliation attitude (great circle) over the study area.

Microfabrics

Microfabrics were not systematically studied during this project, but a few observations on the character of quartz in thin section appear consistent and significant. Many rocks of the Tobacco Root Mountains metamorphic suites are dominated by a granular texture, typical of their grade of metamorphism, in which quartz has an equidimensional habit. In some rocks, however, including some quartzofeldspathic gneisses, ribbon grains of quartz are present (Fig. 3D). Quartz-to-quartz grain boundaries within these ribbons are typically polygonal, even in high-strain zones, but the short length and complex pattern of these polygonal boundaries suggest they may originally have been serrate. From this we conclude first that some rocks in the Indian Creek, Pony–Middle Mountain, and Spuhler Peak Metamorphic Suites may have been mylonitic at early stages of development and, second, either that metamorphic heating outlasted grain-scale deformation or that later annealing occurred.

Fabrics and the MMDS

The metamorphosed mafic dikes and sills (MMDS), found throughout the Indian Creek and Pony–Middle Mountain Metamorphic Suites and dated at 2.06 Ga (Mueller et al., 2004, this volume, Chapter 9), provide an important control on understanding the multistage history by which this seemingly simple pattern of fabric in the Tobacco Root Mountains developed. MMDS cut compositional layering in the Indian Creek and Pony–Middle Mountain Metamorphic Suites (Fig. 5A); they have not been found in the Spuhler Peak Metamorphic Suite (Gillmeister, 1972; Brady et al., 2004a, this volume, Chapter 5). It is, therefore, necessary that widespread production of high-grade gneissic fabric in the Pony–Middle Mountain and Indian Creek Metamorphic Suites predated the intrusion of the MMDS, and it is most likely that juxtaposition of the Spuhler Peak Metamorphic Suite with the two quartzofeldspathic gneiss suites—either depositionally or tectonically—postdated intrusion of the MMDS. Whether or not the Spuhler Peak Metamorphic Suite could have had a pre-MMDS planar fabric depends on the original nature of the contact between it and the quartzofeldspathic gneiss suites. If depositional, all planar fabric in the Spuhler Peak Metamorphic Suite must postdate the MMDS; if tectonic, a pre-MMDS fabric could have been carried with the Spuhler Peak Metamorphic Suite to its present location.

It is also the case that MMDS have an internal foliation and have been boudinaged and folded (Figs. 6A and 6B) (Hanley, 1975; Brady et al., 2004a, this volume, Chapter 5). At least some ductile deformation and fabric production in the Tobacco Root Mountains metamorphic rocks must, therefore, postdate MMDS intrusion. Furthermore, two observations suggest that

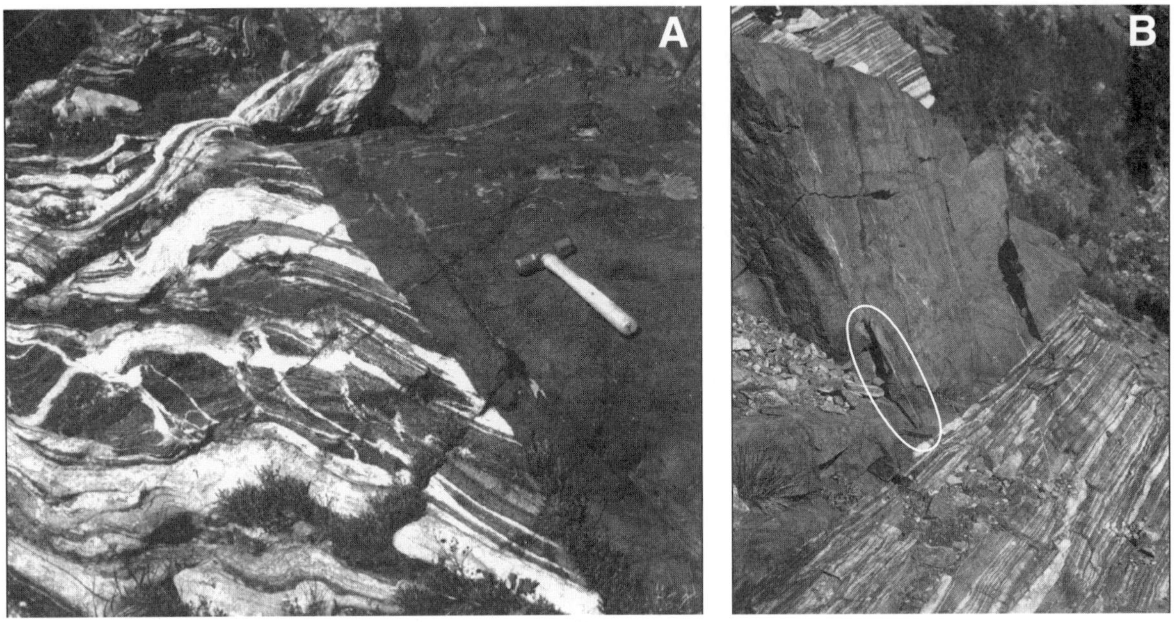

Figure 5. Metamorphosed mafic dikes and sills in crosscutting dike (A) and concordant sill (B) configurations. Note the very low-angle truncation of gneissic banding along the lower sill contact in B. Hammer outlined for scale in B.

many MMDS "sills" are best interpreted as dikes that have been rotated into near parallelism with their host fabric by post-MMDS deformation. First, close examination of such sills shows that their contacts actually crosscut gneissic banding but at a very small angle (Fig. 5B). Second, high-strain zones are present in the quartzofeldspathic gneisses. These are layer-parallel domains in which the typical coarse granular texture of the gneisses is replaced by reduced grain size and exceedingly thin and planar compositional layering (Fig. 7). Localized, intense simple shear within such zones in the host gneisses could rotate embedded high-angle dike contacts into low-angle, apparent sill-like contacts. The preservation of dike geometries in some areas and the existence of "sill" contacts and high-strain zones in others indicate a post-MMDS ductile deformation that profoundly affected fabric but was spatially heterogeneous in intensity. The abundance of sill-like MMDS bodies in the Indian Creek and Pony–Middle Mountain Metamorphic Suites (see Figures 3 and 7 in Brady et al., 2004a, this volume, Chapter 5) indicate the widespread distribution of post-MMDS deformation and reorientation.

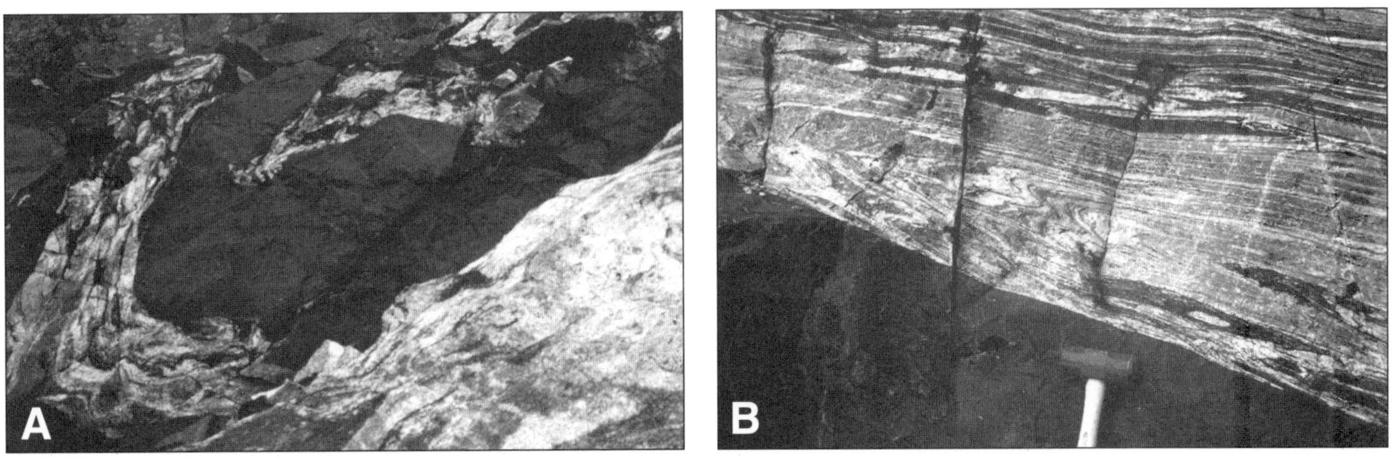

Figure 6. Observed metamorphosed mafic dikes and sills (MMDS) and quartzofeldspathic gneiss fold relationships. A: MMDS body and gneiss bands folded together. Field of view is ~3 m across. B: MMDS dike crosscuts the nose of a tight fold in quartzofeldspathic gneiss bands. See also Figure 6 in Brady et al. (2004a, this volume, Chapter 5).

Figure 7. High-strain zone in the Indian Creek Metamorphic Suite encasing a thin, concordant metamorphosed mafic sill. Note the very fine compositional laminae in the quartzofeldspathic gneiss and their continuity. Compare this to the character of compositional bands in Figures 5A and 3A.

Without the presence of a crosscutting metamorphosed dike, pre-MMDS fabric cannot be distinguished from post-MMDS fabric by appearance, mineralogy, or orientation. The present concordance of planar fabric in the Spuhler Peak Metamorphic Suite with that in the Indian Creek and Pony–Middle Mountain Metamorphic Suites and the parallelism of lineations in all three suites suggests that preexisting planar and linear elements were partially or completely rotated toward and/or recrystallized into a single new, coherent pattern during metamorphism that postdates intrusion of MMDS bodies and developed during or after juxtaposition of the Spuhler Peak Metamorphic Suite with the Indian Creek and Pony–Middle Mountain Metamorphic Suites.

MESOSCOPIC FOLDS

Outcrop-scale folds, from meters to centimeters in wavelength and amplitude, are the most ubiquitous structure in the Precambrian rocks of the Tobacco Root Mountains (Fig. 8). Folds can be found in all four of the metamorphic rock suites, including the MMDS, in all rock types, and in all geologic or geographic domains (see the work of Burger, 1966; Root, 1965; Hess, 1967; Gillmeister, 1972; Cordua, 1973; Hanley, 1975; and Duncan, 1978, for analyses of folds outside the study area of this project). Folds are not, however, evenly distributed. Areas without folds occur as well as fold-rich zones that contain cascades of harmonic and disharmonic folds between enveloping planar foliation.

Mesoscopic folds throughout the metamorphic core of the range fold compositional layering and fold mineral-alignment foliation. Cordua (1973) and Hanley (1975) report axial planar foliation that crosscuts compositional layering in the hinge zones of some folds. Few such fabrics were observed in the course of this study, but are present locally both in the quartzofeldspathic gneisses and in biotite-bearing schists of the Spuhler Peak Metamorphic Suite (Fig. 9A). More commonly, mineral alignment is simply less well developed in the hinge zone of folds.

Folds in the Tobacco Root Mountains exhibit a range of styles (Fig. 8). Fold profiles with thinned limbs and thickened hinges are the most common but near-concentric folds also occur. The latter form is typically associated with rock types such as amphibolites that are likely to have been more competent than surrounding gneisses or micaceous schists at the time of metamorphism and deformation. Hinges are typically subrounded. Folds range from close to isoclinal; interlimb angles of 30°–40° are common. Axial planes vary from moderately inclined to the enveloping compositional layering to fully parallel with the local attitude of layering. Asymmetric folds and isolated isoclines are common. Asymmetric folds are typically intrafolial, with two long limbs concordant with the enveloping foliation and the short limb at an acute angle to that foliation. Both fold profile and interlimb angle may vary between layers of the same fold, suggesting different folding mechanisms at work in different layers.

Because of the lithologic heterogeneity of the Spuhler Peak Metamorphic Suite, erosion along compositional layering is common, and folded surfaces can be beautifully exposed in three dimensions. These surfaces demonstrate the presence of both cylindrical folds and noncylindrical folds with a range of fold axis curvature—from gently undulating hinge lines (Fig. 10A) to tight fold axis bends forming true sheath folds (Fig. 10B). In contrast, the quartzofeldspathic gneisses of the Indian Creek and Pony–Middle Mountain Metamorphic Suites are typically massive, weathering along extensive planar faces, so that fold surfaces are not as commonly revealed. There, eyeball-shaped patterns indicate the presence of highly noncylindrical folds (Fig. 11). That these forms are sheath folds, rather than Type 1 refolded folds (domes or basins), is established by the uniform orientation of mineral lineation on all surfaces of such folds, parallel to the long axis of the paraboloidal fold form.

Folds in the Indian Creek and Pony–Middle Mountain Metamorphic Suites are crosscut by MMDS (Fig. 6B). MMDS are, however, also folded along with compositional layering (Fig. 6A). Clearly, episodes of folding both predate and postdate MMDS emplacement. As is true for pre- and post-MMDS planar fabric, objectively separating individual folds into populations representing the two different phases is not possible. Folds occur with a continuum of fold profile, interlimb angle, symmetry, axial plane attitude, and cylindricity, so that defining different generations on the basis of any of these attributes would be arbitrary. The few folds that have axial planar foliation might represent an early phase of deformation—earlier, at least, than that which produced the alignment of minerals—but, like Root (1965), we believe these folds are best interpreted as the result of differing rheology during folding. And, inasmuch as they have been found in the Spuhler Peak Metamorphic Suite, these folds cannot be associated with the earliest generation of quartzofeldspathic gneissic banding and most likely postdate MMDS emplacement.

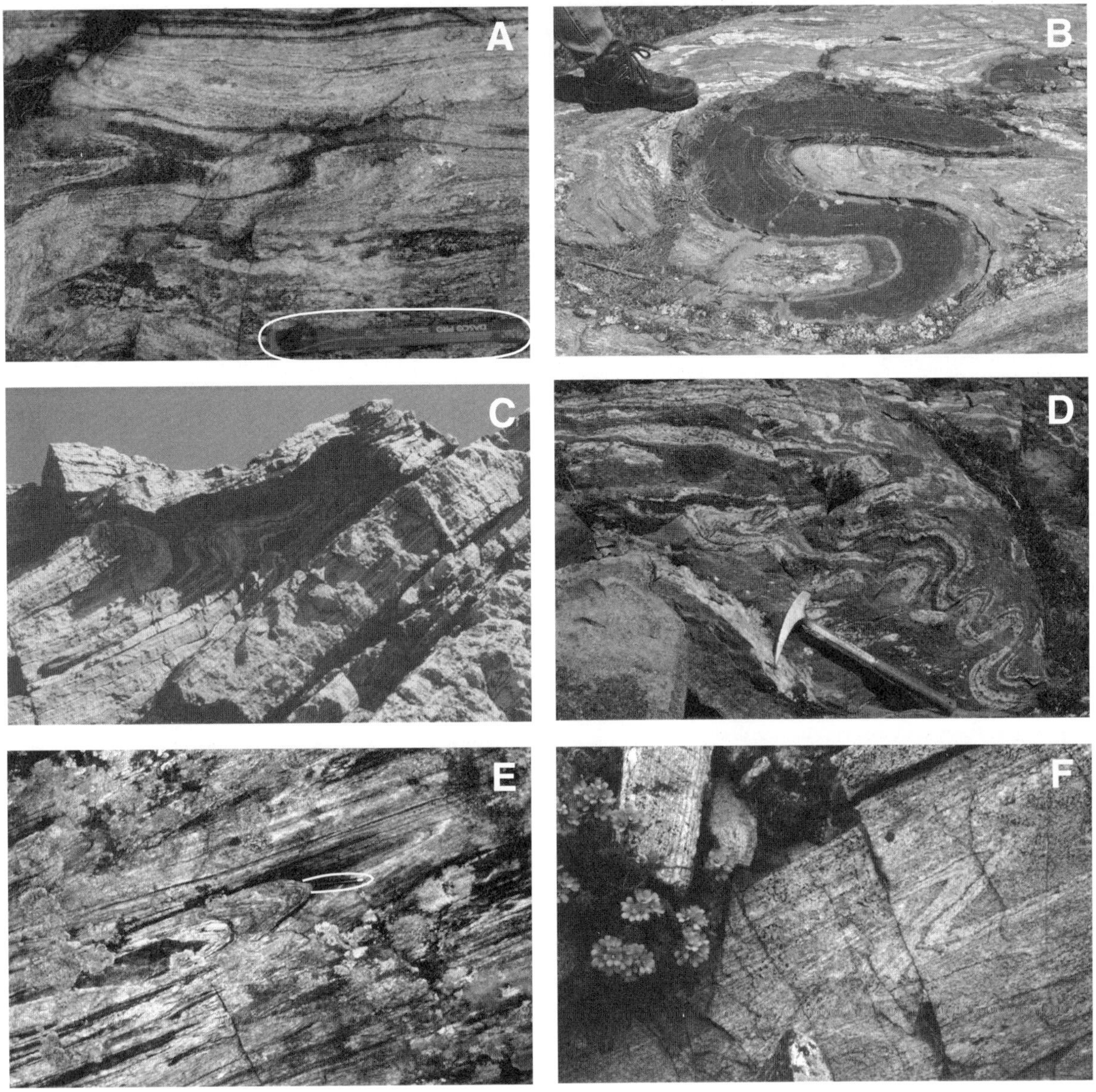

Figure 8. Representative fold styles observed in Precambrian metamorphic rocks of the Tobacco Root Mountains. All fold compositional layering and mineral-alignment foliation, and none have developed an axial planar fabric. A: Tight "M" fold with similar profile in Pony–Middle Mountain Metamorphic Suite gneiss. Chisel in lower right highlighted for scale. B: Isoclinal "S" fold in hornblende-amphibolite layer within biotite-sillimanite schist of the Spuhler Peak Metamorphic Suite. This fold profile is nearly concentric. C: Close, asymmetric, intrafolial fold in the Pony–Middle Mountain Metamorphic Suite. Field of view is ~10 m across. D: Cascade of disharmonic folds in the Spuhler Peak Metamorphic Suite. E: Isolated isoclinal fold in Indian Creek Metamorphic Suite gneiss with thickening of layers in the hinge zone. Pencil along the fold hinge is highlighted for scale. F: Tight, asymmetric, intrafolial "Z" fold in Pony–Middle Mountain Metamorphic Suite quartzofeldspathic gneiss. Note the prominent thickening in the hinge and short limb of the fold. Field of view ~40 cm across.

Finally, hook-shaped or Type 3 refolded folds are not common but can be found in both the quartzofeldspathic gneisses and the Spuhler Peak Metamorphic Suite (Fig. 9B). In this case, too, it seems unlikely that the refolded folds predate MMDS intrusion, and it is equivocal whether or not the earlier fold generation in these interference patterns originated in a distinctly older event. Based on the geometric analysis of folds presented next, we will argue that it is not meaningful to separate pre-MMDS from post-MMDS folds.

Geometric Analysis of Mesoscopic Folds

The orientations of fold axes and linear and planar fabrics were measured across the study area and stereographically

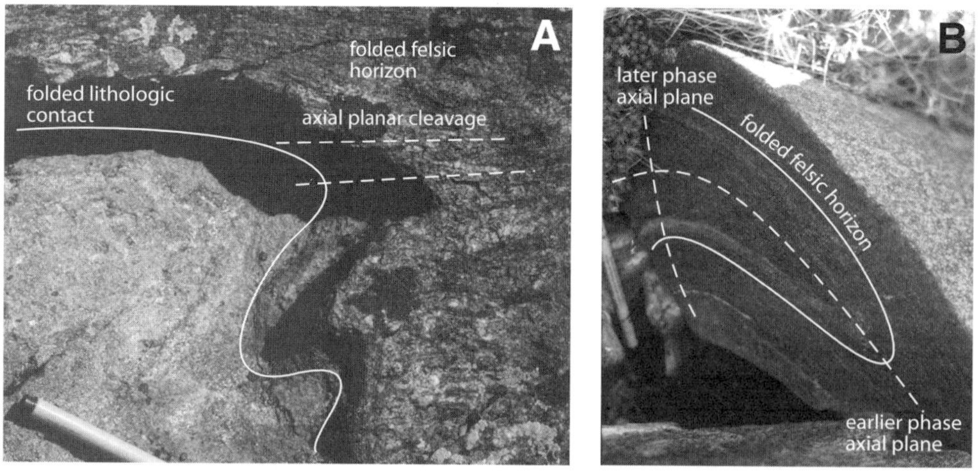

Figure 9. Complex, possibly polyphase, fold geometries observed in the Tobacco Root Mountains metamorphic suites. A: Subtle axial planar mineral-alignment foliation crosses compositional layering in Spuhler Peak Metamorphic Suite gneisses in the hinge zone. B: Close Type 3 refolding of a tight fold in Indian Creek Metamorphic Suite gneiss. Both images have pens for scale to the left of the fold.

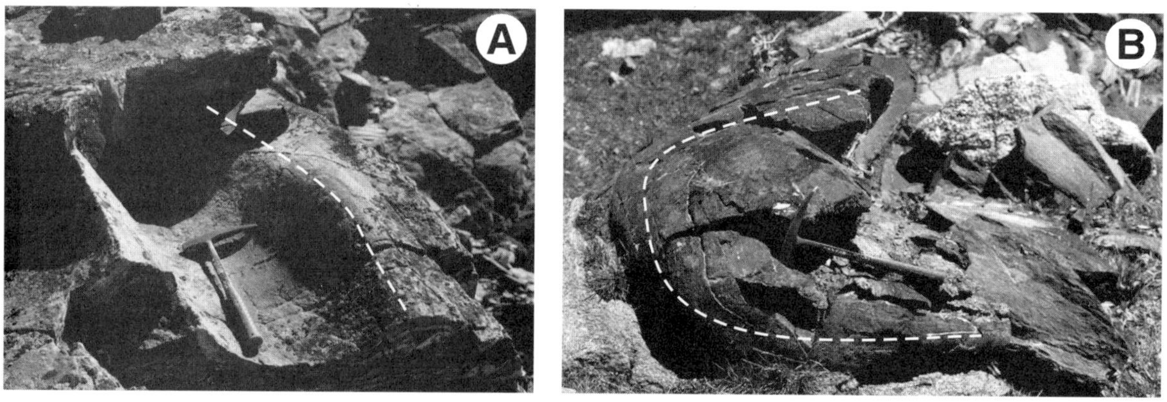

Figure 10. Noncylindrical folds in the Spuhler Peak Metamorphic Suite that range from gently undulating fold hinges (A) to tight sheath folds (B). Hinge lines are highlighted with dashed lines. The view in B is of the bulbous surface of a folded amphibolite layer exposed in three dimensions. A fracture cuts across the folded layer and reveals the fold profile in the upper center of the image.

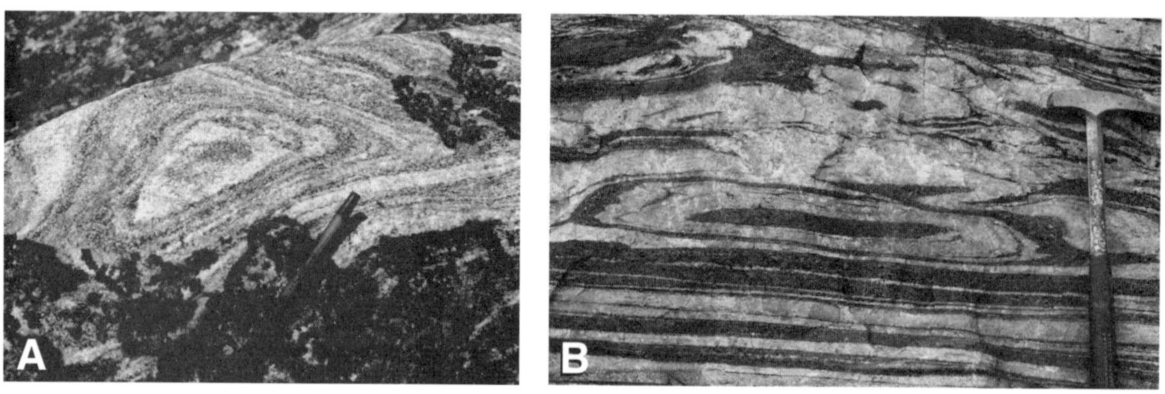

Figure 11. Characteristic "eyeball" shape fold patterns on planar outcrop surfaces indicate noncylindrical folds in quartzo-feldspathic gneisses of the Pony–Middle Mountain Metamorphic Suite (A) and the Indian Creek Metamorphic Suite (B).

analyzed both in local tracts (Figs. 12 and 13) and combined into synopses of the whole study area (Figs. 2, 4, and 14).

A consistent pattern of structures emerges. In each local tract (Figs. 12 and 13), lineations (shape alignment of minerals and/or aggregates of minerals, excluding fold axes) cluster. Poles to foliation lie in a clustered maximum that describes the general local dip of planar fabric, but can be spread into a weak to moderate girdle that reflects rotation of foliation due to folding. The β pole to this girdle is a synoptic fold axis for foliation attitudes. It is near to or coincident with the average mineral lineation, which lies along the plane of the mean foliation.

Fold axes are dispersed about a girdle that is equivalent to the mean foliation plane, that is to say, fold axes lie in the plane of foliation (Figs. 12 to 14). Fold axes have a weak maximum in that girdle, which approximates the mean lineation direction. Asymmetric folds are categorized by their down-plunge sense of rotation as "S" or "Z" folds. Symmetric folds with four exposed limbs are "M" folds. S, Z, and M fold axes are not equally dispersed along the girdle. Although not strongly clustered by type, their means are systematic: S and Z fold axis means fall on opposite sides of the mean lineation (Fig. 12). M folds are not common and are a poorly constrained population, but lie around the mean lineation.

We emphasize that mineral lineations are consistent in orientation within local tracts. Lineations have not been rotated by folding, even where fold axes are not parallel to the mineral lineation.

Thus, in each local tract, the average mineral lineation is the axis of symmetry for the structure set. Foliation, the β pole of foliation, S, Z, and M fold axes, and the mean fold axis are distributed in a predictable pattern about the mean lineation. Regionally, the average lineation plunges gently (~25°) to the north-northeast (Fig. 4). The orientation of each tract-average lineation, however, varies about that general northerly trend, and so the local pattern of linked foliation and fold axes varies in orientation in accordance with the local mean lineation (Fig. 12).

Individual sheath folds mimic the pattern of distribution of fabrics and structures seen in local tracts. One extraordinarily well exposed, large, meter-scale sheath fold in the Spuhler Peak Metamorphic Suite was independently analyzed (Cooper, 2002) and demonstrates the analogy (Fig. 15). Note that mineral lineations are fairly uniform in orientation everywhere on the fold and demonstrate the inherent noncylindrical nature of the fold. S and Z folds are different parts of the strongly curved sheath fold hinge and their axes lie systematically on opposite sides of the lineation within the plane of foliation.

It appears that the Indian Creek, Pony–Middle Mountain, and Spuhler Peak Metamorphic Suites contribute equally to the regional fold pattern. Because of the blocky weathering of the quartzofeldspathic gneisses, fold axes were more difficult to measure in the Pony–Middle Mountain and Indian Creek Metamorphic Suites. In the Sunrise Lake domain, however, comparable numbers of folds in both the Pony–Middle Mountain Metamorphic Suite (Owen, 1996) and the Spuhler Peak Metamorphic Suite were measured and show that the pattern of fold axis distribution is the same in both suites (Fig. 13). Spot studies in the Indian Creek Metamorphic Suite in the headwaters of Mill Creek's Middle Fork and at Ramshorn Mountain, and in the Pony–Middle Mountain Metamorphic Suite near Horse Mountain and around Sailor and Boulder Lakes were selected because they are at some distance (2–5 km) from the Spuhler Peak Metamorphic Suite contact. Orientations in these small areas are consistent with folds and fabrics measured within and along the contact with the Spuhler Peak Metamorphic Suite (Fig. 12). Folded MMDS are compatible with the regional pattern but are too rarely exposed in three dimensions for measurement so that a robust individual population was not obtained.

Whereas the mean orientation of mesoscopic fold axes parallels the average mineral lineation in the metamorphic suites of the Tobacco Root Mountains, individual fold axes—many but not all of which are cylindrical—are dispersed in the plane of foliation (Fig. 14). Mesoscopic folds parallel to lineation (e.g., Escher and Watterson, 1974; Lacassin and Mattauer, 1985; Malavieille, 1987; and Goscombe, 1991) and planar dispersions of folds with vergence sets bisected by the mineral lineation (Borradaile, 1972; Carreras et al., 1977; Bell, 1978; Boyle, 1987; Mies, 1991) have each been described from areas of intense, noncoaxial strain. Mesoscopic sheath folds are considered a key indicator of simple shear (Minnigh, 1979; Berthé and Brun, 1980; Cobbold and Quinquis, 1980; Skjernaa, 1989; Ghosh and Sengupta, 1984). It seems, therefore, that a relatively intense deformation that included a significant component of simple shear strain was imposed on the core of the Tobacco Root Mountains as part of the Big Sky orogeny, during which the observed structure set was formed.

Because some folds and some compositional layers are cut by MMDS, it is likely that a fraction of the folds and fabrics included in this analysis originated well before the Big Sky event. During Big Sky deformation, preexisting planar fabrics would have rotated toward the active flattening plane. Pre-MMDS asymmetric folds would have rotated into the foliation and toward the lineation but would not necessarily show a systematic distribution of S and Z folds on either side of the lineation. This may account for the observed fairly high dispersion of Z and S folds into the opposite side of the girdle from their mean. The coherency and consistency observed in the structural geometry, however, argues that preexisting folds, mineral lineations, and planar fabrics were effectively rotated or recrystallized into the Big Sky strain pattern at the time of that deformation and need not be isolated from that structure set for a viable analysis. The distinctive and uniform pattern of structures and fabrics across the study area is best interpreted as a compatible set that developed as the result of a progressive, noncoaxial regional deformation that produced post-MMDS folds and fabrics and reoriented earlier structures into the Big Sky pattern.

Of those who previously studied fold orientations in the Tobacco Root Mountains, Gillmeister (1971), Hanley (1975), and Duncan (1978) found dispersed girdle patterns of fold axes similar

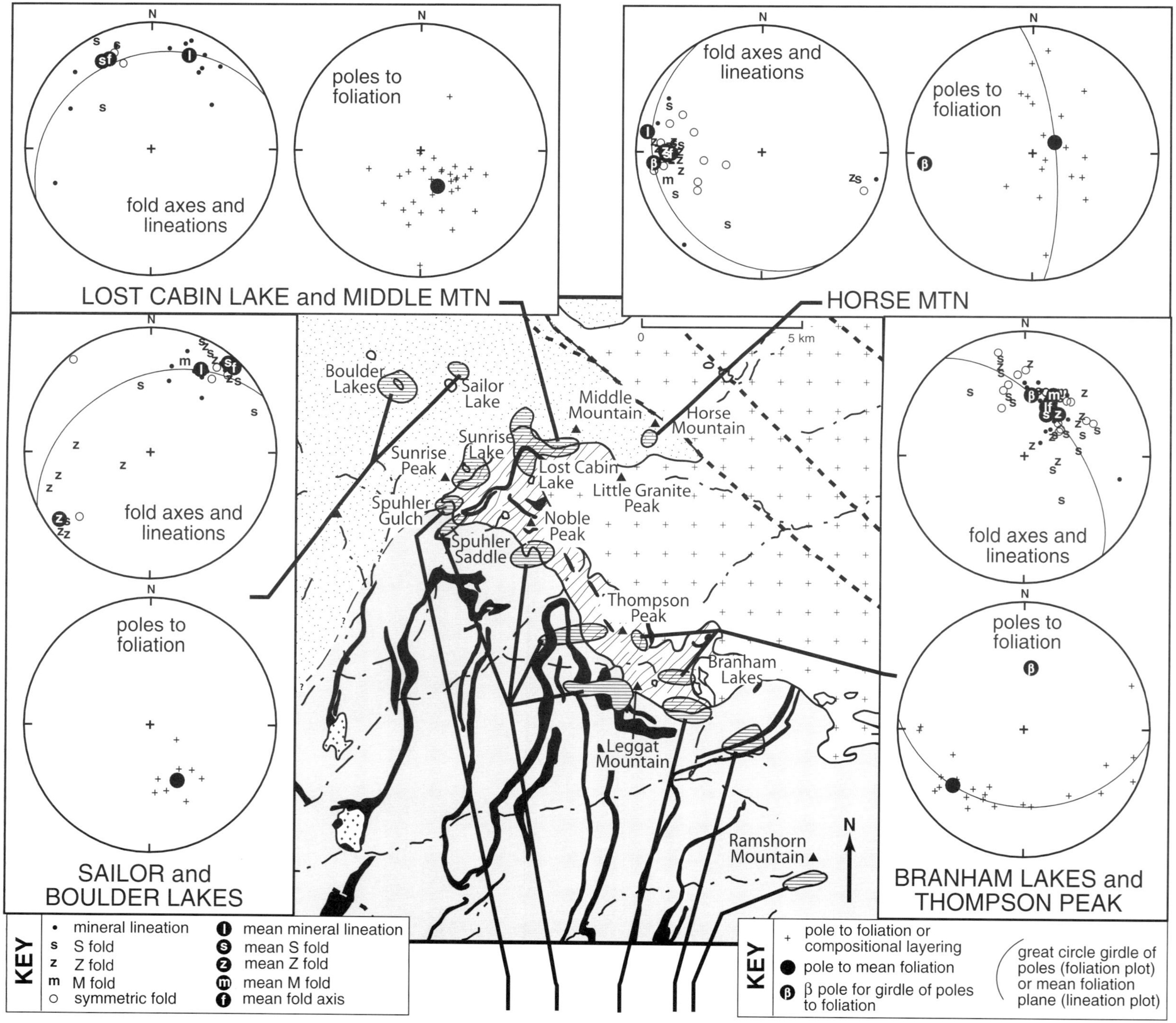

Figure 12. Geometric analysis of structural data from individual tracts within the study area (tracts indicated with the horizontal ruled pattern on the central schematic geologic map). Equal-area, lower-hemisphere projections give foliation, lineation, and fold axis orientations and their calculated averages within those local tracts. These structure data, with those from Sunrise Lake (Fig. 13), are synthesized into Figures 2, 4, and 14. Patterns in the geologic map are defined in Figure 1.

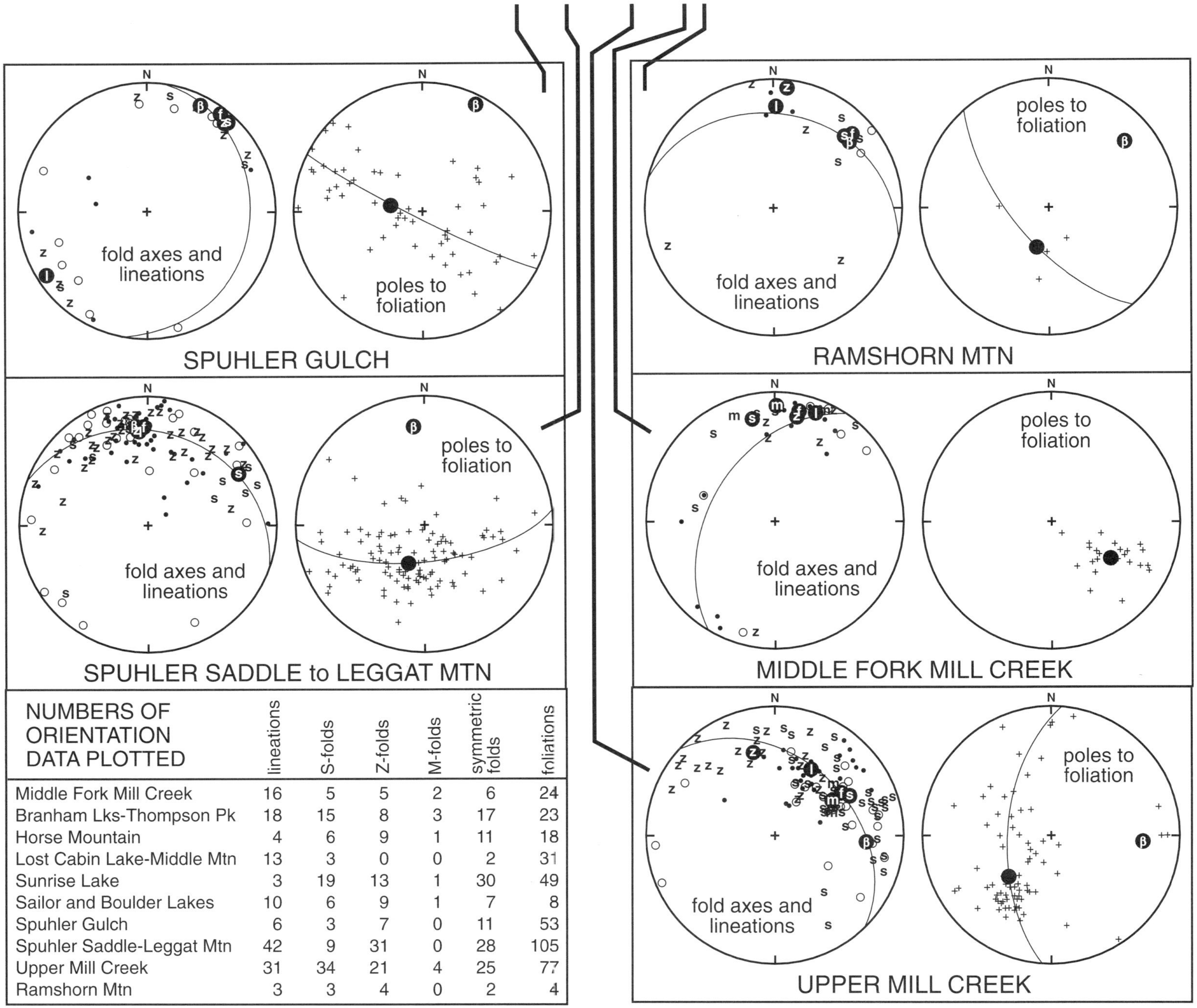

NUMBERS OF ORIENTATION DATA PLOTTED	lineations	S-folds	Z-folds	M-folds	symmetric folds	foliations
Middle Fork Mill Creek	16	5	5	2	6	24
Branham Lks-Thompson Pk	18	15	8	3	17	23
Horse Mountain	4	6	9	1	11	18
Lost Cabin Lake-Middle Mtn	13	3	0	0	2	31
Sunrise Lake	3	19	13	1	30	49
Sailor and Boulder Lakes	10	6	9	1	7	8
Spuhler Gulch	6	3	7	0	11	53
Spuhler Saddle-Leggat Mtn	42	9	31	0	28	105
Upper Mill Creek	31	34	21	4	25	77
Ramshorn Mtn	3	3	4	0	2	4

Figure 12 (*continued*).

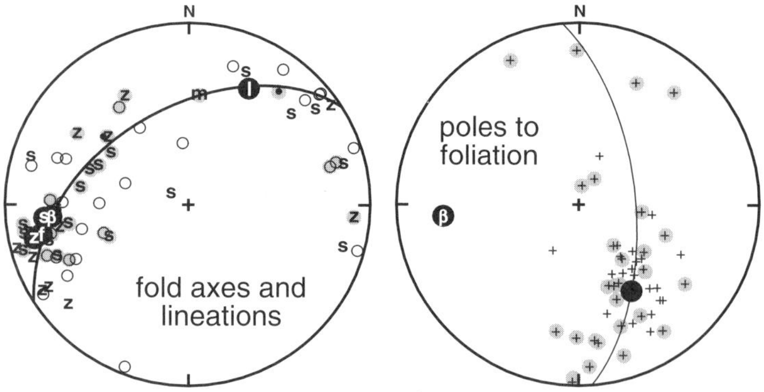

SUNRISE LAKE

NUMBERS OF ORIENTATION DATA PLOTTED	
lineations	3
S-folds	19
Z-folds	13
M-folds	1
symmetric folds	30
foliations	49

Figure 13. Equal-area, lower-hemisphere projections of foliation, lineation, and fold axis orientations from the Sunrise Lake area (shown in the schematic geologic map in Figure 12). Structure symbols superimposed on a gray dot are from the Spuhler Peak Metamorphic Suite; all others are from the Pony–Middle Mountain Metamorphic Suite. See Figure 12 for the key to symbols used in the stereograms.

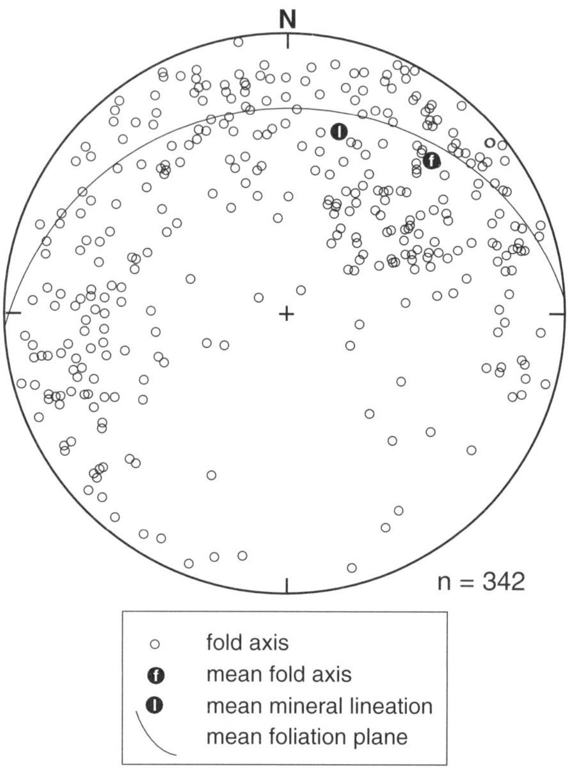

Figure 14. Equal-area, lower-hemisphere projection of orientations of all fold axes measured in all metamorphic suites across the area of the structural study. Fold axes are dispersed along the average foliation (shown by the great circle). Fold axes are slightly concentrated in the northeast quadrant, and a poorly defined mean axis lies at 044°/24°. The average mineral lineation is plotted for comparison.

to those reported here. In contrast, Burger (1966), Root (1965), Hess (1967), and Cordua (1973) report clustered north-northeast trends of mesoscopic folds, parallel to mineral lineation, and to our mean fold axis. This is compatible with the structure set determined in our study and with the proposed intense, progressive noncoaxial deformation. We conclude that the Big Sky orogeny affected all parts of the range but more thoroughly rotated fold axes in some areas than in others.

Kinematic Analysis of Mesoscopic Folds

The pattern of S, Z, and M fold populations and mineral lineations affords the opportunity to assess the direction and sense of displacement during their development according to the model of Hansen (1971), in which the sense of rotation of the S and Z folds indicates the sense of shear. Because of their limited area, the local domains in which structures and fabrics were analyzed can have small populations of any given fold type. More robust kinematic results were derived by grouping similar local domains into synoptic sets in which fold vergence patterns are consistent (Fig. 16).

In our analysis, we assume that the regional mean mineral lineation parallels the axis of maximum elongation during deformation. While not strictly coincident (Lin and Williams, 1992), the lineation approximates the orientation of the slip line along which shear occurred (Hansen, 1971). Because the lineation plunges to the north within north-dipping planar fabric, the displacement recorded by the mesoscopic folds was essentially dip-slip, in present coordinates.

Three internally consistent kinematic domains were recognized (Fig. 16). S folds plunge dominantly to the east of the mineral lineation, indicating a "top-down to the north" sense of shear, in the southwestern Spuhler Peak Metamorphic Suite and along the Spuhler Peak/Indian Creek Metamorphic Suites contact from Leggat Mountain to Spuhler Saddle and southeast to Ramshorn Mountain (kinematic Domain I). Conversely, S folds on average plunge to the west of the mineral lineation and reflect a "top-up to the south" sense of shear in the northern Spuhler Peak Metamorphic Suite around Lost Cabin Lake, along the Spuhler Peak/Pony–Middle Mountain Metamorphic Suites contact from just west of Middle Mountain to the Tobacco Root batholith on the north, in the southern Spuhler Peak Metamorphic Suite in the area of Branham Lakes and Thompson Peak adjacent to the Tobacco Root batholith, and in the Indian Creek Metamorphic Suite in the headwaters of Mill Creek (Domain III). This domain

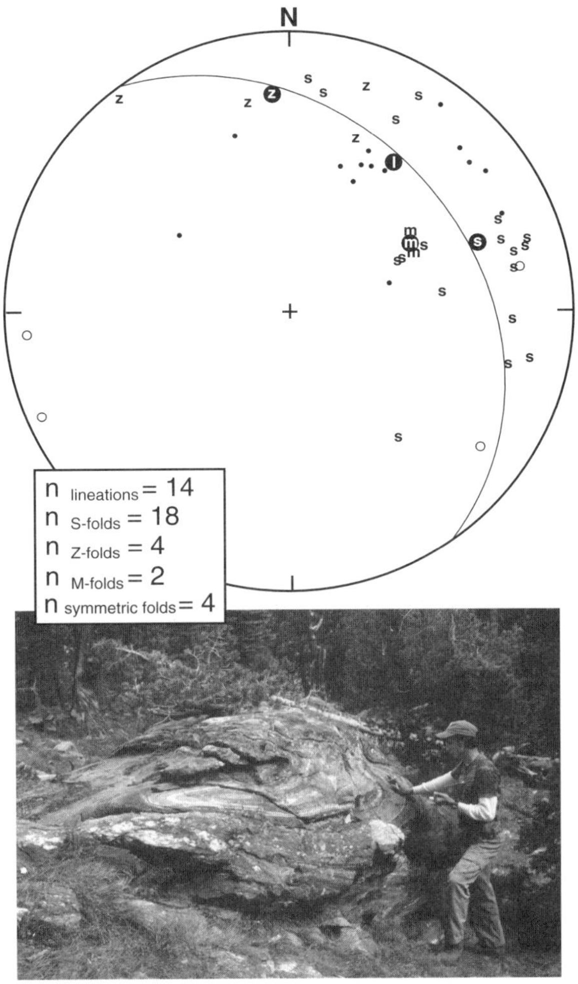

Figure 15. Structural data (upper stereogram) associated with a large sheath fold in the Spuhler Peak Metamorphic Suite (lower photo). The sheath fold is intrafolial; enveloping compositional layering and the plane of flattening of the sheath fold profile dip away from the viewer of the photo. The long axis of the fold plunges directly away from the viewer. The hinge line of the sheath fold is visible in three dimensions on the left side of the outcrop where a thin, dark amphibolite layer outlines a left- (north) plunging Z fold and again in a right- (east) plunging S fold on the right side of the outcrop under the geologist's hand. Equal-area, lower-hemisphere projection shows the cluster of mineral lineations that parallels the long axis of the sheath fold and the distribution of "S" and "Z" fold axes on either side of the calculated mean lineation. All linear elements lie along the average foliation plane. See Figure 12 for the key to symbols used in the stereogram.

was possibly once continuous but subsequently divided by intrusion of the Tobacco Root batholith. Finally, mineral lineations, S folds, and Z folds all cluster and are nearly horizontal, leaving the sense of shear indeterminate and suggesting a distinct strain regime along the Spuhler Peak Metamorphic Suite contact from Spuhler Gulch to Sunrise Lake, and in the Pony–Middle Mountain Metamorphic Suite around Sailor Lake and the Boulder Lakes (Domain II).

In sum, the orientation and vergence patterns of asymmetric mesoscopic folds indicate that the Spuhler Peak Metamorphic Suite moved down and to the north relative to both underlying and overlying quartzofeldspathic gneiss suites. Deformation in the Spuhler Gulch domain was sufficiently intense to have resulted in parallelism of fold axes with the slip line.

Time of Folding

Leucosome is widespread in the metamorphic suites of the Tobacco Root Mountains and is considered to be the product of partial melting associated with the thermal maximum achieved during the Big Sky orogeny (Cheney et al., 2004a, this volume, Chapter 6). A sample of this material collected from between boudins in the limb of an isoclinally folded amphibolite layer in the Spuhler Peak Metamorphic Suite yielded zircon dated at 1.77 Ga (Mueller et al., 2004, this volume, Chapter 9). A smaller amount of similar leucosome also occurs on the convex side of the hinge in the same fold. The location of the crystallized melt was clearly controlled by dilatancy associated with these structures so that the 1.77 Ga age of the melt dates the time of folding and associated deformation.

Geothermobarometry and mineral assemblages have been used to establish a clockwise *P-T* path for Big Sky metamorphism, starting with a high-pressure phase (M1) at >1.0 GPa and >700 °C, followed by decompression while maximum temperatures were achieved, then isobaric cooling, after which the metamorphic suites of the Tobacco Root Mountains were partially reequilibrated (M2) at lower pressures (~0.6–0.8 GPa and ~700 °C) (Cheney et al., 2004a, this volume, Chapter 6). Metamorphic monazite occurs both as inclusions within porphyroblastic minerals such as garnet and kyanite, and within the matrix of the metamorphic rocks (Cheney et al., 2004a, this volume, Chapter 6). These monazites range in age from 1.78 to 1.72 Ga and provide limits on the M1 to M2 metamorphic period of the Big Sky orogeny (Cheney et al., 2004b, this volume, Chapter 8).

M1 and M2 conditions are consistent with the general style of ductile folding observed in all three metamorphic suites, and with minerals such as hornblende and sillimanite after kyanite, which commonly define the mineral lineation that is the key symmetry axis of the Big Sky structure pattern. The zircon date of folding and boudinage falls within the M1 to M2 monazite age range. On this basis, we infer that the pattern of mesoscopic folds observed across the Tobacco Root Mountains developed synchronously with the maximum *P* (deep burial) and maximum *T* (thermal equilibration in response to burial) arc of the clockwise *P-T* path. This would have occurred during shortening, tectonic burial, and crustal thickening, presumably the result of accretion or collision along the northern Wyoming province during the Big Sky orogeny.

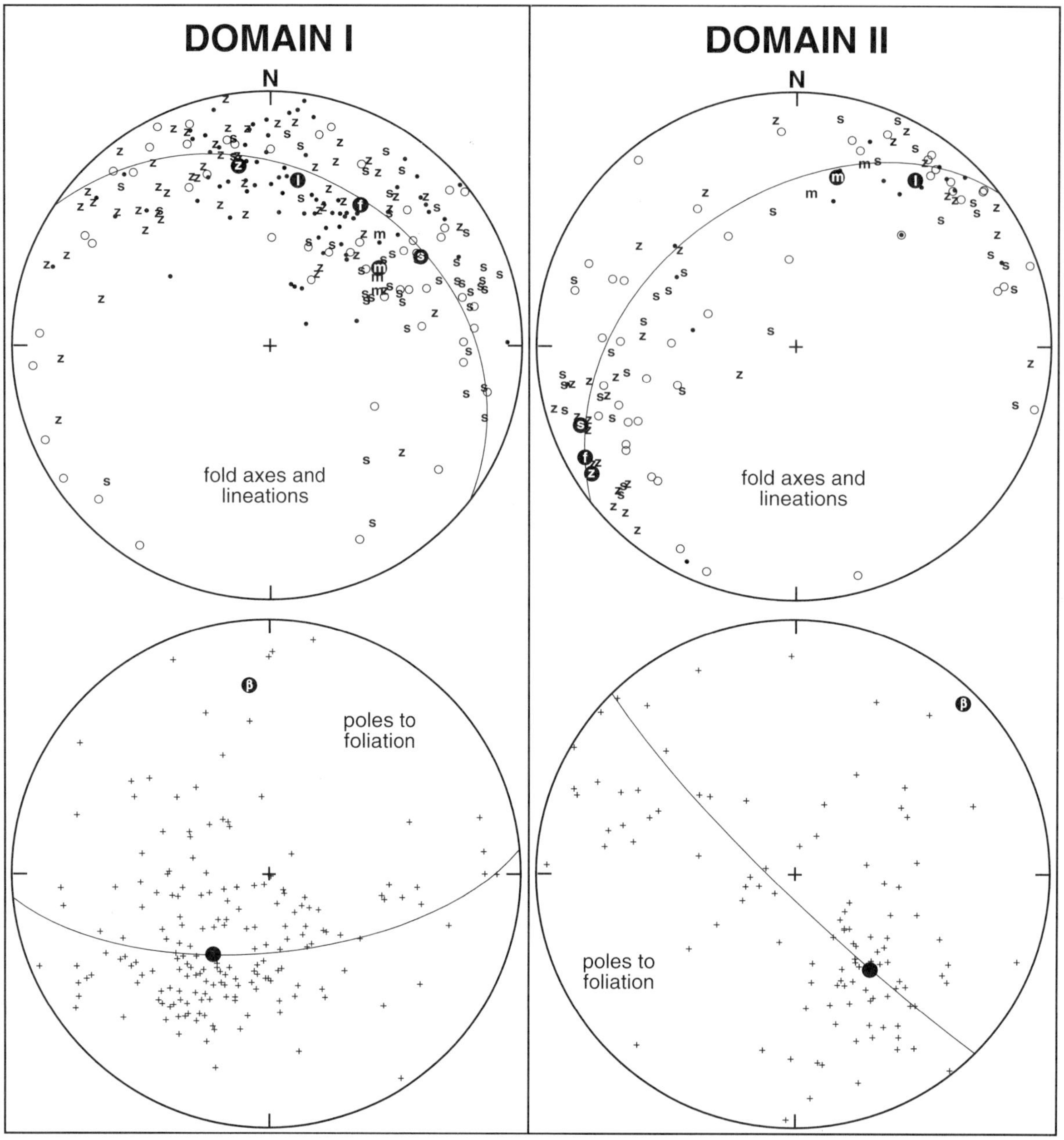

Figure 16. Kinematic analysis of structure data compiled from individual tracts (shown in Figure 12) into synoptic, equal-area, lower-hemisphere projections with consistent distributions of "S" and "Z" folds (as indicated by their calculated average orientation) relative to the calculated mean lineation orientation. Domain I includes data from tracts between Spuhler Saddle and Leggat Mountain, and from the Upper Mill Creek and Ramshorn tracts. Domain II comprises data from the Spuhler Gulch, Sunrise Lake, Sailor Lake, and Boulder Lakes tracts. Domain III incorporates data from the Lost Cabin Lake and Middle Mountain, Branham Lakes and Thompson Peak, and Middle Fork Mill Creek tracts. The geographic distribution of the three domains is shown in the accompanying schematic geologic map. See Figure 12 for the key to symbols used in the stereograms and Figure 1 for patterns used in the geologic map.

Penetrative ductile deformation that produced both fabric and folds, however, was most likely largely complete when delicate symplectite textures that help to define M2 conditions (Cheney et al., 2004a, this volume, Chapter 6) were growing. We did not identify structures that could be unequivocally associated with rapid decompression and quenching at the close of Big Sky metamorphism, which has been determined from metamorphic mineral reactions (Cheney et al., 2004a, this volume, Chapter 6) and $^{40}Ar/^{39}Ar$ cooling dates (Brady et al., 2004b, this volume, Chapter 7).

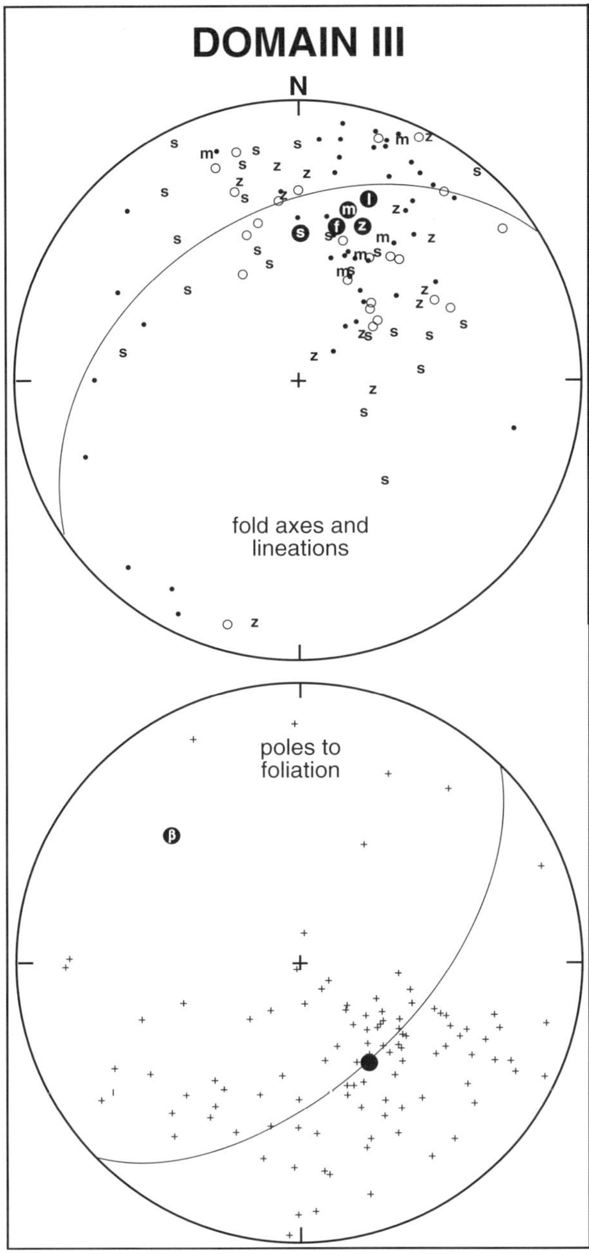

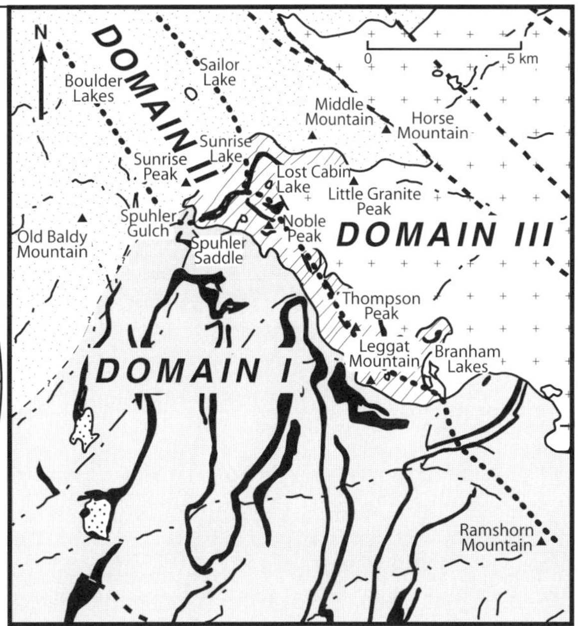

NUMBERS OF ORIENTATION DATA PLOTTED	lineations	S-folds	Z-folds	M-folds	symmetric folds	foliations
DOMAIN I	76	46	56	3	55	186
DOMAIN II	19	28	29	2	44	110
DOMAIN III	47	23	13	5	25	124

Figure 16 (*continued*).

MAP-SCALE FOLDS

Indian Creek Folds

Marble marker horizons outline a series of map-scale antiforms and synforms in the Indian Creek Metamorphic Suite (Burger, 1966; Vitaliano et al., 1979a) (Fig. 1), informally called the Indian Creek folds (Gillmeister, 1972). Burger (1966, 1967, 1969), who first mapped them, used a β plot of compositional layering and foliation in the region to determine a gently north-plunging axis (350°/24°) for the Indian Creek folds, which he used to project a down-plunge transect showing tight to isoclinal, nearly concentric fold profiles overturned to the east. Burger's map-scale fold axis parallels a well-clustered set of 450 mesoscopic fold axes and mineral lineations that he measured over the same area (Burger, 1966) and the mineral lineations documented in the present study.

Burger (1966), Gillmeister (1971), and Duncan (1978) have pointed out the apparent northward truncation of the trend of the Indian Creek folds by the Spuhler Peak Metamorphic Suite contact and have interpreted this as an indication that the folds predate juxtaposition of the Spuhler Peak and Indian Creek Metamorphic

Suites. Our fieldwork shows that compositional layering in the Indian Creek Metamorphic Suite is concordant with the Spuhler Peak Metamorphic Suite north of the folded marble and is not truncated. The orientation of the axis of the Indian Creek folds is compatible with the overall structural geometry of the Big Sky deformation. There is no simple geometric basis on which to determine if these folds are older than the juxtaposition of the Spuhler Peak Metamorphic Suite and were rotated to their present orientation during the Big Sky orogeny or if they originated during Big Sky deformation.

Geologic Relationships in Spuhler Gulch and the Spuhler Peak Metamorphic Suite Fold

In order to constrain relationships between the three major metamorphic units of the Tobacco Root Mountains, we traced out the contact of the Spuhler Peak Metamorphic Suite with the quartzofeldspathic gneisses of the Indian Creek and Pony–Middle Mountain Metamorphic Suites in the field—from Spuhler Saddle, through Spuhler Gulch, to Sunrise Lake (Fig. 17). The contact is readily distinguished by a thin (~1–2 m thick) but distinctive, fuchsitic quartzite in the Spuhler Peak Metamorphic Suite, considered a "basal" clastic unit by Gillmeister (1971, 1972).

We found this contact, and the compositional layering and foliation above and below it, to be continuous and concordant. South of Spuhler Gulch, the contact strikes east-northeast and dips moderately to the north-northwest. In the gulch, the contact shallows to nearly horizontal and, on the north side of the gulch, it steepens from gently south-dipping to the vertical and through the vertical so that north of the gulch the contact again has a moderate north-northwest dip with a northeast strike. The lithologic units and the contact trace out a continuous, tightly folded surface that closes down and to the north. We infer that the hinge of this fold becomes more tightly appressed west of Spuhler Gulch, where compositional layering in the Indian Creek and Pony–Middle Mountain Metamorphic Suites are regionally parallel, and suggest that it lies in a narrow zone of steep to vertical foliation orientations that continues from Spuhler Gulch, just south of Old Baldy Mountain, and along Nugget Creek (Fig. 1).

A β pole derived from orientations of compositional layering and foliation in the Spuhler Gulch area suggests a fold axis at 046°/15° (Fig. 17) for this large isocline (see also Lowell, 1994). The orientation of the Spuhler Gulch fold axis was also independently determined from the outcrop pattern of the Spuhler Peak Metamorphic Suite fuchsitic quartzite that marks the contact of the Spuhler Peak Metamorphic Suite with the quartzofeldspathic gneisses (Kranenburg, 1996). Carefully located sites on the fuchsitic quartzite were used to calculate three-point solutions for short planar segments analogous to tangents to the surface of the map-scale fold. A β plot of poles to these planes defines a girdle that gives a fold axis of 038°/30° (Fig. 17).

The Spuhler Peak Metamorphic Suite lies in the core of this map-scale fold, and quartzofeldspathic gneiss is continuous

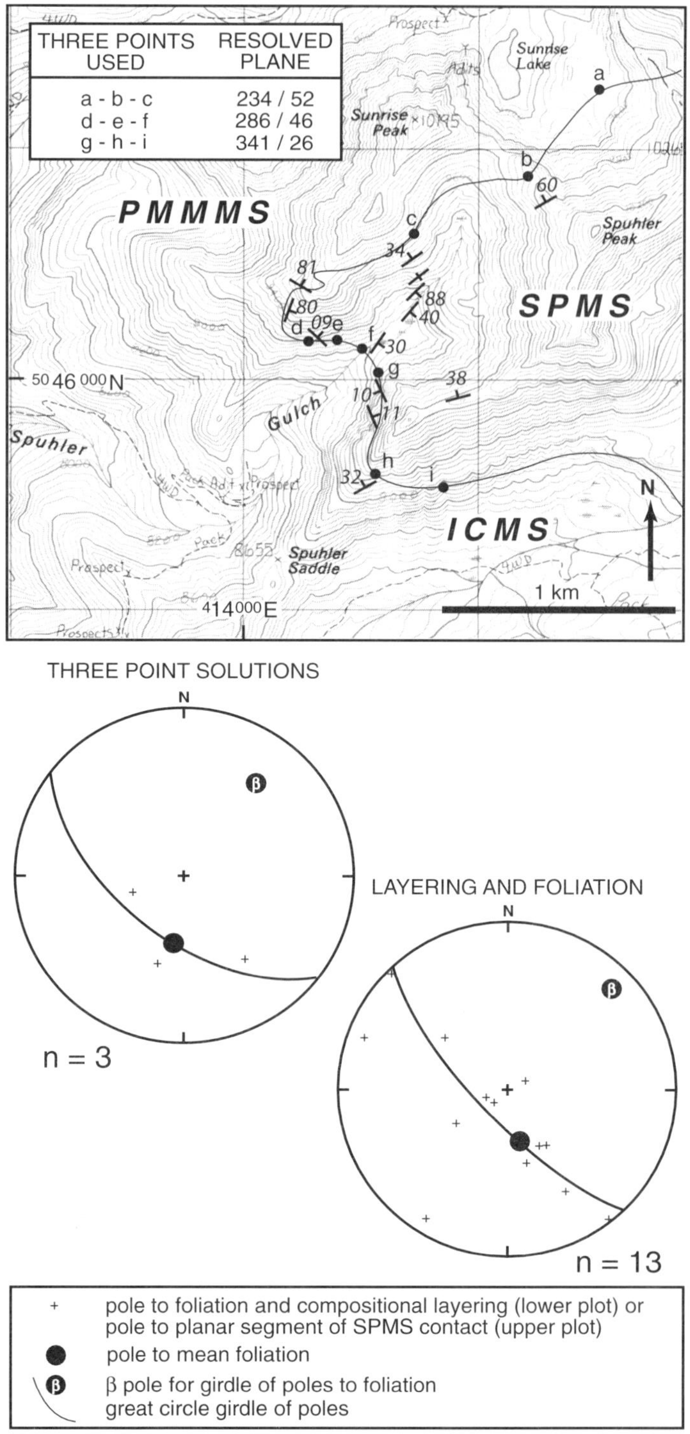

Figure 17. Geologic and structural relationships in Spuhler Gulch. Map shows (1) the trace of the contact of the Spuhler Peak Metamorphic Suite (SPMS) with the Pony–Middle Mountain (PMMMS) and Indian Creek (ICMS) Metamorphic Suites (thin, solid line); (2) representative orientations of compositional layering and mineral foliation measured along the contact; and (3) the location of nine points (a through i) on the contact used for three-point solutions of the attitude of the contact. A northeast-trending fold axis orientation for the isoclinal fold in the SPMS is determined from both the three-point solutions (upper stereogram) and foliation data (lower stereogram). Topographic base taken from the Noble Peak, Montana, 1: 24,000 quadrangle; contour interval 40 feet.

around the outside of the hinge of the fold. Compositional layers in the Indian Creek and Pony–Middle Mountain Metamorphic Suites trend into one another along strike, following the contact with the Spuhler Peak Metamorphic Suite. Thus, at least structurally, the Indian Creek and Pony–Middle Mountain Metamorphic Suites occupy laterally equivalent positions with respect to the Spuhler Peak Metamorphic Suite. If one were to "unfold" the Spuhler Peak Metamorphic Suite isocline, the Indian Creek Metamorphic Suite would neither underlie nor overlie the Pony–Middle Mountain Metamorphic Suite, and the contact between them would run across strike. This might imply that the distribution of marble, quartzite, and other metasupracrustal lithologies in the Indian Creek Metamorphic Suite relative to the Pony–Middle Mountain Metamorphic Suite reflects an original lateral facies transition (see also Vitaliano et al., 1979b; Mogk et al., 2004, this volume, Chapter 2). Conversely, if the Pony–Middle Mountain Metamorphic Suite contains a significant component of metaplutonic rocks (Krogh et al., 1997), the Indian Creek/Pony–Middle Mountain Metamorphic Suites boundary might mark a transposed pluton–host rock contact.

Spuhler Peak Metamorphic Suite Fold as a Map-Scale Sheath Fold

The question then arises as to the relationship between this very large-scale isoclinal fold of the Spuhler Peak Metamorphic Suite and the abundant mesoscopic folds that characterize the Big Sky orogeny. There is substantial obliquity (~35°) between the fold axis orientations determined for the Spuhler Peak Metamorphic Suite isocline in Spuhler Gulch and the synoptic regional β pole of 012°/27° (Fig. 2) or the mean regional lineation at 017°/32° (Fig. 4). Because the lineation is clustered regionally and is parallel on both limbs of the Spuhler Peak Metamorphic Suite isocline (Fig. 12; see data from Lost Cabin Lake versus Spuhler Saddle–Leggat Mountain, for example), it is apparent that the mesoscopic fabric has not been rotated by the Spuhler Peak Metamorphic Suite isoclinal fold, which cannot, therefore, postdate the Big Sky fabric. On the other hand, while the axis of the Indian Creek map-scale folds does parallel the regional lineation, the Spuhler Peak Metamorphic Suite isocline axis in Spuhler Gulch does not and appears uniquely unaffected by the Big Sky strain.

We suggest that the Spuhler Peak Metamorphic Suite isocline is contemporaneous with the Big Sky deformation and that the hinge exposed in Spuhler Gulch is one side of a very large-scale sheath fold of the Spuhler Peak Metamorphic Suite, now partly obliterated by the Tobacco Root batholith. A contemporaneous sheath fold would neither fold Big Sky lineations nor be rotated fully into parallelism with the lineations by that deformation. Map-scale sheath folds of nearly this magnitude have been recognized in other high-strain and high-grade orogens over the past few decades (Henderson, 1981; Lacassin and Mattauer, 1985; Boyle, 1987; Vollmer, 1988; Goscombe, 1991; and Alsop and Holdsworth, 1999).

The following observations support our interpretation of the Spuhler Peak Metamorphic Suite isocline as a regional-scale sheath fold. First, in fact, is the angle between the map-scale fold axis and the regional mean mineral lineation, which is as expected for a highly noncylindrical but not fully tubular fold where the regional lineation parallels the long axis of the sheath fold and the Spuhler Peak Metamorphic Suite isocline is one side of the parabolic hinge of the fold. The obliquity between the mean lineation and the fold axis of the Spuhler Peak Metamorphic Suite isocline gives a hinge line angle of ~70° for the sheath fold, and the direction from the fold axis to the mean lineation indicates that the sheath fold closes downward and to the north (Fig. 18).

Second, regional outcrop distributions and foliation attitudes suggest that the Spuhler Peak Metamorphic Suite may have had a fully closed "eyeball" exposure pattern prior to intrusion of the Tobacco Root batholith. In the area of Branham Lakes and the headwaters of Mill Creek, just west of the Tobacco Root batholith, compositional layers mapped in both the Spuhler Peak and the Indian Creek Metamorphic Suites wrap around from a northwest strike to east and northeast (Fig. 1). This hints that another isoclinal fold of the Spuhler Peak Metamorphic Suite may have existed under the Branham Lakes, in this case with a northwest-plunging hinge that closed to the east. Such a hinge would fold the contact of the Spuhler Peak Metamorphic Suite with the quartzofeldspathic gneisses back to the northwest so that it may once have been continuous from the headwaters of Mill Creek to the area of Little Granite Peak. Had parts of it not been consumed by the Tobacco Root batholith, the Spuhler Peak Metamorphic Suite would thus lie in an elliptical body with its long axis trending northwest from Branham Lakes to Spuhler Gulch and with cross-sectional dimensions of ~10 × 4 km.

Finally, the kinematic analysis of mesoscopic folds in the Spuhler Peak Metamorphic Suite and quartzofeldspathic gneiss units is consistent with contemporaneous large-scale sheath folding. Alsop (1994) and Alsop and Holdsworth (1999) reported reversals in mesoscopic fold asymmetry from one limb to the other of km-scale sheath folds in Great Britain. In the Tobacco Root Mountains, however, S and Z folds are equally numerous both north and south of Spuhler Gulch. It is, instead, the sense of shear recorded by the mesoscopic folds that undergoes a reversal between the two limbs. Kinematic Domain III, where a "top-up to the south" sense of shear accompanied the Big Sky deformation, corresponds to the structurally higher limb of the Spuhler Peak Metamorphic Suite sheath fold. Domain I, with a "top-down to the north" sense of shear, corresponds to the structurally lower, better-preserved limb of the sheath fold. Domain II lies in Spuhler Gulch, along the hinge line of the regional-scale sheath fold. There, both S and Z folds have been rotated into parallelism and lie closer to the east-northeast trend of the Spuhler Gulch fold axis (compare Figs. 16 and 17) than to the north trend of the regional mean lineation. The pattern of top-up on the upper limb and top-down on the lower limb is opposite to the expected slip and parasitic fold distribution associated with a conventional, cylindrical, map-scale synform. We propose that the observed

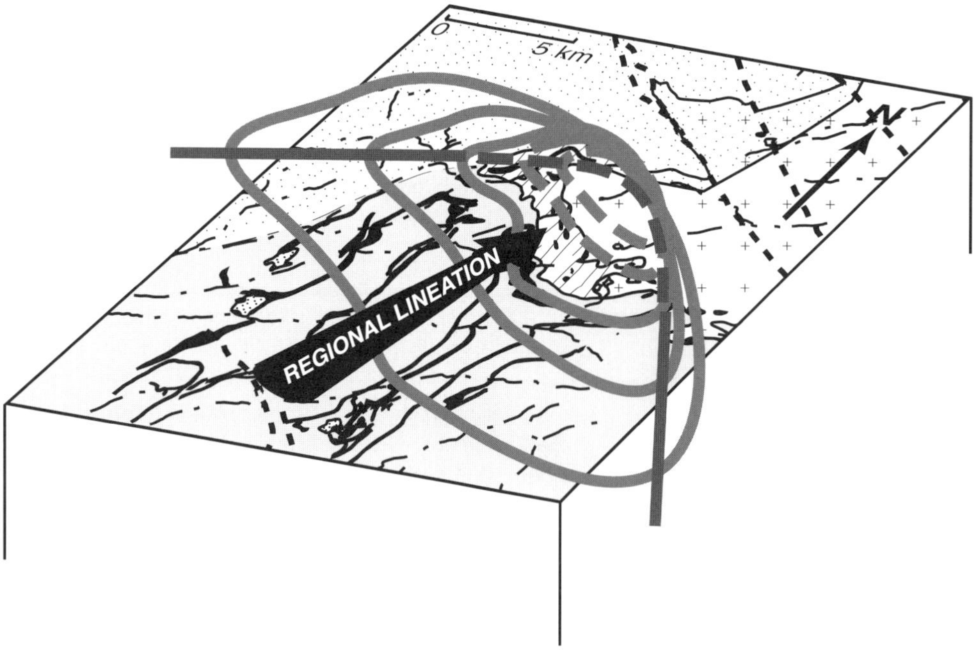

Figure 18. Near down-plunge, perspective view of the map-scale sheath fold superimposed on the geologic map of the area. Parabolic heavy dark grey line represents the hinge line of the fold. It coincides with the hinge of the Spuhler Peak Metamorphic Suite isocline in Spuhler Gulch in the northwest and with the hypothetical fold in the headwaters of Mill Creek in the east. Closed heavy gray lines are serial profiles of the surface of the sheath fold—solid above the present ground surface and dashed below the ground surface. The smallest solid-line profile outlines the possible "eyeball" map pattern of the Spuhler Peak Metamorphic Suite.

kinematic domains record the downward and northward extrusion of a Spuhler Peak Metamorphic Suite tongue relative to the enveloping quartzofeldspathic gneisses during development of the regional-scale sheath fold.

The Indian Creek folds can be integrated into this model and, in doing so, the apparent discontinuity between the northerly trend of the folds and the northwest-striking Spuhler Peak Metamorphic Suite contact versus the conformity of compositional layering in the Indian Creek and Spuhler Peak Metamorphic Suites across that contact can be successfully reconciled. We suggest that the Indian Creek folds may be analogous to "curtain folds" that have been recognized around the rim of salt domes and in modeled diapiric structures (Ramberg, 1981; Talbot and Jackson, 1987; Jackson et al., 1990). Curtain folds result from constrictional flow as material rises from the source bed into the stem of a diapir (Talbot and Jackson, 1987). They are oriented with axes parallel to the flow direction and axial planes at a high angle to the margin of the dome (Fig. 19). The many stunning aerial images of curtain folds surrounding the massive salt domes of the Great Kavir given by Jackson et al. (1990) bear a striking geometric similarity to the map pattern of the Indian Creek folds relative to the Spuhler Peak Metamorphic Suite body, respectively. On this basis, we suggest the Indian Creek folds formed during development of the very large-scale sheath fold as the Spuhler Peak Metamorphic Suite underwent accelerating flow north and downward (in present coordinates) relative to the surrounding quartzofeldspathic gneisses.

The vergence pattern of mesoscopic folds analyzed in this project relates to the development of the Spuhler Peak Metamorphic Suite sheath fold; it differs between the upper and lower limbs of the sheath fold and does not provide a consistent sense of displacement of the Spuhler Peak Metamorphic Suite relative to the Indian Creek and Pony–Middle Mountain Metamorphic Suites. The Spuhler Peak Metamorphic Suite is preserved only in the hinge of the sheath fold for which there is no clear root zone. Rocks that may be equivalent to the Spuhler Peak Metamorphic

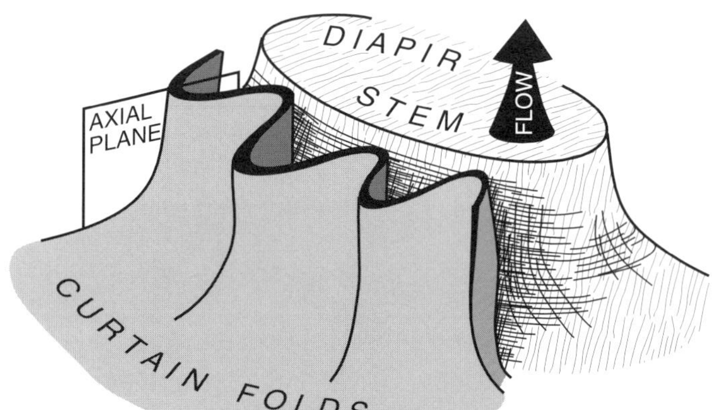

Figure 19. Cut-away view of the relationship of flanking curtain folds to the stem of a diapir. Adapted from Talbot and Jackson (1987). Axes of the folds parallel the flow direction in the diapir, and axial planes of the folds are perpendicular to the rim of the diapir. The top, horizontal cut corresponds to an up-plunge, or worm's-eye, view of the Tobacco Root Mountains map pattern around the Indian Creek folds in the Indian Creek Metamorphic Suite and the elliptical, sheath folded Spuhler Peak Metamorphic Suite, insofar as the flow of the Spuhler Peak Metamorphic Suite in the sheath fold was down relative to the surrounding quartzofeldspathic gneisses.

Suite have been recognized at Cataract Mountain to the north of the map scale sheath fold (Hanley, 1975) and mapped to the southeast of the sheath fold, southeast of Branham Lakes by Vitaliano et al. (1979a). Both of these occurrences are very thin (~200 m according to Hanley, 1975) and have an along-strike continuity of less than 5 km; neither area seems sufficient to accommodate the entire Spuhler Peak Metamorphic Suite rooting into the crust. Consequently, there is no way to determine whether the Spuhler Peak Metamorphic Suite sheath fold is part of a regionally south- or north-vergent system (Fig. 20).

Gentle Folds

Like many others before us (Reid, 1963; Gillmeister, 1971; Cordua, 1973; and Hanley, 1975), we observed one other set of folds in all three metamorphic suites of the Tobacco Root Mountains. These are broad, gentle-to-open folds that produce undulations in metamorphic foliation and compositional layering. At the largest scale, this folding is expressed in two map-scale, north-trending folds (Fig. 1) that warp the regional foliation, the axes of the southeasternmost Indian Creek folds, and the axial plane of the Spuhler Peak Metamorphic Suite sheath fold: the Noble Peak antiform in the north and the Mill Gulch antiform in the south (Vitaliano et al., 1979a). Gentle folds are also seen locally at the outcrop scale, where they warp the axial planes of isoclinal folds as well as compositional layering (Fig. 21).

Vitaliano et al. (1979b) interpreted these folds as different from and younger than the Indian Creek folds, but nevertheless considered them Precambrian in age. As the outcrop-scale gentle folds were largely ignored during this project we cannot shed

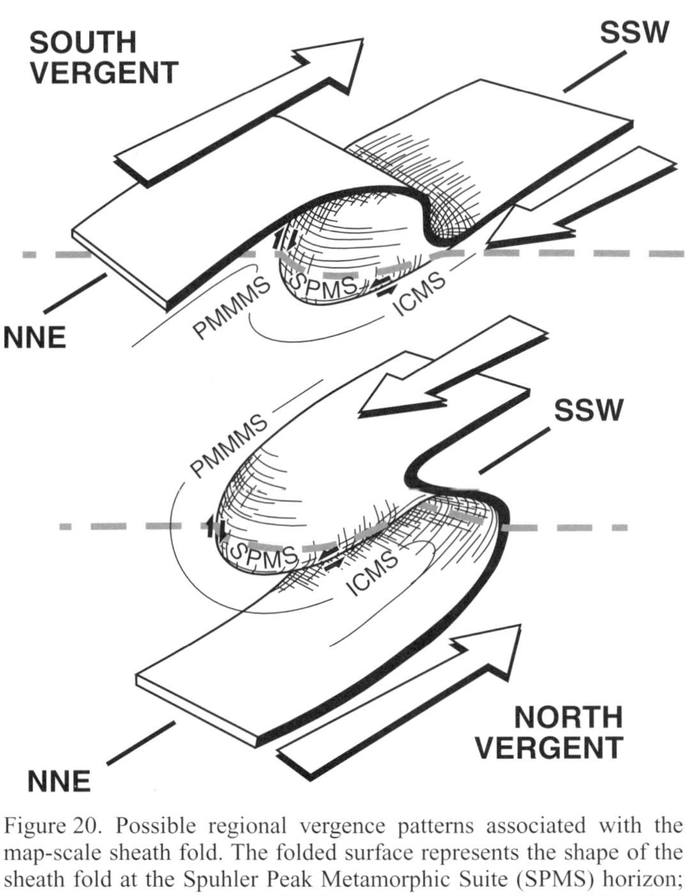

Figure 20. Possible regional vergence patterns associated with the map-scale sheath fold. The folded surface represents the shape of the sheath fold at the Spuhler Peak Metamorphic Suite (SPMS) horizon; the gray dashed line is placed at the present level of the ground surface with respect to the SPMS sheath fold. Small black arrows represent the sense of shear of the SPMS relative to the surrounding Pony–Middle Mountain and Indian Creek gneisses; large white arrows show the regional sense of shear. Both south-vergent (in which the SPMS structurally overlies the Pony–Middle Mountain Metamorphic Suite [PMMMS]–Indian Creek Metamorphic Suite [ICMS] prior to folding) and north-vergent (in which the SPMS underlies the PMMMS-ICMS) scenarios are permitted by the structural data presented in this study.

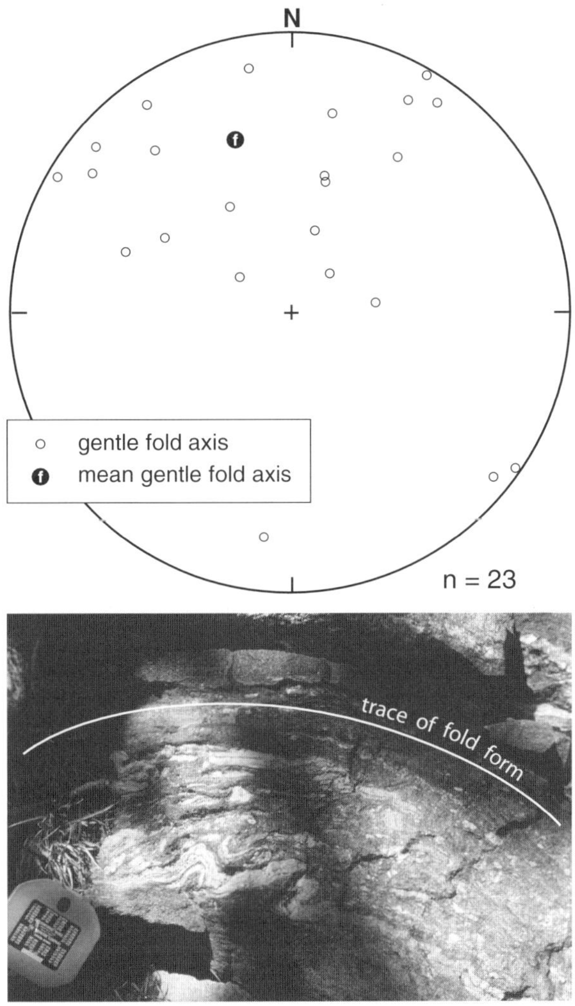

Figure 21. Mesoscopic gentle fold set locally observed in all metamorphic suites of the Tobacco Root Mountains. Upper stereogram of gentle fold axis orientations yields a mean fold axis at 342°/35°. Lower photo shows a representative gentle fold in Spuhler Peak Metamorphic Suite biotite schist.

much light on their origin. Although variable in orientation, the gentle fold axes trend in the northern quadrants (Fig. 21), like the mean orientation of all other outcrop-scale folds in the Tobacco Root Mountains. Consequently, it is geometrically equivocal whether the gentle folds are younger than the Big Sky orogeny or if they simply formed at the closing stages of that orogeny and did not suffer as much tightening or rotation.

Because of the very much larger scale of the Noble Peak and Mill Gulch antiforms in comparison to the Indian Creek folds and the Spuhler Peak Metamorphic Suite sheath fold, and because of their considerably different interlimb angle, we favor a distinctly younger age for the two antiforms and recommend that serious consideration be given to a Laramide origin for them. This is in light of recent recognition that Laramide ranges of southern Montana are carried in basement-involving thrust sheets (Schmidt et al., 1993; Kellogg et al., 1995). West-dipping, east-vergent Laramide thrust faults place Precambrian crystalline rocks over Paleozoic and Mesozoic strata on the westernmost flank of the Tobacco Root Mountains (Hanley, 1975; Vitaliano et al., 1979a; Schmidt and Garihan, 1983) and the westernmost Madison Range (Kellogg et al., 1995) immediately east of the Tobacco Root Mountains (see Figure 1 in Burger, 2004, this volume, Chapter 1). The thrust fault in the western Madison Range rotates foliation in basement rocks as part of a fault bend fold and may project westward under the Tobacco Root Mountains (Kellogg et al., 1995). We believe this presents the distinct possibility that the Noble Peak and Mill Gulch antiforms were produced by ramps associated with a Laramide thrust fault underlying the range.

SYNTHESIS OF FINDINGS

The dominant mesoscopic structural features in Precambrian metamorphic rocks of the Tobacco Root Mountains are compositional layering, mineral-alignment foliation and lineation, and folds—including cylindrical, noncylindrical, and sheath forms. Crosscutting relationships show that gneissic layering developed and was folded prior to emplacement of the MMDS in the Indian Creek and Pony–Middle Mountain Metamorphic Suites. Nevertheless, the present orientations of folds and of planar and linear fabrics throughout the Precambrian core of the range are the result of a post-MMDS deformation that so significantly reworked the Precambrian rocks that all structural elements are compatible with a single strain ellipsoid. Based on the age of zircon in critically located leucosome (Mueller et al., 2004, this volume, Chapter 9), and of monazite in lineated and foliated porphyroblastic metamorphic minerals and their matrix (Cheney et al., 2004a, this volume, Chapter 6), the deformation is interpreted to have occurred between 1.78 and 1.72 Ga as part of the Big Sky orogeny. Despite what may be a much longer history (Harms et al., 2004, this volume, Chapter 11; Mueller et al., 2004, this volume, Chapter 9), the present structural and metamorphic character of Precambrian metamorphic rocks of the Tobacco Root Mountains should be recognized as predominately Proterozoic rather than Archean.

Deformation in the Tobacco Root Mountains constrains the nature of the Big Sky orogeny and, in turn, supports a more robust model for the late history of the northern Wyoming province.

- Big Sky deformation was associated with a significant component of noncoaxial strain. On the mesoscopic scale, this is expressed by high-strain zones, the rotation of metamorphosed dikes to sill-like configurations, and by the widespread occurrence of sheath folds and the planar dispersion of asymmetric cylindrical folds. There is some evidence that, at least locally, the metamorphic rocks were once mylonitized but microfabrics have subsequently been annealed—probably in the late stages of the orogeny, for example, when delicate symplectite textures grew (Cheney et al., 2004a, this volume, Chapter 6). On the map scale, the sheath fold cored by the Spuhler Peak Metamorphic Suite and associated curtain folds in the surrounding quartzofeldspathic gneisses reflect both the intensity and regional dimension of the shear strain.
- Displacement of the Spuhler Peak Metamorphic Suite relative to the quartzofeldspathic gneiss suites during this shear strain is the most likely context for development of the map-scale sheath fold in their contact. This favors a tectonic rather than depositional origin for the juxtaposition of the Spuhler Peak Metamorphic Suite with the Indian Creek and Pony–Middle Mountain Metamorphic Suites.
- Kinematic analysis demonstrates that the Spuhler Peak Metamorphic Suite moved north relative to the surrounding quartzofeldspathic gneisses as the map-scale sheath fold developed, but whether regional transport and associated plate convergence during sheath folding was north- or south-vergent remains indeterminate. But, since Big Sky folding took place at the same time as maximum pressures and temperatures of metamorphism, these structures must have been associated with compressional deformation. Updip displacement, in Proterozoic coordinates, would be expected. Whatever the regional vergence, it is likely that the north-south slip line of simple shear in the Tobacco Root Mountains, defined by the average mineral lineation, identifies the trend of the Big Sky orogen as generally east-west.
- Recognition that the Spuhler Peak Metamorphic Suite cores a map-scale sheath fold places new and different constraints on the mutual relationships between the Indian Creek, Pony–Middle Mountain, and Spuhler Peak Metamorphic Suites. If the Spuhler Peak Metamorphic Suite sheath fold is part of a regionally south-vergent system, "unfolding" the sheath fold restores the Spuhler Peak Metamorphic Suite over the quartzofeldspathic gneisses; if it is a north-vergent system, the Spuhler Peak Metamorphic Suite was under the Indian Creek and Pony–Middle Mountain Metamorphic Suites prior to folding (Fig. 20). This opens the possibility that the quartzofeldspathic gneiss suites were thrust over the Spuhler Peak Metamorphic Suite during the Big Sky orogeny, as well as the more conventionally assumed model in which the suite was

obducted onto the quartzofeldspathic gneisses. Resolving this uncertainty will constrain the geometry of accretion of allochthonous material to the northern margin of the Wyoming province. In either case, "unfolding" the sheath fold places the Indian Creek and Pony–Middle Mountain Metamorphic Suites as lateral equivalents relative to the Spuhler Peak Metamorphic Suite as a datum. This should be a factor in models for the nature of the protoliths of the Indian Creek and Pony–Middle Mountain Metamorphic Suites and their paleotectonic setting.

- The emplacement and sheath folding of the Spuhler Peak Metamorphic Suite was one facet of a larger, regional tectonic stacking during the development of the Big Sky orogen, since the Indian Creek Metamorphic Suite, the Pony–Middle Mountain Metamorphic Suite, the MMDS, and the Spuhler Peak Metamorphic Suite were all under ~40 km of tectonic overburden (Cheney et al., 2004a, this volume, Chapter 6) during metamorphism and simple shear deformation. The metamorphic suites of the Tobacco Root Mountains represent the once deeply buried, ductile core of an orogen of impressive thickness resulting from significant convergence and/or accretion and, presumably, of greater lateral extent.

ACKNOWLEDGMENTS

This work was funded by the W.M. Keck Foundation through the Keck Geology Consortium and by the Trustees of Amherst and Smith Colleges. We thank the residents of Twin Bridges, Montana, for their hospitality and the staff of the U.S. Forest Service in Sheridan, Montana, for their cooperation. The manuscript was greatly improved by the careful review of Robert Bauer, Ernie Duebendorfer, and Mike Williams, to whom we are grateful.

REFERENCES CITED

Alsop, G.I., 1994, Relationships between distributed and localized shear in the tectonic evolution of a Caledonian fold and thrust zone, northwest Ireland: Geological Magazine, v. 131, p. 123–136.

Alsop, G.I., and Holdsworth, R.E., 1999, Vergence and facing patterns in large-scale sheath folds: Journal of Structural Geology, v. 21, p. 1335–1349.

Bell, T.H., 1978, Progressive deformation and reorientation of fold axes in a ductile mylonite zone: The Woodroffe Thrust: Tectonophysics, v. 44, p. 285–320.

Berthé, D., and Brun, J.P., 1980, Evolution of folds during progressive shear in the South Armorican Shear Zone, France: Journal of Structural Geology, v. 2, p. 127–133.

Borradaile, G.J., 1972, Variably oriented co-planar primary folds: Geological Magazine, v. 109, p. 89–98.

Boyle, A.P., 1987, A Model for stratigraphic and metamorphic inversions at Sulitjelma, central Scandes: Geological Magazine, v. 124, p. 451–466.

Brady, J.B., Mohlman, H.K., Harris, C., Carmichael, S.K., Jacob, L.K., and Chaparro, W.R., 2004a, General geology and geochemistry of metamorphosed Proterozoic mafic dikes and sills, Tobacco Root Mountains, Montana, *in* Brady, J.B., et al., eds., Precambrian geology of the Tobacco Root Mountains, Montana: Boulder, Colorado, Geological Society of America Special Paper 377, p. 89–104 (this volume).

Brady, J.B., Kovaric, D.N., Cheney, J.T., Jacob, L.J., and King, J.T., 2004b, $^{40}Ar/^{39}Ar$ ages of metamorphic rocks from the Tobacco Root Mountains region, Montana, *in* Brady, J.B., et al., eds., Precambrian geology of the Tobacco Root Mountains, Montana: Boulder, Colorado, Geological Society of America Special Paper 377, p. 131–149 (this volume).

Burger, H.R., III, 1966, Structure, petrology, and economic geology of the Sheridan District, Madison County, Montana [Ph.D. thesis]: Bloomington, Indiana University, 156 p.

Burger, H.R., III, 1967, Bedrock geology of the Sheridan District, Madison County, Montana: Montana Bureau of Mines and Geology Memoir 41, 22 p.

Burger, H.R., III, 1969, Structural evolution of the southwestern Tobacco Root Mountains, Montana: Geological Society of America Bulletin, v. 80, p. 1329–1342.

Burger, H.R., 2004, General geology and tectonic setting of the Tobacco Root Mountains, Montana, *in* Brady, J.B., et al., eds., Precambrian geology of the Tobacco Root Mountains, Montana: Boulder, Colorado, Geological Society of America Special Paper 377, p. 1–14 (this volume).

Burger, H.R., Peck, W.H., Johnson, K.E., Tierney, K.A., Poulsen, C.J., Cady, P., Lowell, J., MacFarlane, W.A., Sincock, M.J., Archuleta, L.L., Pufall, A., and Cox, M.J., 2004, Geology and geochemistry of the Spuhler Peak Metamorphic Suite, *in* Brady, J.B., et al., eds., Precambrian geology of the Tobacco Root Mountains, Montana: Boulder, Colorado, Geological Society of America Special Paper 377, p. 47–70 (this volume).

Carreras, J., Estrada, A., and White, S., 1977, The effects of folding on the c-axis fabrics of a quartz mylonite: Tectonophysics, v. 39, p. 3–24.

Cheney, J.T., Brady, J.B., Tierney, K.A., DeGraff, K.A., Mohlman, H.K., Frisch, J.D., Hatch, C.E., Steiner, M.L., Carmichael, S.K., Fisher, R.G.M., Tuit, C.B., Steffen, K.J., Cady, P., Lowell, J., Archuleta, L., Hirst, J., Wegmann, K.W., and Monteleone, B., 2004a, Proterozoic metamorphism of the Tobacco Root Mountains, Montana, *in* Brady, J.B., et al., eds., Precambrian geology of the Tobacco Root Mountains, Montana: Boulder, Colorado, Geological Society of America Special Paper 377, p. 105–129 (this volume).

Cheney, J.T., Webb, A.A.G., Coath, C.D., and McKeegan, K.D., 2004b, In situ ion microprobe $^{207}Pb/^{206}Pb$ dating of monazite from Precambrian metamorphic suites, Tobacco Root Mountains, Montana, *in* Brady, J.B., et al., eds., Precambrian geology of the Tobacco Root Mountains, Montana: Boulder, Colorado, Geological Society of America Special Paper 377, p. 151–179 (this volume).

Cobbold, P.R., and Quinquis, H., 1980, Development of sheath folds in shear regimes: Journal of Structural Geology, v. 2, p. 119–126.

Cooper, J.M., 2002, Quartz c-axis studies from a sheath fold in metamorphosed Precambrian rocks, Tobacco Root Mountains, Montana [A.B. thesis]: Amherst, Massachusetts, Amherst College, 55 p.

Cordua, W.S., 1973, Precambrian geology of the southern Tobacco Root Mountains, Madison County, Montana [Ph.D. thesis]: Bloomington, Indiana University, 258 p.

Duncan, D.W., 1978, Structural analysis of the central Tobacco Root Mountains, southwest Montana [M.S. thesis]: University Park, Pennsylvania State University, 79 p.

Escher, A., and Watterson, J., 1974, Stretching fabrics, folds and crustal shortening: Tectonophysics, v. 22, p. 223–231.

Ghosh, S.K., and Sengupta, S., 1984, Successive development of plane noncylindrical folds in progressive deformation: Journal of Structural Geology, v. 6, p. 703–709.

Gillmeister, N.M., 1971, Petrology of Precambrian rocks in the central Tobacco Root Mountains, Madison County, Montana [Ph.D. thesis]: Cambridge, Massachusetts, Harvard University, 201 p.

Gillmeister, N.M., 1972, Cherry Creek Group–Pony Group relationship in the central Tobacco Root Mountains, Madison County, Montana: Northwest Geology, v. 1, p. 21–24.

Goscombe, B., 1991, Intense noncoaxial shear and the development of mega-scale sheath folds in the Arunta Block, Central Australia: Journal of Structural Geology, v. 13, p. 299–318.

Hanley, T.B., 1975, Structure and petrology of the northwestern Tobacco Root Mountains, Madison County, Montana [Ph.D. thesis]: Bloomington, Indiana University, 289 p.

Hansen, E., 1971, Strain facies: New York, Springer Verlag, 207 p.

Harms, T.A., Brady, J.B., Burger, H.R., and Cheney, J.T., 2004, Advances in the geology of the Tobacco Root Mountains, Montana, and their implications for the history of the northern Wyoming province, *in* Brady, J.B., et al., eds., Precambrian geology of the Tobacco Root Mountains, Montana: Boulder, Colorado, Geological Society of America Special Paper 377, p. 227–243 (this volume).

Henderson, J.R., 1981, Structural analysis of sheath folds with horizontal X-axes, northeast Canada: Journal of Structural Geology, v. 3, p. 203–210.

Hess, D.F., 1967, Geology of pre-Beltian rocks in the central and southern Tobacco Root Mountains with reference to superposed effects of the Laramide-age Tobacco Root Batholith [Ph.D. thesis]: Bloomington, Indiana University, 325 p.

Jackson, M.P.A., Cornelius, R.R., Craig, C.H., Gansser, A., Stocklin, J., and Talbot, C.J., 1990, Salt diapirs of the Great Kavir, Central Iran: Geological Society of America Memoir 177, 139 p.

Kellogg, K.S., Schmidt, C.J., and Young, S.W., 1995, Basement and cover-rock deformation during Laramide contraction in the northern Madison Range (Montana) and its influence on Cenozoic basin formation: American Association of Petroleum Geologists Bulletin, v. 79, p. 1117–1137.

Kranenburg, S.R., 1996, Structural relationships among metamorphic suites in the Tobacco Root Mountains of southwestern Montana [B.S. thesis]: Huntsville, Texas, Sam Houston State University, 41 p.

Krogh, T.E., Kamo, S., and Hess, D.F., 1997, Wyoming province 3300+ Ma gneiss with 2400 Ma metamorphism, northwestern Tobacco Root Mountains, Madison County, Montana: Geological Society of America Abstracts with Programs, v. 29, no. 6, p. A-408.

Lacassin, R., and Mattauer, M., 1985, Kilometre-scale sheath fold at Mattmark and implications for transport direction in the Alps: Nature, v. 315, p. 739–742.

Lin, S., and Williams, P.F., 1992, The geometrical relationship between the stretching lineation and the movement direction of shear zones: Journal of Structural Geology, v. 14, p. 491–497.

Lowell, J., 1994, A metamorphic and deformational history of the Spuhler Peak Formation, Tobacco Root Mountains, Montana [B.A. thesis]: Colorado Springs, Colorado College, 100 p.

Malavieille, J., 1987, Extensional shearing deformation and kilometer-scale "a"-type folds in a Cordilleran metamorphic core complex (Raft River Mountains, northwestern Utah): Tectonics, v. 6, p. 423–448.

Mies, J.W., 1991, Planar dispersion of folds in ductile shear zones and kinematic interpretation of fold-hinge girdles: Journal of Structural Geology, v. 13, p. 281–297.

Minnigh, L.D., 1979, Structural analysis of sheath-folds in a meta-chert from the Western Italian Alps: Journal of Structural Geology, v. 1, p. 275–282.

Mogk, D.W., Burger, H.R., Mueller, P.A., Wooden, J.L., D'Arcy, K., Heatherington, A.L., Abeyta, R.L., Martin, J., and Jacob, L.J., 2004, Geochemistry of quartzofeldspathic gneisses and metamorphic mafic rocks of the Indian Creek and Pony–Middle Mountain Metamorphic Suites, Tobacco Root Mountains, Montana, *in* Brady, J.B., et al., eds., Precambrian geology of the Tobacco Root Mountains, Montana: Boulder, Colorado, Geological Society of America Special Paper 377, p. 15–46 (this volume).

Mueller, P.A., Burger, H.R., Wooden, J.L., Heatherington, A.L., Mogk, D.W., and D'Arcy, K., 2004, Age and evolution of the Precambrian crust of the Tobacco Root Mountains, Montana, *in* Brady, J.B., et al., eds., Precambrian geology of the Tobacco Root Mountains, Montana: Boulder, Colorado, Geological Society of America Special Paper 377, p. 181–202 (this volume).

Owen, D.R., 1996, Archean deformation of the Tobacco Root Mountains, southwestern Montana [A.B. thesis]: Amherst, Massachusetts, Amherst College, 70 p.

Pardee, J.T., 1950, Late Cenozoic block faulting in western Montana: Geological Society of America Bulletin, v. 61, p. 359–406.

Ramberg, H., 1981, Gravity, deformation, and the Earth's crust in theory, experiments and geological application: New York, Academic Press, 452 p.

Reid, R.R., 1963, Metamorphic rocks of the northern Tobacco Root Mountains, Madison County, Montana: Geological Society of America Bulletin, v. 74, p. 293–301.

Root, F.K., 1965, Structure, petrology and mineralogy of pre-Beltian metamorphic rocks of the Pony-Sappington area, Madison County, Montana [Ph.D. thesis]: Bloomington, Indiana University, 184 p.

Schmidt, C.J., and Garihan, J.M., 1983, Laramide tectonic development of the Rocky Mountain foreland of southwestern Montana, *in* Lowell, J.D., and Gries, R., eds., Rocky Mountain foreland basins and uplifts: Denver, Rocky Mountain Association of Geologists, p. 271–294.

Schmidt, C.J., Chase, R.B., and Erslev, E.A., editors, 1993, Laramide basement deformation in the Rocky Mountain foreland of the western United States: Geological Society of America Special Paper 280, 365 p.

Skjernaa, L., 1989, Tubular folds and sheath folds: Definitions and conceptual models for their development, with examples from the Grapesvare area, northern Sweden: Journal of Structural Geology, v. 11, p. 689–703.

Talbot, C.J., and Jackson, M.P.A., 1987, Internal kinematics of salt diapirs: American Association of Petroleum Geologists Bulletin, v. 71, p. 1068–1093.

Tansley, W., Schafer, P.A., and Hart, L.H., 1933, A geological reconnaissance of the Tobacco Root Mountains, Madison County, Montana: Montana Bureau of Mines and Geology Memoir 9, 55 p.

Vitaliano, C.J., Burger, H.R., III, Cordua, W.S., Hanley, T.B., Hess, D.F., and Root, F.K., 1979a, Geologic map of southern Tobacco Root Mountains, Madison County, Montana: Geological Society of America Map and Chart Series MC31, scale 1:62,500, 1 sheet, 8 p. text.

Vitaliano, C.J., Burger, H.R., III, Cordua, W.S., Hanley, T.B., Hess, D.F., and Root, F.K., 1979b, Explanatory text to accompany geologic map of southern Tobacco Root Mountains, Madison County, Montana: Geological Society of America Map and Chart Series MC31, scale 1:62,500, 1 sheet, 8 p. text.

Vollmer, F.W., 1988, A computer model of sheath-nappes formed during crustal shear in the Western Gneiss Region, central Norwegian Caledonides: Journal of Structural Geology, v. 10, p. 735–743.

Manuscript Accepted by the Society September 12, 2003

Printed in the USA

Geological Society of America
Special Paper 377
2004

Advances in the geology of the Tobacco Root Mountains, Montana, and their implications for the history of the northern Wyoming province

Tekla A. Harms
Department of Geology, Amherst College, Amherst, Massachusetts 01002, USA

John B. Brady
H. Robert Burger
Department of Geology, Smith College, Northampton, Massachusetts 01063, USA

John T. Cheney
Department of Geology, Amherst College, Amherst, Massachusetts 01002, USA

ABSTRACT

Integrated studies by Keck Geology Consortium participants have generated many new insights into the Precambrian geology of the Tobacco Root Mountains. We have clarified the tectonic setting and origin of two suites of metamorphic rocks: (1) a quartzofeldspathic gneiss complex with associated metasupracrustal rocks (the combined Indian Creek and Pony–Middle Mountain Metamorphic Suites) that originated in a continental arc setting between 3.35 and 3.2 Ga with subsequent sedimentation and (2) mafic metavolcanic rocks with intercalated metasedimentary rocks (the Spuhler Peak Metamorphic Suite) from a suprasubduction zone ophiolite or backarc basin possibly of Proterozoic age. A poorly preserved metamorphic event at 2.45 Ga affected the former but not the latter, as did the intrusion of rift-related mafic dikes and sills at 2.06 Ga. Both suites were amalgamated, metamorphosed to at least upper amphibolite facies, subjected to simple shear strain and folded into map- and outcrop-scale sheath folds, and tectonically unroofed during the period 1.78 to 1.71 Ga. We name this event the Big Sky orogeny.

The Proterozoic geology of the Tobacco Root Mountains can be integrated with coeval features of the geology of the northern Wyoming province to outline a northeast-trending, southeast-vergent belt as the Big Sky orogen. The Big Sky orogen consists of a metamorphic hinterland flanked to the southeast by a foreland of discrete ductile shear zones cutting older basement, and to the northwest by arc-related metaplutonic bodies and the trace of a fossil subduction zone in the upper mantle. Archean blocks to the north of the Big Sky orogen may have been accreted as allochthonous terranes during collision and convergence.

The remarkable synchroneity of collision along the Big Sky orogen with tectonism in the Trans-Hudson orogen along the eastern margin of the Wyoming province and in the Cheyenne belt to the south of the province raise profound but unanswered questions about the process by which the Wyoming province was added to the rest of the ancestral North American craton.

Keywords: Archean, Proterozoic, Wyoming province, Great Falls tectonic zone, Tobacco Root Mountains.

Harms, T.A., Brady, J.B., Burger, H.R., and Cheney, J.T., 2004, Advances in the geology of the Tobacco Root Mountains, Montana, and their implications for the history of the northern Wyoming province, *in* Brady, J.B., Burger, H.R., Cheney, J.T., and Harms, T.A., eds., Precambrian geology of the Tobacco Root Mountains, Montana: Boulder, Colorado, Geological Society of America Special Paper 377, p. 227–243. For permission to copy, contact editing@geosociety.org.

INTRODUCTION

The Wyoming province is a loosely defined craton underlying much of Montana and Wyoming, characterized by its Archean age and seemingly stabilized after a major crust-forming event at 2.6–2.7 Ga (Houston et al., 1993; Frost and Frost, 1993; Frost, 1993; Dutch and Nielson, 1990; Hoffman, 1988). It is flanked to the east and south by Proterozoic orogenic or accretionary belts—the Trans-Hudson orogen (Dahl et al., 1999; Dahl and Frei, 1998; Bickford et al., 1990) and the Cheyenne belt (Chamberlain et al., 1993; Houston, 1993; Premo and Van Schmus, 1989), respectively—that stand in geologic and chronologic contrast to the core of the Archean craton and define the boundaries of the Wyoming province. Knowledge of the Wyoming province comes from discontinuous exposures in basement-cored uplifts of Tertiary age in Wyoming and southern Montana. Even these few windows die out to the north, so that the view necessary to establish the northern limit of the Wyoming province is obscured (Fig. 1).

The Wyoming province consists of three domains that share a characteristic isotopic fingerprint (Frost, 1993; Mueller and Wooden, 1988; Wooden and Mueller, 1988) but are otherwise geologically distinct (Fig. 1). (1) To the south, the Wyoming greenstone province includes greenstone belt sequences like those typical of Archean cratons elsewhere (Hoffman, 1988), whereas the northern Wyoming province consists of (2) the Beartooth-Bighorn magmatic terrane, a largely tonalite-trondhjemite-granodiorite domain intruded by 2.6 Ga, and (3) the Montana metasedimentary terrane, dominated by quartzofeldspathic gneisses in which belts of distinct metasupracrustal sequences are preserved (Mogk and Henry, 1988; Mogk et al., 1992a, 1992b; Mueller et al., 1998; Wooden et al., 1988).

The Tobacco Root Mountains, a Laramide and younger topographic feature in southwest Montana, lie along the northern limit of exposure of Precambrian crystalline basement in the Wyoming province and are cored by metamorphic rocks included in the Montana metasedimentary terrane (Fig. 1). The Tobacco Root Mountains contain several high-grade metamorphic suites: the Pony–Middle Mountain Metamorphic Suite is dominated by quartzofeldspathic gneiss with minor amphibolite; the Indian Creek Metamorphic Suite comprises similar quartzofeldspathic gneisses and amphibolites but also includes mappable sequences of metasupracrustal rocks (marble, quartzite, aluminous schist, and iron formation); the Spuhler Peak Metamorphic Suite is a mineralogically distinctive sequence of mafic rocks with associated quartzite and aluminous schist; and the metamorphosed mafic dikes and sills (MMDS) crosscut and are locally numerous in the Pony–Middle Mountain and Indian Creek Metamorphic Suites but are not present in the Spuhler Peak Metamorphic Suite (Fig. 2).

Building on the work of Charles Vitaliano and his students in the 1970s (Vitaliano et al., 1979a; Vitaliano et al., 1979b; see reprinted map and text accompanying this volume), this Special Paper reports the results of integrated field and instrument-based petrologic, thermobarometric, kinematic, geochronologic, and geochemical studies in the Tobacco Root Mountains conducted under the auspices of the Keck Geology Consortium in the 1990s (Burger, 2004, this volume, Chapter 1). This work clarifies the origin of the suites of Precambrian metamorphic rocks in the Tobacco Root Mountains and, among other findings, demonstrates orogenic activity at both 2.45 and 1.78–1.72 Ga. In contrast to the Late Archean stabilization of the Wyoming province in general, recognition of these Paleoproterozoic events prompts an assessment of how the geologic history of the Tobacco Root Mountains can refine and redefine the nature of the northern Wyoming province.

PRECAMBRIAN GEOLOGIC HISTORY OF THE TOBACCO ROOT MOUNTAINS

A wide range of observations (documented in this volume and summarized in this chapter) indicates extensive reworking of preexisting rocks across the Tobacco Root Mountains between 1.78 and 1.72 Ga. The pressures and temperatures of metamorphism, the widespread distribution and extent of folding, simple shear and transposition of compositional layering, and the degree to which isotopic geochronometers were reset demonstrate an event that was orogenic in character and scope. Because the evidence for this orogeny and most of its known effects are found in Montana, we designate this the Big Sky orogeny.

The Big Sky orogeny so pervasively affected the rocks of the Tobacco Root Mountains that only glimpses of older geologic history are preserved. The vastness of the time span over which at least some metamorphic rocks of the Tobacco Root Mountains existed prior to the Big Sky orogeny (ca. 3.4 to 1.78 Ga, or over a quarter of Earth's history) also contributes to uncertainty. A complete, accurate reconstruction of the evolution of the Tobacco Root Mountains prior to the Big Sky orogeny is out of reach with the presently available observations. We present here the simplest, most straightforward geologic history that is consistent with all data (Fig. 3). Many other interpretations are possible, and future studies may prove some of them more tenable.

The post–1.7 Ga Precambrian history of the Tobacco Root Mountains is comparatively simple and quiescent. Mafic dikes, apparently generated in several different episodes, cut the metamorphic rocks and are not themselves metamorphosed (Wooden et al., 1978). Conglomerates of the LaHood Formation are juxtaposed with Precambrian crystalline rocks across a normal fault and record the initiation of the Belt basin and Belt Supergroup sedimentation in this area (Schmitt, 1988).

Origin of the Indian Creek Metamorphic Suite and the Pony–Middle Mountain Metamorphic Suite

The quartzofeldspathic gneiss and associated amphibolite of the Indian Creek and Pony–Middle Mountain Metamorphic Suites record the earliest stage in Tobacco Root Mountains geologic history. The nature of the protoliths of these metamorphic rocks and the setting in which they arose are of considerable importance to the Archean history of the northern Wyoming province. Yet, because all original contacts have been transposed

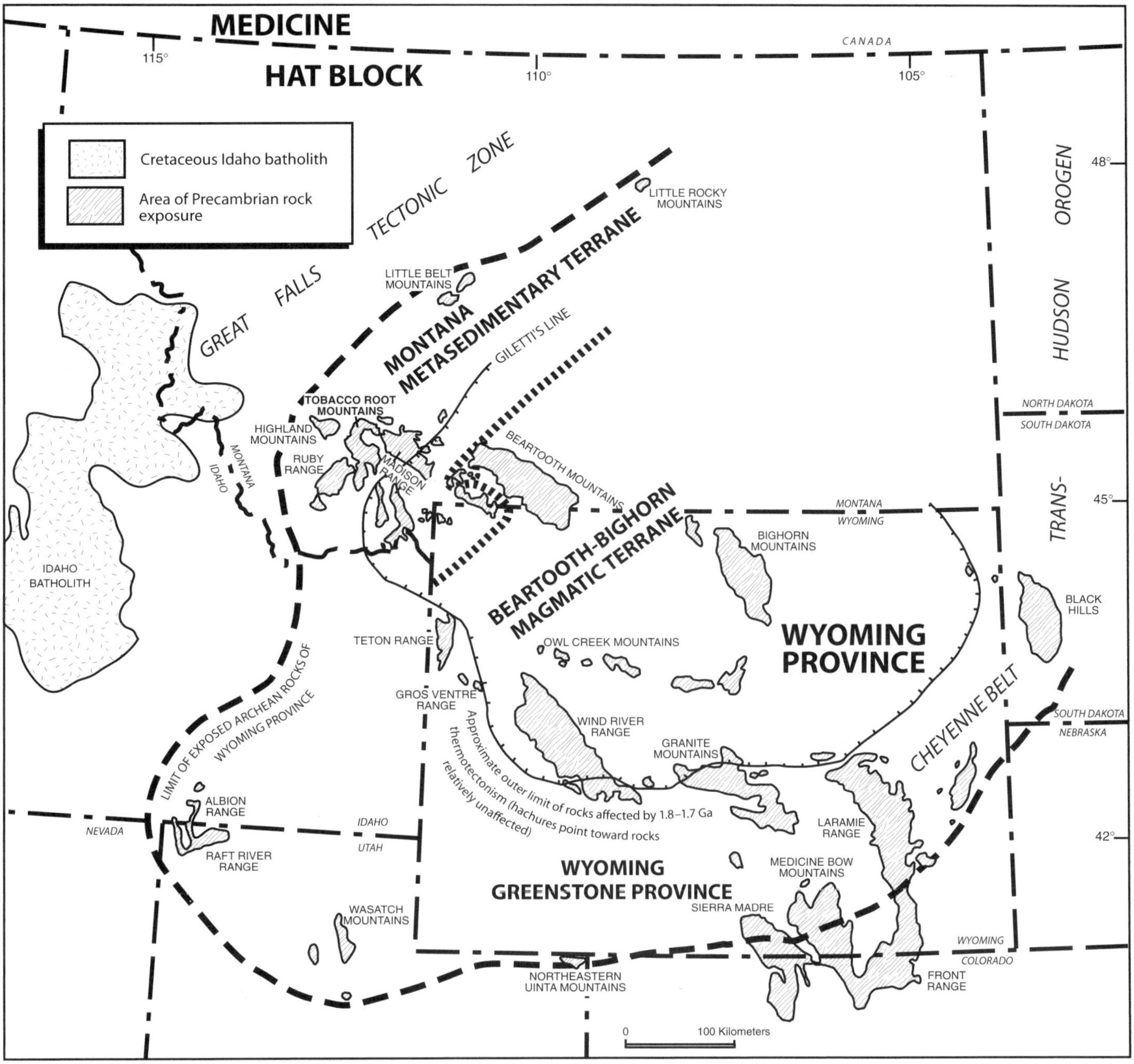

Figure 1. Map of exposed rocks of the Wyoming province and locations referred to in the text. Patterned areas are exposed Precambrian rocks; random dash area is the Idaho batholith; dashed line is the limit of Wyoming province exposure; striped line is the boundary between the Montana metasedimentary terrane and the Beartooth-Bighorn magmatic terrane (after Mogk, 1992); hachured line separates rocks affected by thermotectonism at 1.7–1.8 Ga from undisturbed Archean rocks, and corresponds to Giletti's line (Giletti, 1966) in the northern Wyoming province.

and all primary structures obliterated by extensive high-grade metamorphism both at 1.78 to 1.72 Ga and at 2.45 Ga, direct evidence as to the original character of the Indian Creek and Pony–Middle Mountain Metamorphic Suite rocks has been lost.

Our insights into this problem are provided mainly by major and trace element geochemistry (Mogk et al., 2004, this volume, Chapter 2). Elemental ratios of most Indian Creek and Pony–Middle Mountain Metamorphic Suite samples conform to norms for igneous rocks, but a significant minority indicates a sedimentary origin. Those samples of igneous origin are bimodal in composition: Felsic end members are dominantly calc-alkaline and mafic end members tholeiitic. Discriminant

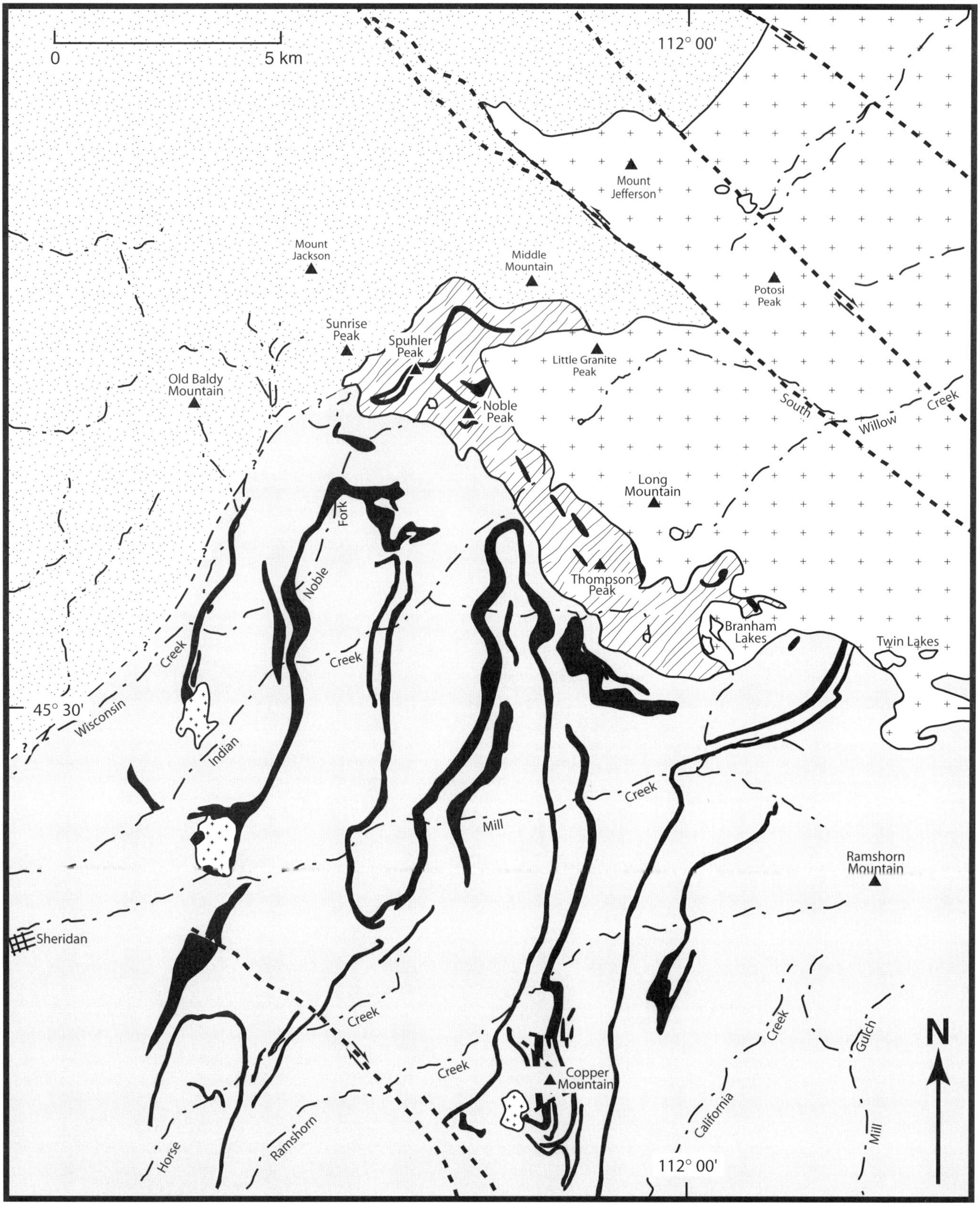

CRETACEOUS

Tobacco Root batholith

PRECAMBRIAN

Spuhler Peak Metamorphic Suite
dominantly amphibolites with
associated quartzite and aluminous schist

Indian Creek Metamorphic Suite
dominantly quartzofeldspathic and hornblende gneisses with
associated marble, quartzite, aluminous schist and iron formation

Pony-Middle Mountain Metamorphic Suite
dominantly quartzofeldspathic and hornblende gneisses

Figure 2. Simplified geologic map of the Tobacco Root Mountains (after Vitaliano et al., 1979b, and O'Neill, 1983). The boundary between the Indian Creek Metamorphic Suite and the Pony–Middle Mountain Metamorphic Suite is schematic and based on the abundance of marble in the Indian Creek Metamorphic Suite.

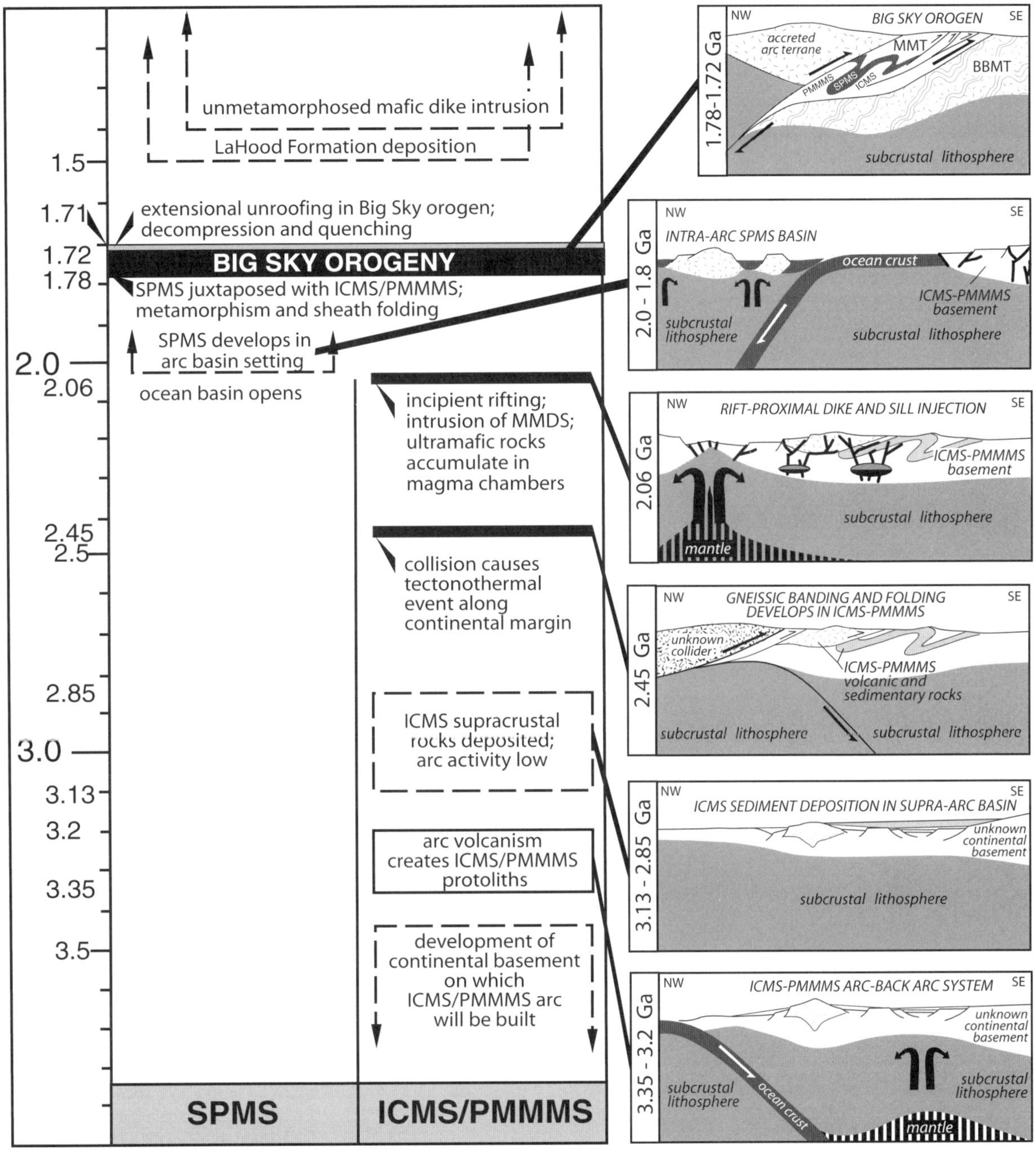

Figure 3. Time line of Precambrian tectonic evolution in the metamorphic suites of the Tobacco Root Mountains. Dashed lines are used where ages are approximate or constrained only by minima and/or maxima. Solid lines and bars indicate known ages. Schematic cross sections illustrate proposed tectonic settings of the Tobacco Root Mountains metamorphic suites at critical points in the time line. PMMMS—Pony–Middle Mountain Metamorphic Suite, SPMS—Spuhler Peak Metamorphic Suite, ICMS—Indian Creek Metamorphic Suite, MMDS—metamorphosed mafic dikes and sills, MMT—Montana metasedimentary terrane, BBMT—Beartooth-Bighorn magmatic terrane

diagrams suggest that the felsic gneisses have an arc affinity, whereas most mafic samples fall in ocean floor fields. Both have strongly negative ε_{Nd} (although the mafic samples are less so) suggesting contamination by older, evolved continental crust (Mueller et al., 2004, this volume, Chapter 9). Finally, samples whose geochemical characteristics indicate a sedimentary protolith are chemically similar to clastic rocks from continental arcs or active continental margin settings (Mogk et al., 2004, this volume, Chapter 2). In sum, these chemical characteristics point to a subduction-related continental arc setting, with intra-arc or backarc extension to generate mid-ocean ridge–like mafic rocks and produce metasomatic alteration that the rocks appear to have experienced (Mogk et al., 2004, this volume, Chapter 2).

The quartzofeldspathic gneisses and amphibolites that dominate the Indian Creek and Pony–Middle Mountain Metamorphic Suites have overlapping ranges of chemical characteristics (Mogk et al., 2004, this volume, Chapter 2). There is no obvious discontinuity between the quartzofeldspathic gneisses of the two suites in the field. Recognition of a map-scale fold involving both suites demonstrates that the two are laterally equivalent along structural strike and that quartzofeldspathic gneiss is continuous from one suite to the other (Harms et al., 2004, this volume, Chapter 10). On this basis, where no distinction need be made between the Indian Creek and Pony–Middle Mountain Metamorphic Suites we will refer to them together simply as the quartzofeldspathic gneiss suite.

We favor an extrusive or volcaniclastic, rather than plutonic, origin for most igneous arc rocks of the quartzofeldspathic suite (Mogk et al., 2004, this volume, Chapter 2). This interpretation is based on the following.

1. The dominance of tabular lithologic units and the interlayering of one unit with another. Strongly crosscutting geometries were not observed in the quartzofeldspathic gneiss. Although the original rock bodies have most certainly been flattened, transposed, and structurally repeated, the ubiquity and fine scale of interlayering argues for it as a primary feature. Interlayering of quartzofeldspathic gneiss of both igneous and sedimentary origin and interlayering of marble, quartzite, aluminous gneiss, and iron formation with the quartzofeldspathic gneiss are easiest to reconcile with a volcanic or volcaniclastic origin for the quartzofeldspathic gneiss.

2. The lithologic heterogeneity on all scales. This is in strong contrast to lithologic homogeneity over large areas that characterize metaplutonic rocks elsewhere in the northern Wyoming province (Mogk et al., 1992b; Wooden et al., 1988). An exception to this interpretation may be the potassium feldspar–megacrystic gneisses of the Pony–Middle Mountain Metamorphic Suite, which we believe are likely candidates for metaplutonic rocks.

At present, our best constraints indicate the quartzofeldspathic gneiss suite originated in a long-standing or repeatedly productive continental arc active between ca. 3.35 and 3.2 Ga. Igneous ages of 3.31 Ga, 3.21 Ga (U-Pb sensitive high-resolution ion microprobe [SHRIMP] dating by Mueller et al., 2004, this volume, Chapter 9) and 3.37 Ga (U-Pb upper-intercept age by Krogh et al., 1997) have been obtained from magmatic zircons separated from Indian Creek Metamorphic Suite samples. These magmatic ages are very similar to the intrusive age of quartzofeldspathic gneiss elsewhere in the Montana metasedimentary terrane (Mueller et al., 1993, 1996c). Whole-rock Rb-Sr ages of 2.75 Ga (James and Hedge, 1980) and 2.67 Ga (Mueller and Cordua, 1976), which were long-standing geochronologic benchmarks for the Tobacco Root Mountains, should now be abandoned in favor of the ages from these more robust dating techniques (see also Roberts et al., 2002). Except through its contribution to the Nd isotope systematics of the arc rocks (Mueller et al., 2004, this volume, Chapter 9) and to detrital zircon populations in Indian Creek Metamorphic Suite metasupracrustal rocks (Mueller et al., 1998), no direct evidence of the older continental basement onto which the 3.35–3.2 Ga continental arc was built has been found.

The obvious metasupracrustal rocks of the Indian Creek and Pony–Middle Mountain Metamorphic Suites—the marbles, quartzites, aluminous schists, and iron formation—represent a different paleoenvironment and, apparently, a different period than the quartzofeldspathic gneisses and amphibolites that envelope them. The presence of rock types such as marble and iron formation, and their lateral continuity, point to a stable depositional environment and periods of tectonic quiescence in and around the basin. Quartzite from the Indian Creek Metamorphic Suite metasupracrustal sequence has yielded detrital zircons ranging in age from 3.13 to 3.93 Ga (Mueller et al., 1998). This 3.13 Ga maximum age for the metasupracrustal rocks is younger than the youngest magmatic age from the quartzofeldspathic gneisses. Mueller et al. (2004, this volume, Chapter 9; 1998) highlight the absence of 2.75 to 2.85 Ga zircons representative of gneisses of that age that are widespread in the northern Wyoming province (Frost, 1993; Frost and Frost, 1993; Mogk and Henry, 1988; Wooden et al., 1988; Montgomery and Lytwyn, 1984). This, they argue, promotes a depositional age older than 2.85 Ga for the quartzite. Alternatively, the zircon age distribution could result from deposition after 2.75 Ga, but in an area either of very restricted provenance or outside the detrital range of the northern Wyoming province.

The Indian Creek and Pony–Middle Mountain Metamorphic Suites, therefore, incorporate two distinct stages in the early history of the Tobacco Root Mountains: long-lived (3.35–3.2 Ga) arc and backarc magmatism on an older continent, and sedimentation after 3.13 Ga. The metasupracrustal rocks may have originated independently and been tectonically juxtaposed with the quartzofeldspathic gneiss, possibly as an aspect of the 2.45 Ga tectonothermal event, or the metasupracrustal rocks may record the transition from an active arc margin to erosion of the arc and deposition in an epicontinental sedimentary basin.

Cryptic Tectonothermal Event at 2.45 Ga

Geochronology provides evidence for what was at least a thermal event in the area occupied by the Pony–Middle Mountain and Indian Creek Metamorphic Suites at 2.45 Ga. Monazites of this age are found in aluminous gneisses in both suites (Cheney et al., 2004b, this volume, Chapter 8). Spot analyses of zircon rims ($^{207}Pb/^{206}Pb$ SHRIMP dating by Mueller et al., 2004, this volume, Chapter 9) and lower-intercept ages from discordant zircon populations (Krogh et al., 1997) in quartzofeldspathic gneisses of the Indian Creek Metamorphic Suite also indicate metamorphic growth or disturbance of zircon at 2.4 Ga. Patchy Ca and Fe zoning in the cores of some Indian Creek and Pony–Middle Mountain Metamorphic Suite garnets may be relicts of this earlier metamorphism (Cheney et al., 2004a, this volume, Chapter 6).

No 2.45 Ga igneous rocks have been identified so far within the quartzofeldspathic gneiss suite (Mueller et al., 2004, this volume, Chapter 9). Inherited zircons that range in age from ca. 2.2 to 3.0 Ga, including several at 2.4 Ga (single zircon, $^{207}Pb/^{206}Pb$ SHRIMP dating), however, have been recovered from the Cretaceous Tobacco Root batholith (Mueller et al., 1996a, 1996b), which cuts the center of the Tobacco Root Mountains (Fig. 2). Mueller et al. (1996a) interpret the zircons as coming directly from magmatic rocks, rather than being second-cycle components of clastic rocks, and point out that plutonic rocks of that age must be preserved at some depth below the range.

Other observations hint that the event at 2.45 Ga was tectonic as well as thermal, and that the arc, backarc, and basinal deposits of the Indian Creek and Pony–Middle Mountain Metamorphic Suites were consolidated into continental crystalline basement by burial, deformation, and metamorphism at that time, implying renewed plate convergence and/or collision. MMDS bodies cut well-developed and folded gneissic fabrics in both the Indian Creek and Pony–Middle Mountain Metamorphic Suites (Brady et al., 2004a, this volume, Chapter 5; Harms et al., 2004, this volume, Chapter 10)—metamorphic fabrics that must, consequently, be older than the 2.06 Ga intrusive age of the MMDS suite (Mueller et al., 2004, this volume, Chapter 9) and have been of high enough grade to survive Big Sky metamorphism at 1.78 Ga. Because it requires the fewest separate metamorphic events, we propose that the 2.45 Ga monazite grew during the development and deformation of the pre-MMDS gneissic banding. This correlation, however, is not strictly required by the data in hand. Both porphyroblasts and matrix in the Pony–Middle Mountain and Indian Creek Metamorphic Suites contain 2.45 Ga monazite, many of which do not have 1.78 Ga rims, but, in the Indian Creek Metamorphic Suite, rocks that have 2.45 Ga monazite most typically also have 1.78 Ga grains. These aspects of the character and distribution of the older monazites suggest that they were profoundly resistant to change even when conditions produced new monazite growth at 1.78 Ga. This being the case, we cannot rigorously or unequivocally relate the 2.45 Ga monazites to the growth of any metamorphic minerals around them, as the latter might or might not have crystallized at 1.78 Ga while the former were clearly unaffected.

Evidence for a thermal or tectonothermal event at 2.45 Ga in the northern Wyoming province is not limited to the Tobacco Root Mountains. Hornblende, biotite, and muscovite $^{40}Ar/^{39}Ar$ plateau ages of 2.4–2.5 Ga in the southern Madison Range are interpreted to reflect the primary cooling history of that area (Erslev and Sutter, 1990). Xenoliths from the Idaho batholith contain zircons with upper-intercept ages of 2.34 Ga (Bickford et al., 1981).

Origin of the Metamorphosed Mafic Dikes and Sills—Mafic Magmatism and Continental Rifting

The Indian Creek and Pony–Middle Mountain Metamorphic Suites are host to metamorphosed mafic dikes and sills (MMDS), many of which retain relict igneous textures (Brady et al., 2004a, this volume, Chapter 5), but all contain metamorphic minerals that reflect the pressure and temperature conditions of the Big Sky orogeny (Cheney et al., 2004a, this volume, Chapter 6), including metamorphic zircons (Mueller et al., 2004, this volume, Chapter 9). Originally there probably were more crosscutting dikes than are seen today, as many have been transposed to layer-parallel geometries by Big Sky deformation (Harms et al., 2004, this volume, Chapter 10).

One representative MMDS body in the Pony–Middle Mountain Metamorphic Suite was sampled for geochronology and yielded a population of prismatic, low-discordance zircons that give what is interpreted to be an intrusive age of 2.06 Ga (Mueller et al., 2004, this volume, Chapter 9). Consistency of major and trace element composition among the many dikes analyzed geochemically argue that all are part of a single magmatic suite intruded into both the Indian Creek and Pony–Middle Mountain Metamorphic Suites together (Brady et al., 2004a, this volume, Chapter 5). However, because multiple sets of dikes of differing Proterozoic ages are known from the Wyoming and other Archean cratons (Frost and Frost, 1993; Krogh et al., 1987; Aspler and Chiarenzelli, 1998, and references therein), and from the post–1.7 Ga history of the Tobacco Root Mountains themselves (Wooden et al., 1978), it is possible that the MMDS suite might include undated bodies of ages other than 2.06 Ga.

Continental rifting at 2.06 Ga is the most likely context for MMDS intrusion. Major element compositions of MMDS rocks indicate a subalkaline, tholeiitic basalt protolith. Their rare earth element (REE) patterns are similar to those of basalts intruded into continental rift settings today (Brady et al., 2004a, this volume, Chapter 5). Discriminant diagrams based on largely immobile elements are inconclusive as to the setting of origin of the MMDS bodies (Hess and Vitaliano, 1990). The sharp contacts of the MMDS bodies, their planar geometries, and the relict chilled margin textures observed indicate that the original dikes and sills cut cool, relatively brittle, continental crust of the Indian Creek and Pony–Middle Mountain Metamorphic Suites. Thus, MMDS-related rifting disrupted the craton that was established or enlarged by collision at 2.45 Ga and stabilized before 2.06 Ga. We speculate that at some distance from the present Tobacco Root Mountains the rifting process went to completion and ocean crust was formed. This new, ca. 2.0 Ga continental margin would subsequently become the locus of collision during the Big Sky orogeny.

Comparison of major element ratios in MMDS rocks indicates that significant fractional crystallization, notably of orthopyroxene but of plagioclase and clinopyroxene as well, occurred in their source reservoir over the course of intrusion of the suite (Brady et al., 2004a, this volume, Chapter 8). Orthopyroxene-rich cumulates created in this way may have been the parents of the relatively silica-rich meta-ultramafic rocks that are present locally in the Indian Creek, Pony–Middle Mountain and Spuhler Peak Metamorphic Suites (Johnson et al., 2004, this volume, Chapter 4).

Origin of the Spuhler Peak Metamorphic Suite

The Spuhler Peak Metamorphic Suite is dominated by hornblende amphibolites and orthoamphibole-garnet gneiss. Geochemical analysis indicates these metamorphosed mafic rocks originated as tholeiitic basalt and hydrothermally altered equivalents, respectively (Burger et al., 2004, this volume, Chapter 3). Trace element chemistry appears arc-related and unlike mid-oceanic-ridge basalt. The scattered metamorphosed ultramafic bodies in the Spuhler Peak Metamorphic Suite do not appear to be ophiolitic either; their geochemistry suggests they are the result of fractional crystallization and were neither komatiites nor mantle peridotites (Johnson et al., 2004, this volume, Chapter 4). A suprasubduction zone ophiolite, backarc, or intra-arc oceanic basin setting of origin for the Spuhler Peak Metamorphic Suite would provide both the observed arc signature and the context for widespread hydrothermal alteration of tholeiitic basalt as required by the prominence of orthoamphibole-garnet rocks.

Interlayered aluminous, siliceous schist and quartzite in the Spuhler Peak Metamorphic Suite reflect the introduction of detritus to—or in the case of quartzite, possibly chemical sedimentation in—the suite's basin (Burger et al., 2004, this volume, Chapter 3). Marbles are notably absent from the suite. These sedimentary components are consistent with a backarc basin setting but do not preclude other interpretations.

The maximum age of the Spuhler Peak Metamorphic Suite is constrained by the few low-discordance, clearly detrital zircon grains that have been recovered from a quartzite in the suite. These range from 3.39 to 3.02 Ga (U-Pb SHRIMP dating); the absence of 2.75–2.85 Ga grains is, again, noteworthy (Mueller et al., 2004, this volume, Chapter 9).

The lithologic contrast between the Spuhler Peak Metamorphic Suite and adjacent quartzofeldspathic gneiss suites (the Indian Creek and Pony–Middle Mountain Metamorphic Suites) has focused attention on the nature of this contact. While the Spuhler Peak Metamorphic Suite experienced all phases of the Big Sky orogeny together with the quartzofeldspathic gneiss suite, there is no evidence that it underwent metamorphism at 2.45 Ga (Cheney et al., 2004b, this volume, Chapter 8) and, like Gillmeister (1972), we observed no MMDS bodies in the Spuhler Peak Metamorphic Suite (Brady et al., 2004a, this volume, Chapter 5). It is clear that the Spuhler Peak Metamorphic Suite was not in contact with the quartzofeldspathic gneiss suite in the period 2.45 to 2.06 Ga and was in contact by 1.78 Ga. But what kind of contact developed in that interval? Two models have been proposed; the implications of each are different and striking.

1. Gillmeister (1972) interpreted the Spuhler Peak Metamorphic Suite to have been deposited on the quartzofeldspathic gneiss suite across an unconformity. This is consistent with the observation that MMDS bodies do not cross the Spuhler Peak Metamorphic Suite contact and with the persistence of a quartzite in the Spuhler Peak Metamorphic Suite along the contact, presumably a basal clastic deposit. One weakness of this model is that it does not account for the lack of feeder dikes within the Indian Creek or Pony–Middle Mountain Metamorphic Suites for the voluminous mafic layers in the Spuhler Peak Metamorphic Suite. Also, in this model, the Spuhler Peak Metamorphic Suite arc and related backarc spreading system would have developed on a basement of quartzofeldspathic gneiss. The absence of a calc-alkaline component in Spuhler Peak Metamorphic Suite metabasalts (Burger et al., 2004, this volume, Chapter 3) makes such a continental setting problematic. Finally, if the Spuhler Peak Metamorphic Suite accumulated unconformably on the quartzofeldspathic gneiss suite, then it must be younger than the 2.06 Ga MMDS intrusions and older than the onset of the 1.78 Ga Big Sky orogeny, that is, wholly Proterozoic. The detrital zircon population of interlayered quartzite requires, in turn, that either sediment supply to the arc-related Spuhler Peak Metamorphic Suite basin was shielded from the nearby Beartooth-Bighorn magmatic terrane, where 2.75–2.85 Ga rocks are common, or the Tobacco Root area, including the Indian Creek and Pony–Middle Mountain Metamorphic Suites, was not adjacent to the Beartooth-Bighorn basement before the Big Sky orogeny.

2. Burger (1969) mapped a fault at the base of the Spuhler Peak Metamorphic Suite because of apparent discordance in structure and metamorphic grade between it and the quartzofeldspathic gneiss suite and because a high-strain zone is coincident with the contact. Our work does not substantiate Burger's original criteria: the grade of Big Sky metamorphism is consistent across all the metamorphic suites of the Tobacco Root Mountains (Cheney et al., 2004a, this volume, Chapter 6); the discordant geometry of Indian Creek Metamorphic Suite folds along the contact may be only an apparent truncation; and high-strain zones are present both along and away from the Spuhler Peak Metamorphic Suite contact (Harms et al., 2004, this volume, Chapter 10). Nevertheless, the Big Sky orogeny is a likely context for tectonic juxtaposition of the Spuhler Peak Metamorphic Suite and the quartzofeldspathic gneiss suite because it was accompanied by substantial simple shear strain across the Tobacco Root Mountains (Harms et al., 2004, this volume, Chapter 10). If the Spuhler Peak Metamorphic Suite was faulted against the quartzofeldspathic gneiss suite, the age of the suite must be older than the Big Sky orogeny but otherwise is constrained only by the detrital zircons recovered to date. The suite could be Archean and older than the 2.75–2.85 Ga Beartooth-Bighorn magmatic terrane crust

not represented by detrital zircons, but we consider the length of time between the origin of the Spuhler Peak Metamorphic Suite in this case and its amalgamation with the rest of the Wyoming province during the Big Sky orogeny over a billion years later to be problematic in light of the transient nature of plate boundaries, arcs, and ocean crust. Conversely, as there is no direct evidence for an Archean age, the Spuhler Peak Metamorphic Suite could be Proterozoic, having developed in an arc-backarc basin system exotic to the northern Wyoming province. The culmination of subduction in this Proterozoic ocean would be the collision that accreted the Spuhler Peak Metamorphic Suite to its present position relative to the quartzofeldspathic gneiss suite and produced the Big Sky orogeny.

Our work in the Tobacco Root Mountains did not, in fact, produce any observation that unequivocally requires or explicitly contradicts either the unconformity or the fault model for the Spuhler Peak Metamorphic Suite contact. A Proterozoic age for the Spuhler Peak Metamorphic Suite is a significant departure from long-standing views as to the antiquity of the rocks in the Tobacco Root Mountains, but we find that the model of tectonic juxtaposition of a Proterozoic Spuhler Peak Metamorphic Suite has the fewest objectionable elements and matches the most observations. For this reason we consider the suite to be an allochthonous terrane within the Big Sky orogen.

The Big Sky Orogeny

Mineral assemblages, mineral textures, and mineral chemistry in all four metamorphic suites in the Tobacco Root Mountains record progressive metamorphism along a clockwise path through pressure-temperature space that reflects stages in the Big Sky orogeny between 1.78 and 1.71 Ga (Fig. 4) (Cheney et al., 2004a, this volume, Chapter 6, 2004b, this volume, Chapter 8; Brady et al., 2004b, this volume, Chapter 7). Metamorphism started with a higher-pressure, higher-temperature (>1.0 GPa and >700 °C) phase (M1). The higher-pressure (M1) start of the metamorphic path is demonstrated predominately in aluminous orthoamphibole-bearing lithologies from the Spuhler Peak Metamorphic Suite, but it is also preserved locally in similar rock types in the Indian Creek Metamorphic Suite, demonstrating that M1 conditions were experienced across the Tobacco Root Mountains at the onset of the Big Sky orogeny. If, as we propose, the Spuhler Peak Metamorphic Suite is a Proterozoic suite with no previous metamorphism, then peak pressures exceeding 1.0 GPa experienced by the suite must have been achieved by tectonic overburden constructed during the Big Sky orogeny, presumably as a consequence of convergence at that time, and therefore suggest crustal thickness in the orogen in excess of 40 to 50 km. Juxtaposition of the Spuhler Peak Metamorphic Suite with the Indian Creek and Pony–Middle Mountain Metamorphic Suites would have occurred as part of this tectonic stacking, although it is not clear from the present geometry in the Tobacco Root Mountains whether the Spuhler Peak Metamorphic Suite was obducted onto the quartzofeldspathic gneisses or was partially subducted beneath them as the Big Sky orogeny commenced (Harms et al., 2004, this volume, Chapter 10).

From the maximum pressures of metamorphism, the Tobacco Root Mountains underwent modest decompression while heating

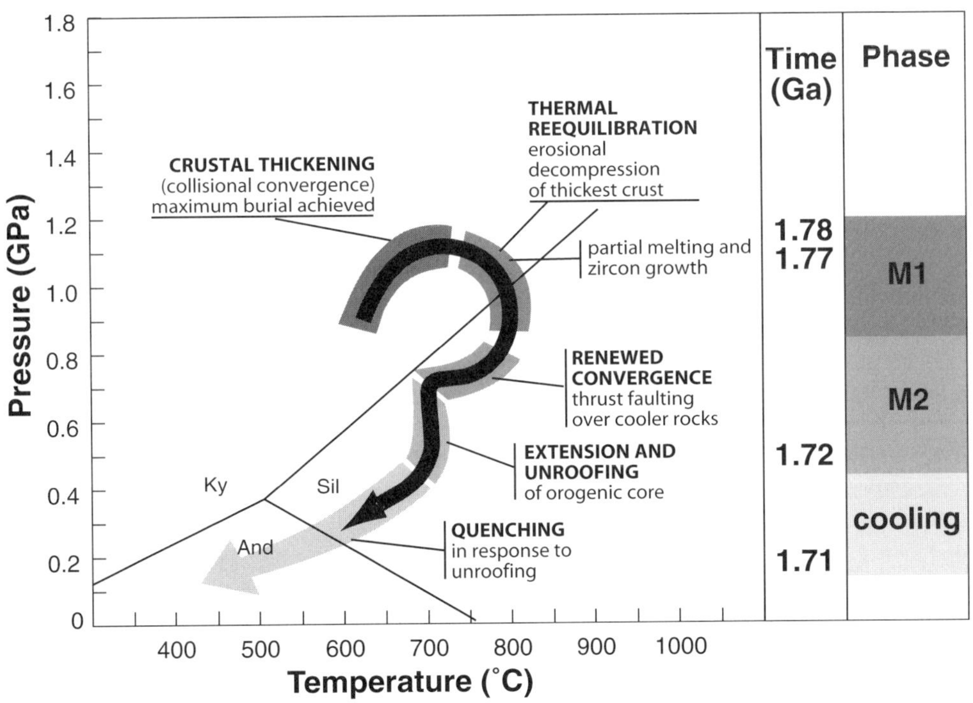

Figure 4. Pressure-temperature-time path for all three metamorphic suites of the Tobacco Root Mountains and the MMDS during the Big Sky orogeny correlated to metamorphic phases and to tectonic stages in the development of the orogen. The numerous mineral stability fields that constrain the pressure-temperature path are given in Cheney et al. (2004a, this volume, Chapter 6; see their Figure 21); only the aluminosilicate stability fields (Holdaway, 1971) are shown here for reference (Ky—kyanite; Sil—sillimanite; And—andalusite).

to a thermal maximum of ~800 °C (Cheney et al., 2004a, this volume, Chapter 6). We suggest this occurred as geotherms reestablished themselves in the thickened crust while erosion was focused at high elevations in the new orogen. No igneous rocks generated during the Big Sky orogeny have been identified in the Tobacco Root Mountains. Leucosome, however, is widely observed. We interpret it to be the result of partial melting that is best linked to the maximum temperatures of metamorphism along this decompressing arc of the pressure-temperature loop (Cheney et al., 2004a, this volume, Chapter 6). Low-discordance, low-Th/U, metamorphic zircon recovered from leucosome gives an age of 1.77 Ga (U-Pb SHRIMP dating) for this stage of the orogeny (Mueller et al., 2004, this volume, Chapter 9). The localization of this leucosome in dilatant zones associated with boudins and folds, the latter of which, in particular, characterize mesoscopic deformation in the Tobacco Root Mountains, links the time of deformation to the maximum temperature and pressure phase (M1) of the Big Sky orogeny (Harms et al., 2004, this volume, Chapter 10; Mueller et al., 2004, this volume, Chapter 9). Mesoscopic folds are part of a pervasive structural pattern of north-dipping fabric, north-plunging lineation, and a map-scale sheath fold of the Spuhler Peak Metamorphic Suite, flanked by the Indian Creek and Pony–Middle Mountain Metamorphic Suites, which we interpret to be the result of intense, regional, simple shear (Harms et al., 2004, this volume, Chapter 10) that accompanied crustal thickening and M1 metamorphism.

The higher-pressure–higher-temperature (M1) arc of the Big Sky metamorphic path was followed by differential reequilibration at lower temperatures and pressures (M2) via nearly isobaric cooling (from ~750 to 700 °C at ~0.7–0.8 GPa), leading to isothermal decompression (from ~0.7 to <0.6 GPa at ~700 °C) (Cheney et al., 2004a, this volume, Chapter 6). The occurrence and conditions of M2 metamorphism are well established by garnet-amphibole thermobarometry (Cheney et al., 2004a, this volume, Chapter 6) in metamorphosed mafic rocks in all four Precambrian suites (Indian Creek, Pony–Middle Mountain, and Spuhler Peak Metamorphic Suites and MMDS). Friberg (1994) and Hess and Vitaliano (1990) also documented pressure and temperature conditions similar to our M2 phase. We suggest isobaric cooling indicates that the Tobacco Root Mountains metamorphic suites and their tectonic overburden were carried piggyback in the hanging wall of a large-scale thrust fault over cooler rocks formerly at higher crustal levels, increasing total crustal thickness in the Big Sky orogen beyond 40–50 km. Isothermal decompression points to gravitational instability and orogenic collapse triggered by this crustal-scale thrust faulting.

Monazite, which is present both as inclusions in various porphyroblasts and within the matrix of metamorphic rocks in the Tobacco Root Mountains, shows dominant growth at 1.78 Ga and continued growth until 1.72 Ga (Cheney et al., 2004b, this volume, Chapter 8). We interpret this age range to represent the M1-to-M2 metamorphic period. Comparison with the 1.77 Ga zircon age of partial melting and pervasive deformation suggests 1.78 Ga monazite growth was associated with the early, higher-pressure portion of the metamorphic path. Monazites as young as 1.753 Ga are present in kyanite (Cheney et al., 2004b, this volume, Chapter 8); the pressure-temperature path must have crossed the kyanite-sillimanite reaction after that time (Fig. 4).

$^{40}Ar/^{39}Ar$ ages of 1.71 Ga from hornblende in the Spuhler Peak and Indian Creek Metamorphic Suites (Brady et al., 2004b, this volume, Chapter 7) confirm cooling through the hornblende closure temperature (~500 °C) immediately after final growth of monazite and upper amphibolite-facies M2 metamorphism. This is a record of rapid quenching following M2 isothermal decompression, which is consistent with tectonic unroofing of the deeper parts of the Big Sky orogen, including the Tobacco Root Mountains. No extensional structures have been identified within the Tobacco Root Mountains (Harms et al., 2004, this volume, Chapter 10). This is not unexpected given the uniformity of metamorphic grade observed throughout the Spuhler Peak, Indian Creek, and Pony–Middle Mountain Metamorphic Suites and MMDS (Cheney et al., 2004a, this volume, Chapter 6) and the consistent time of cooling demonstrated across the Tobacco Root Mountains and adjacent ranges (Brady et al., 2004b, , this volume, Chapter 7, and references therein). The Tobacco Root Mountains must have been positioned in the footwall plate of any detachment systems active during decompression and rapid cooling.

ARCHITECTURE OF THE BIG SKY OROGEN ACROSS THE NORTHERN WYOMING PROVINCE

Nearly four decades ago, Giletti (1966) compiled K-Ar age data for the northern Wyoming province and revealed a domain of ca. 1.6 Ga ages separated from much older, Archean rocks to the south across a northeast-trending line that has become known as "Giletti's line" (Fig. 5). He interpreted the Proterozoic ages as reflecting a regional metamorphism that reset the Ar system in rocks originally metamorphosed much earlier, typically ca. 2.4–2.6 Ga. Later, Erslev and Sutter (1990) and O'Neill (1998) went on to suggest that a northeast-trending Proterozoic orogen underlies the northern Wyoming province and that Giletti's K-Ar boundary separates a reset hinterland to the northwest from a cooler, undisturbed foreland to the southeast in that south-vergent orogen. Evidence for 1.8–1.6 Ga thermal disturbance in much older metamorphic rocks has subsequently been documented in the North (Reid et al., 1975) and South (Montgomery and Lytwyn, 1984) Snowy Blocks of the Beartooth Mountains, along Giletti's line, and in the Stillwater Complex (Houston et al., 1993) south of the line. These additional geochronologic data refine the position of Giletti's line but do not fundamentally change his, or Erslev and Sutter's and O'Neill's vision. The geology of the Tobacco Root Mountains reported in this Special Paper makes clear the importance of a 1.78 to 1.71 Ga orogeny north of Giletti's line and replaces the long-standing view (held since James and Hedge [1980] and Mueller and Cordua [1976]) that metamorphism in this region was entirely or primarily Archean. Despite limited and discontinuous exposures of Precambrian basement, with a wide range of other geologic, geochronologic,

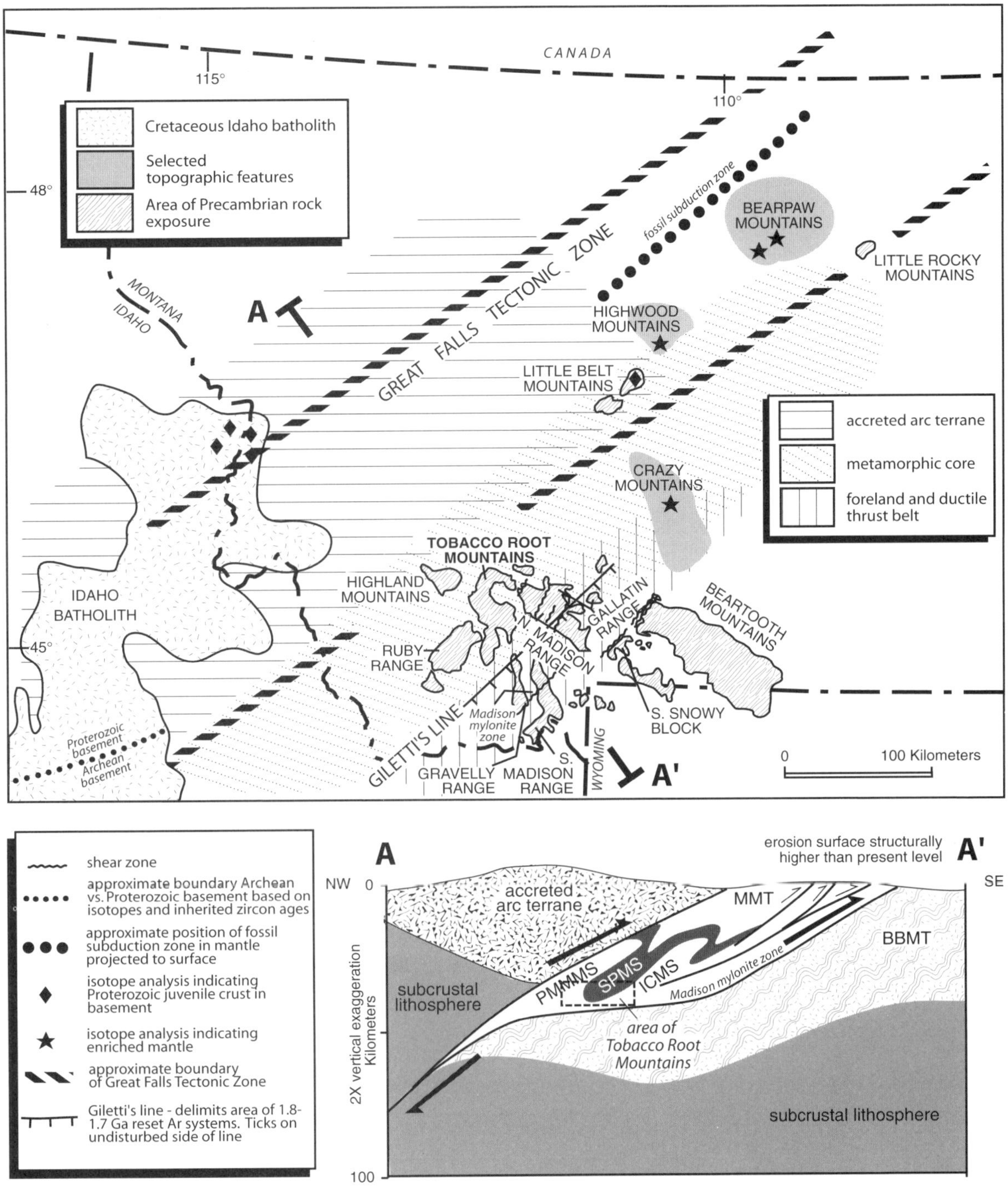

Figure 5. Architecture of the Big Sky orogen in map and cross section. Geochemical, geophysical, and geologic observations cited in the text outline a northeast-trending metamorphic core zone (dotted diagonal ruling, map view) of the orogen that is flanked to the southeast by a foreland of discrete, southeast-vergent, ductile thrust faults (vertical ruling, map view) and to the northwest by a fossil subduction zone and accreted, juvenile, Proterozoic arc terrane (horizontal ruling, map view). See also O'Neill's (1998) Figures 1 and 2. PMMMS—Pony–Middle Mountain Metamorphic Suite, SPMS—Spuhler Peak Metamorphic Suite, ICMS—Indian Creek Metamorphic Suite, MMT—Montana metasedimentary terrane, BBMT—Beartooth-Bighorn magmatic terrane.

geochemical, and geophysical advances in the northern Wyoming province, it is now possible to outline the broad characteristics of this orogen in space and time (see also O'Neill, 1998) (Fig. 5). Major components include

- a ductilely deformed metamorphic core,
- a supracrustal foreland cut by discrete shear zones south of the metamorphic core,
- a north-dipping fossil subduction zone north of surface exposures of Precambrian rocks, and
- a loosely defined domain of juvenile plutonic rocks indicating accretion of an arc along the northern margin of the orogen.

The Tobacco Root Mountains are presently the best-documented part of the metamorphic core zone and provide the best evidence for the evolution of the orogen over time. On this basis, we extend the name "Big Sky orogeny" from the Tobacco Root Mountains across the northern Wyoming province and apply it to the whole Proterozoic orogenic belt.

Metamorphic Core Zone

The core of the Big Sky orogen lies along a northeast-trending belt of ca. 1.8–1.7 Ga upper amphibolite facies metamorphism that not only includes the Tobacco Root Mountains but extends from the Ruby and Highland Ranges to the Little Belt Mountains. It may also encompass metamorphism and/or deformation in the northern Madison and Gallatin Ranges. The ~300 km along-strike exposed length of this belt is modest compared with the high grade of metamorphism and intensity of ductile deformation and shear observed in the Tobacco Root Mountains and with the magnitude of Proterozoic crustal thickening we have proposed. The Big Sky metamorphic core zone may reasonably be expected to continue in the subsurface perhaps as far southwest as the root of the Idaho batholith and northeast to the Trans-Hudson orogen.

Garihan (1979) and O'Neill et al. (1988) report amphibolite- and upper-amphibolite-grade, migmatitic, aluminous gneiss, hornblende gneiss, and quartzofeldspathic gneiss that are mylonitized and folded in the crystalline core of the Ruby and Highland Mountains. Step-leach $^{207}Pb/^{206}Pb$ analysis of garnet, a method that probably yields a maximum age (Dahl et al., 2000), gives an age of metamorphism of younger than 1.82 Ga in the Ruby Range and younger than 1.79 Ga in the Highland Range (Roberts et al., 2002). Cooling through the hornblende and biotite blocking temperatures occurred between 1.74 and 1.71 Ga ($^{40}Ar/^{39}Ar$ dating) (Roberts et al., 2002; Brady et al., 2004b, this volume, Chapter 7) and provides a minimum metamorphic age in the Highland and Ruby Ranges.

To the northeast, in the core of the Little Belt Mountains, an intrusive age of ca. 1.86 Ga has been determined by U-Pb zircon geochronology for various granitoid gneisses and establishes a maximum age for their deformation and metamorphism (Mueller et al., 1997, 2002; Vogl et al., 2002). Catanzaro and Kulp (1964) analyzed concordant monazite of similar age (ca. 1.9 Ga) from a migmatite. A garnet step-leaching age of 1.86 Ga (Dahl et al., 2000) provides another, again maximum age constraint on metamorphism. Hornblende and biotite $^{40}Ar/^{39}Ar$ plateau ages of 1.79–1.77 Ga on the same rocks can be interpreted as the close of deformation and metamorphism (Holm and Schneider, 2002).

The metamorphic history of the Gallatin Range seems very similar to that in the Tobacco Root Mountains. Mogk (1990, 1992) demonstrates a clockwise pressure-temperature path from upper amphibolite or granulite conditions, followed by isothermal decompression. The age of metamorphism and deformation in the Gallatin Range, however, has previously been interpreted as Archean, based on the mid-Archean age of deformed sills interpreted as synkinematic (Mogk, 1990, 1992). Similarly, northeast-striking, northwest-dipping shear zones in the northern Madison and Gallatin Ranges have also been considered Archean because they are crosscut by unfoliated granites dated at 2.55 Ga based on the upper intercept of discordant zircons (Mogk et al., 1992a). We suggest both these ages bear reexamination. The metamorphic facies, structures, and fabrics in the Madison and Gallatin Ranges are parallel to and compatible with those of the 1.78–1.72 Ga Big Sky orogen in the Tobacco Root Mountains, and may also have originated at that time.

Foreland

The Madison mylonite zone in the southern Madison Range (Erslev and Sutter, 1990) is a wide, northeast-striking, northwest-dipping shear zone that roughly parallels Giletti's line (Fig. 5). It is interpreted as a thrust fault because it places higher-grade rocks over lower-grade rocks (Erslev and Sutter, 1990). The Ar system for rocks within the zone appears to have been disturbed at ca. 1.8 Ga, whereas rocks outside the zone give 2.4–2.5 Ga plateau $^{40}Ar/^{39}Ar$ ages (Erslev and Sutter, 1990), demonstrating that the Madison mylonite zone was active during the Big Sky orogeny. Particularly in comparison to the more penetrative deformation in the Tobacco Root Mountains to the northwest, we follow Erslev and Sutter (1990) and interpret the Madison mylonite zone as a component of the cooler, more cratonward part of the Big Sky orogen, where deformation and thermal disturbance was restricted to discrete shear zones.

The Madison mylonite zone is roughly parallel to a number of northeast-striking shear zones within the Montana metasedimentary terrane that juxtapose rocks of different geologic character and/or metamorphic grade, including a mylonite zone in the North Snowy Block that is typically cited as the suture between the Montana metasedimentary terrane and the Beartooth-Bighorn magmatic terrane in the northern Wyoming province (Mogk et al., 1988; Mogk and Henry, 1988; Mogk et al., 1992b; Reid et al., 1975), and shear zones that cut the northern Madison and Gallatin Ranges (the Mirror Lake and Big Brother shear zones of Mogk et al., 1992b). If proven to be Proterozoic, these shear zones constitute a belt of ductile thrust faults between the Big Sky metamorphic core to the northwest and Giletti's line to the southeast that straddles the transition from the infrastructural hinterland to the supracrustal foreland of the orogen.

In outlining his very similar model for a Proterozoic orogen in the northern Wyoming province, O'Neill (1998) suggested that low-grade metasedimentary rocks in the southern Gravelly Range originated as foreland basin deposits localized and overridden by the ductile thrust faults of the southern and northern Madison ranges. As these rocks remain undated, O'Neill's (1998) interpretation is speculative but promising; detrital zircon analysis should substantiate whether or not this basin received sediment shed from the rising core of the Big Sky orogen.

Fossil Subduction Zone

Where it crossed Montana (~110° W longitude), the "Deep Probe" seismic refraction line imaged two north-dipping reflectors between 50 and 100 km in depth, interpreted as the remnants of ancient subduction zones, the more southerly of which projects to the surface along the trend of the Big Sky orogen (Gorman et al., 2002; Henstock et al., 1998). Tomography defines a northeast-striking velocity anomaly in the deeper mantle (100 to 500 km in depth) also below the northern Wyoming province, which may be the result of subduction to that depth (Dueker et al., 2001). While Gorman et al. (2002) interpret the north-dipping mantle reflectors as Archean, we suggest they are better correlated in space and time with the Proterozoic Big Sky orogen (see also Mueller et al., 2004, this volume, Chapter 9). No younger subduction zones in this location and orientation are permissible by the Late Proterozoic through Phanerozoic geologic history of the Cordilleran miogeocline (Bond and Kominz, 1984; Sloss, 1988). The position and attitude of these mantle structures are compatible with an arc collision overriding the northern Wyoming province from north to south as the culmination of north-dipping subduction before 1.78 Ga. The reflectors may mark the location of a cryptic suture, where an ocean that opened following 2.06 Ga rifting associated with MMDS bodies in the Tobacco Root Mountains was consumed. The Spuhler Peak Metamorphic Suite may have originated in or around the arc associated with this subduction zone before being emplaced to the south in the Tobacco Root Mountains as an allochthonous terrane.

Pb, Sr, and Nd isotopic analysis of mantle-derived xenoliths in Eocene minette dikes in the Highwood Mountains (Carlson and Irving, 1994), and of Tertiary alkalic rocks in the Crazy Mountains (Dudas et al., 1987), have been interpreted to reflect a mantle enrichment event, probably brought about by subduction at ca. 1.5–1.8 Ga beneath these areas. The xenoliths from the Highwood Mountains carry 1.8 Ga zircons and 1.78 Ga monazites (Carlson and Irving, 1994). Alternatively, the isotopically enriched character of the Highwood Mountains xenoliths (O'Brien et al., 1995) and of minette bodies and encased xenoliths in the Bearpaw Mountains (MacDonald et al., 1992) have been correlated with what is considered to be a characteristic isotopic profile of the Archean Wyoming province (Frost, 1993; Mueller and Wooden, 1988; Wooden and Mueller, 1988), which suggests that the subcontinental lithosphere of the Wyoming province extends at least that far north of the surface expression of the craton. Either interpretation of the isotope data is consistent with north-dipping subduction closing an ocean basin and drawing the Wyoming craton down to the north below a south-vergent collisional orogen.

Accreted Arc

If the Big Sky orogen is a consequence of north-dipping subduction, a Proterozoic arc should have formed north of the orogen and the subduction zone. Major and trace element geochemistry, isotope geochemistry, and the lack of inherited zircons indicate the arc affinity and juvenile character of granitoid gneisses in the Little Belt Mountains (Mueller et al., 2002). With an intrusive age of ca. 1.86 Ga (U-Pb and Pb-Pb zircon dating by Mueller et al., 1996b, 2002), these rocks would have originated in an arc that was active prior to and at the inception of the Big Sky orogeny, timing that is consistent with an arc-continent collisional orogeny. On this basis, the basement of the Little Belt Mountains should be considered a suspect terrane with respect to the Big Sky orogen.

Precambrian basement in the Little Belt Mountains is the northernmost outlier of the Wyoming province and so may be the only surface exposure of the outboard realm of the Big Sky orogen. Other evidence for the character of this domain must be more indirect. The Mesozoic Idaho batholith straddles the southwest extension of the Big Sky orogen's metamorphic core and any outboard terranes that may be to the north. On the basis of Sm-Nd and Pb isotopes in Mesozoic and Cenozoic igneous rocks, Fleck and others (Fleck and Wooden, 1997; Fleck and Gunn, 1991) and Leeman et al. (1991) locate a northeast-trending boundary between Archean basement to the southeast and Proterozoic basement to the northwest that crosses the central Idaho batholith. Pb isotope ratios in the northern batholith suggest the presence of juvenile igneous crust beneath and around the northern Idaho batholith (Toth and Stacey, 1992). Inherited zircons from the northern Idaho batholith that are largely limited in age to 1.82–1.75 Ga (U-Pb, single zircon SHRIMP dating by Mueller et al. [1995] and Foster and Fanning [1997] that corroborates older, upper-intercept zircon dating by Bickford et al. [1981] and Toth and Stacey [1992]), very similar to the Sm/Nd model age determined for the batholith (Mueller et al., 1995), date that juvenile basement. These studies locate the southwestern extension of the boundary between autochthonous, Wyoming province basement of the Big Sky orogen and an accreted, Proterozoic juvenile arc terrane to the north.

Late Extensional Tectonics

The Snowy shear zone cuts across the north edge of the South Snowy Block of the Beartooth Mountains (Erslev, 1992). It is along strike from the Madison mylonite zone and parallel with the Madison mylonite zone and other ductile shear zones of the foreland belt, which suggests the Snowy shear zone may have been active during the Big Sky orogeny as well (suggested also by Erslev, 1992). The Snowy shear zone, however, has been interpreted as a ductile normal fault based on the offset in metamorphic grade observed across the fault (Erslev, 1992). If these timing and kinematic interpretations are correct, the Snowy shear zone would represent one of the extensional structures on which late-stage collapse and tectonic unroofing in the Big Sky orogen occurred, as recorded by the pressure-temperature history of the Tobacco Root Mountains.

THE GREAT FALLS TECTONIC ZONE AND THE NORTHERN BOUNDARY OF THE WYOMING PROVINCE

The Great Falls tectonic zone in Idaho and Montana was identified by O'Neill and Lopez (1985) on the basis of a linear array of faults that appears to have been active from the Proterozoic to the present and that coincides both with a belt of Mesozoic to Tertiary intrusives and with the trend of geophysical anomalies. The Great Falls tectonic zone is a broad northeast-trending swath that is parallel with and partly overlaps the distribution of the Big Sky orogen, lying along the suture between reworked Wyoming province basement in the Big Sky orogen and accreted juvenile arc rocks (Fig. 5). On this basis, we suggest that the Big Sky orogeny is the Proterozoic stage of the long history of repeated activity that defines the Great Falls tectonic zone (see also O'Neill, 1998).

Since its identification, interest in the Great Falls tectonic zone has focused principally on whether it marks the northern boundary of the Wyoming province (Boerner et al., 1998; Eaton et al., 1999; Holm and Schneider, 2002; Lemieux et al., 2000; Mueller et al., 2002). In assessing the nature of the Great Falls tectonic zone, much attention has been paid to the absence of gravity or electromagnetic signatures expected for a plutonic belt or accretionary sediment wedge that would have been generated by subduction before closure along a boundary. This has led some to look farther north, across the Medicine Hat block to the Vulcan structure, for the limit of the Wyoming province and to attribute the Great Falls tectonic zone to intraplate deformation (Boerner et al., 1998; Eaton et al., 1999; Buhlmann et al., 2000; Gorman et al., 2002; Lemieux et al., 2000). We know of no way other than arc-continent collision to produce a tectonic belt with the characteristics of the Big Sky orogen as we now understand them and conclude that closure of an ocean basin must have occurred (see also Mueller et al., 2002). This model does not, however, obviate the possibility that the Medicine Hat block, or perhaps all of the Hearne province, was related to the Wyoming province prior to rifting at ca. 2.06 Ga so that the northern limit of the Wyoming province was well to the north at that earlier time. The Hearne province, Medicine Hat block, and Wyoming province do share distinct crustal structure (Buhlmann et al., 2000; Gorman et al., 2002; Henstock et al., 1998; Lemieux et al., 2000) and Archean ages (Ross et al., 1991). Nor does the character of the Big Sky orogen eliminate the possibility that the Vulcan structure is also a Proterozoic collision orogen. Both the Medicine Hat block and the Vulcan structure appear to have been tectonically active between 1.7 and 1.85 Ga, as revealed by crustal xenoliths and basement rocks from deep drill cores (Lemieux et al., 2000; Ross et al., 1995, 1991), and the granitoids of the Rimbey belt north of the Vulcan structure are similar in age and arc-character to juvenile plutonic rocks we suggest lie just north of the Big Sky orogen (Ross et al., 1991). If the Big Sky orogen was a broad accretionary orogen, comparable in across-strike width to the Mesozoic Cordillera of western Canada (see also Ross, 2002), it may have ranged from a foreland along Giletti's line, across a metamorphic belt at the collisional suture in the northern Wyoming province, to a series of accreted terranes that could include the Medicine Hat block and remaining Hearne province, with convergent belts both inboard and outboard of each collider. A Proterozoic orogen of this scale would account for the dominance of 1.7–1.9 Ga detrital zircons in the Late Proterozoic Windermere sediments that accumulated in rift basins along the Cordilleran margin in southern British Columbia (Ross and Parrish, 1991).

IMPLICATIONS FOR THE ASSEMBLY OF THE NORTHERN WYOMING PROVINCE AND THE NORTH AMERICAN CRATON

The geologic history of the Tobacco Root Mountains and the Big Sky orogen provides a new context in which to understand the northern Wyoming province but also raises questions that cannot be addressed with our present knowledge. We hope the models introduced here will help focus future work on the following critical issues.

1. What is the relationship of the Big Sky orogen to the Montana metasedimentary terrane? Although rocks older than 1.78 Ga are preserved in many parts of the Montana metasedimentary terrane so that the terrane certainly carries an older history, the Montana metasedimentary terrane and Big Sky orogen are nearly coincident. The Archean history of the Montana metasedimentary terrane is unlike that of the Beartooth-Bighorn magmatic terrane (Mogk et al., 1992a, 1992b; Mogk and Henry, 1988). Juxtaposition of the two is believed to have occurred at ca. 2.55 Ga (Mogk et al., 1992a, 1992b; Mogk and Henry, 1988), but shear zones that define the boundary between the Montana metasedimentary and Beartooth-Bighorn magmatic terranes were, at least in part, active during the Big Sky orogeny and served to transport rocks of the Montana metasedimentary terrane south over the Beartooth-Bighorn magmatic terrane platform. Can any linkages between the Montana metasedimentary and the Beartooth-Bighorn magmatic terranes prior to the Big Sky orogeny be unequivocally demonstrated? The time of accretion of the Montana metasedimentary terrane bears careful reconsideration.

2. What is the relationship of the Big Sky orogen along the northern margin of the Wyoming province to collision and accretion along the eastern and southern margins of this Archean craton? How was the Wyoming province sutured with adjacent Archean cratons? Gravity and magnetic anomaly trends on which the location of the southern Trans-Hudson orogen is

based appear to truncate anomalies associated with the Great Falls tectonic zone (Dutch and Nielson, 1990; Hoffman, 1988; Thomas et al., 1987), and yet the age of the Trans-Hudson belt is traditionally considered to be 1.85–1.80 Ga (Bickford et al., 1990), at least 10–20 million years older than the Big Sky orogen. If Dahl and colleagues (Dahl and Frei, 1998; Dahl et al., 1999) are correct in their interpretation of Proterozoic Black Hills geology, then the time of tectonism in the southern Trans-Hudson orogen may be revised to be contemporaneous with instead of older than the Big Sky orogeny. If not, reinterpretation of the geophysical anomalies is required. In either case, there is remarkable synchroneity between collisional activity along the southern margin of the Wyoming province in the Cheyenne belt (Chamberlain et al., 1993; Houston, 1993; Premo and Van Schmus, 1989) and the development of the Big Sky orogen as we have documented it in the Tobacco Root Mountains (see also Mueller et al., 2002). These timing relationships constrain the process or processes by which the Precambrian basement of the North American craton was assembled, and the orchestration of those processes, and must be of fundamental importance to general models of crustal evolution.

ACKNOWLEDGMENTS

We thank the 35 undergraduate students from the 12 Keck Geology Consortium liberal arts colleges and participating nonconsortium institutions, and all their faculty advisors, for their direct contributions to the research summarized here. Our colleagues Paul Mueller and Dave Mogk have generously shared their extensive expertise in the geology of the northern Wyoming province. This paper benefited immeasurably from reviews by Sam Bowring, Jane Gilotti, and Gerry Ross, but they should not be held responsible for any errors or omissions. We thank them for their input. Finally, we thank Indiana University and the staff of the Indiana University Geologic Field Station, who trained all of us and many others to observe and appreciate field data, and to love the Tobacco Root Mountains. Our work was funded by the W.M. Keck Foundation and the Trustees of Amherst and Smith Colleges.

REFERENCES CITED

Aspler, L.B., and Chiarenzelli, J.R., 1998, Two Neoarchean supercontinents? Evidence from the Paleoproterozoic: Sedimentary Geology, v. 120, p. 75–104.

Bickford, M.E., Chase, R.B., Nelson, B.K., Shuster, R.D., and Arruda, E.C., 1981, U-Pb studies of zircon cores and overgrowths, and monazite: Implications for age and petrogenesis of the northeastern Idaho Batholith: Journal of Geology, v. 89, p. 433–457.

Bickford, M.E., Collerson, K.D., Lewry, J.F., Van Schmus, W.R., and Chiarenzelli, J.R., 1990, Proterozoic collisional tectonism in the Trans-Hudson orogen, Saskatchewan: Geology, v. 18, p. 14–18.

Boerner, D.E., Craven, J.A., Kurtz, R.D., Ross, G.M., and Jones, F.W., 1998, The Great Falls Tectonic Zone: Suture or intracontinental shear zone?: Canadian Journal of Earth Sciences, v. 35, p. 175–183.

Bond, G.C., and Kominz, M.A., 1984, Construction of tectonic subsidence curves for the early Paleozoic miogeocline, southern Canadian Rocky Mountains: Implications for subsidence mechanisms, age of breakup, and crustal thinning: Geological Society of America Bulletin, v. 95, p. 155–173.

Brady, J.B., Mohlman, H.K., Harris, C., Carmichael, S.K., Jacob, L.K., and Chaparro, W.R., 2004a, General geology and geochemistry of metamorphosed Proterozoic mafic dikes and sills, Tobacco Root Mountains, Montana, *in* Brady, J.B., et al., eds., Precambrian geology of the Tobacco Root Mountains, Montana: Boulder, Colorado, Geological Society of America Special Paper 377, p. 89–104 (this volume).

Brady, J.B., Kovaric, D.N., Cheney, J.T., Jacob, L.J., and King, J.T., 2004b, $^{40}Ar/^{39}Ar$ ages of metamorphic rocks from the Tobacco Root Mountains region, Montana, *in* Brady, J.B., et al., eds., Precambrian geology of the Tobacco Root Mountains, Montana: Boulder, Colorado, Geological Society of America Special Paper 377, p. 131–149 (this volume).

Buhlmann, A.L., Cavell, P., Burwash, R.A., Creaser, R.A., and Luth, R.W., 2000, Minette bodies and cognate mica-clinopyroxene xenoliths from the Milk River area, southern Alberta: Records of a complex history of the northernmost part of the Wyoming craton: Canadian Journal of Earth Sciences, v. 37, p. 1629–1650.

Burger, H.R., III, 1969, Structural evolution of the southwestern Tobacco Root Mountains, Montana: Geological Society of America Bulletin, v. 80, p. 1329–1342.

Burger, H.R., 2004, General geology and tectonic setting of the Tobacco Root Mountains, Montana, *in* Brady, J.B., et al., eds., Precambrian geology of the Tobacco Root Mountains, Montana: Boulder, Colorado, Geological Society of America Special Paper 377, p. 1–14 (this volume).

Burger, H.R., Peck, W.H., Johnson, K.E., Tierney, K.A., Poulsen, C.J., Cady, P., Lowell, J., MacFarlane, W.A., Sincock, M.J., Archuleta, L.L., Pufall, A., and Cox, M.J., 2004, Geology and geochemistry of the Spuhler Peak Metamorphic Suite, *in* Brady, J.B., et al., eds., Precambrian geology of the Tobacco Root Mountains, Montana: Boulder, Colorado, Geological Society of America Special Paper 377, p. 47–70 (this volume).

Carlson, R.W., and Irving, A.J., 1994, Depletion and enrichment history of subcontinental lithospheric mantle: An Os, Sr, Nd and Pb isotopic study of ultramafic xenoliths from the northwestern Wyoming craton: Earth and Planetary Science Letters, v. 126, p. 457–472.

Catanzaro, E.J., and Kulp, J.L., 1964, Discordant zircons from the Little Belt (Montana), Beartooth (Montana) and Santa Catalina (Arizona) Mountains: Geochimica et Cosmochimica Acta, v. 28, p. 87–124.

Chamberlain, K.R., Patel, S.C., Frost, R.B., and Snyder, G.L., 1993, Thick-skinned deformation of the Archean Wyoming province during Proterozoic arc-continent collision: Geology, v. 21, p. 995–998.

Cheney, J.T., Brady, J.B., Tierney, K.A., DeGraff, K.A., Mohlman, H.K., Frisch, J.D., Hatch, C.E., Steiner, M.L., Carmichael, S.K., Fisher, R.G.M., Tuit, C.B., Steffen, K.J., Cady, P., Lowell, J., Archuleta, L., Hirst, J., Wegmann, K.W., and Monteleone, B., 2004a, Proterozoic metamorphism of the Tobacco Root Mountains, Montana, *in* Brady, J.B., et al., eds., Precambrian geology of the Tobacco Root Mountains, Montana: Boulder, Colorado, Geological Society of America Special Paper 377, p. 105–129 (this volume).

Cheney, J.T., Webb, A.A.G., Coath, C.D., and McKeegan, K.D., 2004b, In situ ion microprobe $^{207}Pb/^{206}Pb$ dating of monazite from Precambrian metamorphic suites, Tobacco Root Mountains, Montana, *in* Brady, J.B., et al., eds., Precambrian geology of the Tobacco Root Mountains, Montana: Boulder, Colorado, Geological Society of America Special Paper 377, p. 151–179 (this volume).

Dahl, P.S., and Frei, R., 1998, Step-leach Pb-Pb dating of inclusion-bearing garnet and staurolite, with implications for Early Proterozoic tectonism in the Black Hills collisional orogen, South Dakota, United States: Geology, v. 26, p. 111–114.

Dahl, P.S., Holm, D.K., Garder, E.T., Hubacher, F.A., and Foland, K.A., 1999, New constraints on the timing of Early Proterozoic tectonism in the Black Hills (South Dakota), with implications for docking of the Wyoming province with Laurentia: Geological Society of America Bulletin, v. 111, p. 1335–1349.

Dahl, P.S., Hamilton, M.A., Stern, R.A., Frei, R., and Berg, R.B., 2000, In situ SHRIMP investigation of an Early Proterozoic metapelite, with implications for Pb-Pb step leach dating of garnet and staurolite: Geological Society of America Abstracts with Programs, v. 32, no. 7, p. A-297.

Dudas, F.O., Carlson, R.W., and Eggler, D.H., 1987, Regional Middle Proterozoic enrichment of the subcontinental mantle source of igneous rocks from central Montana: Geology, v. 15, p. 22–25.

Dueker, K., Yuan, H., and Zurek, B., 2001, Thick-structured Proterozoic lithosphere of the Rocky Mountain region: GSA Today, v. 11, no. 12, p. 4–9.

Dutch, S.I., and Nielsen, P.A., 1990, The Archean Wyoming province and its relations with adjacent Proterozoic provinces: Geological Association of Canada Special Paper 37, p. 287–300.

Eaton, D.W., Ross, G.M., and Clowes, R.M., 1999, Seismic-reflection and potential-field studies of the Vulcan structure, western Canada: A Paleoproterozoic Pyrenees?: Journal of Geophysical Research, v. 104, p. 23,255–23,269.

Erslev, E.A., 1992, Precambrian geology and ductile normal faulting in the southwest corner of the Beartooth Uplift, Montana, *in* Bartholomew, M.J., et al., eds., Basement tectonics 8: Characterization and comparison of ancient and Mesozoic continental margins—Proceedings of the 8th International Conference on Basement Tectonics (Butte, Montana, 1988): Dordrecht, Netherlands, Kluwer Academic Publishers, p. 313–322.

Erslev, E.A., and Sutter, J.F., 1990, Evidence for Proterozoic mylonitization in the northwestern Wyoming province: Geological Society of America Bulletin, v. 102, p. 1681–1694.

Fleck, R.J., and Gunn, S.H., 1991, The use of integrated geological, geochemical, and isotopic studies in the identification and interpretation of lithospheric boundaries in the northwestern United States: Geological Society of America Abstracts with Programs, v. 23, no. 5, p. A210.

Fleck, R.J., and Wooden, J.L., 1997, Isotopic identification of the Archean/Proterozoic lithospheric boundary in Idaho: Geological Society of America Abstracts with Programs, v. 29, no. 6, p. A70–A71.

Foster, D.A., and Fanning, C.M., 1997, Geochronology of the northern Idaho batholith and the Bitterroot metamorphic core complex: Magmatism preceding and contemporaneous with extension: Geological Society of America Bulletin, v. 109, p. 374–394.

Friberg, L.M., 1994, Geothermobarometry of Precambrian metamorphic rocks from the Tobacco Root Mountains of southwestern Montana: Geological Society of America Abstracts with Programs, v. 26, no. 5, p. 16.

Frost, C.D., 1993, Nd isotopic evidence for the antiquity of the Wyoming province: Geology, v. 21, p. 351–354.

Frost, C.D., and Frost, B.R., 1993, The Archean history of the Wyoming province, *in* Snoke, A.W., et al., eds., The geology of Wyoming: Geological Survey of Wyoming Memoir 5, p. 58–76.

Garihan, J.M., 1979, Geology and structure of the central Ruby Range, Madison County, Montana: Geological Society of America Bulletin, v. 90, p. 323–326 (Part I), p. 695–788 (Part II).

Giletti, B.J., 1966, Isotopic ages from southwestern Montana: Journal of Geophysical Research, v. 71, p. 4029–4036.

Gillmeister, N.M., 1972, Cherry Creek Group–Pony Group relationship in the central Tobacco Root Mountains, Madison County, Montana: Northwest Geology, v. 1, p. 21–24.

Gorman, A.R., Clowes, R.M., Ellis, R.M., Henstock, T.J., Spence, G.D., Keller, G.R., Levander, A., Snelson, C.M., Burianyk, M.J.A., Kanasewich, E.R., Asudeh, I., Hajnal, Z., and Miller, K.C., 2002, Deep probe: Imaging the roots of western North America: Canadian Journal of Earth Sciences, v. 39, p. 375–398.

Harms, T.A., Burger, H.R., Blednick, D.G., Cooper, J.M., King, J.T., Owen, D.R., Lowell, J., Sincock, M.J., Kranenburg, S.R., Pufall, A., and Picornell, C.M., 2004, Character and origin of Precambrian fabrics and structures in the Tobacco Root Mountains, Montana, *in* Brady, J.B., et al., eds., Precambrian geology of the Tobacco Root Mountains, Montana: Boulder, Colorado, Geological Society of America Special Paper 377, p. 203–226 (this volume).

Henstock, T.J., Snelson, C.M., Keller, G.R., Miller, K.C., Harder, S.H., Gorman, A.R., Clowes, R.M., Burianyk, M.J.A., and Humphreys, E.D. (the Deep Probe Working Group), 1998, Probing the Archean and Proterozoic lithosphere of western North America: GSA Today, v. 8, no. 7, p. 1–5, 16–17.

Hess, D.F., and Vitaliano, C.J., 1990, Metabasites: An Indicator of Late Archean geologic history in the Tobacco Root Mountains, Madison County, Montana: Geological Society of America Abstracts with Programs, v. 22, no. 5, p. 13.

Hoffman, P.F., 1988, United Plates of America, the birth of a craton: Early Proterozoic assembly and growth of Laurentia: Annual Review of Earth and Planetary Science, v. 16, p. 543–603.

Holdaway, M.J., 1971, Stability of andalusite and the aluminum silicate phase diagram: American Journal of Science, v. 271, p. 97–131.

Holm, D., and Schneider, D., 2002, $^{40}Ar/^{39}Ar$ evidence for ca. 1800 Ma tectonothermal activity along the Great Falls tectonic zone, central Montana: Canadian Journal of Earth Sciences, v. 39, p. 1719–1728.

Houston, R.S. 1993, Late Archean and Early Proterozoic geology of southeastern Wyoming, *in* Snoke, A.W., et al., eds., The geology of Wyoming: Geological Survey of Wyoming Memoir 5, p. 78–116.

Houston, R.S., Erslev, E.A., Frost, C.D., Karlstrom, K.E., Page, N.J., Zientek, M.L., Reed, J.C., Jr., Snyder, G.L., Worl, R.G., Bryant, B., Reynolds, M.W., and Peterman, Z.E., 1993, The Wyoming province, *in* Reed, J.C., Jr., et al., eds., Precambrian: Conterminous U.S.: Boulder, Colorado, Geological Society of America, Geology of North America, v. C-2, p. 121–170.

James, H.L., and Hedge, C.E., 1980, Age of the basement rocks of southwest Montana: Geological Society of America Bulletin, v. 91, p. 11–15.

Johnson, K.E., Brady, J.B., MacFarlane, W.A., Thomas, R.B., Poulsen, C., and Sincock, M.J., 2004, Precambrian meta-ultramafic rocks from the Tobacco Root Mountains, Montana, *in* Brady, J.B., et al., eds., Precambrian geology of the Tobacco Root Mountains, Montana: Boulder, Colorado, Geological Society of America Special Paper 377, p. 71–87 (this volume).

Krogh, T.E., Corfu, F., Davis, D.W., Dunning, G.R., Heaman, L.M., Kamo, S.L., and Machado, N., 1987, Precise U-Pb isotopic ages of diabase dykes and mafic to ultramafic rocks using trace amounts of baddeleyite and zircon, *in* Halls, H.C., and Farig, W.F., eds., Mafic dyke swarms: Geological Association of Canada Special Paper 34, p. 147–152.

Krogh, T.E., Kamo, S., and Hess, D.F., 1997, Wyoming province 3300+ Ma gneiss with 2400 Ma metamorphism, northwestern Tobacco Root Mountains, Madison County, Montana: Geological Society of America Abstracts with Programs, v. 29, no. 6, p. A-408.

Leeman, W.P., Norman, M.D., and Clarke, C.B., 1991, Geochemical evidence for an Archean/Proterozoic terrane boundary north of the Snake River Plain, central Idaho: Geological Society of America Abstracts with Programs, v. 23, no. 5, p. A210.

Lemieux, S., Ross, G.M., and Cook, F.A., 2000, Crustal geometry and tectonic evolution of the Archean crystalline basement beneath the southern Alberta Plains, from new seismic reflection and potential-field studies: Canadian Journal of Earth Sciences, v. 37, p. 1473–1491.

MacDonald, R., Upton, B.G.J., Collerson, K.D., Hearn, B.C., Jr., and James, D., 1992, Potassic mafic lavas of the Bearpaw Mountains, Montana: Mineralogy, chemistry, and origin: Journal of Petrology, v. 33, p. 305–346.

Mogk, D.W., 1990, A model for the granulite-migmatite association in the Archean basement of southwestern Montana, *in* Vielzeuf, D., and Vidal, Ph., eds., Granulites and crustal evolution: Dordrecht, Netherlands, Kluwer Academic Publishers, p. 133–155.

Mogk, D.W., 1992, Ductile shearing and migmatization at mid-crustal levels in an Archean high-grade gneiss belt, northern Gallatin Range, Montana, USA: Journal of Metamorphic Geology, v. 10, p. 427–438.

Mogk, D.W., and Henry, D.J., 1988, Metamorphic petrology of the northern Archean Wyoming province, southwestern Montana: Evidence for Archean collisional tectonics, *in* Ernst, W.G., ed., Metamorphism and crustal evolution of the western United States, Rubey Volume VII: Englewood Cliffs, New Jersey, Prentice-Hall, p. 362–382.

Mogk, D.W., Mueller, P.A., and Wooden, J.L., 1988, Archean tectonics of the North Snowy block, Beartooth Mountains, Montana: Journal of Geology, v. 96, p. 125–141.

Mogk, D.W., Mueller, P.A., and Wooden, J.L., 1992a, The nature of terrane boundaries: An example from the northern Wyoming province: Precambrian Research, v. 55, p. 155–168.

Mogk, D.W., Mueller, P.A., Wooden, J.L., and Bowes, D.R., 1992b, The northern Wyoming province: Contrasts in Archean crustal evolution, *in* Bartholomew, M.J., et al., eds., Basement tectonics 8: Characterization and comparison of ancient and Mesozoic continental margins—Proceedings of the 8th International Conference on Basement Tectonics (Butte, Montana, 1988): Dordrecht, Netherlands, Kluwer Academic Publishers, p. 283–297.

Mogk, D.W., Burger, H.R., Mueller, P.A., Wooden, J.L., D'Arcy, K., Heatherington, A.L., Abeyta, R.L., Martin, J., and Jacob, L.J., 2004, Geochemistry of quartzofeldspathic gneisses and metamorphic mafic rocks of the Indian Creek and Pony–Middle Mountain Metamorphic Suites, Tobacco Root Mountains, Montana, *in* Brady, J.B., et al., eds., Precambrian geology of the Tobacco Root Mountains, Montana: Boulder, Colorado, Geological Society of America Special Paper 377, p. 15–46 (this volume).

Montgomery, C.W., and Lytwyn, J.N., 1984, Rb-Sr systematics and ages of principal Precambrian lithologies in the South Snowy block, Beartooth Mountains: Journal of Geology, v. 92, p. 103–112.

Mueller, P.A., and Cordua, W.S., 1976, Rb-Sr whole rock ages of gneiss from the Horse Creek area, Tobacco Root Mountains, Montana: Isochron/West, v. 16, p. 33–36.

Mueller, P.A., and Wooden, J.L., 1988, Evidence for Archean subduction and crustal recycling, Wyoming province: Geology, v. 16, p. 871–874.

Mueller, P.A., Shuster, R.D., Wooden, J.L., Erslev, E.A., and Bowes, D.R., 1993, Age and composition of Archean crystalline rocks from the southern Madison Range, Montana: Implications for crustal evolution in the Wyoming craton: Geological Society of America Bulletin, v. 105, p. 437–446.

Mueller, P.A., Shuster, R.D., D'Arcy, K.A., Heatherington, A.L., Nutman, A.P., and Williams, I.S., 1995, Source of the northeastern Idaho batholith: Isotopic evidence for a paleoproterozoic terrane in the northwestern U.S.: Journal of Geology, v. 103, p. 63–72.

Mueller, P.A., Heatherington, A.L., D'Arcy, K.A., Wooden, J.L., and Nutman, A.P., 1996a, Contrasts between Sm-Nd whole-rock and U-Pb zircon systematics in the Tobacco Root batholith, Montana: Implications for the determination of crustal age provinces: Tectonophysics, v. 265, p. 169–179.

Mueller, P., Heatherington, A., Wooden, J., Mogk, D., and Nutman, A., 1996b, Proterozoic evolution of the northwestern Wyoming craton: Geological Society of America Abstracts with Programs, v. 28, no. 7, p. 314.

Mueller, P.A., Wooden, J.L., Mogk, D.W., Nutman, A.P., and Williams, I.S., 1996c, Extended history of a 3.5 Ga trondhjemitic gneiss, Wyoming province, USA: Evidence from U-Pb systematics in zircon: Precambrian Research, v. 78, p. 41–52.

Mueller, P., Wooden, J., Heatherington, A., and Nutman, A., 1997, Distribution of Proterozoic crust along the NW margin of cratonic North America: Evidence from U-Pb zircon ages and isotopic systematics in young granitoids: Geological Society of America Abstracts with Programs, v. 29, no. 6, p. A-70.

Mueller, P.A., Wooden, J.L., Nutman, A.P., and Mogk, D.W., 1998, Early Archean crust in the northern Wyoming province: Evidence from U-Pb ages of detrital zircons: Precambrian Research, v. 91, p. 295–307.

Mueller, P.A., Heatherington, A.L., Kelly, D.M., Wooden, J.L., and Mogk, D.W., 2002, Paleoproterozoic crust within the Great Falls tectonic zone: Implications for the assembly of southern Laurentia: Geology, v. 30, p. 127–130.

Mueller, P.A., Burger, H.R., Wooden, J.L., Heatherington, A.L., Mogk, D.W., and D'Arcy, K., 2004, Age and evolution of the Precambrian crust of the Tobacco Root Mountains, Montana, *in* Brady, J.B., et al., eds., Precambrian geology of the Tobacco Root Mountains, Montana: Boulder, Colorado, Geological Society of America Special Paper 377, p. 181–202 (this volume).

O'Brien, H.E., Irving, A.J., McCallum, I.S., and Thirwall, M.E., 1995, Strontium, neodymium, and lead isotopic evidence for the interaction of post-subduction asthenospheric potassic mafic magmas of the Highwood Mountains, Montana, USA, with ancient Wyoming craton lithospheric mantle: Geochimica et Cosmochimica Acta, v. 59, p. 4539–4556.

O'Neill, J.M., 1983, Geologic map of the Middle Mountain–Tobacco Root roadless area, Madison County, Montana: U.S. Geological Survey Miscellaneous Field Studies Map MF-1590-A, scale 1:50,000.

O'Neill, J.M., 1998, The Great Falls tectonic zone, Montana-Idaho: An Early Proterozoic collisional orogen beneath and south of the Belt basin, *in* Berg, R.B., ed., Belt Symposium III—1993: Montana Bureau of Mines and Geology Special Publication 112, p. 222–228.

O'Neill, J.M., and Lopez, D.A., 1985, Character and regional significance of Great Falls Tectonic Zone, east-central Idaho and west-central Montana: American Association of Petroleum Geologists Bulletin, v. 69, p. 437–447.

O'Neill, J.M., Duncan, M.A., and Zartman, R.E., 1988, An Early Proterozoic gneiss dome in the Highland Mountains, southwestern Montana, *in* Lewis, S.E., and Berg, R.B., eds., Precambrian and Mesozoic plate margins; Montana, Idaho and Wyoming, with field guides for the 8th International Conference on Basement Tectonics: Montana Bureau of Mines and Geology Special Publication 96, p. 81–88.

Premo, W.R., and Van Schmus, W.R., 1989, Zircon geochronology of Precambrian rocks in southeastern Wyoming and northern Colorado: Geological Society of America Special Paper 235, p. 13–32.

Reid, R.R., McMannis, W.J., and Palmquist, J.C., 1975, Precambrian geology of North Snowy block, Beartooth Mountains, Montana: Geological Society of America Special Paper 157, 135 p.

Roberts, H., Dahl, P., Kelley, S., and Frei, R., 2002, New ^{207}Pb-^{206}Pb and ^{40}Ar-^{39}Ar ages from SW Montana, USA: Constraints on the Proterozoic and Archean tectonic and depositional history of the Wyoming province: Precambrian Research, v. 117, p. 119–143.

Ross, G.M., 2002, Evolution of Precambrian continental lithosphere in western Canada: Results from Lithoprobe studies in Alberta and beyond: Canadian Journal of Earth Sciences, v. 39, p. 413–437.

Ross, G.M., and Parrish, R.R., 1991, Detrital zircon geochronology of metasedimentary rocks in the southern Omineca Belt, Canadian Cordillera: Canadian Journal of Earth Sciences, v. 28, p. 1254–1270.

Ross, G.M., Parrish, R.R., Villeneuve, M.E., and Bowring, S.A., 1991, Geophysics and geochronology of the crystalline basement of the Alberta Basin, western Canada: Canadian Journal of Earth Sciences, v. 28, p. 512–522.

Ross, G.M., Mildereit, B., Eaton, D., White, D., Kanasewich, E.R., and Burianyk, M.J.A., 1995, Paleoproterozoic collision orogen beneath the western Canada sedimentary basin imaged by Lithoprobe crustal seismic-reflection data: Geology, v. 23, p. 195–199.

Schmitt, J.G., 1988, Sedimentation and tectonic setting of the middle Proterozoic LaHood Formation, Belt Supergroup, southwestern Montana, *in* Lewis, S.E., and Berg, R.B., eds., Precambrian and Mesozoic plate margins; Montana, Idaho and Wyoming, with field guides for the 8th International Conference on Basement Tectonics: Montana Bureau of Mines and Geology Special Publication 96, p. 89–96.

Sloss, L.L., 1988, Tectonic evolution of the craton in Phanerozoic time, *in* Sloss, L.L., ed., Sedimentary cover—North American craton, U.S.: Boulder, Colorado, Geological Society of America, Geology of North America, v. D-2, p. 25–51.

Thomas, M.D., Sharpton, V.L., and Grieve, R.A.F., 1987, Gravity patterns and Precambrian structure in the North American Central Plains: Geology, v. 15, p. 489–492.

Toth, M.I., and Stacey, J.S., 1992, Constraints on the formation of the Bitterroot lobe of the Idaho batholith, Idaho and Montana, from U-Pb zircon geochronology and feldspar Pb isotopic data: U.S. Geological Survey Bulletin 2008, 14 p.

Vitaliano, C.J., Burger, H.R., III, Cordua, W.S., Hanley, T.B., Hess, D.F., and Root, F.K., 1979a, Geologic map of southern Tobacco Root Mountains, Madison County, Montana: Geological Society of America Map and Chart Series MC31, scale 1:62,500, 1 sheet, 8 p. text.

Vitaliano, C.J., Burger, H.R., III, Cordua, W.S., Hanley, T.B., Hess, D.F., and Root, F.K., 1979b, Explanatory text to accompany geologic map of southern Tobacco Root Mountains, Madison County, Montana: Geological Society of America Map and Chart Series MC31, scale 1:62,500, 1 sheet, 8 p. text.

Vogl, J.J., Foster, D.A., Mueller, P.A., Mogk, D.W., and Wooden, J.L., 2002, Age and character of Precambrian basement in the Little Belt Mountains, Montana: Implications for the role of the Great Falls Tectonic Zone in the Paleoproterozoic assembly of North America: Geological Society of America Abstracts with Programs, v. 34, no. 6, p. 273.

Wooden, J.L., and Mueller, P.A., 1988, Pb, Sr, and Nd isotopic compositions of a suite of Late Archean, igneous rocks, eastern Beartooth Mountains: Implications for crust-mantle evolution: Earth and Planetary Science Letters, v. 87, p. 59–72.

Wooden, J.L., Vitaliano, C.J., Koehler, S.W., and Ragland, P.C., 1978, The Late Precambrian mafic dikes of the southern Tobacco Root Mountains, Montana: Geochemistry, Rb-Sr geochronology and relationship to Belt tectonics: Canadian Journal of Earth Sciences, v. 15, p. 467–479.

Wooden, J.L., Mueller, P.A., and Mogk, D.W., 1988, A Review of the geochemistry and geochronology of the Archean rocks of the northern part of the Wyoming province, *in* Ernst, W.G., ed., Metamorphism and crustal evolution of the western United States, Rubey Volume VII: Englewood Cliffs, New Jersey, Prentice-Hall, p. 383–410.

Manuscript Accepted by the Society September 12, 2003

Printed in the USA

Index of Keywords

Explanatory text to accompany geologic map of southern Tobacco Root Mountains, Madison County, Montana

Charles J. Vitaliano
Department of Geology, Indiana University, Bloomington, Indiana 47405, USA

H. Robert Burger III
Department of Geology, Smith College, Northampton, Massachusetts 01060, USA

William S. Cordua
Department of Plant & Earth Science, University of Wisconsin, River Falls, Wisconsin 54022, USA

Thomas B. Hanley
Department of Earth Science, Columbus College, Columbus, Georgia 31907, USA

David F. Hess
Department of Geology, Western Illinois University, Macomb, Illinois 61445, USA

Forrest K. Root
Ensearch, P.O. Box 4815, Midland, Texas 79701, USA

ABSTRACT

The core of the Tobacco Root Mountains, a domal uplift of southwestern Montana, is composed of regionally metamorphosed Archean rocks that have been intruded by diabase dikes of late Precambrian age and by igneous rocks of late Mesozoic and Cenozoic age. The Archean rocks, originally a stratiform sequence of epiclastic, volcaniclastic, and carbonate sedimentary rocks, have been metamorphosed to amphibolite-granulite rank with attendant migmatization. The distribution of the metamorphic rocks and the later intrusions is shown on the map.

In this report to accompany the map, the nature of the crystalline rocks is described briefly. Two thermal events affected the area in Precambrian time, one about 2.7 b.y. B.P. and the second at about 1.6 b.y. B.P. A third event may have begun in very late Precambrian time, but is most strongly represented by changes of late Mesozoic age.

Because of the structural complexities, correlation of the metamorphic rocks in the Tobacco Root Mountains with Precambrian sequences elsewhere in southwestern Montana is extremely difficult.

INTRODUCTION

Archean metamorphic rocks and associated unmetamorphosed igneous intrusion of Proterozoic and late Mesozoic age in the Tobacco Root Mountains, Montana, crop out in an area 24 km wide and 48 km long. The Archean rocks are bounded on the north and west by Phanerozoic sedimentary rocks and associated Upper Cretaceous intrusions and capped on the south and east by Paleozoic clastic and Tertiary volcanogenic deposits.

Recent studies of these Archean rocks—which are the northernmost Archean rocks in the Wyoming province (Condie and others, 1969)—and their associated Proterozoic intrusions (Wooden and others, 1978) have played an important role in deciphering the geologic history of the Northern Rocky Mountains. Previous studies, because of their restricted areal or topical nature (Hayden, 1872; Merrill, 1895; Peale, 1896; Tansley and others, 1933; Levandowski, 1956; Reid, 1957), have not only failed to produce a unified concept of the distribution, structure,

Reprinted from Vitaliano, C.J., Burger, H.R., III, Cordua, W.S., Hanley, T.B., Hess, D.F., and Root, F.K., 1979, Explanatory text to accompany geologic map of southern Tobacco Root Mountains, Madison County, Montana: Geological Society of America Map and Chart Series MC-31, *for inclusion in* Brady, J.B., Burger, H.R., Cheney, J.T., and Harms, T.A., editors, 2004, Precambrian geology of the Tobacco Root Mountains, Montana: Boulder, Colorado, Geological Society of America Special Paper 377, 256 p.

composition, and history of these rocks, but have also given rise to contradictory concepts of stratigraphy and correlation with Archean rocks in adjacent areas in southwestern Montana. Therefore, it was decided to map the entire area as a unit in order to define the stratigraphy, develop a structural concept of the region consistent with that stratigraphy, elucidate the metamorphic history, and provide an integrated base map for future studies in that part of the range. The accompanying map is the result.

GEOLOGIC SETTING

The stratigraphic section of the Archean rocks in the Tobacco Root Mountains consists of a twofold lithologic association—one, a sequence composed of marbles, sillimanite schists, iron-formation, and quartzites with lesser amounts of amphibolite and quartzofeldspathic gneiss, dominates the western half of the map area; the other, a sequence composed almost entirely of quartzofeldspathic gneiss and amphibolite, dominates the eastern part. Metamorphosed basaltic rocks of Archean age and unmetamorphosed diabase dikes of late Precambrian age are associated with both sequences.

The Archean rock units are described below in the order of abundance and not necessarily in stratigraphic sequence. Thicknesses given represent present conditions only and, owing to deformation, probably do not reflect the original thickness of the protoliths. Furthermore, regionally recognizable structures, textures, or other stratigraphic features such as basal conglomerates and graded bedding are lacking so that assignment of a top or bottom to the stratigraphic section has not been possible.

Most of the Archean rocks in the area have been regionally metamorphosed to the upper amphibolite facies as defined by Turner (1968). However, isolated patches of lower granulite-facies assemblages suggest that a higher rank of metamorphism was attained locally in H_2O-deficient pockets of the quartzofeldspathic gneiss and metabasite igneous rocks in the course of a single prograde metamorphic event. Alternatively, incomplete retrogressive metamorphism may have failed to obliterate entirely the evidence for an earlier metamorphic event of higher grade.

Retrograde metamorphic assemblages developed later in shear zones and at the contacts with intrusions of Proterozoic or Mesozoic age. These commonly are of greenschist-facies rank, but pyroxene-hornfels– and hornblende-hornfels–facies assemblages occur adjacent to large Mesozoic intrusions.

ARCHEAN ROCKS

Quartzofeldspathic Gneiss

Quartzofeldspathic gneiss is the most common unit of the Archean metamorphic assemblages. It consists of at least four rock types: granoblastic to blastomylonitic granitic paragneiss rich in K-feldspar, tonalitic paragneiss rich in plagioclase, quartz-feldspar-biotite gneiss, and plagioclase-quartz-hornblende-biotite gneiss. The quartzofeldspathic gneiss is commonly interlayered with thin layers of amphibolite, quartzite, and sillimanite schist. On the other hand, thin layers of quartzofeldspathic gneiss units are interlayered with most of the other Archean assemblages in the map area, as well. The thickness of the entire unit is estimated to be about 5 km. Migmatization is common throughout, and leucosomes appear to have been mobilized and injected into other parts of the unit. Chemical data, along with the occurrence of crosscutting and concordant pegmatitic veinlets, support the suggestion (Cordua, 1973) that the metamorphic rocks have undergone significant K_2O-metasomatism.

The quartzofeldspathic gneiss generally appears to have behaved competently during deformation, although locally (as along Mill Creek, in the northwest part of the range) migmatite structures suggest extensive flow within the rock mass.

Aluminous and mafic assemblages in the quartzofeldspathic gneisses have been recognized throughout the entire area by Root (1965), Hanley (1975), Burger (1966), Gillmeister (1971), Cordua (1973), and Vitaliano. All the assemblages are consistent with amphibolite-facies metamorphism. In the Mill Gulch cirque area (Cordua, 1973) and at other localities, hypersthene, diopside, rodded quartz, granoblastic antiperthitic plagioclase, and perthite in an assemblage of charnockitic composition may represent prograde metamorphism in H_2O-deficient pockets.

Pervasive metamorphism and intense deformation have obliterated all traces of primary texture and structure in the quartzofeldspathic gneiss except the universally conformable nature of its contact with enclosing units. Consequently, the field relationships do not support a dominantly intrusive origin for its protolith. Grain-shape studies (Hess, 1967; Cordua, 1973) indicate that 72% to 100% of the zircons extracted from the quartzofeldspathic gneisses are rounded; this supports the theory that the parent rocks were chiefly sedimentary; the interlayered mafic units probably represent lava flows or thin pyroclastic layers. Although chemical evidence (Cordua, 1973) indicates that most of the quartzofeldspathic gneiss protoliths have been subjected to alkali metasomatism, some of the rocks which have escaped such metasomatism appear to have been derived from arkose, graywacke, and shale.

Whole-rock Rb-Sr data (Mueller and Cordua, 1976) suggest an age of 2,667 ± 66 m.y. and $^{87}Sr/^{86}Sr$ of 0.707 ± 0.002 for the quartzofeldspathic gneiss in the Horse Gulch area.

Hornblende Gneiss

Hornblende gneiss, the second most common unit in the area, consists mainly of stout prisms of green and brown hornblende; the rock types range from para-amphibolites to hornblende-plagioclase gneisses. Hornblende gneiss is widespread throughout the range but is especially abundant in the western and northeastern part where it forms conspicuous dark gray to black outcrops that may be traced along strike for many kilometers. One prominent band in the southern part of the area is traceable for more than 12 km. Individual hornblende gneiss map units may be as much as 900 m thick, especially where they are associated with thin layers of quartzofeldspathic gneiss and other rock types including thin lenses of marble. The total thickness of the hornblende gneiss is about 2.5 to 3 km.

The hornblende gneiss units are readily distinguished, by their grain size and continuity, from the finely crystalline orthoamphibolite that commonly forms sills, dikes, and boudins in the quartzofeldspathic gneisses. With few exceptions, such as on the noses of folds and at loci of shearing, the contact between the hornblende gneiss and the associated rock types is concordant with the foliation, especially with that of the quartzofelspathic gneiss.

Green or brown hornblende and plagioclase ($An_{25–52}$) are the dominant minerals. Almandine porphyroblasts (as much as 5 cm in diameter), clinopyroxene, orthopyroxene, cummingtonite, and biotite are commonly present, as are accessory epidote, sphene, magnetite, and apatite. Green hornblende assemblages are associated mainly with kyanite assemblages and with sillimanite-bearing rocks in the lowest part of the sillimanite zone. Brown hornblende assemblages are more typically associated with sillimanite-bearing assemblages in the middle or higher parts of the sillimanite zone.

The hornblende gneiss exhibits varying degrees of foliation and banding. Locally it is extensively migmatized, and the leucosomes apparently have been derived by partial melting. In small outcrops scattered throughout the range, a coarse-grained biotite-hornblende diorite has been developed that is similar to a variety ascribed by Mehnert (1968) to extreme anatexis of amphibolite or biotite gneiss.

The nature of the protoliths for the hornblende gneiss and the associated plagioclase-rich migmatites is not clear. The concordant contacts, their gradational character, the frequent alternation with quartzofeldspathic layers, and common association with aluminosilicate schists and marble suggest for some, at least, a sedimentary origin and an initial composition of clay-rich dolomite.

Because all textures and structures of the protoliths, except the concordance with the enclosing units, have been obliterated (in these rocks as well as in the quartzofeldspathic gneisses), further evidence for the nature of the parent rocks was sought in chemical analysis of unmigmatized samples. The analyses show that these rocks have low K_2O values and that, on Niggli diagram plots, they cluster within the trend for Karroo dolerites. This, along with the lack of crosscutting relationships, suggests that much of the hornblende gneiss has been derived from mafic extrusive rocks probably including some volcaniclastic sedimentary rocks.

Anthophyllite-Gedrite Gneiss and Schist

Anthophyllite-gedrite gneiss with or without cummingtonite crops out in many parts of the range. Cummingtonite is, however, more commonly associated with metamorphosed iron-formation; a rare occurrence in quartzofeldspathic gneiss was reported in the northeastern part of the Tobacco Root Mountains (Root, 1965). The anthophyllite- and/or gedrite-bearing rocks are light gray to clove-brown, purplish-brown, or deep golden-brown and weather to dark brown or blackish brown.

Reid (1957, 1963) first recognized anthophyllite- and gedrite-bearing rocks in the northern part of the range. They have since been described from other parts of the range by Root (1965), Burger (1966), Hess (1967), Gillmeister (1971, 1972), Cordua (1973), Hanley (1975, 1976), and Friberg (1976). In the north, anthophyllite- and/or gedrite-bearing rocks crop out sparsely either as thin layers associated with hornblende gneiss or aluminous schist or as pods of meta-ultramafic rock in which anthophyllite is a constituent. In the western part of the area, Burger (1966) reported anthophyllite gneiss assemblages with cummingtonite and cordierite in a 2,000-ft-thick (608-m) complex sequence of amphibolite, hornblende gneiss, and quartzofeldspathic gneiss. In the Cataract Mountain area in the northwestern part of the region, Hanley (1975) described anthophyllite- and/or gedrite-bearing rocks as part of a gneissic complex. In the southern part of the range, Cordua (1973) and Vitaliano found orthoamphibole in schistose rocks associated with garnet gneiss. The garnet gneiss unit of the southern Tobacco Root Mountains is considered equivalent to the gneiss of Cataract Mountain (Hanley, 1975) and the Spuhler Peak Formation of Gillmeister (1971). Finally, in the area immediately bordering the Tobacco Root batholith, Hess (1967) noted anthophyllite-bearing assemblages in the Lost Cabin Lake–Noble Peak region, the Branham Peaks area, and in the Mill Creek drainage. All are closely associated with hornblende gneiss. In the Mill Creek occurrences, cummingtonite is among the associated minerals. Anthophyllite dominates the assemblages in the northern part of the area, and it occurs as lathlike crystals that may reach 15.2 cm in length. In these assemblages, garnet, quartz, plagioclase ($An_{45–50}$), sillimanite, cummingtonite, cordierite, muscovite, and talc are present along with accessory magnetite, apatite, biotite, rutile, and chlorite. In the northwestern part of the range, anthophyllite is associated with garnet, oligoclase, quartz, rutile, ilmenite, chlorite, and talc. Members of the cummingtonite-grunerite series are also found in these rocks associated with actinolite in equilibrium assemblages. Gedrite is the common orthoamphibole in the southern part of the area. Gedrite-bearing rocks crop out as podiform lenses and layers of coarse-grained (poorly foliated) to medium-grained (excellently foliated) habit. Plagioclase ($An_{30–45}$), chlorite, biotite, garnet, cordierite, and enstatite are the primary minerals in the gedrite assemblages. Cummingtonite, where present, generally replaces the magnesian orthoamphiboles. The associated minerals in the cummingtonite-bearing rocks are labradorite, brown hornblende, aluminous antigorite, and accessory rutile and/or ilmenite. In those rocks in which cummingtonite replaces anthophyllite and/or gedrite, the parallel alignment of the cummingtonite grains suggests that it is a retrograde product owing its origin to a fall in temperature following a peak in a regional metamorphic event (Burger, 1966, p. 44). Cordierite, where present, is associated with orthopyroxene and garnet in a symplectic texture apparently overprinted on a dominantly gedrite–aluminosilicate–biotite assemblage.

Studies on the origin of orthoamphibole-bearing rocks in other areas have shed some light on the nature of the parent material for rocks in this region. Grant (1968) and Lal and Moorhouse (1969), for example, suggested that partial melting of a K-feldspar–plagioclase–biotite–sillimanite–cordierite assemblage may yield a granitic melt plus an anthophyllite residuum. Winkler (1968) noted the formation of gedrite in the residuum during melting of a biotite metagraywacke. In samples from the Sheridan district, Burger (1966) found high contents of Ni^{+2} and Cr^{+3} in analyses of the anthophyllite gneiss assemblage. When plotted on a Niggli diagram, analyses of these rocks fall predominantly in the field for the Karroo dolerites; this suggests that their protoliths may have been mafic and ultramafic igneous rocks. The rare association of orthoamphibole-bearing rocks with widespread granitic migmatite lends support to the anatectic model. The orthoamphibole-bearing garnet gneiss assemblage in the Southern Tobacco Root Mountain may have been derived from a complex interlayered sequence of sedimentary and igneous rocks dominated by pelititc and mafic igneous rock types. According to Auboin (1965), such a sequence is reminiscent of marine sequences consisting of ferruginous shales, radiolarian cherts, mafic to ultramafic volcanic rocks, and thin, intercalated carbonate.

Quartzite

Quartzite occurs in bands about 100 m thick, although locally thicknesses of as much as 400 m have been encountered. The layers characterized as quartzite contain more than 60% quartz; the remainder consists of green mica (fuchsite), biotite, feldspar, sillimanite, kyanite, and, rarely, graphite. The quartzite layers and lenses form erosion-resistant ridges and low hills, respectively. The quartzite is white, red, pink, or yellow. Quartzite layers not mapped individually are also found in association with marble, metamorphosed iron-formation, amphibolite, and garnet gneiss.

Aluminous Schists

Aluminous schists crop out in many parts of the range. Typical assemblages with kyanite or sillimanite or both are prevalent in the west-central and northeastern part of the map area. The mineral assemblages consist of kyanite, sillimanite, quartz, red-brown biotite, garnet, rutile, and, locally, graphite in various proportions. Such rocks are associated with marble, quartzite, gedrite-garnet schist, metamorphosed iron-formation, hornblende-plagioclase gneiss, hornblende granulite, and quartzofeldspathic gneiss. Kyanite is the dominant aluminosilicate in the northern part of the area; sillimanite is prevalent in the northeastern, southeastern, and western parts. Locally, as along the fault in the Pony–Carmichael Creek valley, both sillimanite and kyanite are present. Structural state studies of K-feldspars from aluminous assemblages in the northwestern part of the Tobacco Root Mountains (Hanley, 1975) yielded Spencer B (intermediate obliquity) and orthoclase types (Wright and Stewart, 1968). According to Guidotti and others (1973), such structures are representative of granulite-facies assemblages. Kyanite and sillimanite also form clusters and radiating aggregates in pegmatite and quartz veins. Kyanite locally is replaced by fibrolite sillimanite. This may be the result of an earlier granulite-facies assemblage that was replaced by a sillimanite-bearing amphibolite-facies assemblage during retrogressive metamorphism, as suggested by Reid (1957, 1958, 1963). However, the kyanite-bearing assemblages may also be products of kyanite-almandine–subfacies metamorphism in which the sillimanite assemblages reflect conditions near the univariant sillimanite-kyanite boundary. We favor the latter explanation.

The aluminous schists appear to have been derived from a shale protolith, but may in part have been derived from subgraywackes. The close association of these schists with quartzite and marble and the presence of graphite in some of them suggest that the protoliths were carbonaceous and that the associated quartzites were derived from sandstone.

Iron-Formation

Small, relatively thin iron-bearing bands crop out in the Copper Mountain region (east of Sheridan) and in the Mineral Hill district on the north slope of Carmichael Creek. Lesser outcrops are present in the Meadow Creek area near Sailor Lake, Dry Boulder Creek, Curly Lake cirque (Hanley, 1975; Immega and Klein, 1976), and in scattered small outcrops throughout the range. The iron-formation is composed of magnetite-bearing schists and gneisses, grunerite-magnetite schists, and garnet-quartz rocks. The minerals include magnetite, members of the cummingtonite-grunerite series, blue-green hornblende, orthopyroxenes (ferrohypersthene to eulite compositions), green diopside, garnet, quartz, and plagioclase. In the Copper Mountain occurrence, thin layers of iron-formation are associated with amphibolite, biotite–garnet–sillimanite schist, and massive quartzite. The Meadow Creek unit is similar to that at Copper Mountain. In the Mineral Hill district, iron formation is closely associated with cummingtonite-bearing rocks, discontinuous lenses and pods of garnet-quartz rocks, garnetiferous gneiss and amphibole-kyanite schist. Ferrohypersthene and diopside are also present in both the Mineral Hill and the Copper Mountain deposits. In the Sailor Lake region the iron-formation, which contains magnetite and orthopyroxene, is associated with hornblende gneiss and quartzite. The small scattered outcrops in various parts of the range are associated with pegmatite and kyanite or sillimanite schist. On the basis of electron microprobe studies of pyroxene-rich assemblages, Immega (1976) and Immega and Klein (1976) concluded that the iron-formation assemblages of the Tobacco Root Mountains are of granulite-facies rank. The iron-formation assemblages fall into the magnetite-iron silicate type of James (1954). According to Högberg (1960), the association of iron-formation with cherty-dolomitic marbles, schists, and quartzofeldspathic gneiss suggests a shallow-marine or brackish water origin. Although we favor such an origin for the deposits of the Tobacco Root Mountains, we realize that a deeper-water origin may be possible for those iron-formations associated with amphibolites.

Marble

Metamorphosed carbonate rocks form prominent buff to dark brown ridges in the western, southwestern, and south-central parts of the mapped area. The major carbonate layers range in thickness from 0.3 to 244 m and form excellent marker beds that clearly depict the folded nature of the metamorphic rocks despite local complication by faulting. Elsewhere in the Tobacco Root Mountains, they are sparse and represented by relatively thin discontinuous layers commonly difficult to discern because of their association with amphibolite layers. An exception is the very tight isoclinally folded carbonate layers in the northwestern part of the range (Hanley, 1976).

Calcite and dolomite, in a 3:1 ratio, are the dominant minerals in the carbonate layers. The freshly broken surface of the rock is white and laced with bands and lenses of quartz and calc-silicates, such as partly serpentinized forsterite and large diopside porphyroblasts replaced in part by fine accicular tremolite. The silicate minerals are present in accessory quantities only and probably represent the metamorphic products of chert-rich layers and lenses in the carbonate protoliths. The chert nodules or beds have been severely disrupted and deformed into boudins during metamorphism. Where SiO_2 was deficient, forsterite developed, but elsewhere diopside formed. In the vicinity of plutons of Late Cretaceous age, such as the Tobacco Root batholith, hydrothermal contact metamorphism developed albite–epidote–hornfels–facies assemblages in which serpentine replaces forsterite and tremolite replaces diopside.

Pegmatite

Lenses and streaks of pegmatite, mostly of small dimensions, crop out in many parts of the Tobacco Root Mountains. Some are concordant, others are discordant to the foliation of the metamorphic host rock; nevertheless, they are all considered to be of Precambrian age. Pegmatites that crop out around the margins of the Tobacco Root batholith are of late Mesozoic age.

Pegmatites of the concordant group occur as pods in the crests of subisoclinal folds and are clearly of magmatic origin. They are the oldest pegmatites in the area, although a few of those that locally depart from conformity and crosscut the foliation may be slightly younger. Their mineral content corresponds to that of the host rock—those occurring in the plagioclase-rich gneisses are composed of quartz, plagioclase, garnet, magnetite, and rare muscovite; those in the granitic gneisses consist of quartz and microcline. Allanite, tourmaline, graphite, kyanite, and sillimanite may be accessory minerals. Such pegmatites predate the latest metamorphic episode and, like the migmatites elsewhere in the range, are considered to be the product of anatectic melting.

The discordant pegmatites were preferentially emplaced either along joints and faults or subparallel to the axial surfaces of open folds. They are coarser grained and consist of quartz, plagioclase, and microcline. Veinlets of epidote, clots of biotite, and local chlorite-sericite alteration contained in them may be products of retrograde metamorphism. Although unfoliated, the presence of these pegmatites as cobbles in the conglomerate at the base of the LaHood facies of the Proterozoic Belt Supergroup to the north of the map area clearly establishes their Precambrian age.

Orthoamphibolites

Metamorphosed rocks of intermediate to mafic igneous parentage (metabasites) crop out as sills, dikes, and phacoliths throughout the map area. The sills are especially well exposed in cirque walls and are concordant to the foliation of the enclosing rocks. Commonly they show pinch-and-swell features as well as boudinage. The dikes crop out as connecting links between two or more sills or as tabular bodies 1–50 m wide that clearly transect the regional foliation. Folded sills (phacoliths) occupy the crests of subisoclinal folds; the best example is that in the Ramshorn Creek drainage in the western part of the mapped area.

In contrast to the hornblende gneisses, the orthoamphibolites are fine- to medium-grained and dark brown to black; they have granoblastic, clustered granoblastic, and microflaser textures. Green or brown hornblende, plagioclase (An_{15-41}, locally antiperthitic), clinopyroxene (augite or salite), and almandine are the major constituents and are accompanied by accessory amounts of quartz, apatite, ilmenomagnetite, blue-green hornblende, biotite, hypersthene, epidote, chlorite, pyrite, and antiperthite.

The orthoamphibolites provide important clues to at least part of the metamorphic history of the region. Many of the dikes and sills have been sheared, folded, and intruded by pegmatite dikes. Furthermore, the dikes crosscut the foliation of the host rock, and the presence of sharp contacts with chill effects indicates that their protoliths were emplaced during or sometime after an earlier metamorphic event.

The orthoamphibolites have been metamorphosed to almandine–amphibolite–subfacies grade. However, hornblende-granulite–facies assemblages in these rocks crop out in widely scattered localities within the map area, for example, in the Pony region (Root, 1965), in Bumby Gulch (Reid, 1957), and throughout the southern Tobacco Root Mountains (Cordua, 1973). Pyroxene granulite assemblages have also been reported to crop out near Sheridan in the Indian Creek drainage (Burger, 1966). These granulite-facies rocks, which range from granoblastic to nematoblastic, consist of pink pleochroic hypersthene (in many it constitutes as much as 40% to 55% of a thin section), clinopyroxene (augite or salite), plagioclase (An_{65-68}), garnet, antiperthite, and quartz. Not all the hypersthene is of primary metamorphic origin; some, especially where altered and in association with altered plagioclase, is evidently relict from the original igneous protoliths that have been metamorphosed to almandine-amphibolite–facies grade, but under conditions close to those of hornblende granulite facies.

Meta-ultramafic Rocks

Small lenses, pods, and stocklike masses of dark green to black, coarse-grained meta-ultramafic rocks crop out in scattered localities throughout the range. They weather to black or mottled brown and consist mainly of serpentinized olivine, hornblende, orthopyroxene, and clinopyroxene in various assemblages, associated with lesser quantities of spinel, phlogopite, paragasite, cummingtonite, anthophyllite, talc, tremolite, carbonate, magnetite, brucite, and chlorite (see Tendall, 1978). The meta-ultramafites occur in close association with metabasites. They seem to have been intruded at about the same time and to have evolved along similar trends.

The origin of the meta-ultramafic rocks and the nature of their protoliths have been investigated by Tendall (1978), who has considered three possible models for their origin—tectonically emplaced slices of upper mantle, refractory residues from partially melted solid upper mantle, and cumulates of a differentiating melt. The sharp contacts, the lack of evidence for migmatization, the nonlinear distribution of the meta-ultramafic bodies, and the dominance of orthopyroxene over olivine appear to favor an origin from a deep-seated magma. The lack of visible layering is attributed to destruction by internal deformation and recrystallization during diapiric uprise. The lack of metasomatic effects, the presence of narrow shear rinds around the bodies, as well as the presence of occasional fractures in the pyroxene crystals, appear to support emplacement of the protoliths as solid bodies.

PROTEROZOIC DIABASE DIKES

During late Precambrian time three sets of mafic dikes were emplaced in the Tobacco Root Mountains region (Koehler, 1972, 1976; Wooden and others, 1978). The oldest dikes, of low-K tholeiite composition, were emplaced in the southern Tobacco Root Mountains about 1,455 m.y. ago. The other two sets, both of which are high-K quartz-normative types, were intruded sometime between 1,120 and 1,130 m.y. ago. One of the two is strongly enriched in Fe and is similar to ferrobasalt or ferrogabbro in composition; the other is low in Fe and has differentiated along strong alkali and silica enrichment trends. The 1,455-m.y.–old dikes and the Fe-enriched 1,120-m.y.–old dikes have initial $^{87}Sr/^{86}Sr$ ratios in the range of 0.7020 to 0.7030, which indicate probable derivation from mantle material that has maintained a low Rb/Sr ratio (0.024) for much of the Earth's history. The Fe-poor 1,130-m.y.–old set of dikes has an initial $^{87}Sr/^{86}Sr$ ratio of 0.709, which suggests contamination by crustal Sr.

POST-PRECAMBRIAN ROCKS

Sedimentary Rocks

The Archean rocks are overlain by undifferentiated sedimentary rocks (limestone, shale, siltstone, and sandstone) of Paleozoic to Mesozoic age. Upper Mesozoic and Cenozoic igneous rocks either overlie the Archean rocks or have been intruded into them. These younger rocks crop out in the northern, northwestern, and western parts of the map area. In addition, a small area of sandstone, presumed to be equivalent to the Cambrian Flathead Formation, is preserved in the southeastern part just south of the old Virginia City highway (State Highway 34, now abandoned).

Intrusions

In late Mesozoic time the Tobacco Root batholith, a large sill, and a number of smaller plutons were emplaced within and at or near the periphery of the area of the Archean rocks. All are granitic in composition. The Tobacco Root batholith is the most important of these plutons. It is a zoned, composite pluton 120 mi^2 (310 km^2 in area emplaced in the central to east-central part of the Archean rock outcrop area. The units of the batholith range from hornblende diorite to granite with intermediate facies of tonalitic, granodioritic, quartz monzonitic, and monzonitic composition.

Modal and chemical analyses (Smith, 1970) show a gradational concentric zoning from the periphery to the core of the batholith. Hornblende diorite is restricted to satellite stocks near the batholith's border and to the margin directly adjacent to mafic and ultramafic Archean metamorphic rocks. Inward toward the southeastern core of the pluton, the rocks become progressively more silicic. The most sialic concentrate in the batholith occupies an oval area of some 60 km^2 in the southeastern part of the pluton.

Studies of the compositional variation of K-feldspar, plagioclase, and biotite (Smith, 1970; Klusman, 1969) confirm the gradual change from the periphery to the center of the batholith. This has led us to conclude that the pluton, emplaced as a single injection of magma, was differentiated in situ, partly by reaction with the host rock at its outer margins but mainly by crystal fractionation. The lack of preserved primary flow features within the pluton, except at its margins, is interpreted as further evidence that it is the product of a single injection followed by crystallization in situ. K-Ar dating (Vitaliano and others, in prep.) has indicated an age of 74 to 71 m.y. for the rocks of the pluton.

The marginal-facies pegmatites of the Tobacco Root batholith crop out both within the batholith and in the adjacent host rocks. Microcline and quartz are the dominant minerals; zircon, tourmaline, ilmenite, rutile, rare albite (in miarolitic cavities), and magnetite occur in accessory quantities. The pegmatites of Late Cretaceous age contain sulfide minerals and are thus distinguished from those of Precambrian age. The younger pegmatites are considered to be the product of fluid-rich late-stage differentiates of the granitic magma of the batholith, emplaced in latest Cretaceous or earliest Eocene time. Aplite dikes and quartz veins are commonly associated with, and occasionally replace, the pegmatite.

The positioning of the Tobacco Root batholith was controlled, for the most part, by previously existing structures in the Archean rocks, mainly foliation and a set of northwest-trending major faults. As a result, despite local minor crosscutting such as along the southern margin of the batholith, the foliation of the Archean rocks is conformable to the batholith periphery for considerable distances. Furthermore, near the southern contact of the batholith, the dip of the foliation in the Archean rocks steepens considerably so as to conform, for some depth at least, to the outer surface of the batholith. In the northeastern part of the batholith near Norris, the conformity between the granitic rocks of the batholith and the foliation in the Archean rocks is well exposed in outcrops on the west wall of the valley near the head of Burnt Creek. Post-batholith movement along major faults, notably on the northwestern margin of the batholith, has offset the contact between the igneous rock and the Archean host.

A major sill of quartz monzonitic composition crops out in Brown Back Gulch, near the northwestern part of the mapped area, where it has intruded in part along the contact between the Archean and Paleozoic rocks and in part within the Lower Cambrian rocks adjacent to the contact. The sill is nearly 5 mi (8 km) long and attains a maximum width of more than 0.8 km. Stocks, bosses, and lesser sills and dikes of felsic composition are scattered throughout the western part of the map area. The majority are quartz monzonitic in composition, but one small pluton associated with the stock near Copper Mountain is composed of gabbro and diorite.

The location of the sill in Brown Back Gulch and of many of the smaller bodies appears to have been controlled by foliation or bedding planes. Precious and nonferrous metal mines, now mostly worked out, are associated with the Tobacco Root batholith and with some of the stocks and bosses. Scattered occurrences of andesite and basalt that crop out in the Archean terrane are probably of late Tertiary age and may even be related to the volcanism described below. Several small plugs and sills of latite and quartz latite, which crop out on the Middle Mountain in the eastern part of the map area, may be of similar age.

Volcanic Rocks

Volcanic rocks of Cenozoic age crop out north of Virginia City, in the North Meadow Creek drainage northwest of McAllister, and west of Norris in the Canadian Gulch drainage. The Virginia City unit forms a broad plateau capping the Archean rocks in the southern tip of the mapped area. The rocks range in composition from basalt to rhyolite and compose a series of lava flows and pyroclastic as well as volcaniclastic sedimentary units. An andesite at the base of the section is reported by K.L. Weir (1972, personal commun. to Cordua) to have yielded a radiometric date of 51 m.y. B.P. A basalt higher in the section was dated at 34 m.y. B.P. In the McAllister region, the volcanic rocks are rhyodacitic and are associated with tuffaceous clastic sedimentary rocks.

Sedimentary rocks of the Tertiary Bozeman Group adjoin the Precambrian rocks on the south and east parts of the Tobacco Root Mountains. These rocks include volcaniclastic sandstones, shales, conglomerates, bentonites, and, locally, freshwater and tuffaceous limestones.

Unconsolidated Deposits

Alluvium and colluvium representing debris derived from local outcrops have accumulated on some of the lower slopes of the mountains, on many of the valley floors, and on some of the higher plateau areas. Sediments of ground, lateral, and a few terminal moraines have been deposited in many higher parts of the mountains by glaciers that carved a residual cirque- and arête-like topography found on the highest peaks.

STRUCTURE

The geologic structure of the area is dominated by the Mill Gulch antiform, a north-trending broad open fold that is flanked on the west and east by a series of subisoclinal folds of nearly parallel trend. The strata of the Mill Gulch antiform are actually refolded, tightly folded, recumbent isoclinal folds. The Mill Gulch antiform is best developed and exposed south of the Tobacco Root batholith. North of the batholith, the stretching required to accommodate the batholith appears to have modified but not obliterated the antiformal nature of this fold. Here the axial trend of the tight subisoclinal folds is easterly to southeasterly. The plunge of the folds north of the batholith varies from north-northeast to east-northeast and east.

In the southwestern part of the map area the subisoclinal folds flanking the Mill Gulch antiform trend N10°W, plunge 25°N, and are overturned to the east. In the southeastern part of the area, the subisoclinal folds trend N42°E, their plunge averages 22°NE, and the folds are overturned to the west. The fold sets described above are clearly of Precambrian age. A variety of lesser folds are found throughout the area; the most important of them is a major refolded marble horizon that crops out in the vicinity of California Creek and terminates at a major fault.

Later, although probably still during Precambrian time, the recumbent, subisoclinally folded strata of the earlier folding epoch were deformed into a set of northeast-trending open folds. The antiform whose axis is near Noble Peak belongs to this set.

The Archean rocks were folded again, along with the overlying Phanerozoic rocks, during the Laramide orogeny. Recognizable evidence from this period of folding is concentrated along major transcurrent faults (for example, the fault along Carmichael Canyon) in the form of drag folds. A similar drag fold exposed in the northeastern part of the map area was also formed during this period. Other evidence of folding during this period may be present within the area of Archean rocks proper, but if so, it has not been recognized as yet.

FAULTS

The major faults in the area can be classified into two sets: faults that trend northwest and those that trend northeast. The majority of these faults formed during Precambrian time. In the northwestern part of the area, near Dead Mountain, a major northwest-trending fault has controlled the late Precambrian emplacement of diabase dikes. In the southern part of the area the similar northwest trend of the diabase dike swarm and most of the faults suggests that the faults may have controlled the late Precambrian emplacement of the dikes. In the northwestern part of the batholith the contact between the Archean rocks and the batholith has been offset along three faults—one passing through the Bismark Mine, one near Skylake, and one near the Mammoth Mine. Movement, at least along the northwest-trending faults, apparently continued after emplacement of the Tobacco Root batholith.

The northeast-trending faults, though not as prominent as those of northwest trend, also are of probable Precambrian age. Movement along some of these faults continued in post-Precambrian time. In the vicinity of the Atlantic and Pacific Mine in the north-central part of the map, a major northwest-trending fault has been offset in a right-lateral sense by a small northeast-trending fault. In the vicinity of the Bertha Mine in the southeastern part of the map, a northwest-trending fault has been offset left laterally by a major northeast-trending one. Along Willow Creek and south of Norris, the border of the batholith has been offset along similarly trending faults.

Finally, lesser faults of variable trend are scattered throughout the map area. These also may date from Precambrian time, but the evidence is less conclusive. In the northwestern part of the map area in the vicinity of Hell Roaring Creek, a thrust fault has brought Archean garnet gneiss on top of Mississippian Mission Canyon Limestone.

METAMORPHISM

We conclude that the Archean rocks in the Tobacco Root Mountains have probably undergone two episodes of metamorphism (M_1 and M_2) in Precambrian time and a third one (M_3) at least partly in post-Precambrian (Laramide) time.

After deposition of the metamorphic protoliths, the region experienced an early Precambrian regional thermal event that may have reached granulite facies intensity. Relicts of this event may be represented by the local occurrence of hypersthene and clinopyroxene in quartzofeldspathic gneiss, enstatite in gedrite schist, forsterite and hypersthene in marble, two-pyroxene assemblages in iron-formation, and fibrolite in aluminous schist.

A second Precambrian regional metamorphic event (M_2) 1.6 b.y. ago (Giletti, 1966) equilibrated within the almandine-amphibole facies and apparently obliterated much of the evidence for M_1. In the early stages Of M_2, kyanite-muscovite–bearing assemblages developed in the aluminous schists and quartzofeldspathic gneisses. In the later stages, quartz–K-feldspar–sillimanite assemblages stabilized in the aluminous schists. Gedrite replaced enstatite in the gedrite schists, and augite and labradorite in the metabasites were replaced by hornblende, diopside, and andesine. Locally, some granulite-facies assemblages were produced by M_2. For example, in the metabasites, hornblende and diopside are replaced by garnet; then garnet was replaced by hypersthene. In gedrite schists, gedrite and biotite are locally replaced by an orthopyroxene-cordierite symplectite. In the meta-ultramafic bodies, orthopyroxene porphyroblasts enclose a variety of amphiboles, and in the stratiform hornblende gneisses, hornblende is replaced locally by hypersthene. These granulite-facies assemblages probably developed in H_2O-deficient environments, while the surrounding rocks were equilibrating in the almandine-amphibolite facies.

Alternatively, our M_1 and M_2 could be represented by a single, multiphase metamorphic event. This possibility has been pointed out by Immega and Klein (1976); on the basis of mineral assemblages in the iron-formations, they suggested equilibration conditions of metamorphism at 650 to 750 °C and 4 to 6 kbar. The fact that the metamorphic intensity during M_2 seems to have been higher in the central part of the Tobacco Root Mountains (where sillimanite–K-feldspar assemblages are ubiquitous in aluminous schists) but a bit lower in the northern and extreme southeastern parts (where kyanite-bearing assemblages appear in the aluminous schists) would appear to support this alternative.

The duration and timing of M_3 are moot. The effects of this event are concentrated along major faults, shear zones, and at the contacts of igneous intrusions. Many of the products of M_3 are characterisitic of greenschist facies. Plagioclase, hornblende, biotite, garnet, kyanite, sillimanite, K-feldspar, forsterite, and orthopyroxene are replaced by sericite, paragonite, zoisite, clinozoisite, chlorite, epidote, actinolite, serpentine, biotite, talc, tremolite, carbonate, sphene, and magnetite (see especially, Friberg, 1976; Tendall, 1978).

Archean rocks in contact with the Upper Cretaceous–lower Eocene Tobacco Root batholith and related granitic rocks were overprinted with pyroxene- and hornblende-hornfels–facies assemblages. Gedrite and biotite in gedrite schists have been replaced by an orthopyroxene-cordierite symplectite (Hess, 1967).

PONY SERIES–CHERRY CREEK GROUP PROBLEM

The name Cherry Creek Group was first applied to staurolite-kyanite schists, marbles, gneisses, and phyllites exposed in the vicinity of Cherry Creek, about 18 mi (29 km) south-southwest of Ennis (Runner and Thomas, 1928). In subsequent studies, other workers have applied the name to Precambrian marbles, hornblende gneisses, sillimanite schists, anthophyllite gneisses, and green quartzites in the Tobacco Root Mountains, despite the lack of precise correlation with the units at the Cherry Creek type locality. The problem has been further complicated by the selection of a green quartzite as the base of the Cherry Creek Group (Reid, 1957). This green quartzite is in a section in the South Meadow Creek area described by Tansley and others (1933) and other early workers. However, sillimanite schists and green quartzites (admittedly sparse) have been found (Hess, 1967) immediately south of Pony, not far from the massive microcline-rich paragneisses that Tansley and others (1933) apparently had in mind when they first applied the name Pony Series. Furthermore, green quartzites have been described elsewhere in the vicinity of the Tobacco Root Mountains (Heinrich, 1949; McThenia, 1960; Spencer and Kozak, 1965) in rocks labeled Pony Series. Sillimanite paragneisses and green quartzite have also been found just north of Ennis Lake (Hess, 1967). Although rocks in the vicinity of the Indiana University Geologic Field Camp have been placed in the Pony Series (Reid, 1957), F.K. Root (1965, unpub. data) found nearly every rock type known from the "Cherry Creek" assemblage with the exception of marble in that area.

The data presented above illustrate the nature of the Cherry Creek Group–Pony Series problem and are not meant to obscure the fact that very essential differences in composition of the protoliths for these rocks must have existed from place to place in the southwestern Montana Precambrian province. As mentioned above, the stratigraphic section in the Tobacco Root Mountains consists of one sequence dominated by carbonate-pelite assemblages and another by coarse-grained clastic assemblages. If we assume that the section is right side up and that the western sequence is younger, the quartzofeldspathic gneisses and amphibolites on the east would correlate with the pre-Cherry Creek gneiss of Heinrich (1960). If the reverse is assumed and overturning has occurred, the western sequence is older, and the quartzofeldspathic gneiss and amphibolite correlate instead with the Pony gneiss of Tansley and others (1933). However, the structural complexities in the region require caution in stating whether a now highly metamorphosed succession is older or younger than another adjacent one. The presence of large-scale isoclinal recumbent or overturned folds in the Tobacco Root Mountains, as suggested by Reid (1963) and confirmed by this report, make repetition of sequences throughout the southwestern Montana Precambrian province almost certain. Therefore, the designation of Precambrian rocks in any one area as Cherry Creek Group or Pony Series may be unsound. In addition to original differences in the protoliths, the presence of inverted and repeated successions in various outcrop belts of the Montana Precambrian section is bound to have added to the confusion in age relations. Thus, although rocks like those of the Cherry Creek Group have been found underlying rocks like those of the Pony Series (Reid, 1957), the latter are not necessarily younger than the former, a possibility also pointed out by Reid (1957, 1966). Therefore, on the basis of present knowledge, the evidence is insufficient to justify application of the terms Cherry Creek Group and Pony Series to the sequence of the Archean rocks in the Tobacco Root Mountains. Although the evidence to date has failed to provide support for a definite answer to the problem, it has suggested a third possibility; namely, that the entire section is part of a single sequence derived from a landmass to the east and deposited in a sedimentary basin. The detrital sediments thus derived were poured into the basin and tended to concentrate mainly in its eastern part. The finer clastic material was transported farther west before being deposited together with carbonates. The eastward thinning and the gradual disappearance of the carbonate beds in this area at least appear to support this thesis. Despite the high rank of metamorphism and intense deformation experienced by the rocks of the Tobacco Root Mountains area, the general distribution of depositional facies tended to be retained.

REFERENCES CITED

Auboin, J., 1965, Geosynclines: New York, Elsevier Scientific Publishing Company, 335 p.

Burger, H.R., 1966, Structure, petrology, and economic geology of the Sheridan district, Madison County, Montana [Ph.D. thesis]: Bloomington, Indiana University, 156 p.

Condie, K.C., Leech, A.P., and Baadsgaard, H., 1969, K-Ar ages of Precambrian mafic dikes in Wyoming: Geological Society of America Bulletin, v. 80, p. 899–906.

Cordua, W.S., 1973, Precambrian geology of the southern Tobacco Root Mountains, Madison County, Montana [Ph.D. thesis]: Bloomington, Indiana University, 247 p.

Friberg, L.M., 1976, Petrology of a metamorphic sequence of upper-amphibolite facies in the central Tobacco Root Mountains, southwestern Montana [Ph.D. thesis]: Bloomington, Indiana University, 147 p.

Giletti, B., 1966, Isotopic ages from southwestern Montana: Journal of Geophysical Research, v. 71, p. 4029–4036.

Gillmeister, N.M., 1971, Petrology of Precambrian rocks in the central Tobacco Root Mountains, Madison County, Montana [Ph.D. thesis]: Cambridge, Harvard University, 201 p.

Gillmeister, N.M., 1972, Cherry Creek–Pony Group relationship in the central Tobacco Root Mountains, Madison County, Montana: Northwest Geology, v. 1, p. 21–24.

Grant, J.A., 1968, Partial melting of common rocks as a possible source of cordierite-anthophyllite bearing rocks: American Journal of Science, v. 266, p. 908–931.

Guidotti, C.V., Herd, H.H., and Tuttle, C.L., 1973, Composition and structural state of K-feldspars from K-feldspar + sillimanite grade rocks in northwestern Maine: American Mineralogist, v. 58, p. 705–716.

Hanley, T.B., 1975, Structure and petrology of the northwestern Tobacco Root Mountains, Madison County, Montana [Ph.D. thesis]: Bloomington, Indiana University, 289 p.

Hanley, T.B., 1976, Stratigraphy and structure of the central fault block, northwestern Tobacco Root Mountains, Madison County, Montana, in Tobacco Root Geological Society, 1976 Field Conference Guidebook: Montana Bureau of Mines and Geology Special Publication 73, p. 7–14.

Hayden, F.V., 1872, Preliminary report of the U.S. Geological Survey of Montana and portions of adjacent territories: U.S. Geological Survey of Territories, 5th Annual Report, p. 33–43.

Heinrich, E.W., 1949, Pegmatite mineral deposits in Montana: Montana Bureau of Mines and Geology Memoir 28, 56 p.

Heinrich, E.W., 1960, Pre-Beltian geology of the Cherry Creek and Ruby Mountains areas, southwestern Montana—Part 2, Geology of the Ruby Mountains and nearby areas in southwestern Montana: Montana Bureau of Mines and Geology Memoir 38, p. 23–40.

Hess, D.F., 1967, Geology of pre-Beltian rocks in the central and southern Tobacco Root Mountains, with reference to superposed effects of the Laramide-age Tobacco Root batholith [Ph.D. thesis]: Bloomington, Indiana University, 333 p.

Högberg, R.K., 1960, Geology of the Ruby Creek iron deposit, Madison County, Montana, in West Yellowstone earthquake area, 11th Annual Field Conference Guidebook: Billings Geological Society, p. 268–272.

Immega, I., 1976, Mineralogy and petrology of some Pre-Cambrian iron-formations in southwestern Montana [Ph.D. thesis]: Bloomington, Indiana University, 97 p.

Immega, I., and Klein, C., 1976, Mineralogy and petrology of some metamorphic Precambrian iron-formations of southwestern Montana: American Mineralogist, v. 6 1, p. 1117–1144.

James, H.L., 1954, Sedimentary facies of iron-formation: Economic Geology, v. 49, p. 235–293.

Klusman, R., 1969, Electron microprobe analysis of feldspars [Ph.D. thesis]: Bloomington, Indiana University, 270 p.

Koehler, S.W., 1972, Petrology and petrogeny of the diabase dikes of the Tobacco Root Mountains, southwestern Montana [M.A. thesis]: Bloomington, Indiana University, 60 p.

Koehler, S.W., 1976, Petrology of the diabase dikes of the Tobacco Root Mountains, Montana, *in* Tobacco Root Geological Society, 1976, Field Conference Guidebook: Montana Bureau of Mines and Geology Special Publication 73, p. 27–36.

Lal, R., and Moorhouse, W., 1969, Cordierite-gedrite rocks and the associated gneisses of Fishtail Lake, Harcourt Township, Ontario: Canadian Journal of Earth Sciences, v. 6, p. 145–165.

Levandowski, D.W., 1956, Geology and mineral deposits of the Sheridan-Alder area, Madison County, Montana [Ph.D. thesis]: Ann Arbor, Michigan, University Microfilms 19.707.

McThenia, A.W., Jr., 1960, Geology of the Madison River Canyon area, north of Ennis, Madison County, Montana, in West Yellowstone earthquake area, 11th Annual Field Conference, Guidebook: Billings Geological Society, p. 155–164.

Mehnert, K.R., 1968, Migmatites and the origin of the granitic rocks: Amsterdam, Elsevier Publishing Company, 393 p.

Merrill, G.P., 1895, Notes on some eruptive rocks from Gallatin, Jefferson, and Madison Co., Montana: U.S. National Museum Proceedings, v. 17, p. 637–673.

Mueller, P., and Cordua, W.S., 1976, Rb-Sr whole rock age of gneisses from the Horse Creek area, Tobacco Root Mountains, Montana: Isochron/West, no. 16, p. 33–36.

Peale, A.C., 1896, Geologic atlas of the United States, no. 24, Three Forks Folio, Montana: U.S. Geological Survey, 7 p., 3 maps, 2 secs.

Reid, R.R., 1957, Bedrock geology of the north end of the Tobacco Root Mountains, Madison County, Montana: Montana Bureau of Mines and Geology Memoir 36, 25 p.

Reid, R.R., 1958, Origin of the pre-Beltian metamorphic rocks of the Tobacco Root Mountains, southwestern Montana [abs.]: Geological Society of America Bulletin, v. 69, p. 1741.

Reid, R.R., 1963, Metamorphic rocks of the northern Tobacco Root Mountains, Madison County, Montana: Geological Society of America Bulletin, v. 74, p. 293–306.

Reid, R.R., 1966, Pony–Cherry Creek age relation in southwestern Montana: Geological Society of America Bulletin, v. 66, p. 1315–1316.

Root, F.K., 1965, Structure, petrology, and mineralogy of pre-Beltian metamorphic rocks of the Pony–Sappington area, Madison County, Montana [Ph.D. thesis]: Bloomington, Indiana University, 184 p.

Runner, J.J., and Thomas, L.C., 1928, Stratigraphic relationships of the Cherry Creek Group in the Madison Valley, Montana: Geological Society of America Bulletin, v. 39, p. 202.

Smith, J.L., 1970, Petrology, mineralogy and chemistry of the Tobacco Root batholith, Madison County, Montana [Ph.D. thesis]: Bloomington, Indiana University, 164 p.

Spencer, E.W., and Kozak, S.J., 1965, Tectonic style in the Spanish Peaks, Madison Range, Montana: Washington and Lee University, Geological Department Report 1.

Tansley, W., Shafer, A., and Hart, L., 1933, A geologic reconnaissance of the Tobacco Root Mountains, Madison County, Montana: Montana Bureau of Mines and Geology Memoir 9, 55 p.

Tendall, B.A., 1978, Mineralogy and petrology of Precambrian ultramafic bodies from the Tobacco Root Mountains, Madison County, Montana [M.A., thesis]: Bloomington, Indiana University, 126 p.

Turner, F.J., 1968, Metamorphic petrology: New York, McGraw-Hill Publishing Company, 403 p.

Winkler, H.G.F., 1968, Petrogenesis of metamorphic rocks, third edition: New York, Springer-Verlag, 320 p.

Wooden, J.L., Vitaliano, C.J., Koehler, S.W., and Ragland, P.C., 1978, The late Precambrian mafic dikes of the southern Tobacco Root Mountains, Montana: Geochemistry, Rb-Sr geochronology, and relationship to Belt tectonics: Canadian Journal of Earth Sciences, v. 15, p. 467–479.

Wright, T.C., and Stewart, D., 1968, X-ray and optical study of the alkali-feldspar: I. Determination of composition and structural state from refined unit cell parameters and 2*V*: American Mineralogist, v. 53, p. 38–87.

ACKNOWLEDGMENTS

Field work on this project was begun in 1962 and ended in 1972. Funds for the field work were provided by the National Science Foundation (Grants GA-570, GA-1550, and GA-21369 to C.J. Vitaliano) and the Geological Society of America (grants to H.R. Burger III and to W.S. Cordua). Indiana University through Judson Mead, Director of Indiana University Geologic Field Station, kindly made available the field station as a base and provided logistical support as well. C. Klein and D.B. Vitaliano kindly reviewed the manuscript and made many useful suggestions. Publication of the geologic map has been made possible by National Science Foundation Grant EAR-78-12955 and by a grant from the Indiana University Foundation. Final drafting of the map was done by James R. Tolen under the direction of William Moran, both of the Indiana Geological Survey. All of this assistance is gratefully acknowledged.

Manuscript Received April 24, 1978
Revised Manuscript Received February 27, 1979
Manuscript Accepted May 21, 1979